Foreword

After nicotine and the bases of solanaceous plants, the isoquinoline alkaloids take pride of place in the classical series.

We recall the pioneering researches of Goldschmidt on papaverine, the syntheses of Pictet, Perkin's studies of berberine, the U.V. spectral work of Hartley and Dobbie, the biogenetic intuition of Winterstein and Trier, and many other outstanding investigations in the field that were veritable milestones of progress.

Chemistry of the morphine-thebaine group was developed in parallel, though these bases received recognition as isoquinoline derivatives relatively late in their chemical history. The names of Rabe, Knorr and Pschorr are outstanding in the earlier period during which the foundations of our present understanding were well and truly laid.

The isoquinoline alkaloids were, in my own student days, relatively limited in number. Now, due largely to vastly improved methods of isolation and characterisation, they have become a multitude. Professor Kametani's work gives eloquent testimony to this fact and it is therefore very important that his systematic arrangement of the data is so helpful by facilitating the quick identification of the section of the book relevant to the matter in which the reader is interested. We have here, therefore, an excellent work of reference. But equally, the explanatory sections are well designed to be read through, and they will be found to be scholarly essays of great value to the student, and let it be said, to the teacher also.

As one who has been concerned in the study of isoquinoline alkaloids, I may be allowed to say that Professor Kametani has given accounts of these chapters in the history of our science, as befits a pupil of Shigehiko Sugasawa, that have afforded me great satisfaction. His contribution to further developments, by supplying accurate information and stimulating interest, must be most significant.

19th July, 1968

Robert Robinson

Robert Robinson

Foreword

During these odd one hundred and sixty years, since the first discovery in 1805 of the alkaloid, morphine, by the German pharmacist Serthürner, there has been found a host of alkaloids, especially during the last couple of decades. In my student days, around 1920, all natural alkaloidal compounds could be compiled in a single volume of a book of several hundred pages, e. g. Winterstein-Trier, Wolfenstein, Henry, only to mention a few. But in present days more than ten volumes of similar sized books are necessary to make a complete description of the known alkaloids, e. g. Manske-Alkaloids.

Of these numerous natural bases the so-called isoquinoline alkaloids are probably the most abundant in number. Incidentally the first discovered base, morphine, belongs to this group as was established by Professor Sir Robert Robinson in 1923 through his subtle insight into the biogenesis of this important compound.

I may well say that Professor Kametani is a leading isoquinoline chemist, who has been working in this field for more than fifteen years, especially specializing in the synthesis of isoquinoline bases and I am confident that he is a well-qualified person to be the author of "Chemistry of Isoquinoline Alkaloids." In this book more than six hundred isoquinoline alkaloids are well classified and compiled; in the introductory section general synthetic methods of isoquinoline derivatives are described and moreover, current biogenetic views for the important isoquinoline bases are sensibly summarized, making this book a valuable instructor for research chemists as well as for students.

It is my expectation and sincere desire that Professor Kametani with his collaborators will render further contributions in this area through their works.

August, 1968

Shigehiko Sugasawa

Shigehiko Sugasawa

Foreword

Prof. Kametani has compiled a Monograph which will be welcomed by organic chemists, specifically by those interested in alkaloids, and more specifically by those whose researches include the isoquinolines. An enormous amount of information is condensed into a readily accessible form by an author who has had extensive experience, particularly in many brilliant syntheses of these alkaloids.

Most of the transformations and metabolic changes that appear possible to ring on tyrosine can be recognized in the separate types of isoquinoline alkaloids and many of the essential stages are now understood. The chemistry involved in the exhaustive study of these alkaloids includes most of the reactions known to organic chemistry and much of it is included in this Monograph.

I am personally indebted to Prof. Kametani not only for providing me with an excellent reference work, full of useable information, but on at least one occasion he had corrected an error of mine in this field of alkaloid chemistry and on another occasion he has achieved a synthesis (of cularine) which had baffled me. I congratulate him.

August 23, 1968

R. H. F. Manske

Preface

The opium poppy, *Papaver somniferum*, has been greatly valued for its ability to remove pain from the human being for many centuries. Papaverine is used as a muscle relaxant, while codeine is also used as an analgesic. All of them belong to alkaloids of the isoquinoline group.

My work with isoquinoline derivatives has been continuing over the last fifteen years since 1950, with particular emphasis on the total syntheses of isoquinoline alkaloids.

The alkaloids of the isoquinoline group may be subdivided into several families. In this work, an attempt has been made to explain the chemical and physical characteristics and the origins of all the alkaloids which have been reported until 1966.

I have also attempted to explain the biogenetic concept and total syntheses of some representative alkaloids in this work.

I am extremely grateful to my first readers, Prof. Shigehiko Sugasawa, Prof. Sir Robert Robinson and Dr. R. H. F. Manske, for their valuable suggestions and for special forewards.

Sendai, August 1968

Tetsuji Kametani

Acknowledgements

This book could not be made without the help of many colleagues, who extremely kindly informed me about their new results and observations. I therefore wish to express my sincere gratitude for their collaboration. I want to thank Dr. Keiichiro Fukumoto (Tohoku University), who provided very helpful and constructive comments on the manuscripts. Without his earnest help, this book would not be born. I should like to express my warmest thanks to Seiichi Takano, Ph. D., Kunio Ogasawara, Ph. D., Shiroshi Shibuya, Ph. D., Haruhiko Yagi, Ph. D., Isao Noguchi, Ms. D., Kazumi Ohkubo, Ms. D., Masataka Ihara, Ms. D., and Mitsuo Satoh, Ms. D. of our laboratories, who informed me their recent observations and examined recent literatures. I also express my special gratitude to Emeritus Professor Shigehiko Sugasawa (University of Tokyo), Professor Sir Robert Robinson (London) and Dr. R. H. F. Manske (UNIROYAL Ltd., Research Laboratories, Guelph, Ontario, Canada), who not only read the entire book and gave a number of suggestions, but also informed me of their more recent observations. I was also grateful to Dr. Betty McFarland (NIH postdoctral fellow at Tohoku University), who read the entire book and offered kindest suggesions. Finally, I am indebted to President Genji Hirokawa and Vice-president Setsuo Hirokawa of Hirokawa Publishing Company who cooperated in publishing this book.

CONTENTS

Contributors to this book

Seiichi Takano
Keiichiro Fukumoto
Kunio Ogasawara
Shiroshi Shibuya
Haruhiko Yagi
Isao Noguchi
Kazumi Ohkubo
Toyohiko Kikuchi
Masataka Ihara
Mitsuo Satoh
Shigeo Kaneda
Atsuto Kozuka
Hideo Agui
Hideo Nemoto
Fumio Satoh
Kazuya Yamaki
Tomoko Yoshida
Yoshiko Suzuki

Note to the reader

Classification of the isoquinoline alkaloids in this book depends mainly upon the biogenetical view and structural formula. The isoquinoline alkaloid of interest can be seen at glance. In each column you can understand the plant origins, physical constants of the free base and its salts, structural determination, total and partial syntheses, biogenesis and biosynthesis, data of modern physical methods of analysis (UV, IR, ORD,CD, NMR, Mass, X-ray, and chromatographical values), and pertinent literature references. Furthermore, the details are as follows.

(1) Arrangement: The compounds are arranged with respect to their molecular weight.

(2) Nomenclature: The alkaloids are generally listed under the name used in the original literature. The compounds having two (or more) names are listed by their older name, and the new name is shown in parentheses.

(3) Details of structure and synthesis: In the structural determination, the references are shown without distinguishing the planar and stereoisomeric structures, and in the column of synthesis the total synthesis mainly is shown but in some cases the synthesis of simple derivatives such as O-methyl is also listed. The physical data mainly are listed as those of the alkaloid itself but the data of simple derivatives are shown in some cases.

(4) References: The abbreviations used in the references depend upon those of "Chemical Abstracts", and the initial page of all the literatures is cited.

(5) The abbreviations: Various abbreviations are represented by the following symbols.

Ac: acetate
Amorph: amorphous
aq: aqueous
B: free base
BG: biogenesis
BS: biosynthesis
Bz: benzoate
CD: circular dichroism
CR: colour reaction
Cryst: crystalline form
C & S: crystal and solubility
d: decomposition
2,4-D: 2,4-dinitrophenylhydrazone
Et: ethyl
IR: infrared spectrum
Insol: insoluble
Mass: mass spectrum
Me: methyl
MeI: methiodide
Mod: moderate
NMR: nuclear magnetic resonance spectrum
org. solv: organic solvent
Oxal: oxalate
OP: original plants
ORD: optical rotatory dispersion
Petr.: petroleum
Ph: physical data
Pic: picrate
Picrol: picrolonate
Py: pyridine
Rf: Rf value in T. L. C. (thin layer chromatogram) or P. C. (paper chromatogram)
Semicarb: semicarbazone
S. & B.: synthesis and biosynthesis
Sa. & D.: salt and derivative
S. D.: structural determination
Sol: soluble
Spar: sparingly
Styph: styphnate
UV: ultraviolet spectrum

Chapter 1 Introduction

(I) Isoquinoline alkaloids and their classification

Sertürner, a German pharmacist, isolated the active constituent of opium and named it "morphine" in 1805, when many people in the world, especially in Europe, indulged in opium spread by Arabians. Although the complete constitution of this base was too complicated to elucidate until 1925, it has been well known that morphine, as well as other compounds found in opium, has an isoquinoline structural unit which belongs to a large family of natural bases, "alkaloids".

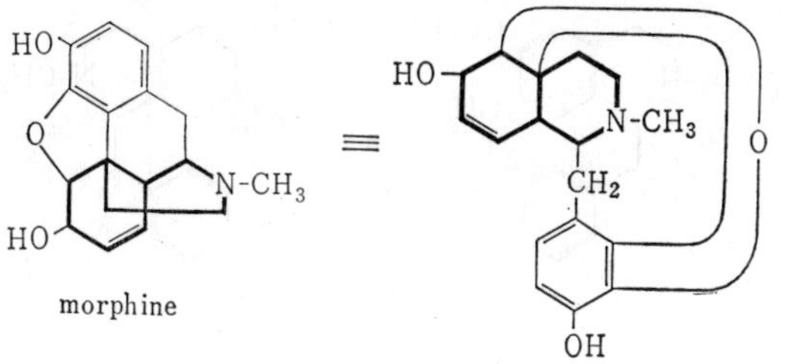

morphine

The isoquinoline alkaloids are further structurally and/or biogenetically subdivided into small groups as follows;

(1) Simple isoquinoline alkaloids
(2) 1-Benzylisoquinoline alkaloids
(3) Pavine and isopavine alkaloids
(4) 2-Benzylisoquinoline alkaloids
(5) Bisbenzylisoquinoline alkaloids
(6) Cularine and related alkaloids
(7) Proaporphine alkaloids
(8) Aporphine alkaloids
(9) Protoberberine alkaloids
(10) Protopine alkaloids
(11) Phthalide alkaloids
(12) Ochotensine and related alkaloids
(13) Rhoeadine and related alkaloids
(14) Morphine and related alkaloids
(15) Hasubanonine and related alkaloids (Hasubanan alkaloids)
(16) Emetine and related alkaloids
(17) Erythrina alkaloids
(18) Protostephanine alkaloids
(19) Tyrophorine and cryptopleurine alkaloids
(20) Amaryllidaceae alkaloids
(21) Benzophenanthridine alkaloids
(22) Dibenzopyrrocoline alkaloids
(23) Indoloisoquinoline alkaloids*
(24) Phenethylisoquinoline and colchicine alkaloids

(II) Structural Determination and Total Synthesis

There are many kinds of structures among the isoquinoline alkaloids, in which not only simple anhalonine but also complicated morphine are included, and the structural determination methods must necessarily be different. Since there is no determination procedure peculiar to isoquinoline alkaloids, zinc

*) This group is not discussed in this book. For a complete survey of indoloisoquinoline alkaloids, the reader is referred to "The Indole Alkaloids" by Manfred Hesse.

dust distillation, dehydrogenation with selenium or sulphur, oxidation and reduction, and Hofmann and Emde degradation have been used as for other alkaloids. These reactions reduce the complicated compound to a more simple moiety, whose structure indicates the basic nucleus and determines the position of the substituents.

For instance, the structure of magnoline (I), which was isolated from the leaves of *Magnolia fuscata* Andr., was assigned by oxidation as follows.[1] This alkaloid has the composition, $C_{36}H_{40}O_6N_2$, and contains two methoxyl and three hydroxyl groups. Oxidation of trimethyl ether gives the acid (III: $R=CH_3$), and the isoquinoline (II: $R=CH_3$), whereas oxidation of the triethyl ether gives the corresponding ethoxy compounds (II: $R=C_2H_5$) and (III: $R=C_2H_5$). Assuming that the carbonyl and carboxyl groups of (II) and (III) represent the residues obtained by fission of benzylisoquinoline systems, the structure (I) follows logically for magnoline.

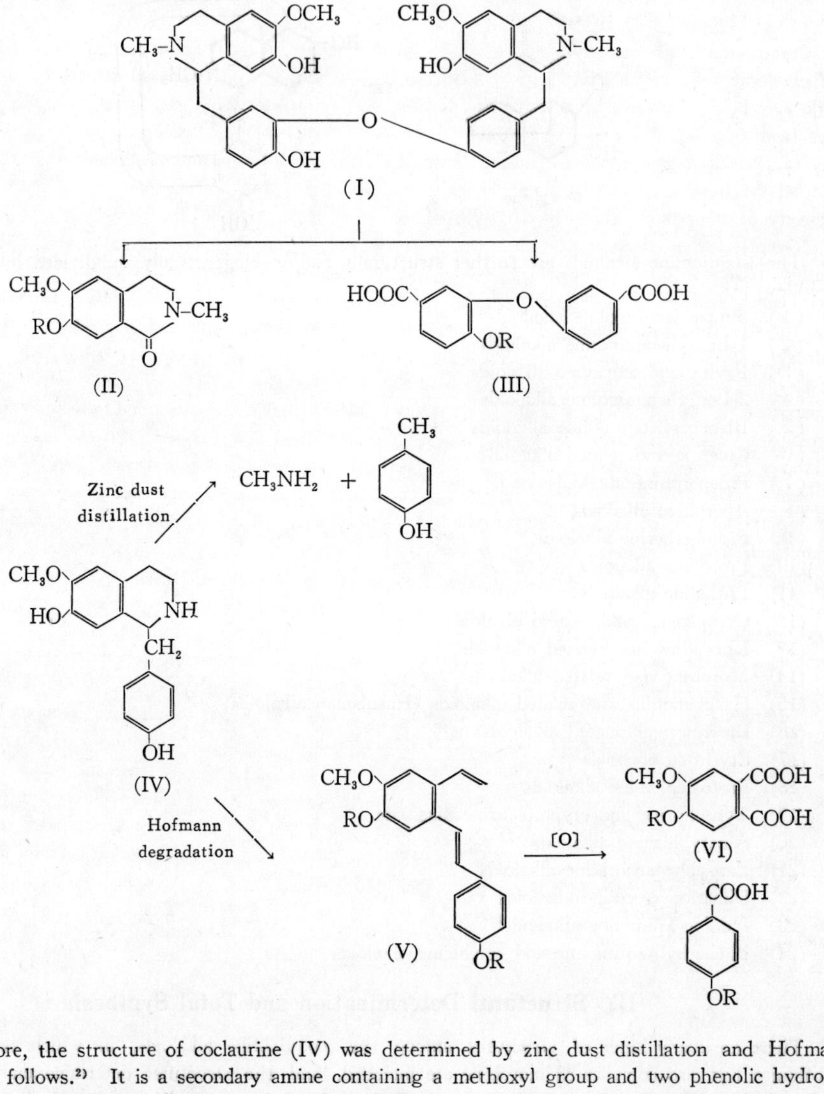

Furthermore, the structure of coclaurine (IV) was determined by zinc dust distillation and Hofmann degradation as follows.[2] It is a secondary amine containing a methoxyl group and two phenolic hydroxyl groups. Zinc dust distillation affords *p*-cresol and methylamine, and Hofmann degradation of the fully methylated base yields a methine (V: $R=CH_3$) which on oxidation gives anisic and *m*-hemipic acids (VI: $R=CH_3$). Hofmann degradation of coclaurine diethyl ether gives an analogous methine (V: $R=C_2H_5$), which on oxidation breaks down into 5-ethoxy-4-methoxyphthalic acid (VI: $R=C_2H_5$) and 4-ethoxybenzoic acid. This would be consistent with the formulation of this methine as (V), and coclaurine as (IV).

The structure above is a planar one showing only the intramolecular relationships, but the stereochemical structure of the molecule must be examined. In general, the absolute configuration is determined by conversion of the alkaloids to an amino acid, whose configuration is known. For instance, the absolute configuration of (−)-salsolidine (VII) has been elucidated by relation to the configuration of amino acid (VIII), which was obtained by degradation of N-formyl-(−)-salsolidine with ozone and peracetic acid, followed by hydrolysis of the resultant N-formyl-L-alanine derivative.[3]

CH_3O, CH_3O, NH, H CH_3 (VII) —3 steps→ HOOC, HOOC, NH, H CH_3 (VIII)

Comparison of the natural product with a synthetic sample must be carried out, from the physical and chemical point of view in order to ascertain the validity of the proposed structure, but the synthetic work may be a long and difficult task. For instance, after morphine was isolated in the pure crystalline state from opium by Sertürner[4] in 1805, its correct structure was assigned by Sir Robert Robinson[5] only in 1923 following extensive investigations by chemists in Germany, England, and Austria. Finally, its total synthesis was accomplished by the American chemist, Professor Gates in 1952.[6] Between the discovery of morphine and its synthesis one hundred forty-seven years intervened.

Recently, physical methods, for instance, ultraviolet, infrared, nuclear magnetic resonance, optical rotatory dispersion, and circular dichroism spectrophotometry and mass spectroscopy and X-ray analysis, have been applied to the structural elucidation of many isoquinoline alkaloids, surprisingly shortening the time for structural determination. It is interesting and valuable that the position of the methoxyl and hydroxyl groups in the benzene ring may be determined by nuclear magnetic resonance spectrum. According to older procedures, the product, obtained by ethylation of hydroxyl group with diazoethane, was decomposed in order to determine the position of the substitutents, and the resultant product was identified by comparison with an authentic sample alternatively synthesised. In certain cases the position of ethoxy or methoxy substituents could not be characterised by chemical procedures. At present the chemical shifts of the aromatic ring protons and the methoxy methyl protons are known with considerable exactness, and therefore the position of methoxy groups in an unknown compound can be determined by nuclear magnetic resonance measurement. Furthermore, application of optical rotatory dispersion and circular dichroism is an important tool for elucidating the absolute configuration of hydrogen at the C_1-position in the isoquinoline nucleus. Furthermore, application of X-ray analysis has been successful for determination of structure as shown in case of ochotensine,[7] but as yet there have been only a few such examples. In the near future, X-ray analysis is expected to increase in usefulness as a method for the structural determination.

Battersby's structural determination of melanthioidine (IX: R=H) by physical methods is explained as follows.[8] High resolution mass spectrometry established the molecular formula $C_{38}H_{42}O_6N_2$(m/e 622) for this base. The presence of two phenolic hydroxyl groups was proved by the infrared spectrum, by the shift of the ultraviolet spectrum in base, and by the formation of an O,O-diacetate (m/e 706; $\nu_{max.}$ 1760 cm^{-1}) and O,O-dimethyl ether (m/e 650). Melanthioidine was shown to be a symmetrical molecule by the striking simplicity of its nuclear magnetic resonance spectrum, showing two N-methyl groups (singlet, 7.56), two O-methyl groups (singlet, 6.21), ten aryl protons (3.1–3.5), and unresolved signals corresponding to eighteen protons (6.0–9.0), similar to that of 1-phenethylisoquinoline. Furthermore, the ultraviolet spectrum was closely similar to that of curine, a bisbenzylisoquinoline alkaloid. Consideration of these facts together with the foregoing information suggested that melanthioidine is a bisphenethylisoquinoline alkaloid. Reductive cleavage of the O,O-dimethyl ether base with sodium in liquid ammonia afforded a phenolic compound (M^+, m/e 327; base peak, m/e 192) (X). This product has the R-configuration, deduced from its negative Cotton effect.

The symmetry of melanthioidine (IX: R=H) and all the experimental data can be accommodated by the structure (IX: R=CH_3) for the O,O-dimethyl ether base. The foregoing evidence does not rigorously exclude linkage of the ether bridge *meta* to the methoxyl groups on ring C and C′ of O,O-dimethyl base (IX: R=CH_3). However, the illustrated arrangement is strongly favoured on biogenetic grounds and this structure would be expected to yield the compound (X) as the main reductive product. Major fragment ions appear at m/e 312 and 310 in the mass spectrum of melanthioidine. These correspond to the favoured

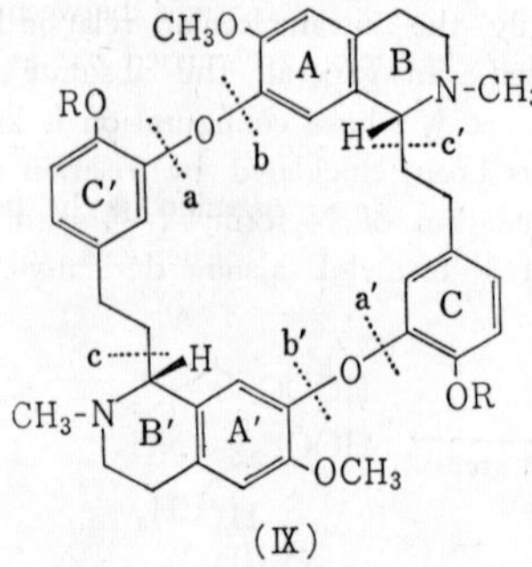

(IX) 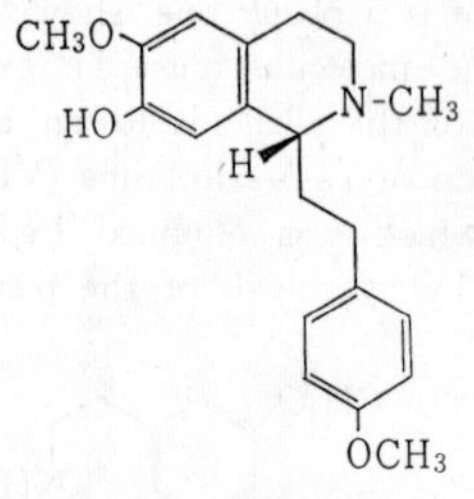 (X)

cleavage at c and c′ with hydrogen transfer. A further important fragment at m/e 485 corresponds to the loss of $C_8H_9O_2$ from the parent ion. This fragment supports the structure (IX: R=H) for melanthioidine since its formation can be explained by fission at b–c (=b′–c′), again with hydrogen transfer. The fission at b–c was confirmed by the appearance of strong peaks at m/e 499 and 529, respectively, in the mass spectra of the O, O-dimethyl ether and O, O-diacetyl bases. The two hydroxyl groups thus are proved to be located on rings C and C′ and the structure of the alkaloid can be assigned the formula (IX: R=H).

The structure proposed by chemical and physical methods also should be considered from the biogenetical point of view. A very interesting example is the structure of emetine, first determined by biogenetic mechanism and then confirmed by its total synthesis.

After Pyman[9] had proposed the structure (XI) for emetine by assuming Späth's experimental results of various degradation reactions, Sir Robert Robinson[10] proposed the structure (XII) for it merely from the biogenetical point of view in 1948.

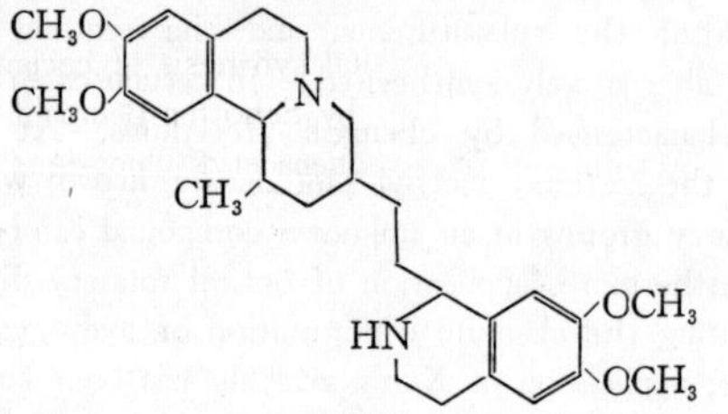

(XI) 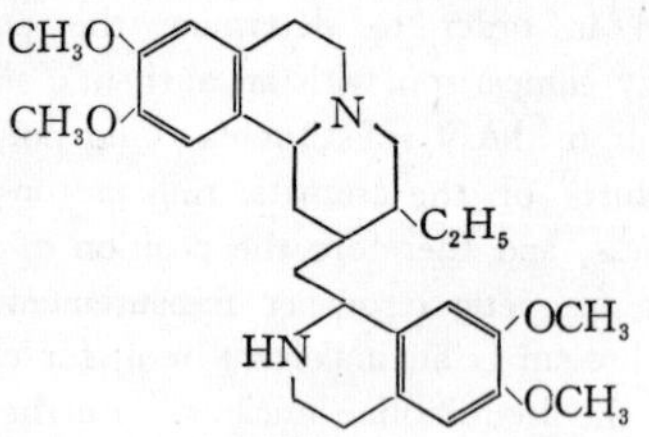 (XII)

By making use of the so-called "Woodward fission" applied by Woodward in the structural determination of strychnine, Robinson has considered that the cleavage of the D ring in the protoberberine type compound (XIII) gives the formyl derivative (XIV) in two steps. Subsequent Mannich reaction of the latter compound (XIV) with dopamine or its biogenetic equivalents affords the emetine (XII). Robinson's structure was assigned without experiment, but, afterwards, Pailer,[11] Battersby,[12] and Preobrazhenski[13] proved it to be correct by chemical degradations and total syntheses.

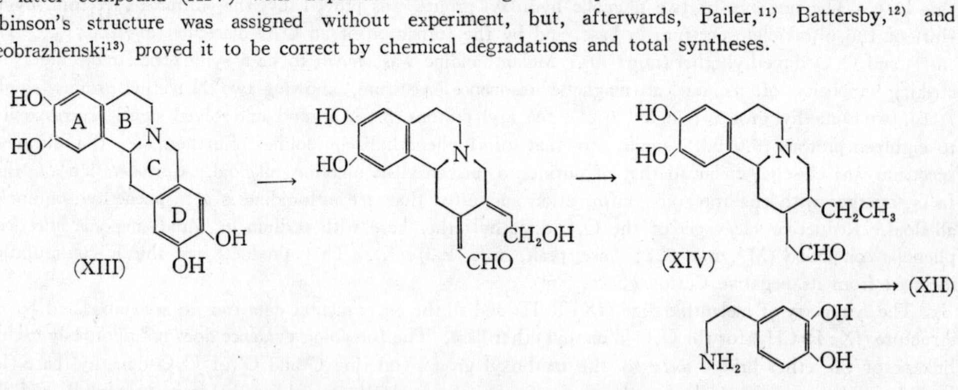

(XIII) → → (XIV) → (XII)

The final stage of structural determination of isoquinoline alkaloids is total synthesis. Three main reactions have been used to synthesise the isoquinoline alkaloids: (1) Bischler-Napieralski reaction[14], (2)

Pictet-Spengler reaction[15], and (3) Pomeranz-Fritsch reaction.[16] Cyclisations between C_1 and N_2-position, N_2 and C_3-positions, and C_3 and C_4-positions in isoquinoline nuclei, are carried out as special cases in isoquinoline syntheses.

1. Bischler-Napieralski Reaction

Dehydrative cyclisation of the acyl-derivative (XV) of β-phenethylamine in the presence of a Lewis acid such as phosphoryl chloride in an inactive solvent, for example, benzene, gives a 3,4-dihydroisoquinoline (XVI), which must be reduced to the 1,2,3,4-tetrahydroisoquinoline (XVII) since natural products exist as the tetrahydro-derivatives in most cases. For this purpose, 3,4-dihydroisoquinoline hydrochloride can be directly reduced with sodium borohydride to give tetrahydro-derivative (XVII). When the N-methyl-derivative (XVIII) is desired, the Eschweiler-Clarke reaction of (XVII) with formalin and formic acid gives an expected N-methyl-derivative (XVIII). Reduction of the methiodide (XIX) of a 3,4-dihydro-derivative (XVI) with sodium borohydride to (XVIII) is also recommendable.

CH_3O NH CO R (XV) $\xrightarrow{H^+}$ CH_3O N R (XVI) $\xrightarrow{NaBH_4}$ CH_3O NH R (XVII)

(XVII) $\xrightarrow{CH_2O-HCO_2H}$ (XVIII); (XVI) ⟶ CH_3O $\overset{+}{N}-CH_3$ I^- R (XIX) ⟶ CH_3O $N-CH_3$ R (XVIII)

For instance, the Bischler-Napieralski reaction is a key step in the total synthesis of coclaurine (IV).[3] Bischler-Napieralski reaction of the amide (XXII), which was obtained by Schotten-Baumann reaction of 4-benzyloxy-3-methoxyphenethylamine (XX) with 4-ethoxycarbonyloxyphenacetyl chloride (XXI), gave

CH_3O, $PhCH_2O$, NH_2 (XX) + COCl CH_2 C_2H_5OOCO (XXI) ⟶ CH_3O $PhCH_2O$ NH CO CH_2 C_2H_5OOCO (XXII) $\xrightarrow{POCl_3}$ CH_3O $PhCH_2O$ N CH_2 $OCOOC_2H_5$ (XXIII)

$\xrightarrow[PtO_2]{H_2}$ CH_3O $PhCH_2O$ NH CH_2 $OCOOC_2H_5$ (XXIV) $\xrightarrow{HCl}$ CH_3O HO NH CH_2 OH (IV)

1-benzyl-3,4-dihydroisoquinoline (XXIII). The total synthesis of coclaurine has thus been accomplished after further reduction of (XXIII) and hydrolysis of (XXIV).

On the other hand, mild dehydrogenation of 3,4-dihydroisoquinoline (XXVIII), which was obtained by Schotten-Baumann reaction between (XXV) and (XXVI) and then by Bischler-Napieralski reaction of (XXVII), gives an expected isoquinoline, papaverine (XXIX).[19]

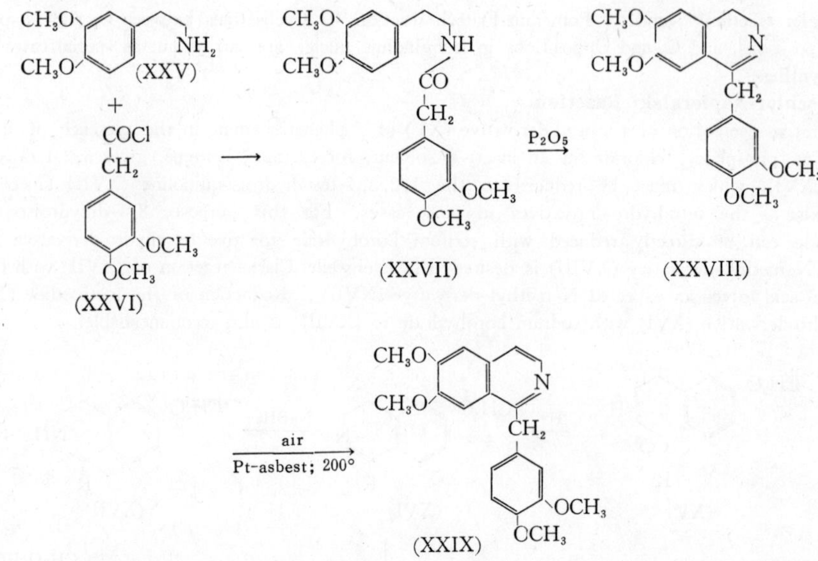

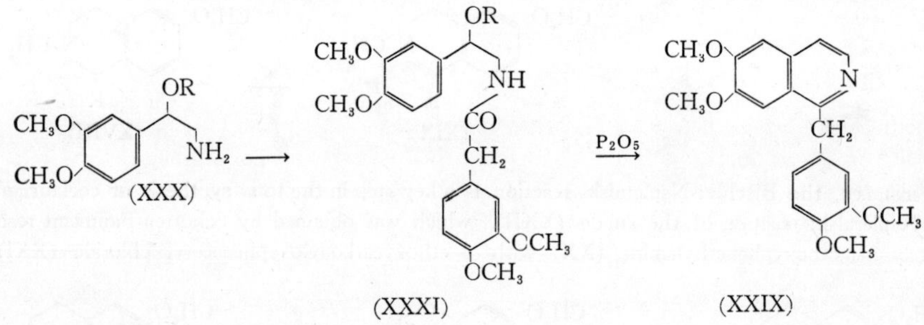

The Pictet-Gams modification[20] is used for the purpose of obtaining an isoquinoline derivative itself. Namely, if β-hydroxy- (XXX: R=H) and β-methoxyphenethylamine (XXX: R=CH_3) instead of β-phenethylamine (XXV) is used in the above reaction with acid chloride, the isoquinoline (XXIX) will be obtained *via* the amide (XXXI).

2. Pictet-Spengler Reaction

Treatment with acid of the Schiff base (XXXIII), prepared by condensation of a phenethylamine (XXV) with a carbonyl compound (XXXII), gives a 1,2,3,4-tetrahydroisoquinoline (XXXIV). This reaction is considered a special case of Mannich reaction.

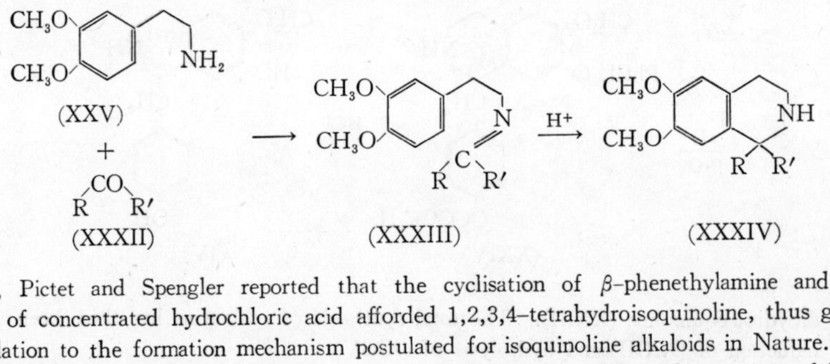

In 1911, Pictet and Spengler reported that the cyclisation of β-phenethylamine and formalin in the presence of concentrated hydrochloric acid afforded 1,2,3,4-tetrahydroisoquinoline, thus giving experimental foundation to the formation mechanism postulated for isoquinoline alkaloids in Nature.[22] Furthermore, Späth and Berger succeeded in a biogenetic type synthesis by Mannich reaction of homoveratrylamine (XXV) with homoveratraldehyde (XXXV) to give norlaudanosine (XXXVI). Dehydrogenation of the resultant norlaudanosine (XXXVI) gave papaverine (XXIX).[23]

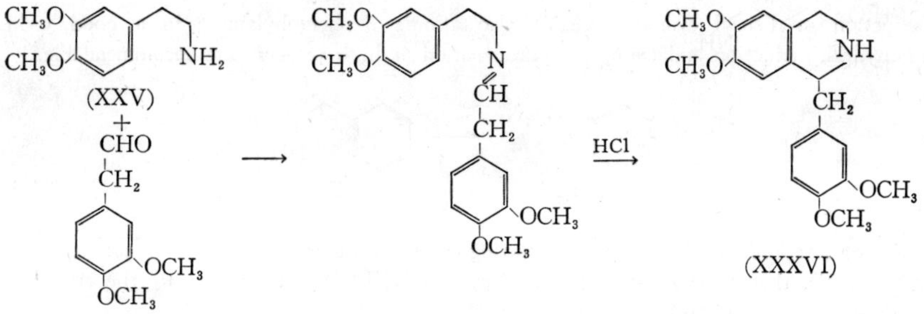

CH_3O CH_3O NH_2 (XXV) + CHO CH_2 OCH_3 OCH_3 (XXXV) ⟶ CH_3O CH_3O N CH CH_2 OCH_3 OCH_3 $\xrightarrow{HCl}$ CH_3O CH_3O NH CH_2 OCH_3 OCH_3 (XXXVI)

The berberine bridge can be also formed very easily by application of the Pictet-Spengler reaction. Cyclisation of 1,2,3,4-tetrahydroisoquinoline (XXXVII) with formalin in the presence of hydrochloric acid or formic acid[24] affords protoberberine type alkaloids such as (±)-coreximine (XXXVIII) in good yield as follows.[25]

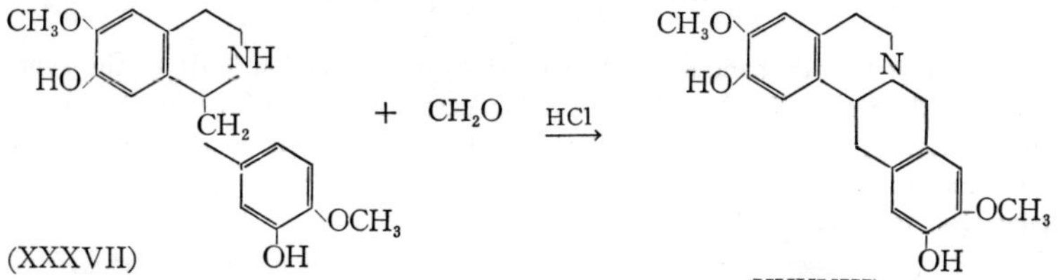

CH_3O HO NH CH_2 OCH_3 OH (XXXVII) + CH_2O $\xrightarrow{HCl}$ CH_3O HO N OCH_3 OH (XXXVIII)

In both Bischler-Napieralski and Pictet-Spengler reactions it is characteristic that the cyclisation of the amide (XXXIX) occurs predominantly at the position *para* to the methoxyl group (XL) as follows.[14,15] Accordingly, the total syntheses of the alkaloids cularine and petaline, having the ether linkage in the C_7 and C_8-positions, are impossible by these reactions.

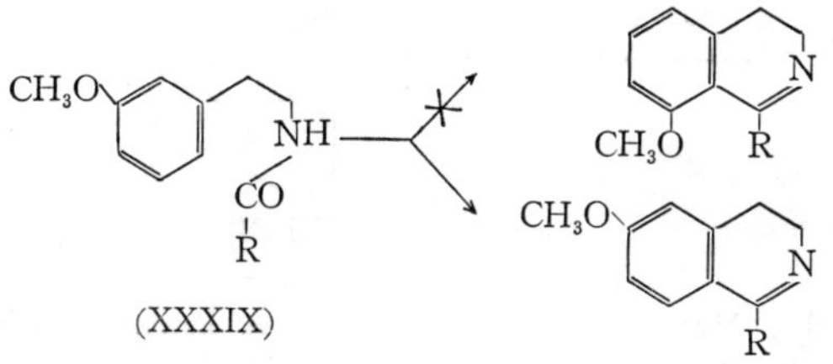

CH_3O NH CO R (XXXIX) ⟶ N CH_3O R ; CH_3O N R (XL)

3. Preparation of β-Phenethylamines[26]

The important starting materials in both reactions above are β-phenethylamine and its acyl derivative, and therefore various synthetic methods have hitherto been reported.

(a) β-Phenethylamine: β-Phenethyl bromide (XLI) can be converted into the corresponding β-phenethylamines (XLII) by application of the Gabriel synthesis[27] or by the reaction of the bromides with urotropine.[28] The oximes of arylacetaldehydes can be reduced to β-phenethylamines.[29] Closely related

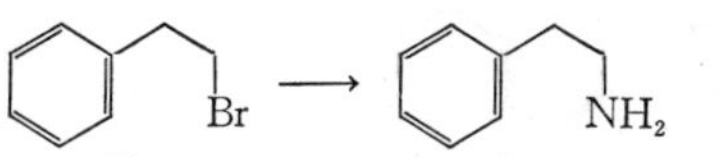

Br (XLI) ⟶ NH_2 (XLII)

reactions may be found in the reduction of ω-nitrostyrene with lithium aluminium hydride or Tafel electrolytic reduction to β-phenethylamines.[30] In this case the nitro compounds (XLIII) are prepared by the condensation of aromatic aldehydes with nitromethane in the presence of base.

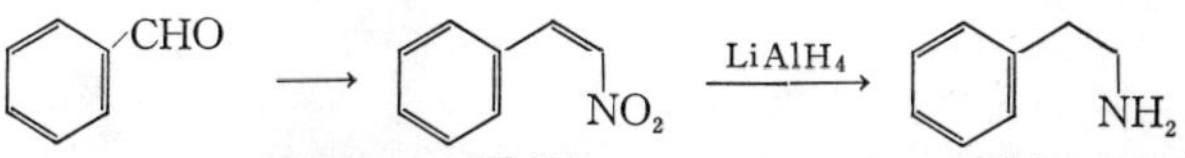

CHO ⟶ NO_2 (XLIII) $\xrightarrow{LiAlH_4}$ NH_2 (XLII)

Reduction of benzyl cyanide (XLIV) also affords β-phenethylamine in a good yield; lithium aluminium hydride reduction and catalytic hydrogenation at high pressure are recommendable.[31]

CH_2CN → NH_2

(XLIV) (XLII)

A series of reactions has been reported starting from aromatic aldehydes and proceeds first to the cyanhydrins (XLV), then to the α-acetoxybenzyl cyanides (XLVI) and, finally, to the arylethylamines by hydrogenolysis and hydrogenation over palladium.[32]

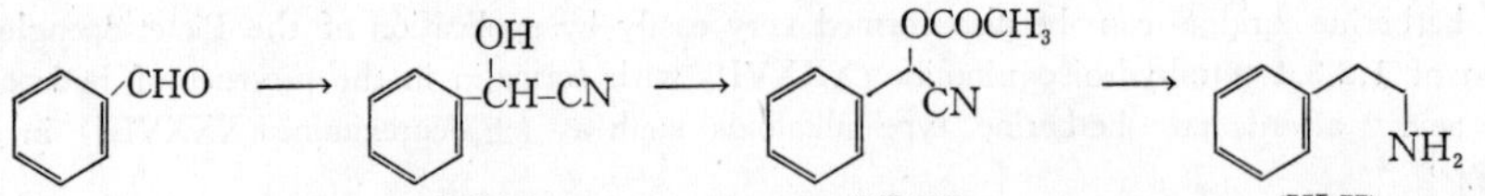

CHO → OH CH-CN → $OCOCH_3$ CN → NH_2

(XLV) (XLVI) (XLII)

The application of the Hofmann reaction or of the Curtius reaction leads to the syntheses of the corresponding β-phenethylamines.[32,33,34] An interesting extension of the latter method is use of an isocyanate (XLVIII), which formed by rearrangement of the azide (XLVII). As soon as the isocyanate (XLVIII) is generated, it combines with the carboxylic acid, thereby yielding the amide (XLIX) suitable for the Bischler-Napieralski reaction.[35]

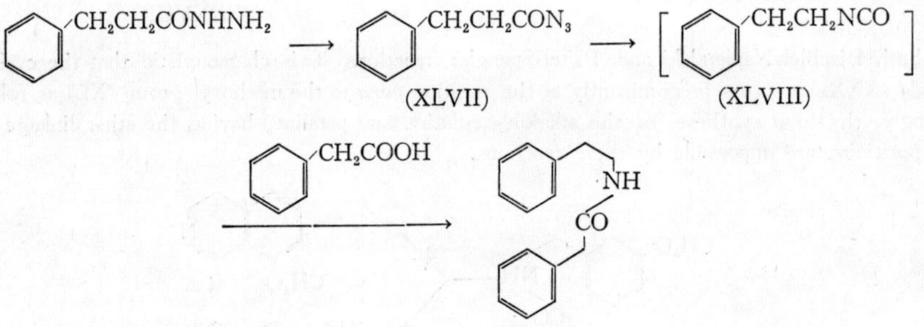

$CH_2CH_2CONHNH_2$ → $CH_2CH_2CON_3$ → [CH_2CH_2NCO]

(XLVII) (XLVIII)

CH_2COOH → NH CO

(XLIX)

(b) β-Hydroxy- and β-Methoxyphenethylamine: A ketone is converted into the isonitroso compound (L), reduced to the aminoketone (LI) and then reduced to amino alcohol (LII).[36]

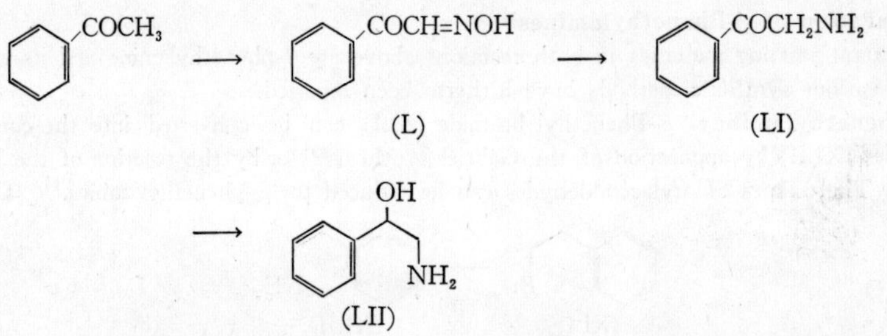

$COCH_3$ → COCH=NOH → $COCH_2NH_2$

(L) (LI)

→ OH NH_2

(LII)

Treatment of the propenylbenzene (LIII) with sodium nitrite and ethereal sulphuric acid gives the pseudonitrosite (LIV), which is converted into the acetoxynitro derivative (LV). Reduction of these compounds affords β-hydroxyphenethylamines (LVI) by the following two methods.[37]

$$\text{Ar-CH=CH-CH}_3 \ (\text{LIII}) \rightarrow \text{Ar-CH(NO)-CH(NO}_2\text{)-CH}_3 \ (\text{LIV}) \rightarrow \text{Ar-CH(CH}_3\text{COO)-CH(NO}_2\text{)-CH}_3 \ (\text{LV})$$

$$\text{Ar-CH(OH)-CH(NO}_2\text{)-CH}_3 \rightarrow \text{Ar-CH(RCOO)-CH(NO}_2\text{)-CH}_3$$

$$\text{Ar-CH(HO)-CH(NHCOCH}_3\text{)-CH}_3 \rightarrow \text{Ar-CH(OH)-CH(NH}_2\text{)-CH}_3 \ (\text{LVI}) \rightleftarrows \text{Ar-CH(HO)-CH(NHCOR)-CH}_3$$

β–Aryl–β–methoxyphenethylamines (LVIII) are prepared by way of the base-catalysed addition of methanol to the nitrostyrene (XLIII), followed by reduction of the nitro group.[38)]

(XLIII) → (LVII) → (LVIII)

(c) **Pomeranz-Fritsch Reaction:** Cyclisation of the benzalaminoacetal (LIX), obtained by condensation of an aromatic aldehyde with an aminoacetal, in the presence of acidic catalysts such as sulphuric and polyphosphoric acids, affords an isoquinoline (LX).

$$\text{C}_6\text{H}_5\text{CHO} + \text{H}_2\text{NCH}_2\text{CH(OC}_2\text{H}_5)_2 \longrightarrow (\text{LIX}) \xrightarrow{\text{H}^+} (\text{LX})$$

In general, the yield of this reaction is variable and depends upon the starting materials, but it is characteristic that the isoquinolines having methoxyl and hydroxyl groups in either of the C_5, C_6, C_7, or C_8–positions of an isoquinoline nucleus can be synthesised by this reaction. Accordingly, this reaction is important as a synthetic compliment to the Bischler-Napieralski and Pictet-Spengler reactions. For instance, although the cularine (LXI) having the ether bridges at C_7 and C_8–positions could not be obtained by Bischler-Napieralski reaction, the Pomeranz-Fritsch reaction of dibenzketo-oxepine (LXII) gave the isoquinoline derivative (LXIV) *via* Schiff base (LXIII). Methylation of (LXIV), followed by reduction of its methiodide (LXV), gave (±)–cularine (LXI) successfully.[39)]

$$(\text{LXII}) \xrightarrow{\text{H}_2\text{NCH}_2\text{CH(OC}_2\text{H}_5)_2} (\text{LXIII}) \xrightarrow{\text{H}_2\text{SO}_4}$$

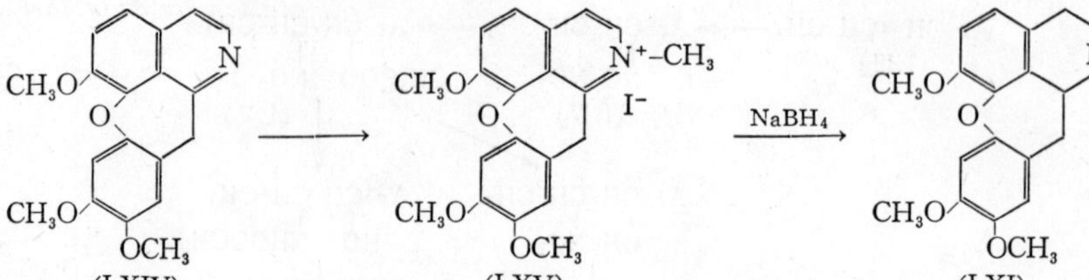

(LXIV) → (LXV) —$NaBH_4$→ (LXI)

Furthermore, Schlittler and Müller reported that cyclisation of the Schiff base (LXVI) derived from benzylamine and glyoxal semiacetal gave the expected isoquinoline (LXVII) in the presence of acid. This reaction is also important as a modified procedure of Pomeranz-Fritsch reaction.[40]

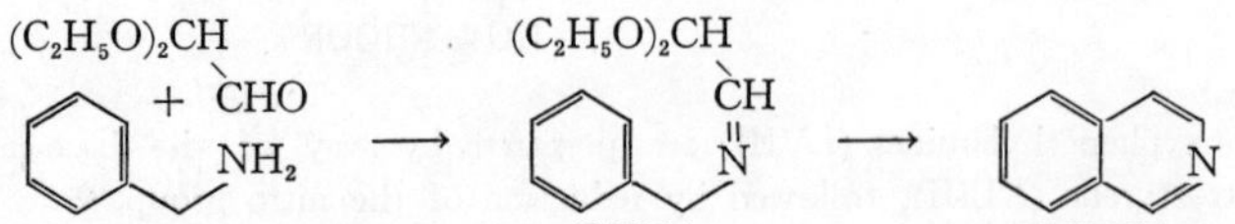

$(C_2H_5O)_2CH-CHO$ + benzylamine → (LXVI) → (LXVII)

4. Miscellaneous Synthetic Methods

(a) Isocarbostyrils from Isocoumarins: The starting materials in this reaction are isocoumarins, homophthalic acid derivatives, and phthalimidoacetic acid derivatives; from each, isocarbostyrils are formed. Gabriel reported the conversion of 3-phenylisocoumarin (LXVIII) into 3-phenylisocarbostyril (LXIX) with an alcoholic ammonia solution at 100°.[41]

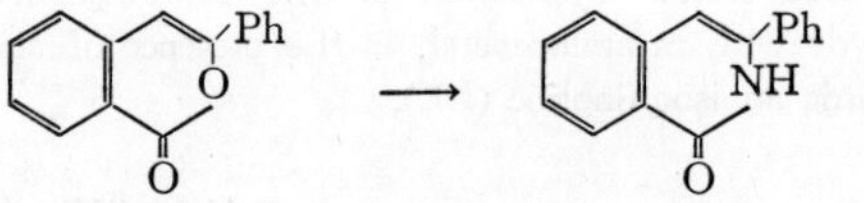

(LXVIII) → (LXIX)

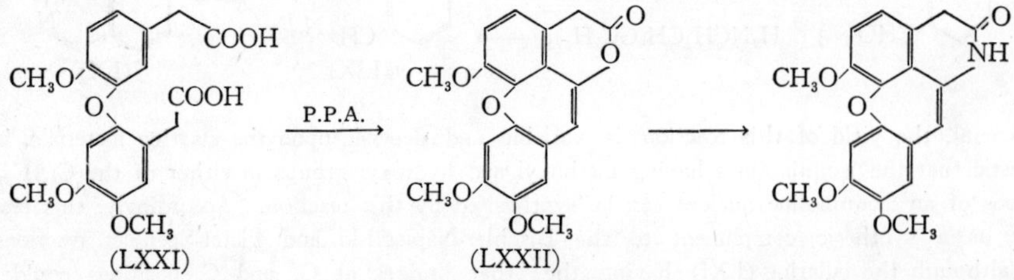

(LXXI) —P.P.A.→ (LXXII) → lactam —1) H_2/Pt 2) $LiAlH_4$→ (LXX)

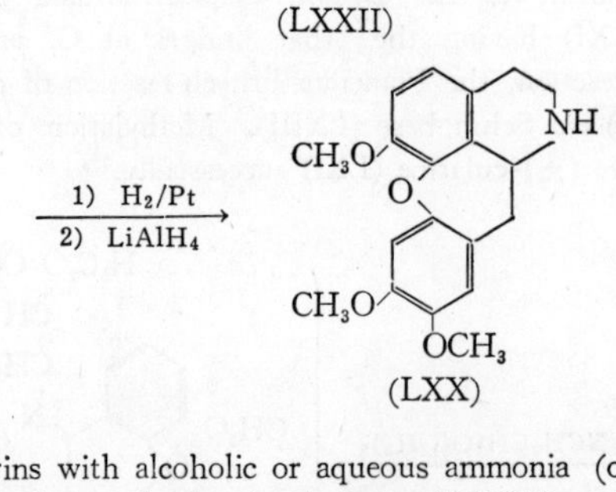

The reaction of isocoumarins with alcoholic or aqueous ammonia (or with primary amines) is quite general and has been applied to the synthesis of cularimine (LXX)[42] and benzophenanthridine alkaloids,[46] the former of which is shown as above. Cyclisation of the acid (LXXI) with polyphosphoric acid gave the lactone (LXXII), which was converted into the lactam by ammonolysis with ethanolic ammonia. Catalytic hydrogenation of the lactam, followed by reduction with lithium aluminium hydride, gave (±)-cularimine (LXX).

(b) Homophthalimide Synthesis: The homophthalimides (LXXIII) can be prepared in an excellent yield by heating the homophthalic acid diammonium salts.[43] Other methods which give these compounds include cyclisation by heating of homophthalamic acid[44] and the treatment of homophthalonitrile with strong acid.[45] Gates succeeded in the total synthesis of morphine by application of this reaction, as described later.[6]

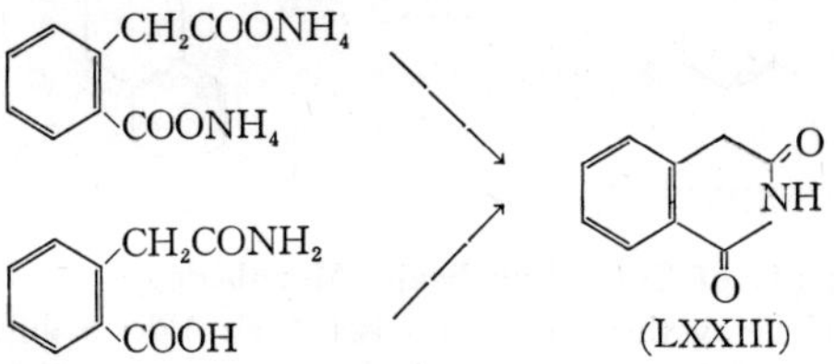

(LXXIII)

(c) Synthesis by Phenolic Oxidative Coupling: Total syntheses of the proaporphine (LXXIV), aporphine[47] (LXXV), and morphine type alkaloids,[48] (LXXVI) and (LXXVII), have been achieved recently

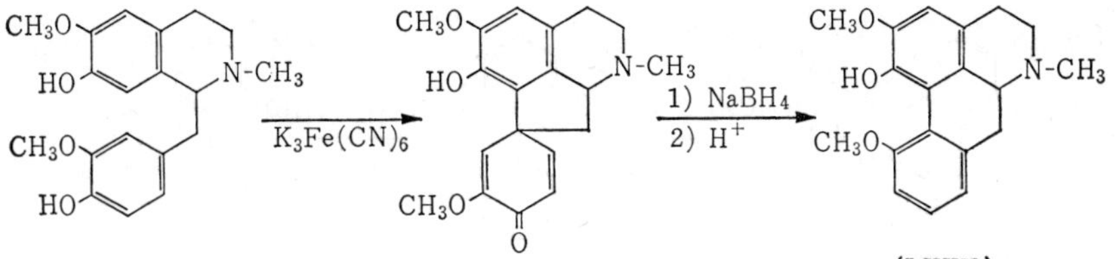

(LXXIV) (LXXV)

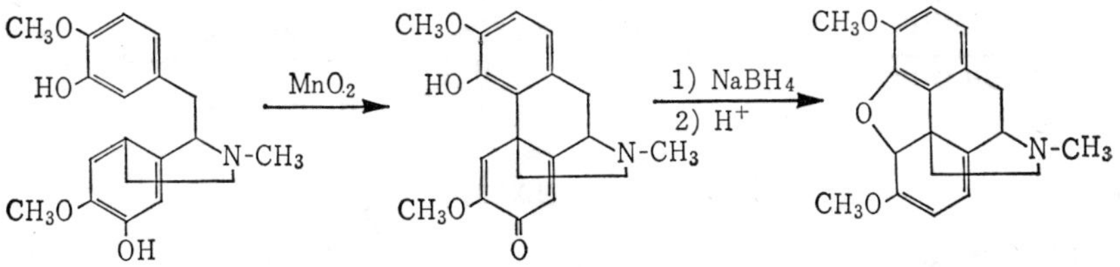

(LXXVI) (LXXVII)

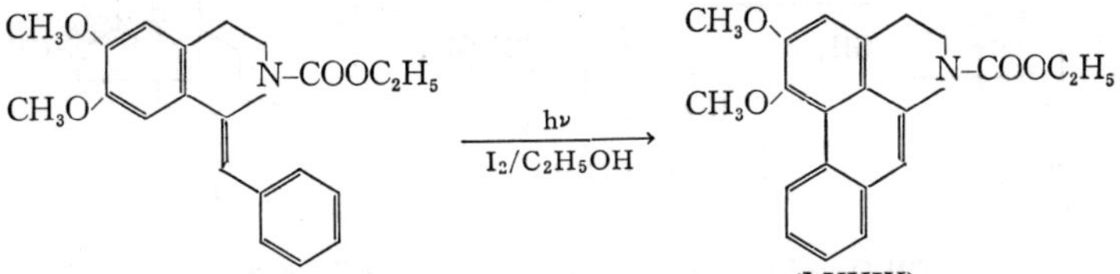

(LXXVIII) (LXXIX)

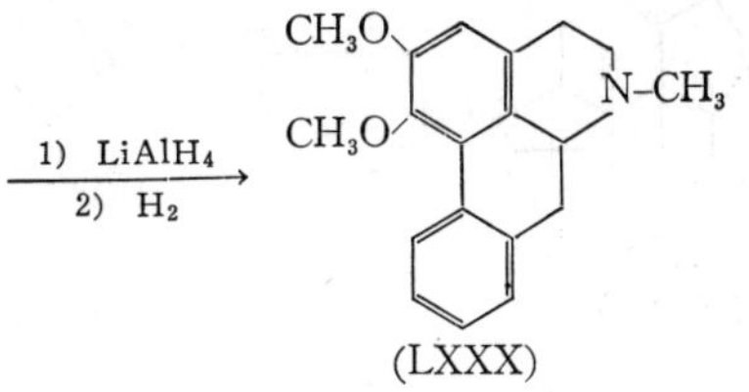

(LXXX)

as is shown. This reaction will be important from the synthetic and biogenetic points of view.

Furthermore, the formation of the aporphine nucleus by photooxidation[49] and photocyclisation[50] has recently been reported. Photooxidation of the isoquinoline (LXXVIII) in ethanol in the presence of iodine gave the compound (LXXIX), whose reduction with lithium aluminium hydride and catalytic hydrogenation afforded the aporphine derivative (LXXX).[49] Photocyclisation of (LXXXI) likewise gave the aporphine derivative (LXXX).[50]

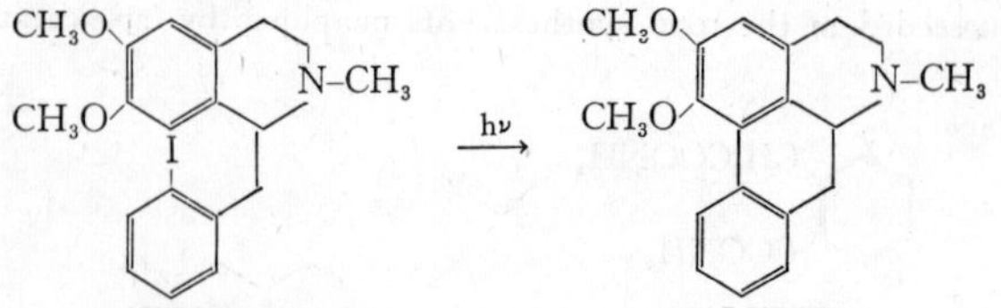

(LXXXI) (LXXX)

5. Representative Example of Total Synthesis (Morphine)

The key step in the morphine synthesis by Gates is a Diels-Alder addition of butadiene to 4-cyano-methyl-1,2-naphthoquinone (LXXXII). On catalytic hydrogenation the adduct (LXXXIII) undergoes ring closure to the keto-lactam (LXXXIV), the carbonyl group of which was removed by Wolff-Kishner reduction. The amide carbonyl group was removed by further reduction with lithium aluminium hydride. N-Methylation then gave the product (LXXXV). The compound thus obtained had the correct ring

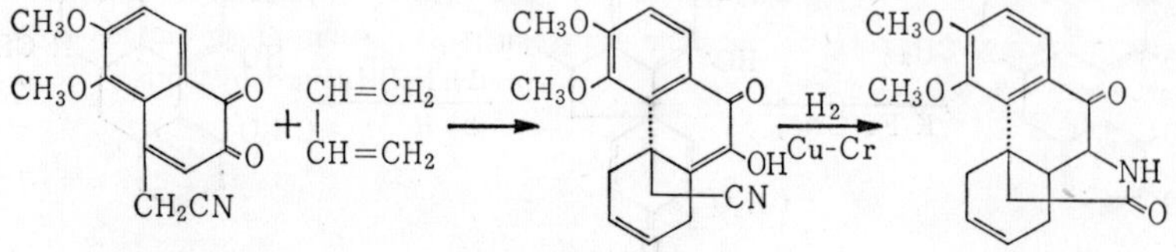

(LXXXII) (LXXXIII) (LXXXIV)

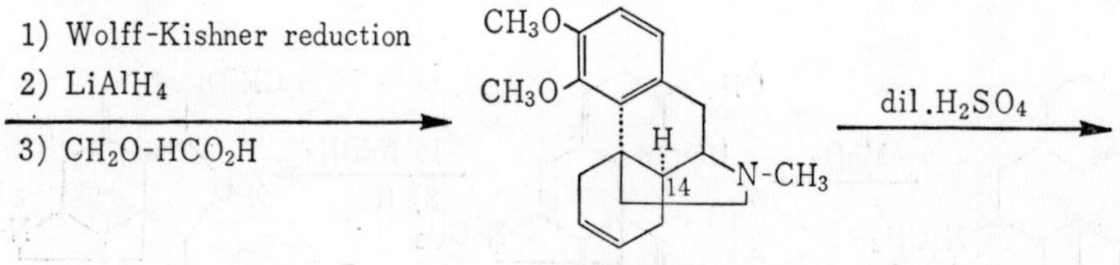

1) Wolff-Kishner reduction
2) $LiAlH_4$
3) CH_2O-HCO_2H

(LXXXV)

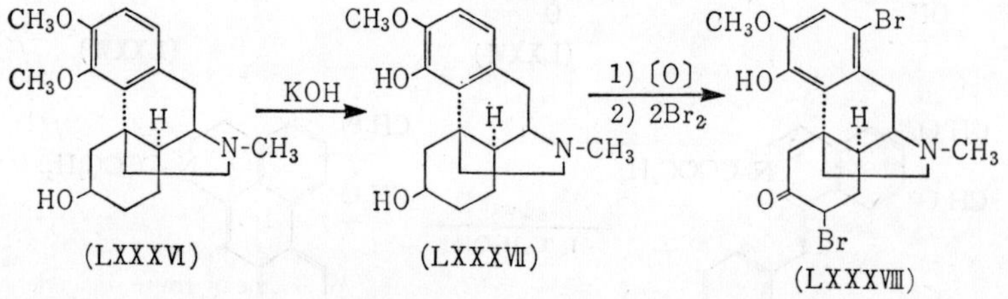

(LXXXVI) (LXXXVII) (LXXXVIII)

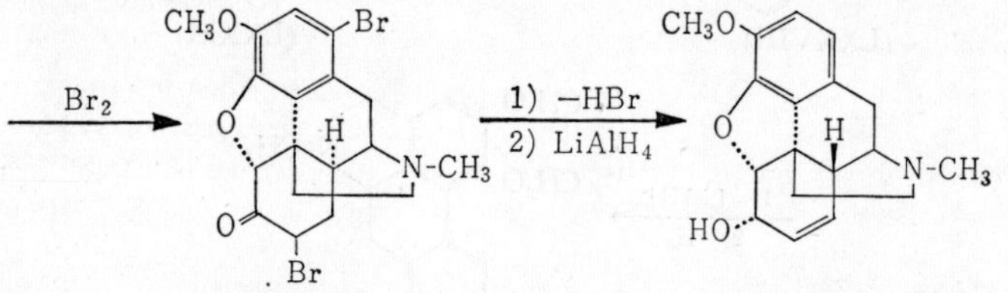

(LXXXIX) codeine (XC)

structure but the unnatural configuration at the C_{14}-position. However, this fault, discovered by a fortunate rearrangement, was corrected near the end of synthetic sequence. The double bond of (LXXXV) was hydrated to the compound (LXXXVI), the ether group at the C_4-position was selectively demethylated to (LXXXVII), and oxidation and bromination gave the 6-ketone (LXXXVIII), which was further brominated to give the compound (LXXXIX). In this case bromine is introduced first at C_7 and C_1-positions, and one mole of bromine then formed the oxide bridge by ring closure to give (LXXXIX). In the course of dehydrobromination of (LXXXIX) with 2, 4-dinitrophenylhydrazine to produce the α, β-unsaturated ketone structure of codeinone, inversion occurred at the centre adjacent to the conjugated system to make the natural orientation at C_{14}-position. Since the optical resolution had been effected in an intermediate stage (LXXXV), reductive removal of the bromine atom at C_1-position gave codeine (XC) which was convertible into morphine by cleavage of the ether group.

Since a synthetic sample, without optical resolution, is a racemate, it is different from the natural product, and therefore, optical resolution is necessary. But even in case of the racemate, comparison of synthetic and natural products can be made by R_F values in paper and thin-layer chromatogram and infrared (in chloroform), ultraviolet, nuclear magnetic resonance, and Mass spectra.

(III) Biogenesis and Biosynthesis

Sir Robert Robinson[51)] proposed the biogenesis of many alkaloids at the beginning of this century[52)]; his proposal was based on the idea that hypothetical simple, active precursors were condensed by Mannich type reaction to give the complicated natural products. Although the precursors might be present in the plant cell, they could not be identified at that time.

In 1950's Barton proposed the so-called "phenol oxidation mechanism" in the biogenesis of alkaloids[53)]; the biphenyl ether bonds in biscoclaurine type alkaloids and the biphenyl bond in morphine or aporphine type alkaloids were formed by oxidative coupling of phenols by action of an enzyme present in plants. The biogenesis of all isoquinoline alkaloids seems to be rationally explained by both these theories.

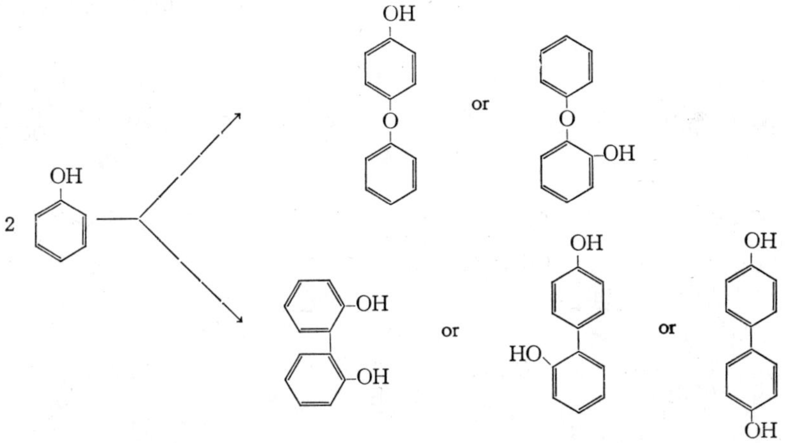

OH
2
OH
O
or
O
OH
OH
OH
or
HO
or
OH
OH
OH
OH

There are many benzylisoquinoline alkaloids, and related alkaloids such as morphine, Amaryllidaceae alkaloids, and colchicine type alkaloids, which are produced by Mannich condensation of aromatic compounds derived from amino acids, followed by their oxidative coupling. For some of these, biogenetic mechanisms have been proved true by tracer methods. For example, a labelled papaverine (XCII) was obtained by feeding tyrosine-2-C^{14} (XCI) in *Papaver somniferum*.[54)]

The mechanistic pathway has been illustrated in the following equation proceeding by way of (XCIII) and (XCV); but the step of hydroxylation in the aromatic ring system has not yet been confirmed.[54)]

The position of labelled carbon (C*) was proved as follows; laudanosine (XCVI) derived from active papaverine (XCII) was degradated by the Hofmann method to give compound (XCVII). Further oxidation of (XCVII) gave a mixture of the inactive veratric acid (XCVIII) and active amino acid (XCIX), whose Hofmann degradation afforded a doubly labelled carboxylic acid (C).[54)]

Battersby[55,56)] found that an aqueous solution of tyrosine-2-C^{14} was incorporated into morphine (CI) in opium and labelled carbon was located specifically at C_9 and C_{16}-positions as demonstrated by chemical

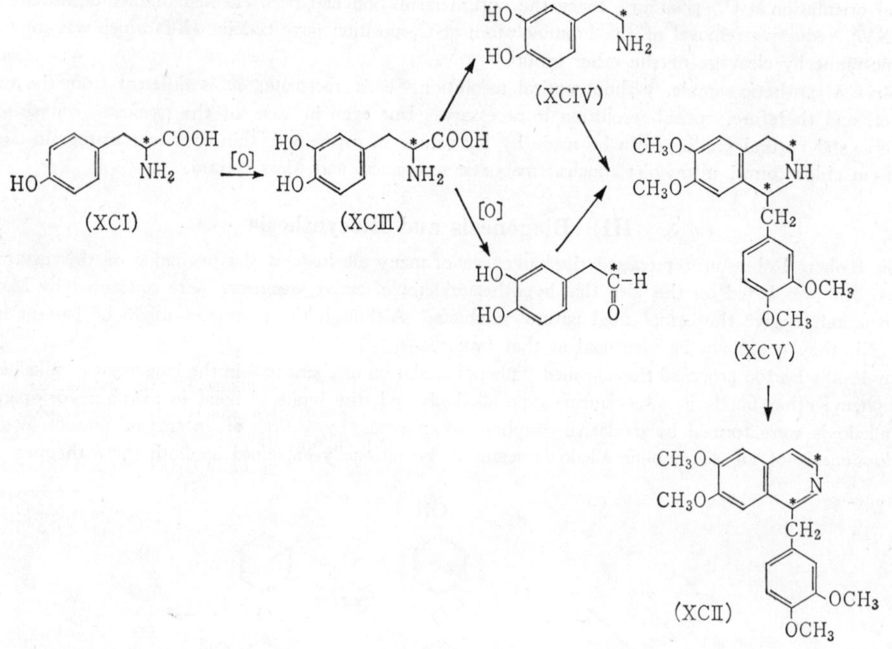

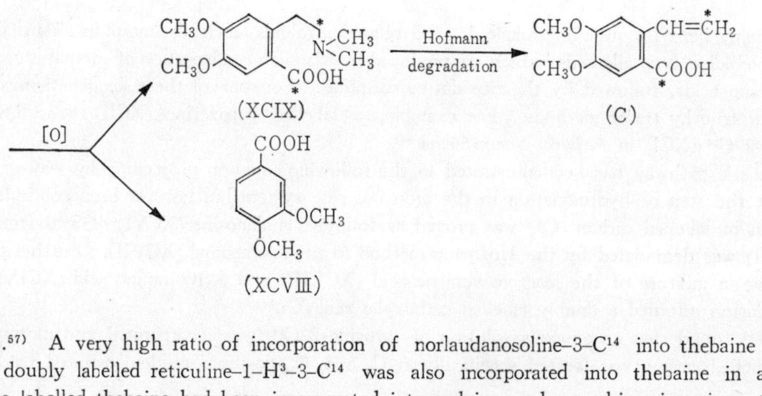

(XCI) (XCIII) (XCIV) (XCV) (XCII)

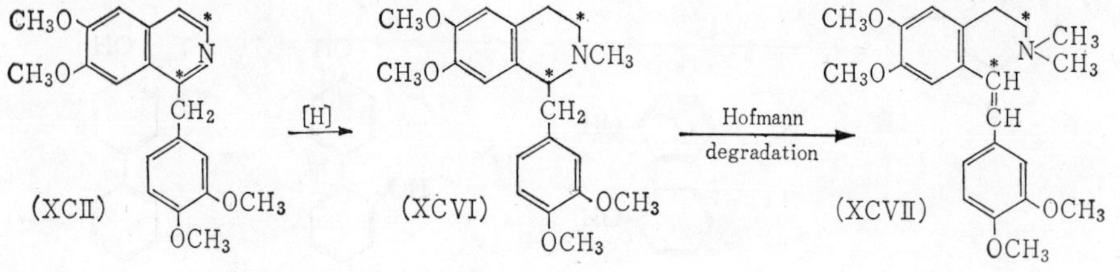

[O] [O]

(XCII) [H] (XCVI) Hofmann degradation (XCVII)

(XCIX) Hofmann degradation (C) [O] (XCVIII)

methods.[57] A very high ratio of incorporation of norlaudanosoline-3-C^{14} into thebaine was observed[58] and the doubly labelled reticuline-1-H^3-3-C^{14} was also incorporated into thebaine in a high ratio.[59] Since the labelled thebaine had been incorporated into codeine and morphine in opium, the biosynthetic pathway to morphine from simple precursors was shown as follows.

tyrosine ⟶ norlaudanosoline (or its dimethyl ether) ⟶ reticuline ⟶ thebaine ⟶ morphine

Although the hydrophenanthrene ring is formed by phenolic oxidative coupling of reticuline or its biogenetic equivalents in this process, there are two mechanisms for formation of the second ethereal oxide bridge as shown in pathways (1)[53] and (2)[58,60]. Sinomenine (CII), one of the hydrophenanthrene alkaloids having no ethereal oxide bridge, has an oxygen group at the C_7-position, whereas morphine alkaloids, having no oxygen group at the C_7-position, possess an ethereal ring. Therefore, Ginsburg[60] and Battersby[58] independently proposed the biogenesis of these groups as follows; a carbonyl group at the C_7-position in the dienone system (CIII) is reduced to a carbinol as in (CIV), and then allylic elimination of this hydroxyl group gives an ether bridge compound (CV) as shown in pathway (2).

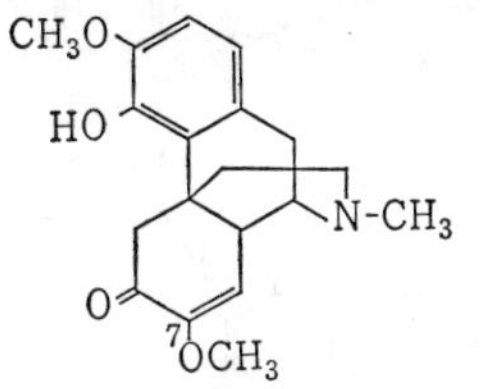

(CII)

Barton converted H^3-labelled thebaine (CV) into the dienone type alkaloid, salutaridine (CIII), also found in opium, and the latter alkaloid was reduced with sodium borotritiride to give salutaridinols-I and II (stereoisomers) (CIV) doubly labelled with tritium (H^3). The salutaridinol-I (CIV) was converted into thebaine (CV) with an incorporation ratio of 7%. In this case the position of isotope and the ratio of activity were the same as those of precursor, but the other precursor showed very poor incorporation ratio. These experiments proved that the correct pathway was route (2).[61]

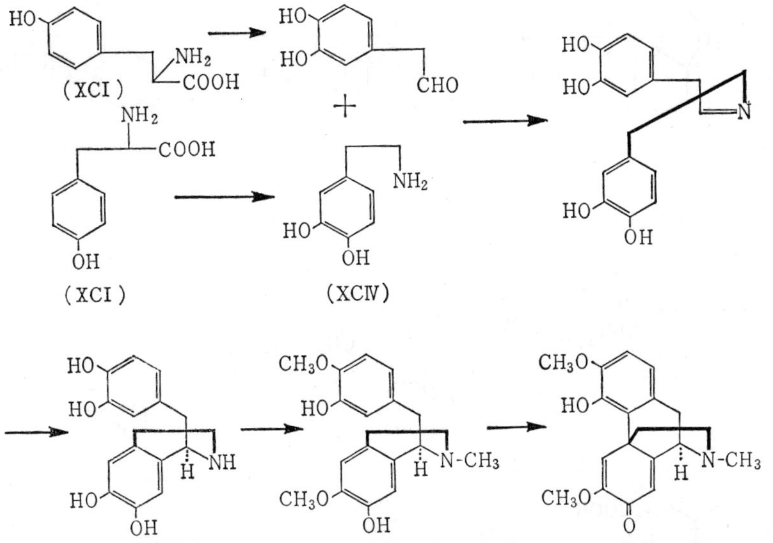

(XCI) (XCI) (XCIV) (CIII)

Narcotine and hydrastine (CVI), phthalide isoquinoline alkaloids in opium, and cryptopine (CVII), berberine and protoberberine alkaloids in *Coptis japonica*, are derived biogenetically from norlaudanosoline (CVIII).[51] Spencer[62] succeeded in transformation of tyrosine-2-C^{14} into hydrastine (CVI) with incorporation of activity being 39% at C_1 and 51% at the C_3-position, but hydrastine (CVI) from 3,4-dihydroxyphenethylamine-1-C^{14} was labelled only at the C_3-position. Battersby[63] also showed tyrosine-2-C^{14} (XCI) as well as norlaudanosoline-1-C^{14} (CVIII) and formate-1-C^{14} to be precursors of narcotine (CIX).

The so-called "berberine bridge"[51] in berberine and the carbon of the lactone carbonyl group in phthalide alkaloids were derived from C_1-units, the precursor being the methyl group of methionine rather than formate or its biogenetic equivalents.

Reticuline (CXI), having activities of 5.9% at O-$\overset{*}{C}H_3$ group and 94.1% at N-$\overset{*}{C}H_3$ group, was transformed into berberine (CX) showing activities of 5.7% at the methylenedioxy group and 91% at the berberine bridge. Thus the carbon in the berberine bridge was derived from the N-methyl group.[64,65] These facts were also proved in the case of protopine.[66]

path 1 ← (CIII) → path 2

(CIV)

(CV)

morphine (CI)

hydrastine (CVI)

cryptopine (CVII)

(CVIII)

(CIX)

(CXI)

(CX)

3, 4, 5, 6– and 2, 3, 5, 6–tetrasubstituted aporphine alkaloids, (CXIII) and (CXIV) were formed by oxidative cyclisation of benzyltetrahydroisoquinoline (CXII).[51] However, 5, 6–disubstituted (CXVI) and 4, 5, 6–trisubstituted aporphine alkaloids could not be interpreted by the simple biogenetic mechanism above. In 1957, Barton[53] presumed "dienone-phenol rearrangement" (later revised to "dienol-benzene rearrangement") to take place in these biogeneses. This hypothesis[61] has been proved recently by the incorporation of isoreticuline–1–H^3 or labelled (+)–orientaline into isothebaine[67] and the isolation of crotonosine having a dienone structure from *Croton linearis*.[68]

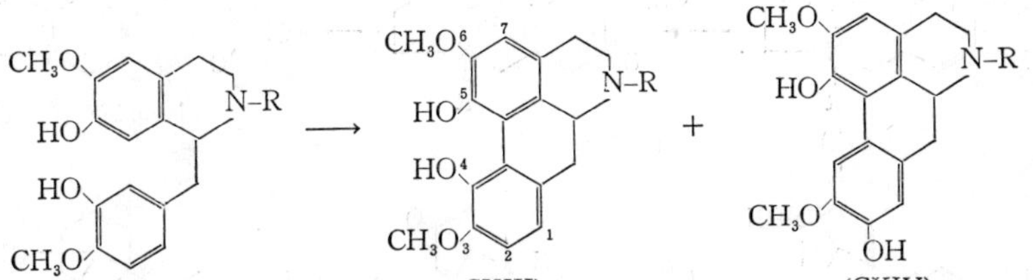

(CXII) (CXIII) (CXIV)

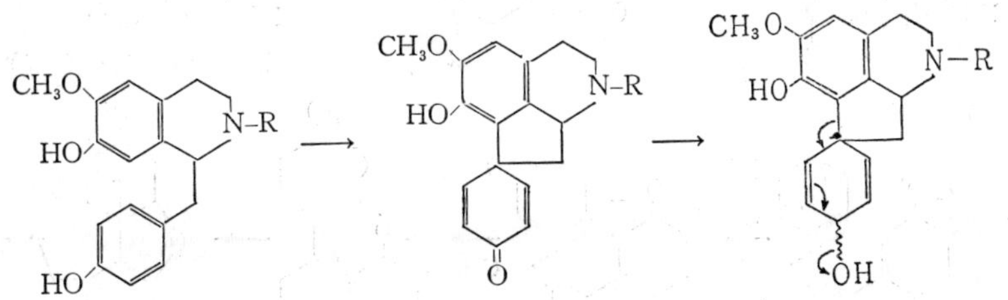

(CXV)

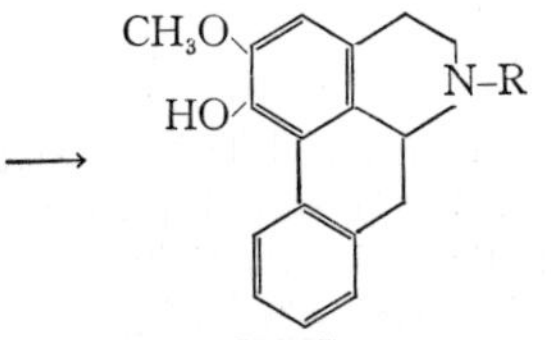

(CXVI)

When this type of oxidation occurs between carbon and oxygen in the plant, there are two possibilities; one is intramolecular carbon-oxygen bond formation and the other is intermolecular carbon-oxygen bond formation. The former example occurs in the formation of cularine alkaloids (LXI)[69] and the latter explains the formation of biscoclaurine (bisbenzylisoquinoline) alkaloids (CXVII).[51,70]

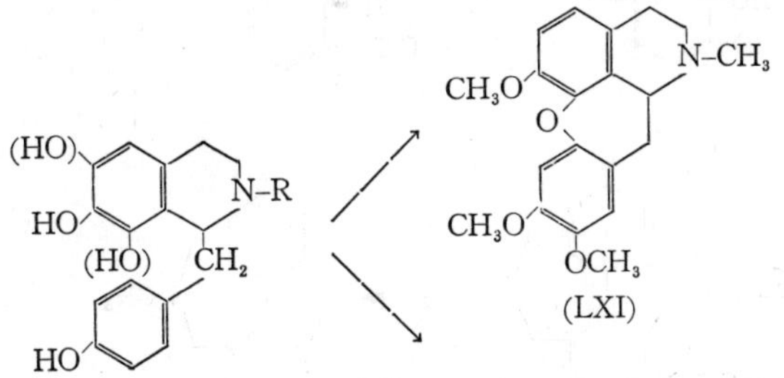

(LXI)

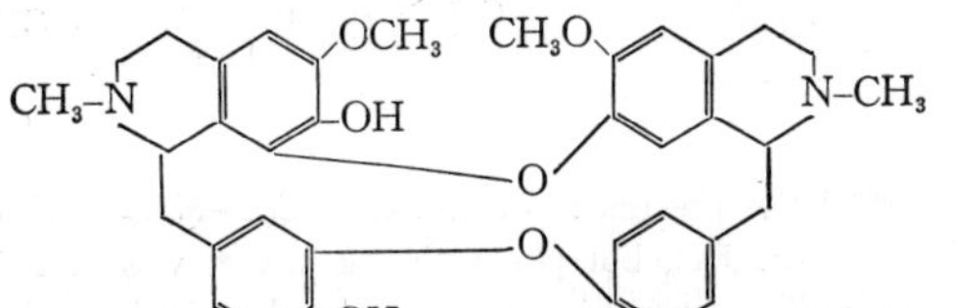

(CXVII)

Barton,[53] who had observed the existence of two C(6)–C(2) and C(6)–C(1) units in all Amaryllidaceae alkaloids, proposed a biogenesis based on phenol oxidation of a C(6)–C(1)–N–C(2)–C(6) intermediate such as O–methylnorbelladine (CXVIII : R=H). This was proved by many tracer experiments using O–methylnorbelladine, or its biogenetic equivalent (CXXII), by the Barton[71] and Battersby[58] schools.

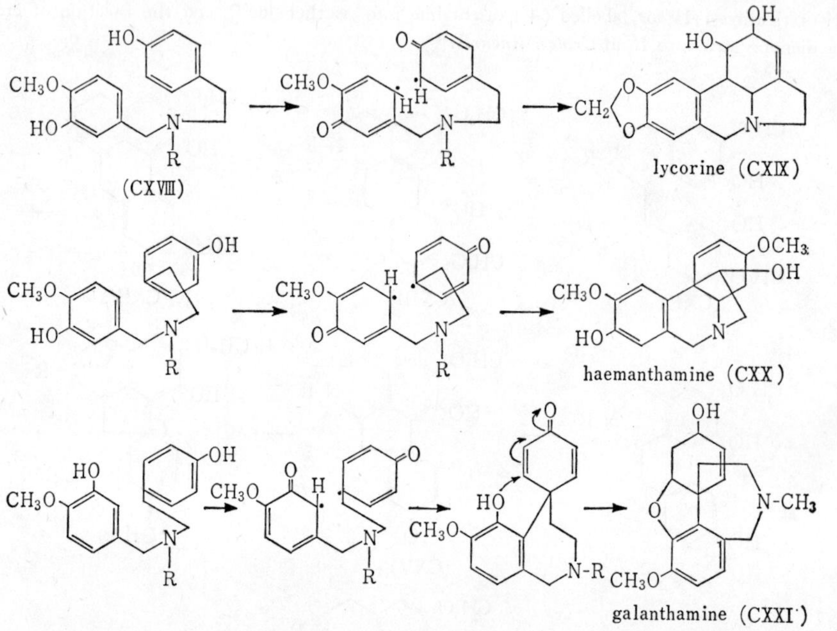

lycorine (CXIX)

(CXVIII)

haemanthamine (CXX)

galanthamine (CXXI)

Norbelladine (CXXII), N–methylnorbelladine, and O, N–dimethylnorbelladine (methylated at the C_4–hydroxyl group), which were labelled with C^{14} in several positions, were incorporated into galanthamine (CXXI), but O-methylnorbelladine–C^{14} (methylated at the C_3-hydroxyl group) was not incorporated. These facts show that the biosynthetic route for the formation of galanthamine (CXXI) is as below.[72]

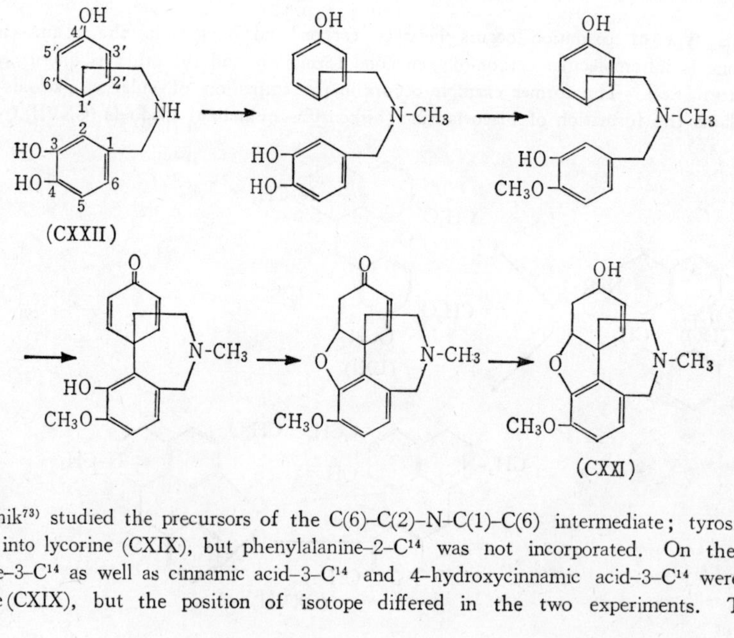

(CXXII)

(CXXI)

Suhadolnik[73] studied the precursors of the C(6)–C(2)–N–C(1)–C(6) intermediate; tyrosine–3–C^{14} was incorporated into lycorine (CXIX), but phenylalanine–2–C^{14} was not incorporated. On the other hand, phenylalanine–3–C^{14} as well as cinnamic acid–3–C^{14} and 4–hydroxycinnamic acid–3–C^{14} were incorporated into lycorine (CXIX), but the position of isotope differed in the two experiments. Therefore, he

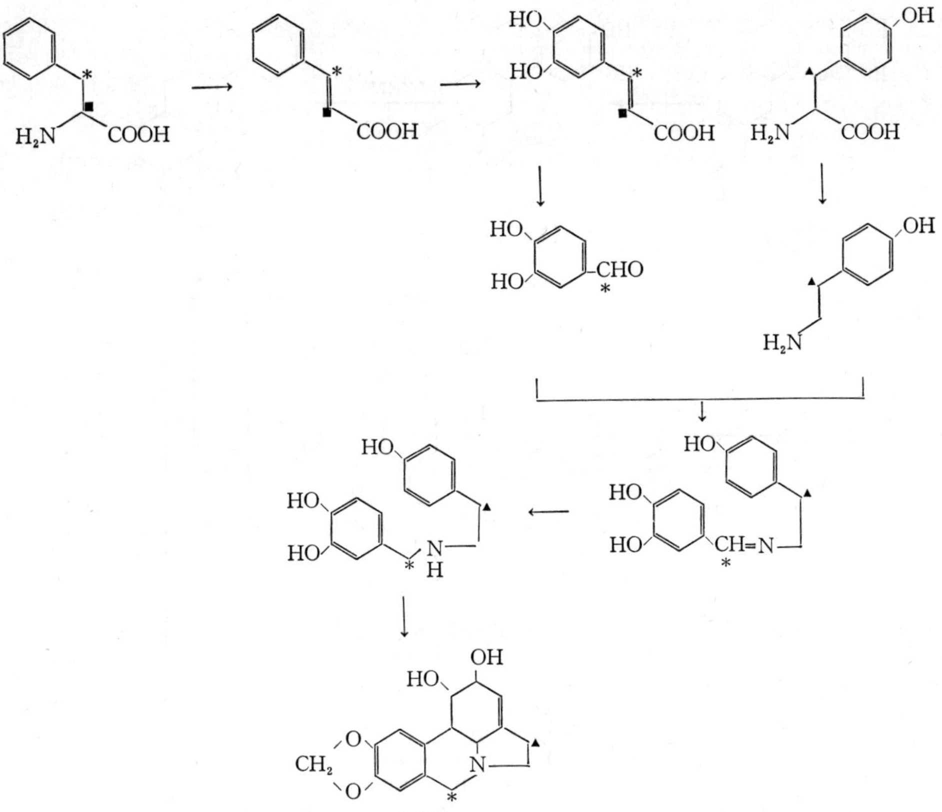

CXIX

concluded that the C(6)–C(1) unit was derived from the latter three compounds and C(6)–C(2) unit was based on the tyrosine–3–C^{14} precursor.

There were four types of biogeneses suggested for emetine (XII) and related alkaloids. Namely, Robinson[51] supposed "Woodward fission" of protoberberine type compounds (XIII), followed by Mannich type condensation with dopamine (XCIV) (or its equivalents). Leete[75] presumed that emetine was constructed by condensation of 2 moles of dopamine with the tricarboxylic acid (CXXIII) which was derived from 3 moles of acetic acid, 1 mole of malonic acid and 1 mole of formic acid. Furthermore, Wenkert[76] supposed the hypothesis that emetine was formed from prephenic acid (CXXIV) *via* shikimic acid. On the other hand, Thomas[77] proposed the "terpenoid theory" that the precursors of emetine were 2 moles of tyrosine or its equivalent and the "cyclopentanoid (CXXVII)" derived from mevalonic acid (CXXV) *via* geraniol (CXXVI). Scott[78] proved this latter hypothesis indirectly in the biosynthesis of the indole alkaloid, vindoline.

After the biosynthesis of emetine (XII) and related alkaloids had been studied by Battersby, the "Woodward fission theory" of D–ring in protoberberine (XIII),[51] the "acetic acid-malonic acid theory" of Leete,[75] and the "prephenic acid theory" of Wenkert[76] were denied, and he proposed the terpenoid theory[77,78] as follows.[74]

The erythrina alkaloids seem plausibly to be derived by oxidative coupling of the C(6)–C(2)–N–C(2)–C(6) intermediate, which forms from two moles of tyrosine (XCI).[51,53,79,80,81,82] Leete has recently proved one part of this hypothesis by incorporation of tyrosine–2–C^{14} (XCI) into erythroidine (CXXIX).[84] Addition of the amino group to the quinoid system (CXXX) leads to the spiran type compound (CXXXI). Unexceptional steps then yield the aromatic erythrina alkaloids which are represented by erysopine (CXXXII). The non-aromatic alkaloids of this family as represented by α–erythroidine (CXXIX) may be formed by oxidative fission of the catechol ring to yield the dicarboxylic acid (CXXXIII), followed by appropriate modifications of the side chains and ultimate lactonization. On the other hand, Wenkert favoured the formation of these alkaloids from prephenic acid (CXXIV).[83]

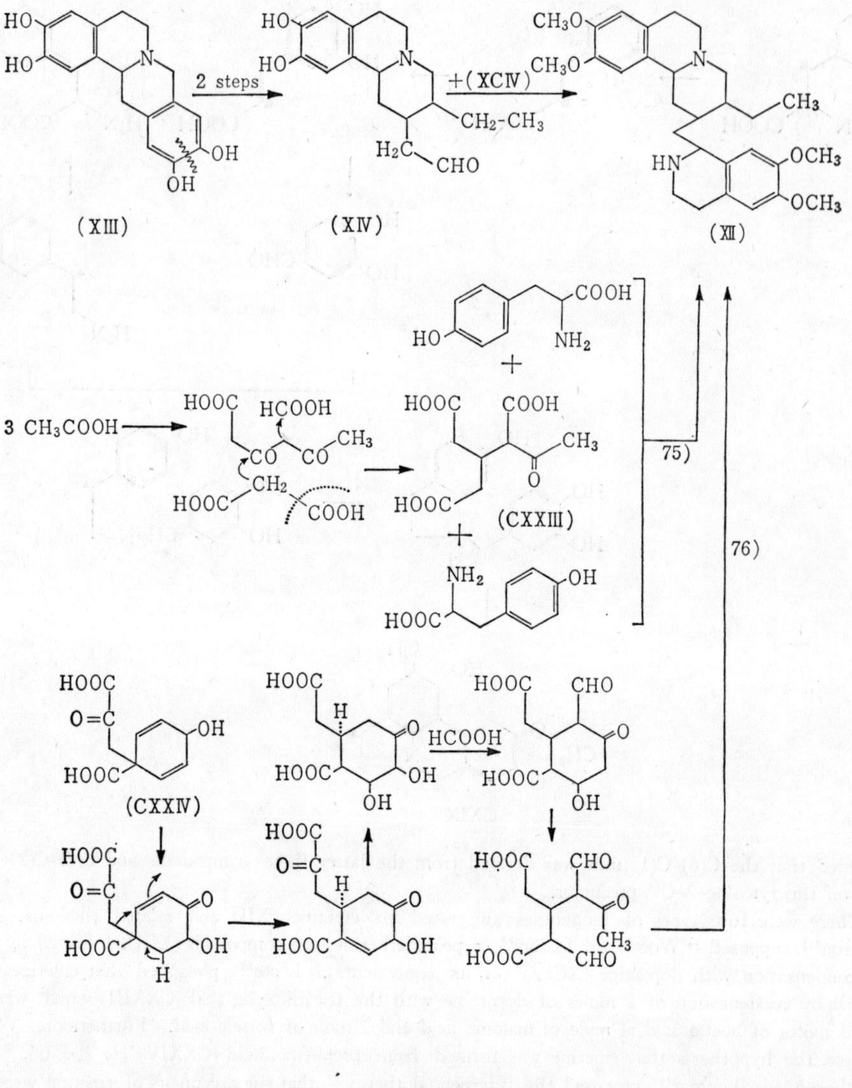

(XIII) —2 steps→ (XIV) —+(XCIV)→ (XII)

3 CH_3COOH → → (CXXIII) + tyrosine (×2) —75)→ (XII)

(CXXIV) → … —HCOOH→ … —76)→ (XII)

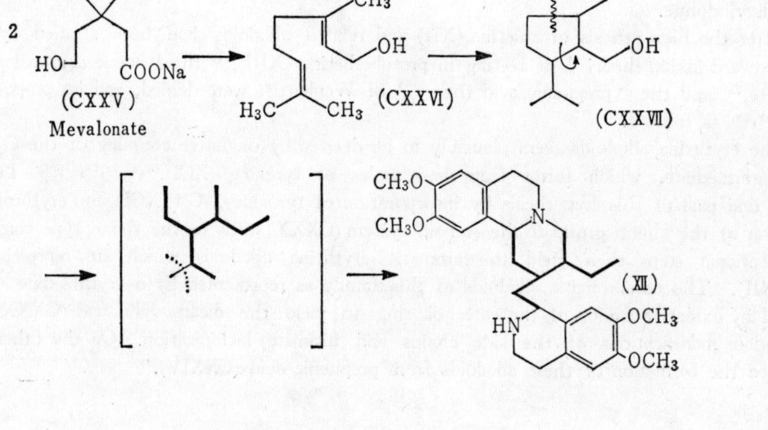

2 (CXXV) Mevalonate → (CXXVI) → (CXXVII) → [ ] → (XII)

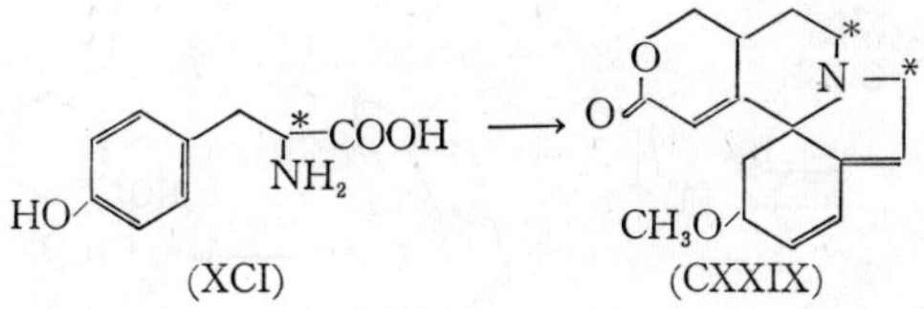

(XCI) (CXXIX)

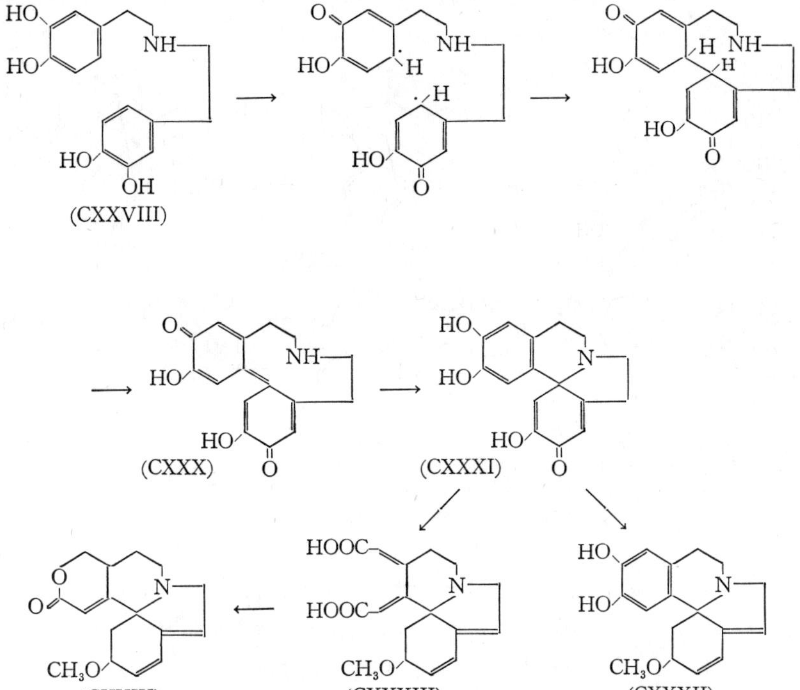

(CXXVIII)

(CXXX) (CXXXI)

(CXXIX) (CXXXIII) (CXXXII)

Furthermore, Prelog pointed out another cyclisation pathway as below.[75]

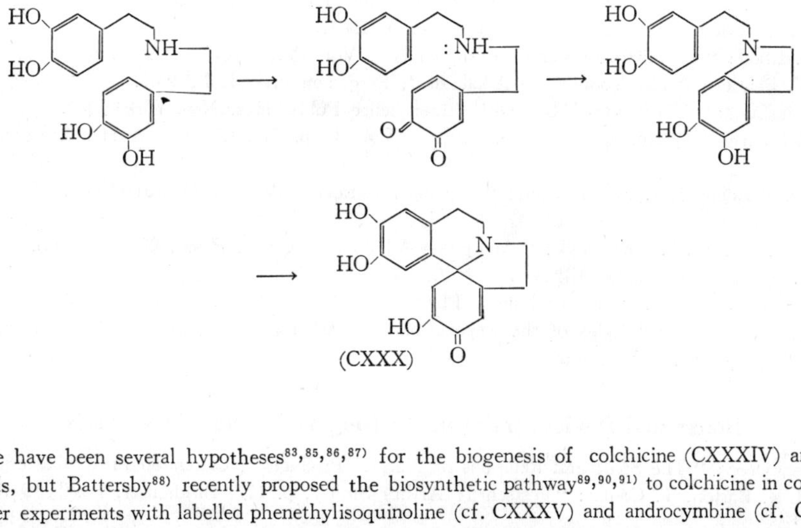

(CXXX)

There have been several hypotheses[83,85,86,87] for the biogenesis of colchicine (CXXXIV) and related compounds, but Battersby[88] recently proposed the biosynthetic pathway[89,90,91] to colchicine in considerable detail after experiments with labelled phenethylisoquinoline (cf. CXXXV) and androcymbine (cf. CXXXVI).

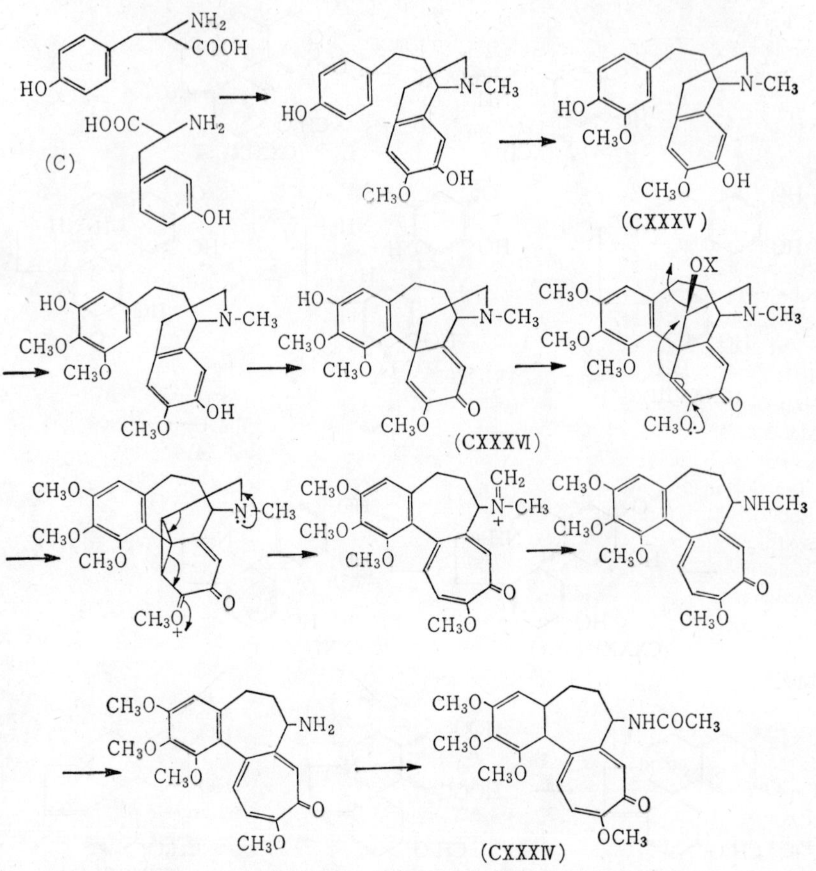

Books and Reviews relating to Structural Determinations and Total Syntheses

1. R. H. F. Manske, H. L. Holmes: "The Alkaloids" Vol. I, II, III, IV, VI, VII, Academic Inc., New York.
2. H.-G. Boit: "Ergebnisse der Alkaloid-Chemie bis 1960 under Besonderer Berücksichtung der Fortschritte seit 1950", Akademie-Verlag, Berlin, 1961.
3. E. H. Rodd: "Chemistry of Carbon Compounds", Vol. IVc, Elsevier Publishing Company, 1960.
4. K. W. Bentley: "The Isoquinoline Alkaloids", Pergamon Press Ltd., 1965.
5. D. Ginsburg: "The Opium Alkaloids", Interscience Publishings, New York, 1962.
6. R. C. Elderfield: "Heterocyclic Compounds", Vol. 4, p. 344, Chapman & Hall Ltd., New York, 1962.
7. W. M. Whaley, T. R. Govindachari: "Organic Reactions", Vol. **6**, 74, 151 (1951).
8. W. J. Gensler: *Ibid.*, **6**, 191 (1951).
9. M. Shamma, W. A. Slusarchyk: "Aporphine Alkaloids", Chem. Revs., **63**, 59 (1963).
10. A. R. Pinder: "Phthalide Alkaloids", *Ibid.*, **64**, 551 (1964).
11. S. Pfeifer, L. Kühn: "Proaporphine", Pharmazie, **20**, 659 (1965).
12. T. Kariyone: "Anual Index of the Reports on Plant Chemistry in 1957–1961", Hirokawa Publishing Company, Inc. Tokyo, Japan.

Books and Reviews relating to Biogenesis and Biosynthesis

1. R. Robinson: "The Structural Relations of Natural Products", Clarendon Press, Oxford, 1955.
2. D. H. R. Barton, T. Cohen: "Festschrift Arthur Stoll", p. 117, Birkhauser, Basel, 1957.

3. D. H. R. Barton: "Chemistry of Natural Products, Vol. III, p. 35 (1964).
4. P. Bernfeld: "Biosynthesis of Natural Compounds", Pergamon Press, 1963.
5. D. H. R. Barton: Proc. Chem. Soc., **1963**, 293.
6. A. R. Battersby: Quart. Revs., **15**, 259 (1961); Proc. Chem. Soc., **1963**, 189.
7. K. Mothes, H. R. Schutte: Angew. Chem., **72**, 357 (1963).

References

1) N. F. Proskurnina, A. P. Orechoff: Zhur. Obshchei. Khim., **10**, 707 (1940).
2) H. Kondo, T. Kondo: J. Pharm. Soc. Japan, **45**, 876 (1925); **46**, 104 (1926); **48**, 55, 106 (1928); **50**, 63 (1930); cf. J. prakt. Chem., **126**, 24 (1930).
3) A. R. Battersby, T. P. Edwards: J. Chem. Soc., **1960**, 1214.
4) F. W. Sertürner: Trommsdorff's Journal der Pharmazie, [i] **13**, 234 (1805).
5) J. M. Gulland, R. Robinson: Nature, **45**, 625 (1925).
6) M. Gates, G. Tschudi: J. Am. Chem. Soc., **74**, 1109 (1952); **78**, 1380 (1956).
7) S. McLean, M.-S. Lin, A. C. Macdonald, J. Trotter: Tetrahedron Letters, **1966**, 185.
8) A. R. Battersby, R. B. Herbert, F. Santavy: Chem. Comm., **1965**, 415.
9) W. H. Brindley, F. L. Pyman: J. Chem. Soc., **1927**, 1067.
10) R. Robinson: Nature, **162**, 524 (1948).
11) E. Späth, M. Pailer: Monatsh., **79**, 128 (1948); M. Pailer, L. Bilek: Monatsh., **79**, 135 (1948); M. Pailer: Monatsh., **79**, 331 (1948); M. Pailer, K. Porschinski: Monatsh., **80**, 94 (1949).
12) A. R. Battersby, H. T. Openshaw: J. Chem. Soc., **1949**, 59, 3207.
13) R. P. Eustigneeva, N. A. Preobrazhensky: Tetrahedron, **4**, 223 (1958); cf. N. A. Preobrazhensky, *et. al.*: Zhur. Obshchei Khim., **22**, 1467 (1952).
14) W. M. Whaley, T. R. Govindachari: Organic Reactions, **6**, 74 (1951).
15) W. M. Whaley, T. R. Govindachari: Organic Reactions, **6**, 151 (1951).
16) W. J. Gensler: Organic Reactions, **6**, 191 (1951).
17) M. L. Moore: Organic Reactions, **5**, 301 (1949).
18) J. Finkelstein: J. Am. Chem. Soc., **73**, 550 (1951).
19) A. Pictet, M. Finkelstein: Ber., **42**, 1979 (1909); E. Späth, A. Burger: Ber., **60**, 704 (1927).
20) A. Pictet, A. Gams: Ber., **43**, 2384 (1910); cf. F. W. Bergstrom: Chem. Revs., **35**, 220 (1944).
21) A. Pictet, A. Gams: Ber., **44**, 2480 (1911).
22) A. Pictet, T. Spengler: Ber., **44**, 2030 (1911).
23) E. Späth, F. Berger: Ber., **63**, 2098 (1930).
24) T. Kametani, M. Ihara: J. Chem. Soc. (C), **1966**, 2010.
25) M. Tomita, J. Kunitomo: J. Pharm. Soc. Japan, **80**, 1238 (1960).
26) W. J. Gensler: Heterocyclic Compounds, Vol. IV, Chap. 2 (1952).
27) H. R. Ing, R. H. F. Manske: J. Chem. Soc., **1926**, 2348.
28) A. Galat, G. Elion: J. Am. Chem. Soc., **61**, 3585 (1939).
29) D. H. Hey: J. Chem. Soc., **1930**, 18.
30) R. F. Nystrom, W. G. Brown: J. Am. Chem. Soc., **70**, 3738 (1948).
31) J. C. Robinson Jr., H. R. Snyder: Organic Syntheses, Coll. Vol. III, 720 (1955).
32) K. Kindler: Arch. Pharm., **269**, 70 (1931).
33) R. H. F. Manske, H. L. Holmes: J. Am. Chem. Soc., **67**, 95 (1945).
34) W. S. Ide, J. S. Buck: J. Am. Chem. Soc., **62**, 425 (1940).
35) C. Schöpf, W. Salzer: Ann., **544**, 1 (1940); C. Schöpf, H. Perry, I. Jäckh: Ann., **497**, 47 (1932).
36) A. Pictet, A. Gams: Ber., **42**, 2943 (1909); **43**, 2384 (1910); J. L. Bills, C. R. Naller: J. Am. Chem. Soc., **70**, 957 (1948).
37) V. Bruckner, G. Foder, J. Kiss, J. Kovacs: J. Chem. Soc., **1948**, 885.
38) C. Mannich, M. Falber: Arch. Pharm., **267**, 601 (1929); K. W. Rosenmund, M. Nothnagel, H. Riesenfeldt: Ber., **60**, 392 (1927).
39) T. Kametani, K. Fukumoto: J. Chem. Soc., **1963**, 4289.
40) E. Schlittler, J. Müller: Helv. Chim. Acta, **31**, 914 (1948).
41) S. Gabriel; Ber., **18**, 2433, 3470 (1885).
42) T. Kametani, S. Shibuya, S. Seino, K. Fukumoto: J. Chem. Soc., **1964**, 4146.
43) B. R. Harriman, R. S. Shelton, M. G. Van Campen, M. R. Warren: J. Am. Chem. Soc., **67**, 1481

(1945); cf. U.S. pat. 2,351,391 [C.A., **38**, 5228 (1944)].
44) R.D. Haworth, H.S. Pink: J. Chem. Soc., **127**, 1368 (1925).
45) R.D. Haworth, W.H. Perkin Jr.: J. Chem. Soc., **1926**, 1769.
46) A.S. Bailey, C.R. Worthing: J. Chem. Soc., **1956**, 4535.
47) A.R. Battersby, T.H. Brown, J.H. Clements: J. Chem. Soc., **1965**, 4550.
48) D.H.R. Barton, G.W. Kirby, W. Steglich, G.M. Thomas: Proc. Chem. Soc., **1963**, 203.
49) M.P. Cava, S.C. Havlicek, A. Lindert, R.J. Spangler: Tetrahedron Letters, **1966**, 2937.
50) S.M. Kupchan, R.M. Kanojia: Tetrahedron Letters, **1966**, 5353.
51) R. Robinson: "The Structural Relations of Natural Products", Clarendon Press, Oxford, (1955).
52) R. Robinson: J. Chem. Soc., **111**, 762, 876 (1917).
53) D.H.R. Barton, T. Cohen: "Festschrift Arthur Stoll", p. 117. Brikhauser, Basel (1957).
54) A.R. Battersby, B.J.T. Harper: Proc. Chem. Soc., **1959**, 152; J. Chem. Soc., **1962**, 3526.
55) A.R. Battersby, B.J.T. Harper: Chem. & Ind., **1958**, 364.
56) A.R. Battersby, B.J.T. Harper: Tetrahedron Letters, **1960**, No. 27, 21.
57) A.R. Battersby, R. Binks, B.J.T. Harper: J. Chem. Soc., **1962**, 3534.
58) A.R. Battersby: Proc. Chem. Soc., **1963**, 189.
59) D.H.R. Barton, G.W. Kirby, W. Steglich, G.M. Thomas, A.R. Battersby, T.A. Dobson, H. Ramuz: J. Chem. Soc., **1965**, 2423.
60) D. Ginsburg: "The Opium Alkaloids", p. 91. Interscience Publishers, New York, (1962).
61) D.H.R. Barton: "Chemistry of Natural Products", Vol. III, p. 35. International Symposium, Kyoto (1964).
62) J.R. Gear, I.D. Spenser: Can. J. Chem., **41**, 783 (1963).
63) A.R. Battersby, D.J. McCaldin: Proc. Chem. Soc., **1962**, 365.
64) D.H.R. Barton, R.H. Hesse, G.W. Kirby: J. Chem. Soc., **1965**, 6379.
65) R.N. Gupta, I.D. Spenser: Can. J. Chem., **43**, 133 (1965).
66) A.R. Battersby, R.J. Francis, E.A. Ruveda, J. Staunton: Chem. Comm., **1965**, 89.
67) A.R. Battersby, R.T. Brown, J.H. Clements, G.G. Iverach: Chem. Comm., **1965**, 230.
68) L.J. Haynes, K.L. Stuart, D.H.R. Barton, G.W. Kirby: Proc. Chem. Soc., **1963**, 280; J. Chem. Soc. (C), **1966**, 1676.
69) R.H.F. Manske: J. Am. Chem. Soc., **72**, 55 (1950).
70) D.H.R. Barton, G.W. Kirby, A. Wiechers: Chem. Comm., **1966**, 266.
71) D.H.R. Barton: Proc. Chem. Soc., **1963**, 293.
72) D.H.R. Barton, G.W. Kirby: Proc. Chem. Soc., **1960**, 392; J. Chem. Soc., **1962**, 806; D.H.R. Barton, G.W. Kirby, J.B. Taylor, G.M. Thomas: Proc. Chem. Soc., **1961**, 254.
73) R.J. Suhadolnik, A.G. Fisher, J. Zulaliar: J. Am. Chem. Soc., **84**, 4348 (1962).
74) A.R. Battersby, R. Binks, W. Lawrie, G.V. Parry, B.R. Webster: J. Chem. Soc., **1965**, 7459.
75) E. Leete, S. Ghosal, P.N. Edwards: J. Am. Chem. Soc., **84**, 1068 (1962).
76) E. Wenkert, N.V. Bringi: J. Am. Chem. Soc., **81**, 1474 (1959); E. Wenkert: J. Am. Chem. Soc., **84**, 98 (1962).
77) R. Thomas: Tetrahedron Letters, **1961**, 544.
78) T. Money, I.G. Wright, F. McCapra, A.I. Scott: Proc. Natl. Acad. Sci. U.S., **53**, 901 (1965).
79) B. Witkop, S. Goodain: Experientia, **8**, 377 (1952).
80) E. Wenkert: Chem. & Ind., **1953**, 1088.
81) V. Boekelheide, J. Weinstock, M.G. Grundon, G.L. Sauvage, E.J. Agnello: J. Am. Chem. Soc., **75**, 2550 (1953).
82) V. Prelog: Angew. Chem., **69**, 33 (1957).
83) E. Wenkert: Experientia, **15**, 169 (1959).
84) E. Leete, A. Ahmad: J. Am. Chem. Soc., **88**, 4722 (1966).
85) F.A.L.v. Anet, R. Robinson: Nature, **166**, 924 (1950).
86) B. Belleau: Experientia, **9**, 178 (1953).
87) A.I. Scott: Nature, **186**, 556 (1960).
88) A.R. Battersby, R.B. Herbert, E. McDonald, R. Ramage, J.H. Clements: Chem. Comm., **1966**, 603.
89) E. Leete: Tetrahedron Letters, **1965**, 333.
90) A.R. Battersby, J.J. Reynolds: Proc. Chem. Soc., **1960**, 346.
91) E. Leete, P.E. Nemeth: J. Am. Chem. Soc., **82**, 6055 (1960).

Chapter 2 Simple Isoquinoline Alkaloids

Hydrohydrastinine $C_{11}H_{13}O_2N \equiv 179.21$

≪mp≫ 66°
≪$[\alpha]_D$≫ 0°

≪Ph.≫
≪OP.≫ Corydalis tuberosa[1]
≪C. & So.≫
≪Sa. & D.≫ B-HCl: 276~278°
B-HI: 240~242° B-Pic: 175~176°
≪SD.≫ Degradative product of hydrastine and cotarnine[2)3)4)]
≪S. & B.≫ 5), 6), 7)

Corypalline $C_{11}H_{15}O_2N \equiv 193.24$

≪mp≫ 168° (MeOH-Et_2O)
≪$[\alpha]_D$≫ 0°

≪Ph.≫
≪OP.≫ Corydalis aurea[8], C. pallida[8]
≪C. & So.≫ Prisms
≪Sa. & D.≫ B-Pic: 178° B-OMe: 65°
≪SD.≫ 8)
≪S. & B.≫ 8)

***d*-Salsoline** $C_{11}H_{15}O_2N \equiv 193.24$

≪mp≫ 215~216°
≪$[\alpha]_D$≫ B-HCl +40.1° (H_2O)

≪Ph.≫ UV[95]
≪OP.≫ Salsola kali[9], S. richteri[10)11)12)13)14)15)16)18)], S. ruthenica[9], S. soda[9]
≪C. & So.≫ Sol. in EtOH, $CHCl_3$. Spar. Sol. in H_2O, C_6H_6. Insol. in petr. ether.
≪Sa. & D.≫ B-HCl: 171~172°
≪SD.≫ 17), 18)
≪S. & B.≫ 17), 19), 94), BG: 20)

***dl*-Salsoline**

≪mp≫ 224~226°
≪$[\alpha]_D$≫ 0°
≪Ph.≫
≪OP.≫ S. richteri[10)11)12)13)14)15)16)]

≪C. & So.≫
≪Sa. & D.≫ B-HCl: 141~152°
B-NBz: 172~174° B-N,O-DiBz: 166~168°
≪SD.≫ cf. *d*-salsoline
≪S. & B.≫

***l*-Salsolidine** $C_{12}H_{17}O_2N \equiv 207.26$

≪mp≫ 69~70°
≪$[\alpha]_D$≫ −53° (EtOH)

≪Ph.≫ UV[95]
≪OP.≫ Salsola richteri[10)11)12)13)14)15)16)]
≪C. & So.≫ Plates. cf. bp ca. 120°/0.01
≪Sa. & D.≫ B-HCl: 229~231°
B-Pic: 194~195° B-Picrol: 220~221°
≪SD.≫ 18), 19)
≪S. & B.≫ 19), 21), 92), BG: 22)

***d*-Salsolidine**

≪mp≫ 69~70°
≪$[\alpha]_D$≫
≪Ph.≫
≪OP.≫ Genista (Cytisus) purgans[23]

≪C. & So.≫
≪Sa. & D.≫
≪SD.≫ cf. *l*-salsolidine
≪S. & B.≫

_dl_-Salsolidine

≪mp≫
≪[α]$_D$≫
≪Ph.≫
≪OP.≫ Genista (Cytisus) purgans[23)]
≪C. & So.≫
≪Sa. & D.≫
≪SD.≫ cf. _l_-salsolidine
≪S. & B.≫

Carnegine[24)] **(Pectenine)**[25)] $C_{13}H_{19}O_2N \equiv 221.29$

CH_3O, CH_3O, N–CH_3, CH_3

≪mp≫
≪[α]$_D$≫ +20° (H_2O)
≪Ph.≫
≪OP.≫ Carnegia gigantea[24)], Cereus pectenabroginum (Pachycereus pecten-abroginum)[25)26)]
≪C. & So.≫ Colourless oil. bp 170°/1
≪Sa. & D.≫ B–HCl: 210～211° B–HBr: 228°
B–Pic: 212～213° (222～222.5° (d))
B–MeI: 210～211°
≪SD.≫ 26), 27)
≪S. & B.≫ 27), 28), BG: 22)

_d_-Calycotomine $C_{12}H_{17}O_3N \equiv 223.26$

CH_3O, CH_3O, NH, H CH_2OH

≪mp≫ 139～141°
≪[α]$_D$≫ +21° (H_2O)
≪Ph.≫
≪OP.≫ Calycotome spinosa[29)], Cytisus proliferus[30)]
≪C. & So.≫
≪Sa. & D.≫ B–HCl: 193° B–$HClO_4$: 176～177°
B–Pic: 163～166° B–$HgCl_2$: 118～119°
≪SD.≫ 18), 29), 31)
≪S. & B.≫ 32), 33), 34), 35), 36), 37), 38), 93), BG: 93)

_dl_-Calycotomine

≪mp≫
≪[α]$_D$≫ 0°
≪Ph.≫
≪OP.≫ Calycotome spinosa[29)], Cytisus nigricans[29)31)], C. proliferus[30)]
≪C. & So.≫
≪Sa. & D.≫ B–HCl: 181° B–$HClO_4$: 172°
≪SD.≫ cf. _d_-calycotomine
≪S. & B.≫

Anhalamine $C_{11}H_{15}O_3N \equiv 209.34$

CH_3O, CH_3O, NH, HO

≪mp≫ 189～191° (EtOH)
≪[α]$_D$≫ 0°
≪Ph.≫ UV[45)] IR[45)] NMR[45)]
≪OP.≫ Anhalonium lewinii (Lophophora williamsii)[39)40)]
≪C. & So.≫ Needles, Sol. in H_2O, alkali. Spar. sol. in Et_2O, C_6H_6, EtOH
≪Sa. & D.≫ B–HCl: 257～258° (vacuum)
B–Pic: 273～274° B–OMe=anhalinine
B–NMe=anhalidine B–NBz: 167～168°
B–O, N–DiBz: 128～129°
≪SD.≫ 40), 41), 42), 43), 44)
≪S. & B.≫ 44), 45), 46)

Hydrocotarnine $C_{12}H_{15}O_3N \equiv 221.25$

CH_2, O, O, N–CH_3, CH_3O

≪mp≫ 55～56° [B+$^1/_2$ H_2O] (EtOH)
≪[α]$_D$≫ 0°
≪Ph.≫
≪OP.≫ Opium[47)]
≪C. & So.≫ Colourless plates, Sol. in org. solv. Insol. in H_2O
≪Sa. & D.≫ B–HBr: 236～237°
B–HI: 206～207° B–MeI: 206～207°, cf. ref. 48)
≪SD.≫ ≪S. & B.≫ reduction product of cotarnine[2)3)], cf. ref. 5), 6), 7), 49), 50), 51), 52)

Anhalonine

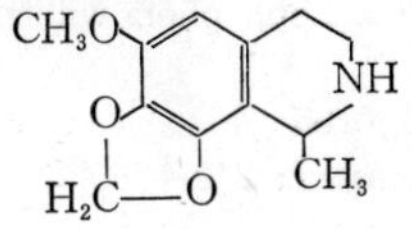
CH3O
NH
O
H2C—O
CH3

≪mp≫ 83～85° (petr. ether)
≪$[\alpha]_D$≫ −56° ($CHCl_3$)

$C_{12}H_{15}O_3N \equiv 221.25$

≪Ph.≫
≪OP.≫ Anhalonium lewinii (Lophophora williamsii)[53)54)55)]
≪C. & So.≫ Needles, Sol. in H_2O, EtOH, $CHCl_3$, Et_2O
≪Sa. & D.≫ B–HCl: 254～256°
B–NMe=lophophorine
≪SD.≫ 18), 41), 56), 57), 58)
≪S. & B.≫ 56), 59)

Anhalidine

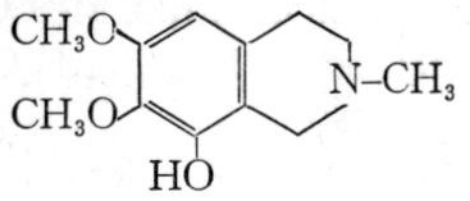
CH3O
N–CH3
CH3O
HO

≪mp≫ 131～133° (sublime)
≪$[\alpha]_D$≫ 0°

$C_{12}H_{17}O_3N \equiv 223.26$

≪Ph.≫ UV[45)] IR[45)] NMR[45)]
≪OP.≫ Anhalonium lewinii (Lophophora williamsii)[57)]
≪C. & So.≫
≪Sa. & D.≫ B–OMe–MeI: 211.5～212.5°
≪SD.≫ 57), 58)
≪S. & B.≫ 46), 45)

Anhalinine

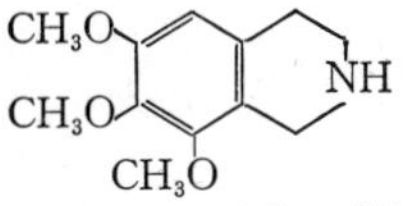
CH3O
NH
CH3O
CH3O

≪mp≫ 61～63° (sublime)
≪$[\alpha]_D$≫ 0°

$C_{12}H_{17}O_3N \equiv 223.26$

≪Ph.≫
≪OP.≫ Anhalonium lewinii (Lophophora williamsii)[41)]
≪C. & So.≫ bp 144～5°/0.1
≪Sa. & D.≫ B–HCl: 242～243°
B–Pic: 184～185° B–1/2 H_2PtCl_6: 207～208°
B–$HAuCl_4$: 139～140° B–MeI: 211.5～212.5°
≪SD.≫ 41), 57), 58)
≪S. & B.≫ 42), 43), 60)

Anhalonidine

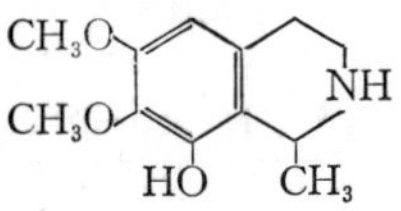
CH3O
NH
CH3O
HO
CH3

≪mp≫ 160～161°
≪$[\alpha]_D$≫ 0°

$C_{12}H_{17}O_3N \equiv 223.26$

≪Ph.≫ UV[45)] IR[45)] NMR[45)]
≪OP.≫ Anhalonium lewinii (Lophophora williamsii)[61)], Echinocactus lewinii[62)], Lamaireocereus weberi[63)]
≪C. & So.≫ Small octahedra, Sol. in H_2O, Et_2O. Insol. in petr. ether
≪Sa. & D.≫ B–Pic: 201～208° B–1/2 H_2PtCl_6: 152° B–NMe=pellotine B–NBz: 189°
B–O, N–DiBz: 125～126°
≪SD.≫ 40), 42), 62), 64), 65)
≪S. & B.≫ 45), 62), BS: 66)

Lophophorine

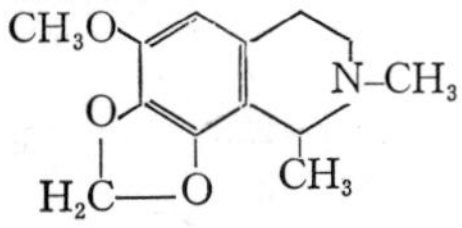
CH3O
N–CH3
O
H2C—O
CH3

≪mp≫
≪$[\alpha]_D$≫ −47° ($CHCl_3$)

$C_{13}H_{17}O_3N \equiv 235.27$

≪Ph.≫
≪OP.≫ Anhalonium lewinii (Lophophora williamsii)[61)]
≪C. & So.≫ Oil
≪Sa. & D.≫ B–Pic: 162～163° B–MeI: 223°
≪SD.≫ 18), 41), 56)
≪S. & B.≫ 56), 59)

O-Methyl-d-anhalonidine $C_{13}H_{19}O_3N \equiv 237.29$

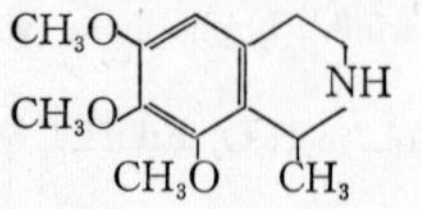

≪mp≫

≪[α]D≫ +20.7° (MeOH)

≪Ph.≫

≪OP.≫ Anhalonium lewinii (Lophophora williamsii)[67]

≪C. & So.≫ Oil, bp 140°/0.05

≪Sa. & D.≫ B–Trinitrobenzoyl: 259~260°

≪SD.≫ 67)

≪S. & B.≫ 42), 67)

Pellotine $C_{13}H_{19}O_3N \equiv 237.29$

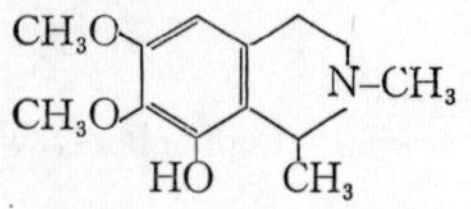

≪mp≫ 111~112° (EtOH)

≪[α]D≫ 0°

≪Ph.≫ UV[45] IR[45] NMR[45]

≪OP.≫ Anhalonium lewinii (Lophophora williamsii)[39], A. (Echinocactus) williamsii[62][68]

≪C. & So.≫ Plates, Sol. in EtOH, Et_2O, Me_2CO, $CHCl_3$. Mod. sol. in petr. ether. Insol. in H_2O

≪Sa. & D.≫ B–HI: 125~130°
B–Pic: 172~173° B–$HAuCl_4$: 147~148°
B–MeI: 199° B–OMe–MeI: 226~227°

≪SD.≫ 42), 62), 64), 65), 69), 70)

≪S. & B.≫ 28), 45), 62), 69), BS: 71)

Lophocerine $C_{15}H_{23}O_2N \equiv 249.33$

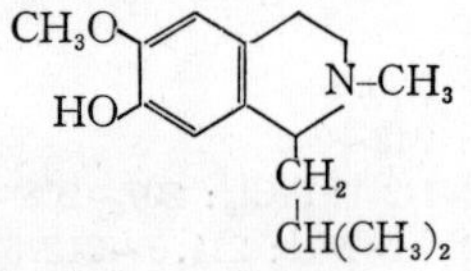

≪mp≫

≪[α]D≫ 0°

≪Ph.≫ UV[73]

≪OP.≫ Lophocerus schottii[72]

≪C. & So.≫ Oil

≪Sa. & D.≫ B–Pic: 191.5~193° (172°)
B–Styph: 212~214° (171~172°)
B–OMe–Pic: 180~182°
B–OMe–Styph: 210~212°

≪SD.≫ 19)

≪S. & B.≫ 73), 74), 75), BG: 71), 72)

Pilocereine $C_{45}H_{65}O_6N_3 \equiv 743.996$

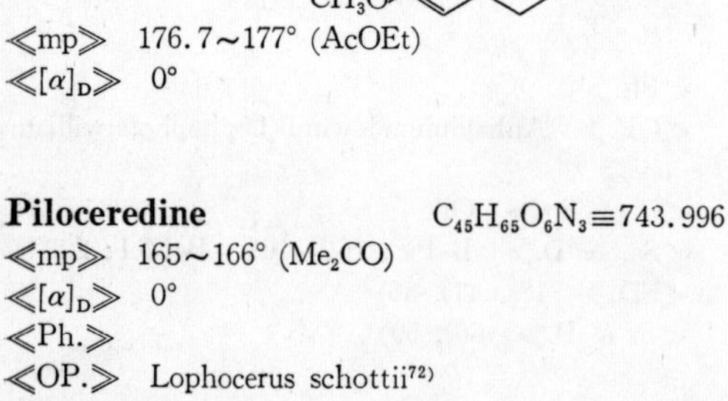

≪mp≫ 176.7~177° (AcOEt)

≪[α]D≫ 0°

≪Ph.≫ UV[75][76] IR[72][75][85][86][89] NMR[79][86] Mass[79] Rf[87]

≪OP.≫ Lophocerus australis[63], L. gatesii[63], L. schottii[25][76], Pachycereus maginatus[63], Pilocereas (Cereus) sargentianus[25]

≪C. & So.≫

≪Sa. & D.≫ B–3HCl: 233°
B–$3HClO_4$: 214~217° B–3Oxal: 158~163°(d) or 198~200°(d) B–OMe: 92~105°, 153~154°, 133~135°

≪SD.≫ 76), 77), 78), 79), 80), 81), 82), 83), 84), 85), 72)

≪S. & B.≫ 86), 87), 88), 89), BG: 72)

Piloceredine $C_{45}H_{65}O_6N_3 \equiv 743.996$

≪mp≫ 165~166° (Me_2CO)

≪[α]D≫ 0°

≪Ph.≫

≪OP.≫ Lophocerus schottii[72]

≪C. & So.≫

≪Sa. & D.≫ B–$3HClO_4$: 221~222°
B–OMe: 142° B–OAc: 133~134°

≪SD.≫ Piloceredine is a diastereoisomer of pilocereine

≪S. & B.≫

Salsamine $C_{26}H_{36}O_4N_2 \equiv 408.56$

CH_3O, CH_3O, N, $(CH_2)_2$, N, OCH_3, OCH_3, CH_3, CH_3

≪mp≫ 165～167°
≪[α]D≫

≪Ph.≫
≪OP.≫ Salsola richteri[14]
≪C. & So.≫
≪Sa. & D.≫
≪SD.≫ 90)
≪S. & B.≫

Spartocytisine $C_{12}H_{17}O_3N \equiv 223.26$

CH_3O, CH_3O, NH, OH, CH_3

≪mp≫ 158～159° (heptane–C_6H_6)
≪[α]D≫ +11.4°

≪Ph.≫ CR[91] Rf[91]
≪OP.≫ Spartocytisus filipes[91]
≪C. & So.≫
≪Sa. & D.≫ B–Pic: 174～175° B–Styph: 212°
B–O, N–DiAc: 125～126°
≪SD.≫ 91)
≪S. & B.≫

1) E. Späth, P.L. Julian: Ber., **64**, 1131 (1931).
2) D.B. Clayson: J. Chem. Soc., **1949**, 2016.
3) V.M. Rodionov, M.G. Chentsova: Zhur. Obshchi Khim., **21**, 321 (1951).
4) M. Freund, W. Will: Ber., **20**, 2400 (1887).
5) F.L. Pyman, F.G.P. Remfry: J. Chem. Soc., **101**, 1595 (1912).
6) Y. Tanaka, T. Midzuno, T. Okami: J. Pharm. Soc. Japan, **50**, 559 (1930).
7) K. Topchiev: Zhur. Priklado. Khim. (USSR), **6**, 529 (1933).
8) R.H.F. Manske: Can. J. Res., **B 15**, 159 (1937).
9) B. Borkowski, K. Drost, B. Pasichowa: Acta polan. pharm., **16**, 57 (1959). [C. A., **53**, 11533 (1959)].
10) A. Orechoff, N. Proskurnina: Ber., **66**, 841 (1933).
11) A. Orechoff, N. Proskurnina: Ber., **67**, 878 (1934).
12) A. Orechoff, N. Proskurnina: Khim. Farm. Prom., **1934**, No. 2, 8～10. [C. A., **28**, 5460 (1934)].
13) A. Orechoff, N. Proskurnina: Bull. acad. sci. U.S.S.R., Sér. Chim., **1936**, 957. [C. A., **31**, 5365 (1937)].
14) N. Proskurnina, A. Orechoff: Bull. Soc. Chim. France, [5] **4**, 1265 (1937).
15) N. Proskurnina, A. Orechoff: Bull. Soc. Chim. France, [5] **6**, 144 (1939).
16) A.A. Konovalova, T.F. Platonova, R.A. Konovalova: Zhur. Priklado. Khim. (USSR), **23**, 927 (1950).
17) E. Späth, A. Orechhoff, F. Kuffner: Ber., **67**, 1214 (1934).
18) A.R. Battersby, T.P. Edward: J. Chem. Soc., **1960**, 1214.
19) Ö. Kovacs, G. Fodor: Ber., **84**, 795 (1951).
20) R. Robinson: "The Structural Relations of Natural Products", Clarendon Press, Oxford, (1955).
21) E. Späth, F. Dengel: Ber., **71**, 113 (1938).
22) C. Schöpf, H. Bayerle: Ann., **513**, 190 (1934).
23) R. Barca, J. Dominguez, I. Ribas: Anales real soc. españ, fés. y. quím, **55B**, 717 (1959). [C. A., **54**, 14289 (1960)].
24) G. Heyl: Arch. Pharm., **266**, 668 (1928).
25) G. Heyl: Arch. Pharm., **239**, 451 (1901).
26) E. Späth, F. Kuffner: Ber., **62**, 2242 (1929).
27) E. Späth: Ber., **62**, 1021 (1929).
28) T. Nakada, K. Nishihara: J. Pharm. Soc. Japan, **64**, 74 (1934).
29) E.P. White: New Zealand J. Sci. Technol., **B 25**, 152 (1944). [C. A., **38**, 6494 (1944)].
30) E.P. White: New Zealand J. Sci. Technol., **B 25**, 103 (1944). [C. A., **38**, 3690 (1944)].
31) E.P. White: New Zealand J. Sci. Technol., **B33**, 38 (1951). [C. A., **46**, 6656 (1952)].
32) L. Dúbravkoνá, I. Ježo, P. Šefcorič, Z. Votický: Chem. Zvesti., **12**, 459 (1958). [C.A., **53**, 5314 (1959)].
33) A.R. Battersby, T.P. Edward: J. Chem. Soc., **1959**, 1909.
34) K. Kazka, J. Kobor: Szegedi Pedagogiai Foiskola Evkonyve., **1962**, 207. [C. A., **62**, 2761 (1965)].
35) F. Benington, R.D. Morin: J. Org. Chem., **26**, 194 (1961).
36) A. Chatterjee, N. Adityachaudhury: J. Org. Chem., **27**, 309 (1962).
37) H.W. Gibson, E.D. Popp, A. Catala: J. Heterocyclic Chem., **1**, 251 (1964).
38) A. Chatterjee, N.A. Chaudhury: Sci. Cult., **25**, 389 (1959).
39) E. Kauder: Arch. Pharm., **237**, 190 (1899).

40) A. Heffter: Ber., **34**, 3004 (1901).
41) E. Späth, F. Becke: Ber., **68**, 501 (1935).
42) E. Späth: Monatsh., **42**, 97 (1921).
43) E. Späth, F. Becke: Ber., **67**, 2100 (1934).
44) E. Späth, H. Röder: Monatsh., **43**, 93 (1922).
45) A. Brossi, F. Schenker, W. Leimgruber: Helv. Chim. Acta, **47**, 2089 (1964).
46) T. Kametani, N. Wagatsuma, F. Sasaki: J. Pharm. Soc. Japan, **86**, 913 (1966).
47) O. Hesse: Ann. (Supple), **8**, 261 (1872).
48) M. Semonsky: Coll. Czech. Chem. Comm., **15**, 1024 (1951).
49) G.H. Beckett, C.R.A. Wright: J. Chem. Soc., **28**, 577 (1875).
50) Bandow, Wolffenstein: Ber., **31**., 1577 (1898).
51) Steiner: Compt. rend., **176**, 224 (1923).
52) Dey, Kantan: J. Indian Chem. Soc., **12**, 421 (1935).
53) L. Lewin: Ber. deut. botan. Ges., **12**, 283 (1894).
54) L. Lewin: Arch. exptl. Pathol. Pharmakol., **24**, 401 (1888).
55) L. Lewin: Arch. exptl. Pathol. Pharmakol., **34**, 374 (1894).
56) E. Späth, J. Gangl: Monatsh., **44**, 103 (1923).
57) E. Späth, F. Becke: Ber., **68**, 944 (1935).
58) E. Späth, F. Becke: Monatsh., **66**, 327 (1935).
59) E. Späth, F. Kesztler: Ber., **68**, 1663 (1935).
60) J.A. Castrillón: J. Am. Chem. Soc., **74**, 558 (1952).
61) A. Heffter: Ber., **29**, 216 (1896).
62) E. Späth: Monatsh., **43**, 477 (1922).
63) C. Djerassi, C.R. Smith, S.P. Marfey, R.N. McDonald, A.J. Lemin, S.K. Figdor, H. Estrada: J. Am. Chem. Soc., **76**, 3215 (1954).
64) E. Späth: Ber., **65**, 1778 (1932).
65) E. Späth, F. Boschan: Monatsh., **63**, 141 (1933).
66) E. Leete: J. Am. Chem. Soc., **88**, 4218 (1966).
67) E. Späth, J. Bruck: Ber., **72**, 334 (1939).
68) A. Heffter: Ber., **27**, 2975 (1894).
69) E. Späth, F. Becke: Ber., **67**, 266 (1934).
70) E. Späth, F. Kesztler: Ber., **69**, 755 (1936).
71) A.R. Battersby: Quart. Revs., **15**, 259 (1961).
72) C. Djerassi, T. Nakano, J.M. Bobbitt: Tetrahedron, **2**, 58 (1958).
73) C. Djerassi, J.J. Beereboom, S.P. Marfey, S.K. Figdor: J. Am. Chem. Soc., **77**, 484 (1955).
74) J.M. Bobbitt, T.-t. Chou: J. Org. Chem., **24**, 1106 (1959).
75) M. Tomita, T. Kikuchi, K. Bessho, Y. Inubushi: Chem. Pharm. Bull., **11**, 1477 (1963).
76) C. Djerassi, N. Frick, L.E. Geller: J. Am. Chem. Soc., **75**, 3632 (1953).
77) C. Djerassi, S.K. Figdor, J.M. Bobbitt, F.X. Markley: J. Am. Chem. Soc., **78**, 3861 (1956).
78) C. Djerassi, S.K. Figdor, J.M. Bobbitt, F.X. Markley: J. Am. Chem. Soc., **79**. 2203 (1957).
79) C. Djerassi, H.W. Brewer, C. Clarke, L.J. Durham: J. Am. Chem. Soc., **84**, 3210 (1962).
80) M. Tomita, T. Kikuchi, K. Bessho, Y. Inubushi: Tetrahedron Letters, **1963**, 127.
81) K. Bessho: Chem. Pharm. Bull., **11**, 1504 (1963).
82) K. Bessho: Chem. Pharm. Bull., **11**, 1507 (1963).
83) M. Tomita, T. Kikuchi, K. Bessho, T. Hori, Y. Inubushi: Chem. Pharm. Bull., **11**, 1484 (1963).
84) K. Bessho: Chem. Pharm. Bull., **11**, 1491 (1963).
85) K. Bessho: Chem. Pharm. Bull., **11**, 1500 (1963).
86) B. Franck, G. Blaschke: Tetrahedron Letters **1963**, 569.
87) B. Franck, G. Blaschke, K. Lewejokann: Ann., **685**, 207 (1965).
88) Ger. Pat. 1,207,932. [C.A., **64**, 12653 (1966)].
89) J.M. Bobbitt, R. Ebermann, M. Schubert: Tetrahedron Letters, **1963**, 575.
90) N.F. Proskurnina: Zhur. Obshchei Khim., **28**, 256 (1958). [C.A., **52**, 12879 (1958)].
91) A.G. Gonzáles, C. Casanova: Anales Real Soc. Espan. Fis. Quim., **58 B**, 483 (1962).
92) I. Zezo, M. Karvas, K. Tihlárik: Chem. Zvesti., **14**, 38 (1960). [C.A., **54**, 22697 (1960)].
93) A. Chatterjee, N.A. Chaudhury: Naturwiss., **47**, 207 (1960).
94) O. Kovacs, G. Fodor: Hung. Pat. 141,779. [C.A., **59**, 2782 (1963)].
95) R.M. Pinyazhko: Farmatsevi. Zh. (Kiev), **19**, 12 (1964). [C.A., **64**, 6701 (1966)].

Chapter 3 Benzylisoquinoline Alkaloids

Papaverine $C_{20}H_{21}O_4N \equiv 339.38$

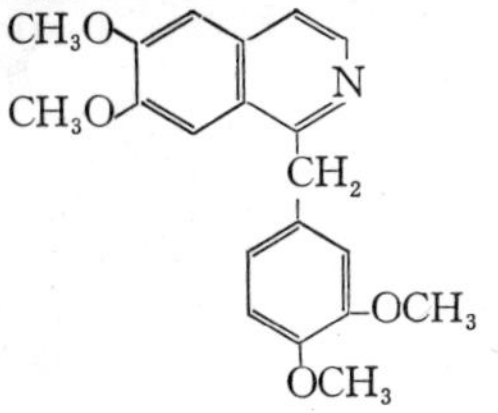

≪mp≫ 147～148° (EtOH or C_6H_6 or $CHCl_3$–petr. ether)
≪[α]D≫ 0°

≪Ph.≫ CR[2)8)24)] UV[62)63)64] IR[54)] NMR[65)] Mass[66)]
≪OP.≫ Papaver floribundum[1)], P. somniferum[1)], P. somniferum var. album[2)3)4)5)]. isolation→6)～15)
≪C. & So.≫ Needles, Sol. in hot EtOH, Et_2O. Spar. sol. in $CHCl_3$, CCl_4. Sublime at 135～140°/11
≪Sa. & D.≫ B–HCl: 220～221°(d)
B–HBr: 213～214° B–HI: 200°
B–Oxal: 201～202° B–Pic: 186°
B–Picrol: 220° B–MeI: 195°
≪SD.≫ 2), 7), 16)～33)
≪S. & B.≫ 28), 34～55)
BS: 56), 57), 58), 59)
BG: 60), 61)

Escholamine $C_{19}H_{16}O_4N^{\oplus} \equiv 322.33$

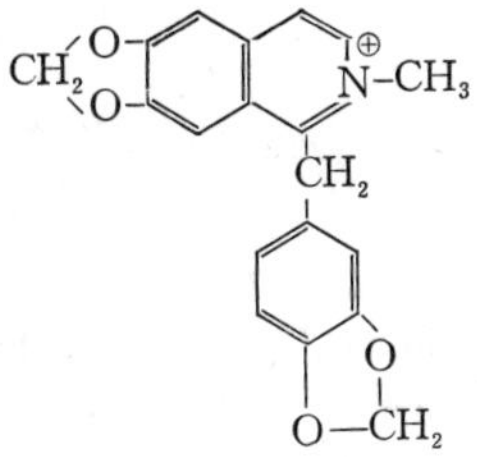

≪mp≫
≪[α]D≫ B⊕–I⊖: 0±3° (H_2O)

≪Ph.≫ UV[169)] IR[169)] Mass[169)]
≪OP.≫ Eschschlotzia glauca[169)], E. oregana[169)]
≪C. & So.≫
≪Sa. & D.≫ B⊕–I⊖: 265～266°
Tetrahydro–Der.: 96～97°
≪SD.≫ 169)
≪S. & B.≫

Takatonine $C_{21}H_{24}O_4N^{\oplus} \equiv 354.41$

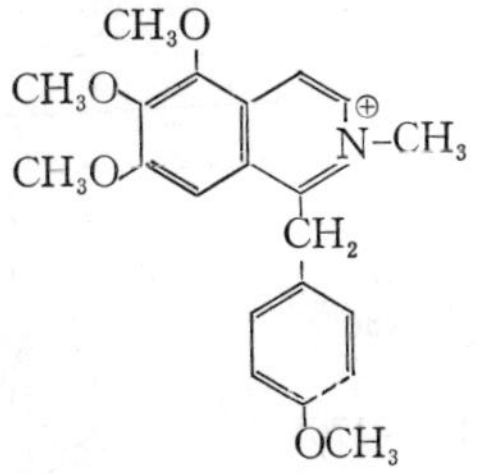

≪mp≫
≪[α]D≫ 0°

≪Ph.≫ UV[68)69)70)] IR[68)69)70)] NMR[69)70)] Rf[69)70)]
≪OP.≫ Thalictrum thunbergii(minus)[68)]
≪C. & So.≫
≪Sa. & D.≫ B⊕–I⊖: 192～193°
Tetrahydro–Der. B–HCl: 186～187°
B–HBr: 184～185°
B–Pic: 141～143°
≪SD.≫ 68), 69), 70)
≪S. & B.≫ 69), 70)

Xanthaline (Papaveraldine) $C_{20}H_{19}O_5N \equiv 353.36$

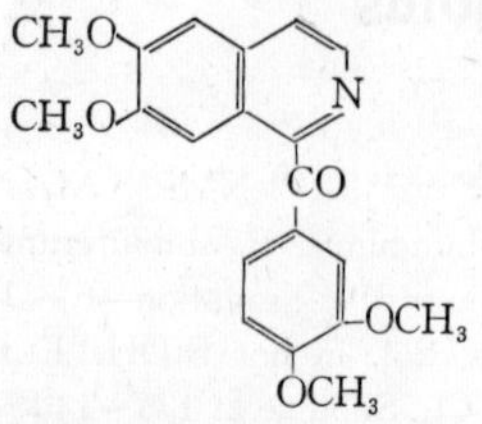

≪mp≫ 147～148° or 209～211°
≪[α]D≫ 0°

≪Ph.≫
≪OP.≫ Opium (natural product?)
≪C. & So.≫ Sol. in C_6H_6, $CHCl_3$. Spar. sol. in EtOH, petr. ether
≪Sa. & D.≫ B–HCl: 200° B–Oxal: 202°
B–Pic: 208～209° B–MeI: 133～135°
≪SD.≫ 71)
≪S. & B.≫ 48), 51), 54)

***N*-Methylxanthaline** $C_{21}H_{22}O_5N^{\oplus} \equiv 368.38$

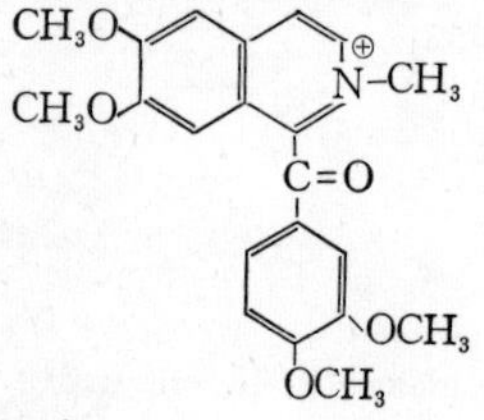

≪mp≫
≪[α]D≫ 0°

≪Ph.≫ UV[72] IR[72] NMR[72] Rf[72]
≪OP.≫ Stephania sasakii[72]
≪C. & So.≫
≪Sa. & D.≫ $B^{\oplus}-Cl^{\ominus}$: 159.5～160°
B–Pic: 129.5～131° Tetrahydro–Der.: 108～109°
≪SD.≫ 72)
≪S. & B.≫ 72)

Coclaurine (Machiline)[79] $C_{17}H_{19}O_3N \equiv 285.33$

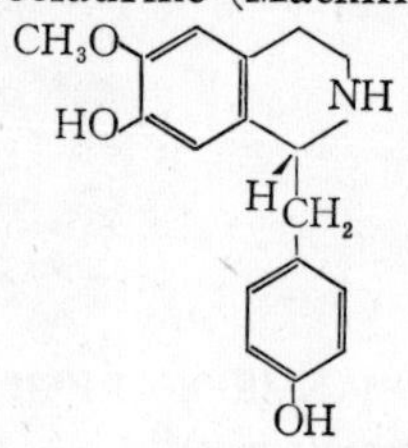

≪mp≫ 220～221° (Et_2O or MeOH)
≪[α]D≫ +0.9° (MeOH)

≪Ph.≫ UV[76] IR[88,93,94] NMR[94,95]
≪OP.≫ Cocculus laurifolius,[73,74,75,76] C. hirsutus[77], Machilus kusanoi[76,78]
≪C. & So.≫ Plates
≪Sa. & D.≫ B–HCl: 263～264°
B–HI: 249～250° B–MeI: 155°(d)
B–NMe: 161～162° B–O, O–DiMe: 200～203°
B–O, O, N–TriMe: 62° B–O, O, N–TriAc: 174°
≪SD.≫ 80), 81)
≪S. & B.≫ 82)～94)

Isococlaurine $C_{17}H_{19}O_3N \equiv 285.33$

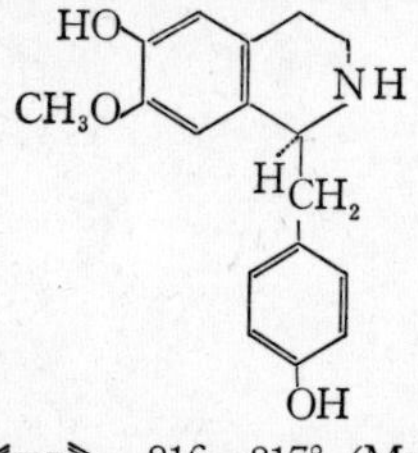

≪mp≫ 216～217° (MeOH)
≪[α]D≫ B–HCl +24°(H_2O)

≪Ph.≫ CR[97] IR[88]
≪OP.≫ Radix Pareirae bravae[96]
≪C. & So.≫ Plates
≪Sa. & D.≫ B–HCl: 175～176°
cf. (±)–B–HCl: 215°
≪SD.≫ 81), 96)
≪S. & B.≫ 97), cf. 88)

N-Methylcoclaurine $C_{18}H_{21}O_3N \equiv 299.36$

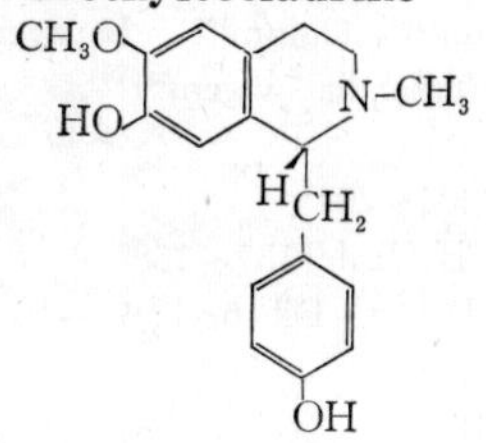

CH_3O HO N–CH_3 H CH_2 OH

≪mp≫ 184~185° (154~155°) (C_6H_6–MeOH)
≪$[\alpha]_D$≫ −96° (EtOH)

≪Ph.≫ UV[98] IR[98] NMR[95,98,104] pKa[98]
≪OP.≫ Phylica rogersii[98]
≪C. & So.≫
≪Sa. & D.≫ B–HCl: 252~254°
B–HBr: 235~238° B–$HClO_4$: 232~234°
B–MeI: 200~202° B–O, O–DiAc: 70~73°
B–O, O–DiMe: 58.5~61°
≪SD.≫ 98)
≪S. & B.≫ 74), 86), 91), cf. 88)

L-(−)-N-Norarmepavine $C_{18}H_{21}O_3N \equiv 299.36$

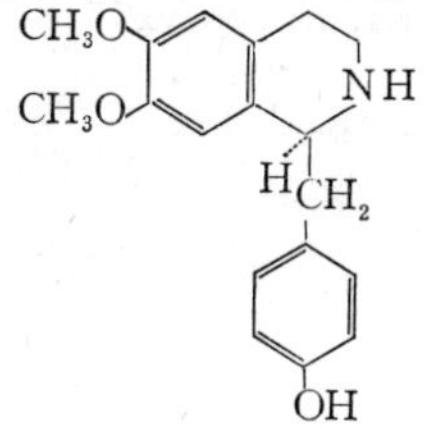

CH_3O CH_3O NH H CH_2 OH

≪mp≫ 152~153° (Me_2CO or MeOH)
≪$[\alpha]_D$≫ −23° ($CHCl_3$)

≪Ph.≫ CR[99,100] UV[64,78,99,101,102] IR[78,101]
NMR[95,99,104,103] ORD[100]
≪OP.≫ Machilus kusanoi,[76,78] M. tunbergii, Nelumbo lutea[99]
≪C. & So.≫ Needles
≪Sa. & D.≫ B–HCl: 153~155°
B–HI: 242~245° B–$HClO_4$: 212~215°
B–Oxal: 228~229° cf. 237°~238°
B–NAc: 237~238° B–NMe=armepavine
≪SD.≫ 78), 99), 100)
≪S. & B.≫ 78)

D-(+)-N-Norarmepavine
≪mp≫ 157~158° (MeOH)
≪$[\alpha]_D$≫ +31.5°
≪Ph.≫ UV[101] IR[101]
≪OP.≫ Magnolia kachirachirai[78,101]

≪C. & So.≫ Needles
≪Sa. & D.≫ B–Oxal: 228~229°
B–NMe: 145~146°
≪SD.≫ 78), 101)
≪S. & B.≫

Armepavine $C_{19}H_{23}O_3N \equiv 313.38$

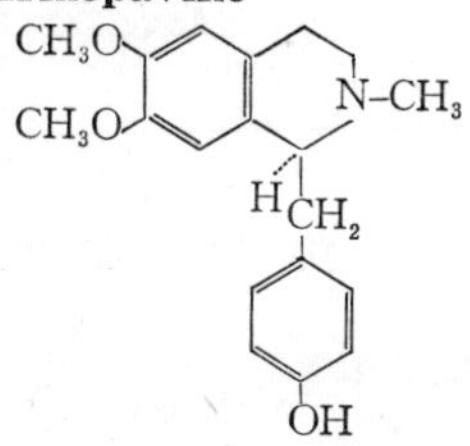

CH_3O CH_3O N–CH_3 H CH_2 OH

≪mp≫ ca. 100°, solidified and remelted at 148~149° (Me_2CO–Et_2O)
≪$[\alpha]_D$≫ +92.8° ($CHCl_3$)

≪Ph.≫ UV[78,102] IR[99,101,115] NMR[104] Mass[103]
ORD[100,116,117,118] Rf[111]
≪OP.≫ Machilus kusanoi,[78] Magnolia kachirachirai,[101] Nelumbo nucifera[105], Papaver caucasicum (P. armeiacum),[4] P. fugax[106]
≪C. & So.≫ Colourless needles, Sol. in EtOH, $CHCl_3$. Mod. sol. in Et_2O
≪Sa. & D.≫ B–HCl: 151~152°
B–Oxal: 211~212° B–MeI: 199~200°
B–OMe: 63~64° B–OMe–MeI: 135~136°
≪SD.≫ 4), 74), 81), 100), 107), 108), 109), 110)
≪S. & B.≫ 105), 107), 109), 110), 111), 112), 113) cf. 84), 114)

***dl*-Armepavine**
≪mp≫ 167~168° (Me_2CO–Et_2O)
≪$[\alpha]_D$≫ ±0°
≪Ph.≫ CR[115] UV[115] IR[115]
≪OP.≫ Nelumbo nucifera[115]
≪C. & So.≫ Needles

≪Sa. & D.≫ B–Oxal: 210~211°
B–Picrol: 214~215° B–MeI: 200~202°
B–OMe–Pic: 170~172°
≪SD.≫ 115)
≪S. & B.≫ 115)

Reticuline $C_{19}H_{23}O_4N \equiv 329.38$

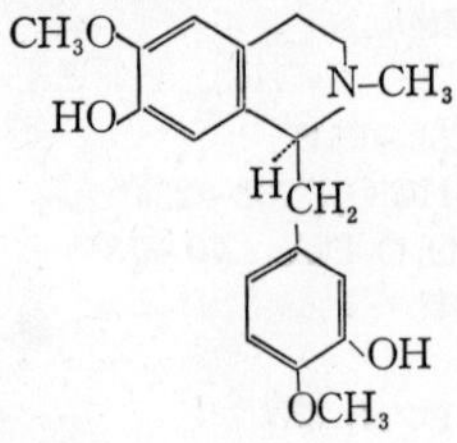

≪mp≫

≪$[\alpha]_D$≫ +85.0° (EtOH)

≪Ph.≫ UV[64) 119)] IR[120) 124)] NMR[120) 126)]
≪OP.≫ Anona reticulata[119)], Opium[120)], Papaver (Phylica) rogersii[121)], Hernandia ovigera[181)]
≪C. & So.≫ Oil
≪Sa. & D.≫ B–HCl: 203~204° B–$HClO_4$: 203~204° B–Oxal: 157~158° B–OMe=laudanosine B–O, O–Dibenzyl: 90~91° B–O, O–DiAc: Oil
≪SD.≫ 119), 120), 122)
≪S. & B.≫ 87), 123), 124), 125), BS: 122)

Remneine $C_{20}H_{23}O_4N \equiv 341.39$

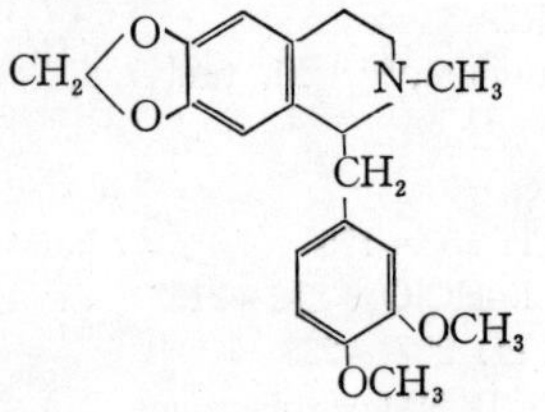

≪mp≫

≪$[\alpha]_D$≫

≪Ph.≫ UV[127) 181)] IR[127) 181)] NMR[127) 181)] Mass[127)]
≪OP.≫ Remneya coulteri var. trichocalyx[127)]
≪C. & So.≫ Oil
≪Sa. & D.≫ B–HBr: 224~225°
≪SD.≫ 127)
≪S. & B.≫ 127)

Codamine $C_{20}H_{25}O_4N \equiv 343.41$

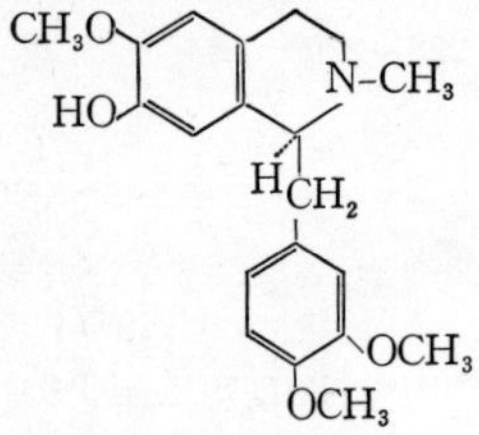

≪mp≫ 126~127°

≪$[\alpha]_D$≫ +75.5° (EtOH)

≪Ph.≫ CR[128)] IR[128)] NMR[128)] Rf[128)]
≪OP.≫ Papaver somniferum[7) 128)]
≪C. & So.≫ Sol. in Et_2O, $CHCl_3$, C_6H_6
≪Sa. & D.≫ B–OMe=laudanosine
≪SD.≫ 7), 16), 129), 130), 131)
≪S. & B.≫ 131), 132), 133), 134), 135)

Thalifendlerine $C_{20}H_{25}O_4N \equiv 343.41$

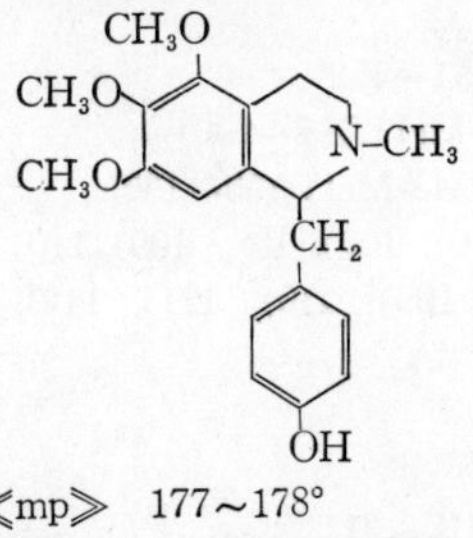

≪mp≫ 177~178°

≪$[\alpha]_D$≫ −108° (MeOH)

≪Ph.≫ CR[136)] UV[136)] IR[136)] NMR[136)] Mass[136)] Rf[136)]
≪OP.≫ Thalictrum fendleri[136)]
≪C. & So.≫
≪Sa. & D.≫ B–OMe: 195~197°
≪SD.≫ 136)
≪S. & B.≫ cf. 136)

Laudanine $C_{20}H_{25}O_4N \equiv 343.41$

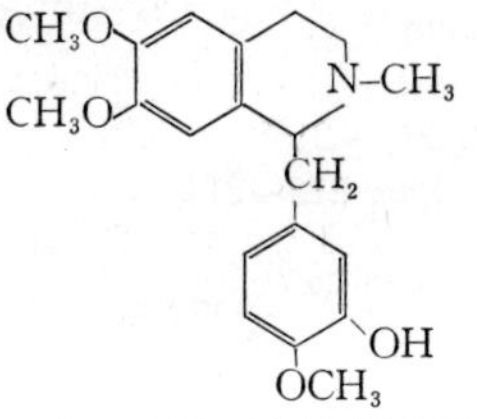

≪mp≫ 166～167° (EtOH)
≪$[\alpha]_D$≫ 0°

≪Ph.≫ CR[7,16] UV[65,102,110,139] ORD[100]
≪OP.≫ Papaver somniferum[7,16]
≪C. & So.≫ Prisms, Sol. in $CHCl_3$, C_6H_6. Spar. sol. in EtOH, Et_2O
≪Sa. & D.≫ B-Pic: 176～177°
B-Oxal: 110° B-OAc: 40° B-OBz: 156～157°
≪SD.≫ 7), 16), 110), 129), 131), 137), 138), 139), 140), 141), 142)
≪S. & B.≫ 110), 131), 135), 142), 143), 144), 145), 146)

Laudanidine (Tritopine) $C_{20}H_{25}O_4N \equiv 343.41$

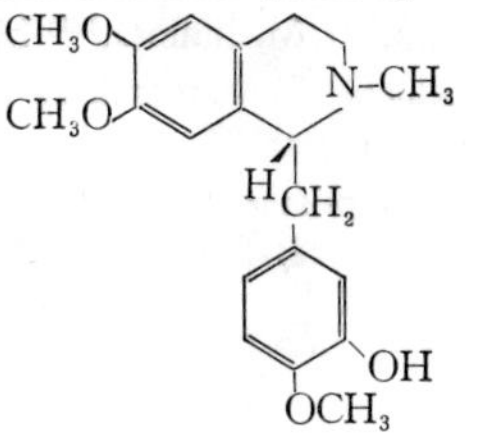

≪mp≫ 184～185° (EtOH)
≪$[\alpha]_D$≫ −100.6° (EtOH)

≪Ph.≫
≪OP.≫ Opium[7,16]
≪C. & So.≫ Sol. in C_6H_6, H_2O. Spar. sol. in EtOH, Et_2O, petr. ether
≪Sa. & D.≫ B-OAc: 98°
≪SD.≫ cf. laudanine
≪S. & B.≫

Laudanosine $C_{21}H_{27}O_4N \equiv 357.43$

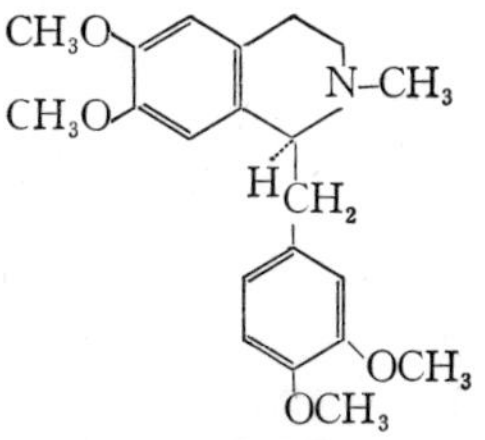

≪mp≫ 89° (C_6H_6)
≪$[\alpha]_D$≫ +52° ($CHCl_3$), +100° (EtOH)

≪Ph.≫ UV[64,102,157,166] IR[157,158,166] NMR[126] Mass[103] ORD[102,118] CD[102]
≪OP.≫ Papaver somniferum[7,16]
≪C. & So.≫ Needles, Sol. in EtOH, Et_2O, $CHCl_3$, C_6H_6. Insol. in H_2O
≪Sa. & D.≫ B-MeI: 215～217°, cf. (±)-B-Pic: 174～175°
≪SD.≫ 7), 16), 131), 139), 147)～160), 182)
≪S. & B.≫ 7), 16), 20), 24), 87), 132), 143), 151), 154), 155), 158), 161～166)

Latericine $C_{24}H_{31}O_8N \equiv 461.50$

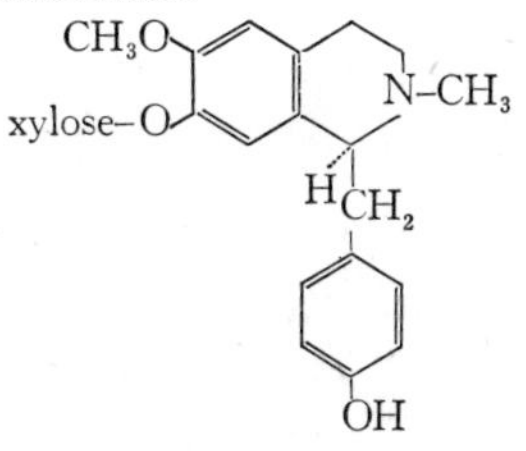

≪mp≫ 216～223° (MeOH)
≪$[\alpha]_D$≫ +93±3° (EtOH), +40±3° (Py)

≪Ph.≫ UV[167] IR[167] Mass[167] ORD[167]
≪OP.≫ Various Papaver species[167]
≪C. & So.≫
≪Sa. & D.≫ B-OMe: 202～203°
aglycone: 134～136°
≪SD.≫ 167)
≪S. & B.≫

Lotusine $C_{19}H_{24}O_3N^{\oplus} \equiv 314.39$

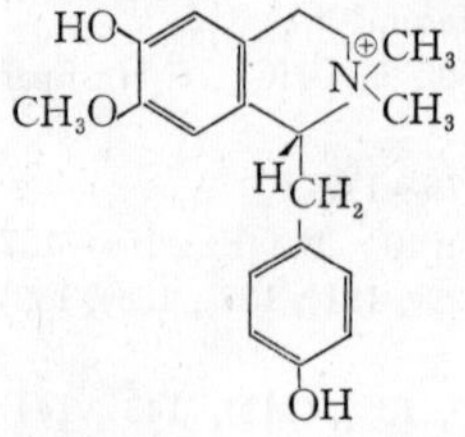

≪mp≫
≪$[\alpha]_D$≫ B⊕-Cl⊖ −15° (MeOH)
≪Ph.≫ CR[169] UV[169] IR[169] NMR[169] Rf[169]
≪OP.≫ Nelumbo nucifera[169]
≪C. & So.≫
≪Sa. & D.≫ B⊕-Cl⊖ : 213~215°
B⊕-I⊖ : 202~204° B-Pic : 212~214°
≪SD.≫ 169)
≪S. & B.≫

***l*-Magnocurarine** $C_{19}H_{24}O_3N^{\oplus} \equiv 314.39$

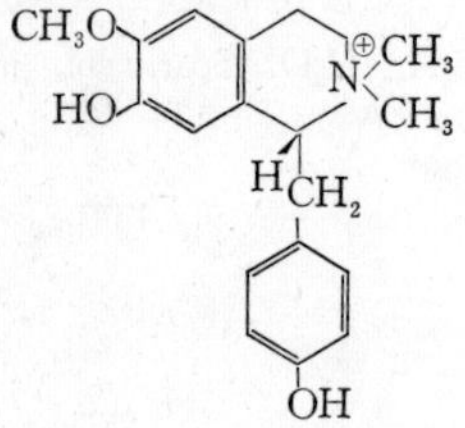

≪mp≫ 199~200°
≪$[\alpha]_D$≫ −91° (H_2O)
≪Ph.≫ CR[172] UV[64,172,176] ORD[118] Rf[173]
≪OP.≫ Magnolia denudata,[170] M. liliflora,[171] M. obovata,[172,173,174] M. officinalis[174], M. parviflora[75], M. salicifolia,[173] Michelia fuscata[175]
≪C. & So.≫ Orange-yellow prisms
≪Sa. & D.≫ B⊕-I⊖ : 93~95° B-Pic : 182~183°
B⊕-O, O-DiMe-I⊖ : 112°
≪SD.≫ 81), 144), 172)
≪S. & B.≫ 176)

***d*-Magnocurarine**

≪mp≫ 199~200°
≪$[\alpha]_D$≫ +91° (H_2O)
≪Ph.≫
≪OP.≫ Gyrocarpus armenicanus[144]
≪C. & So.≫
≪Sa. & D.≫
≪SD.≫ cf. *l*-magnocurarine
≪S. & B.≫

Petaline $C_{20}H_{26}O_3N^{\oplus} \equiv 328.42$

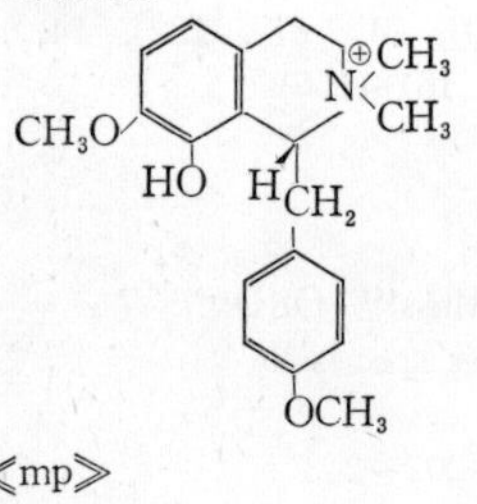

≪mp≫
≪$[\alpha]_D$≫ B⊕-Cl⊖ +11.3°
≪Ph.≫ UV[102,116,179] IR[179] NMR[179] Mass[179] ORD[102] CD[102]
≪OP.≫ Leontice leontopetalum[177,178]
≪C. & So.≫
≪Sa. & D.≫ B⊕-Cl⊖ : 140~143°
Reineckate : 179~181.5° B-Pic : 165.5~166°
Methine base : 123°
≪SD.≫ 102), 116)
≪S. & B.≫ 179)

Tembetarine $C_{20}H_{26}O_4N^{\oplus} \equiv 344.42$

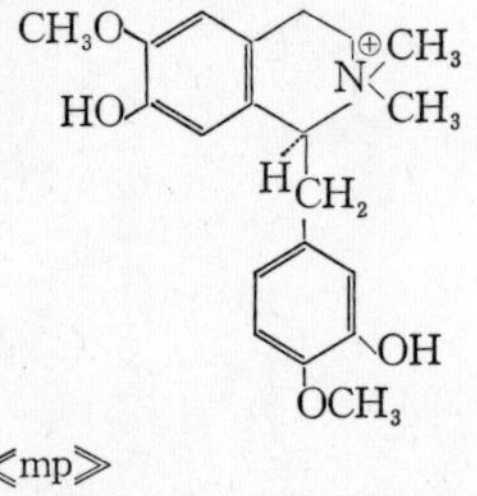

≪mp≫
≪$[\alpha]_D$≫ B⊕-Cl⊖ +123.3° (EtOH)
≪Ph.≫ CR[180] UV[180] IR[180] NMR[180] Rf[180]
≪OP.≫ Fagara hyemalis[180], F. naranjillo[180], F. nigerescens[180], F. pterota[180], F. rhoifolia[180]
≪C. & So.≫
≪Sa. & D.≫ B⊕-Cl⊖ : 237~238°
B-Pic : 190~191° B-Styph : 177~178°
B-O, O-Di Me = laudanosine-MeI
≪SD.≫ 180)
≪S. & B.≫

1) R. Konowalowa, S. Yu. Yunusov, A. Orechoff: Ber., **68**, 2277 (1935).
2) G. Merck: Ann., **66**, 125 (1848); **73**, 50 (1850).
3) P. C. Plugge: Arch. Pharm., [3] **25**, 343 (1887); Analyst, **12**, 197 (1887); Rec. trav. chim., **6**, 167 (1887); Z. anal. Chem., **30**, 385 (1891).
4) R. Konowalowa, S. Yu. Yunusov, A. Orechoff: Ber., **68**, 2158 (1935). cf. Zhur. Obshchei Khim., **10**, 641 (1940).
5) H. Asahina, T. Kawatani, M. Ono, S. Fujita: Bull. Narcotics, **9**, No. 2, 20 (1957). [C.A., **51**, 17094 (1957)].
6) G. Dragendorff: Die Heilpflanzen, 1898, p. 249.
7) O. Hesse: Ann., **153**, 47 (1870); Ber., **4**, 693 (1871); Ann. (Supple), **8**, 261 (1872).
8) A. Pictet, G. H. Kramers: Ber., **43**, 1329 (1910).
9) A. H. Allen: Commercial Organic Analysis, Vol. VI, p. 372 (1912).
10) E. Machiguchi: J. Pharm. Soc. Japan, **529**, 185 (1926).
11) L. A. Alyavdina: Russian Pat. 39,109. [C. A., **30**, 3591 (1936)].
12) G. R. Levi, F. Castelli: Gazz. chim. ital., **68**, 459 (1938). [C. A., **33**, 808 (1939)]; Arquiv. biol. (São Paulo), **23**, 263 (1939); Anales farm. bioquím (Buenos Aires), **11**, 6 (1940). [C. A., **34**, 7532 (1940)].
13) S. Busse, V. Busse: Khim. Farm. Prom., **1933**, 127. [C. A., **28**, 478 (1934)].
14) H. Willstaedt: Swed. Pat. 98,873. [C. A., **40**, 990 (1946)].
15) G. E. Foster: Analyst, **71**, 139 (1946).
16) O. Hesse: Ann., **176**, 189 (1875).
17) G. Goldschmiedt: Monatsh., **4**, 704 (1883); **6**, 372, 667, 954 (1885); **7**, 485, 504 (1886); **8**, 510 (1887); **9**, 42, 327, 349, 762, 778 (1888); **10**, 673 (1889); **19**, 321 (1898).
18) G. Goldschmiedt, H. Strache: Monatsh., **10**, 156, 692 (1889).
19) A. Hirsch: Monatsh., **12**, 486 (1891).
20) G. Goldschmiedt, F. Schranzhofer: Monatsh., **13**, 697 (1892).
21) F. Schranzhofer: Monatsh., **14**, 521, 597 (1893).
22) G. Goldschmiedt, A. Kirpal: Monatsh., **17**, 491 (1896).
23) R. Wegscheider: Monatsh., **18**, 418 (1897), **23**, 369 (1902).
24) A. Pictet, B. Athanasescu: Ber., **33**, 2346 (1900).
25) L. Stuchlik: Monatsh., **21**, 813 (1900).
26) G. Goldschmiedt, O. Hönigschmidt: Monatsh., **24**, 681 (1903).
27) J. Gadamer, W. Schulemann: Arch. Pharm., **253**, 284 (1915).
28) J. S. Buck, R. D. Haworth, W. H. Perkin, Jr.: J. Chem. Soc., **125**, 2176 (1924).
29) J. S. Buck, W. H. Perkin, Jr., T. S. Stevens: J. Chem. Soc., **127**, 1462 (1925).
30) A. Müller, M. Dorfman: J. Am. Chem. Soc., **56**, 2787 (1934).
31) M. I. Kabachnik, A. I. Zitser: Zhur. Obshchei Khim., **7**, 162 (1937).
32) J. Büchi, H. Welti: Pharm. Acta Helv., **16**, 67 (1941).
33) K. N. Menon: Proc. Indian Acad. Sci., **A 19**, 21 (1944).
34) T. and H. Smith Co.: Pharm. J. and Trans., [3] **23**, 793 (1893).
35) A. Pictet, A. Gams: Compt. rend., **149**, 210 (1909); Ber., **44**, 2480 (1911).
36) A. Pictet, M. Finkelstein: Ber., **42**, 1979 (1909).
37) B. Dobson, W. H. Perkin, Jr.: J. Chem. Soc., **99**, 135 (1911).
38) K. W. Rosenmund: Ber., **46**, 1034 (1913).
39) K. W. Rosenmund, M. Nothnagel, H. Riesenfeldt: Ber., **60**, 392 (1927).
40) H. Späth, A. Burger: Ber., **60**, 704 (1927).
41) C. Mannich, O. Walther: Arch. Pharm., **265**, 1 (1927).
42) E. Späth, F. Berger: Ber., **63**, 2098 (1930).
43) C. Schöpf, H. Perroy, I. Jäckh: Ann., **497**, 47 (1932).
44) K. Kindler, W. Peschke: Arch. Pharm., **272**, 236 (1934).
45) F. Boedecker, A. Heymans: Ger. Pat. 674, 400 (1937).
46) E. Kaufman, E. E. Eliel, J. Rosenkranz: Ciencia (Mex.), **7**, 136 (1946).
47) L. E. Craig, D. S. Tarbell: J. Am. Chem. Soc., **70**, 2783 (1948).
48) A. Galat: J. Am. Chem. Soc., **72**, 4436 (1950); **73**, 3654 (1951).
49) H. Wahl: Bull. Soc. Chim. France, **1950**, 680; **1951**, D. 1.
50) A. Dobrowsky: Monatsh., 82, 122, 140 (1951).
51) J. Gardent: Compt. rend., **234**, 1374 (1952); Am. Chim., (12) **10**, 413 (1955).
52) D. A. Guthrie, A. W. Frank, C. B. Purves: Can. J. Chem., **33**, 729 (1955).
53) V. R. Shrinivasan, F. Turba: Biochem. Z., **327**, 362 (1956).
54) F. D. Popp, W. E. McEwen: J. Am. Chem. Soc., **79**, 3773 (1957).
55) N. A. Chaudhury, A. Chatterjee: J. Indian Chem. Soc., **36**, 585 (1959).

56) A. R. Battersby, B. J. T. Harper: Proc. Chem. Soc., **1959**, 152.
57) A. R. Battersby, R. Binks: Proc. Chem. Soc., **1960**, 360.
58) A. R. Battersby, R. Binks, B. J. T. Harper: J. Chem. Soc., **1962**, 3534.
59) G. Kleinschmidt, K. Mothes: Z. Naturforsch., **14 b**, 52 (1959).
60) R. Robinson: "The Structural Relations of Natural Products," Clarendon Press, Oxford (1955).
61) E. Wenkert: Experientia, **10**, 346 (1954); **15**, 165 (1959).
62) R. H. Pinyazhko: Farmatsevii Zh. (Kiev), **19**, 12 (1964). [C. A., **64**, 6701 (1966)].
63) J. Knabe: Pharmazie, **20**, 741 (1965).
64) A. W. Sangster, K. L. Stuart: Chem. Revs., **65**, 69 (1965).
65) L. Pohl, W. Wiegrebe: Z. Naturforsch., **20 b**, 1032 (1965).
66) B. Ohashi, J. M. Wilson, H. Budzikiewicz, M. Shamma, W. A. Slusarchyk, C. Djerassi: J. Am. Chem. Soc., **85**, 2807 (1963).
67) W. Debska: Doznan. Towarz. Przyjaciol Nauk Wydzial Lekar., Prace. Komisji Farm., **2**, 23 (1962). [C. A., **59**, 6451 (1963)].
68) E. Fujita, T. Tomimatsu: J. Pharm. Soc. Japan, **79**, 1082 (1959).
69) S. Kubota, T. Matsui, E. Fujita, S. M. Kupchan: Tetrahedron Letters, **1965**, 3599.
70) S. Kubota, T. Matsui, E. Fujita, S. M. Kupchan: J. Org. Chem., **31**, 516 (1966).
71) Small, Lutz: "Chemistry of the Opium Alkaloids" U. S. Public Health Reports, Supple. No. 103, Washington (1932).
72) J. Kunitomo, E. Yuge, Y. Nagai: J. Pharm. Soc. Japan, **86**, 456 (1966).
73) H. Kondo, T. Kondo: J. Pharm. Soc. Japan, **524**, 3 (1925).
74) M. Tomita, F. Kusuda: J. Pharm. Soc. Japan, **72**, 793 (1952).
75) T. Nakano, M. Uchiyama: Pharm. Bull., **4**, 408 (1956).
76) S-T. Lu: J. Pharm. Soc. Japan, **83**, 19 (1963).
77) K. V. Rao, L. R. Row: J. Sci. Ind. Res., **20 B**, 125 (1961).
78) M. Tomita, T.-H. Yang, S-T. Lu: J. Pharm. Soc. Japan, **83**, 15 (1963).
79) M. Tomita, T.-H. Yang, K. H. Gaind, S. K. Baveja: J. Pharm. Soc. Japan, **83**, 218 (1963).
80) H. Kondo, T. Kondo: J. Pharm. Soc. Japan, **524**, 876 (1925), **538**, 1029 (1926), **554**, 324 (1928), **562**, 1156, 1163 (1928); J. prakt. Chem., **126**, 24 (1930).
81) M. Tomita, J. Kunitomo: J. Pharm. Soc. Japan, **82**, 734 (1962).
82) J. Finkelstein: J. Am. Chem. Soc., **73**, 550 (1951).
83) K. Kratzl, G. Billek: Monatsh., **82**, 568 (1951).
84) M. Tomita, K. Nakaguchi, S. Takagi: J. Pharm. Soc. Japan, **71**, 1046 (1951).
85) M. Tomita, F. Kusuda: J. Pharm. Soc. Japan, **72**, 280 (1952).
86) D. A. A. Kidd, J. Walker: J. Chem. Soc., **1954**, 609.
87) M. Tomita, I. Kikkawa: Pharm. Bull., **4**, 230 (1956).
88) H. Yamaguchi: J. Pharm. Soc. Japan, **78**, 733 (1956).
89) H. Yamaguchi, K. Nakano: J. Pharm. Soc. Japan, **79**, 1106 (1959).
90) H. Yamaguchi, K. Sempuku, E. Morokawa: J. Pharm. Soc. Japan, **80**, 1469 (1960).
91) H. Hellmann, W. Elser: Ann., **639**, 77 (1961).
92) H. Yamaguchi, C. Tanaka, F. Nagatani: J. Pharm. Soc. Japan, **82**, 522 (1962).
93) T. Kametani, S. Takano, K. Masuko, S. Kuribara: J. Pharm. Soc. Japan, **85**, 166 (1965).
94) M. Tomita, T. Ibuka: J. Pharm. Soc. Japan, **85**, 557 (1965).
95) M. Tomita, T. Shingu, H. Furukawa: J. Pharm. Soc. Japan, **86**, 373 (1966).
96) H. King: J. Chem. Soc., **1940**, 737.
97) M. Tomita, H. Yamaguchi: J. Pharm. Soc. Japan, **72**, 1219 (1952).
98) R. R. Arndt: J. Chem. Soc., **1963**, 2547.
99) S. M. Kupchan, B. Dasgupta, E. Fujita, M. L. King: Tetrahedron, **19**, 227 (1963).
100) J. C. Craig, S. K. Roy: Tetrahedron, **21**, 401 (1965).
101) T.-H. Yang, S.-T. Lu: J. Pharm. Soc. Japan, **83**, 22 (1963).
102) J. C. Craig, M. Martin-Smith, S. K. Roy, J. B. Stenlake: Tetrahedron, **22**, 1335 (1966).
103) M. Tomita, H. Furukawa, T. Kikuchi, A. Kato, T. Ibuka: Chem. Pharm. Bull., **14**, 232 (1966).
104) M. Tomita, T. Shingu, K. Fujitani, H. Furukawa: Chem. Pharm. Bull., **13**, 921 (1965).
105) J. Kunitomo, C. Yamamoto, T. Otuki: J. Pharm. Soc. Japan, **84**, 1141 (1964).
106) S. Yu. Yunusov, V. A. Mnatsakanyan, S. T. Akramov: Dokl. Akad. Nauk Uz. SSR., **1961**, No. 8, 43. [C. A., **57**, 9900 (1962)].
107) S. Yu. Yunusov, R. Konowalowa, A. Orechoff: Zhur. Obshchei Khim., **10**, 641 (1940). [C., **1941**, I, 2530].

108) L. Marion, L. Lemay, V. Portelance: J. Org. Chem., **15**, 216 (1950).
109) M. Tomita, H. Yamaguchi: J. Pharm. Soc. Japan, **72**, 793 (1952).
110) C. Ferrai, V. Deulofeu: Tetrahedron, **18**, 419 (1962).
111) M. Tomita, J. Niimi: J. Pharm. Soc. Japan, **78**, 1232 (1958).
112) G. J. Kapadia, N. J. Shah, R. I. Highet: J. Pharm. Sci., **53**, 1431 (1964).
113) H. W. Gibson, F. D. Popp, A. C. Noble: J. Heterocyclic Chem., **3**, 99 (1966).
114) H. Yamaguchi: J. Pharm. Soc. Japan, **78**, 692 (1958).
115) M. Tomita, Y. Watanabe, H. Furukawa: J. Pharm. Soc. Japan, **81**, 1644 (1961).
116) N. J. McCorkindale, D. S. Magrill, M. Martin-Smith, S. J. Smith, J. B. Stenlake: Tetrahedron Letters, **1964**, 3841.
117) A. R. Battersby, I. R. C. Bick, W. Klyne, J. P. Jennings, P. H. Scopes, M. J. Vernengo: J. Chem. Soc., **1965**, 2239.
118) S. M. Albonico, J. Comin, A. M. Kuck, E. Sanchez, P. H. Scopes, R. J. Swan, M. J. Vernengo: J. Chem. Soc., **1966**, 1340.
119) K. W. Gopinath, T. R. Govindachari, B. R. Pai, N. Viswanathan: Ber., **92**, 776 (1959).
120) E. Brochmann-Hansen, B. Nielsen: Tetrahedron Letters, **1965**, 1271.
121) R. R. Arndt, W. H. Waarschers: J. Chem. Soc., **1964**, 2244.
122) A. R. Battersby, G. W. Evans, R. O. Martin, M. E. Warren, Jr., H. Rapoport: Tetrahedron Letters, **1965**, 1275,
123) K. W. Gopinath, T. R. Govindachari, N. Viswanathan: Ber., **92**, 1657 (1959).
124) J. Kunitomo: J. Pharm. Soc. Japan, **81**, 1253 (1961).
125) W. W. C. Chen, P. Maitland: J. Chem. Soc.(C), **1966**, 753.
126) D. H. R. Barton, R. H. Hesse, G. W. Kirby: J. Chem. Soc., **1965**, 6379.
127) F. R. Stermitz, L-Chen, J. I. White: Tetrahedron, **22**, 1095 (1966).
128) F. Brochmann-Hansen, B. Nielsen, G. E. Utzinger: J. Pharm. Sci., **54**, 1531 (1965).
129) E. Späth, H. Epstein: Ber., **59**, 2791 (1926).
130) W. Leithe: Ber., **63**, 1498 (1930); **64**, 2827 (1931).
131) C. Schöpf, K. Thierfelder: Ann., **497**, 22 (1932).
132) A. Burger: Dissertation, Vienna, 1927.
133) E. Späth, H. Epstein: Ber., **61**, 334 (1928).
134) G. Billek: Monatsh., **87**, 106 (1956).
135) A. Chatterjee, S. K. Kundu: Ber., **99**,1764 (1966).
136) M. Shamma, M. A. Greenberg, B. S. Dudock: Tetrahedron Letters, **1965**, 3595.
137) E. Kauder: Arch. Pharm., **228**, 419 (1890).
138) O. Hesse: J. prakt. Chem., **65**, 42 (1902).
139) J. J. Dobbie, A. Lauder: J. Chem. Soc., **83**, 626 (1903).
140) E. Späth: Monatsh., **41**, 297 (1920).
141) O. Hesse: Ann., **282**, 209 (1894).
142) E. Späth, N. Lang: Monatsh., **42**, 273 (1921).
143) E. Späth, A. Burger: Monatsh., **47**, 733 (1926).
144) A. W. McKenzie, J. R. Price: Austral. J. Chem., **6**, 180 (1953).
145) B. Frydman, R. Bendisch, V. Deulofeu: Tetrahedron, **4**, 342 (1958).
146) A. M. Kuck, B. Frydman: J. Org. Chem., **26**, 5253 (1961).
147) H. Decker, L. Galatty: Ber., **42**, 1179 (1909).
148) F. L. Pyman: J. Chem. Soc., **95**, 1266 (1909).
149) J. Gadamer, H. Kondo: Arch. Pharm., **253**, 281 (1915).
150) W. Leithe: Ber., **63**, 1498 (1930); **64**, 2827 (1931).
151) T. Kondo, N. Mori: J. Pharm. Soc. Japan, **51**, 615 (1931).
152) V. K. Baghwat, D. K. Moore, F. L. Pyman: J. Chem. Soc., **1931**, 443.
153) R. Robinson, S. Sugasawa: J. Chem. Soc., **1932**, 789.
154) F. E. King, P. L'Ecuyer, F. L. Pyman: J. Chem. Soc., **1936**, 731.
155) F. E. King, P. L'Ecuyer: J. Chem. Soc., **1937**, 427.
156) H. Corrodi, E. Hardegger: Helv. Chim. Acta, **39**, 889 (1956).
157) J. Knabe: Arch. Pharm., **292**, 416 (1959).
158) J. M. Z. Gladyeh: J. Chem. Soc., **1963**, 733.
159) M. Ohta, H. Tani, S. Morozumi, S. Kodaira, K. Kuriyama: Tetrahedron Letters, **1963**, 1875.
160) J. Knabe, G. Grund: Arch. Pharm., **296**, 854 (1963).
161) F. L. Pyman: J. Chem. Soc., **95**, 1610 (1909).
162) F. L. Pyman, W. C. Reynolds: J. Chem. Soc., **97**, 1320 (1910).
163) E. Späth, H. Epstein: Ber., **59**, 2721 (1926).
164) W. Awe, H. Unger: Ber., **70**, 472 (1937).
165) Shionogi & Co. Ltd.: Japan Pat., 781 (1959). [C. A., **54**, 8879 (1960)].
166) M. K. Jain: J. Chem. Soc., **1962**, 2203.
167) V. Preininger, A. D. Cross, F. Santavy: Coll. Czech. Chem. Comm., **31**, 3345 (1966).
168) L. Slavikova, J. Slavik: Coll. Czech. Chem. Comm., **31**, 3362 (1966).
169) H. Furukawa, T.-H. Yang, T.-J. Lin: J.

Pharm. Soc. Japan, **85**, 472 (1965).

170) M. Tomita, T. Nakano: J. Pharm. Soc. Japan, **72**, 1260 (1952).

171) T. Nakano: Pharm. Bull., **1**, 29 (1953).

172) M. Tomita, Y. Inubushi, M. Yamagata: J. Pharm. Soc. Japan, **71**, 1069 (1951).

173) M. Tomita, T. Nakano: J. Pharm. Soc. Japan, **72**, 197 (1952).

174) T. Nakano: Pharm. Bull., **3**, 234 (1955).

175) K. Ito, I. Uchida: J. Pharm. Soc. Japan, **79**, 1108 (1959).

176) M. Tomita, H. Yamaguchi: J. Pharm. Soc. Japan, **73**, 495 (1953).

177) N. J. McCorkindale, D. S. Magrill, M. Martin-Smith, J. B. Stenlake: Tetrahedron Letters, **1964**, 3841.

178) J. McShefferty, P. F. Nelson, J. L. Paterson, J. B. Stenlake, J. P. Todd: J. Pharm. Pharmacol., **8**, 1117 (1956).

179) G. Grethe, M. Uskokovic, A. Brossi: Tetrahedron Letters, **1966**, 1599.

180) S. M. Albonico, A. M. Kuck, V. Deulofeu: Chem. & Ind., **1964**, 1580; Ann., **685**, 200 (1965).

181) H. Furukawa: J. Pharm. Soc. Japan, **86**, 1143 (1966).

182) R. L. Gay, T. F. Crimmins, C. R. Hauser: Chem. & Ind., **1966**, 1635.

Chapter 4 Pavine and Isopavine Alkaloids

Crychine $C_{19}H_{17}O_4N \equiv 323.33$

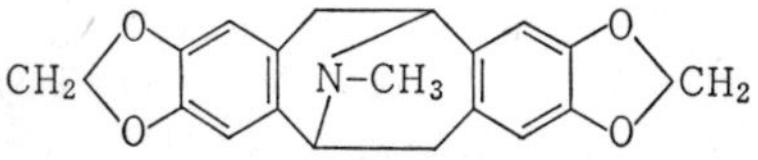

≪mp≫
≪[α]$_D$≫ −184.5° ($CHCl_3$), −220.2° (EtOH)
≪Ph.≫ UV[1] IR[1)2)] NMR[1] Mass[3] Rf[1]
≪OP.≫ Cryptocarya chinensis[1]
≪C. & So.≫ Colourless oil
≪Sa. & D.≫ B–Pic: 177° B–MeI: 286°(d)
≪SD.≫ 1)
≪S. & B.≫

Eschscholtzine $C_{19}H_{17}O_4N \equiv 323.33$

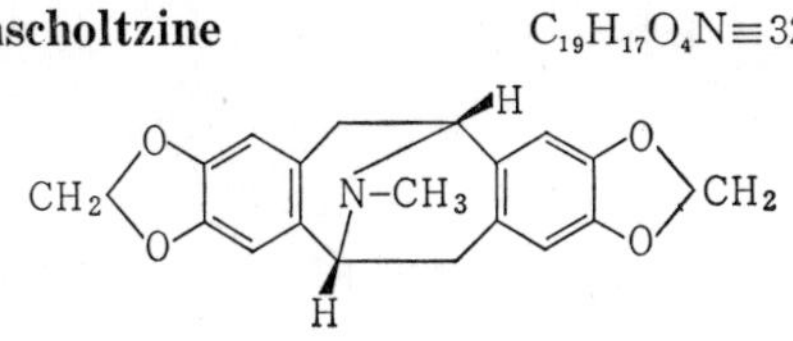

≪mp≫ 128°
≪[α]$_D$≫ −202° (MeOH)
≪Ph.≫ UV[5] NMR[5]
≪OP.≫ Eschscholtzia californica[4]
≪C. & So.≫
≪Sa. &–D.≫ B–HCl: 246° (225°)
B–Oxal: 140° (sint.)
≪SD.≫ 5)
≪S. & B.≫

***l*-Caryachine** $C_{19}H_{19}O_4N \equiv 325.33$

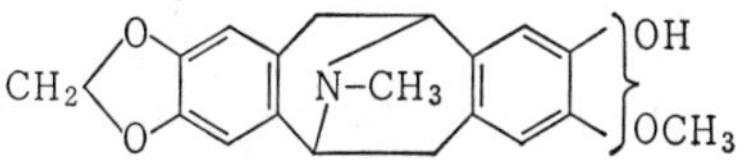

≪mp≫ 174∼175° (Me_2CO)
≪[α]$_D$≫ −269.6° (EtOH)
≪Ph.≫ CR[1] UV[1] IR[1] NMR[1] Mass[3]
≪OP.≫ Cryptocarya chinensis[1]
≪C. & So.≫ Colourless needles
≪Sa. & D.≫ B–OMe–MeI: 246° (225°)
B–HBr: 295∼296° (d)
≪SD.≫ 1)
≪S. & B.≫

***dl*-Caryachine**
≪mp≫ 241∼242°
≪[α]$_D$≫
≪Ph.≫ CR[1] UV[1] NMR[1]
≪OP.≫ Cryptocarya chinensis[1)2)]
≪C. & So.≫
≪Sa. & D.≫
≪SD.≫ 1)
≪S. & B.≫

Munitagine $C_{19}H_{21}O_4N \equiv 327.37$

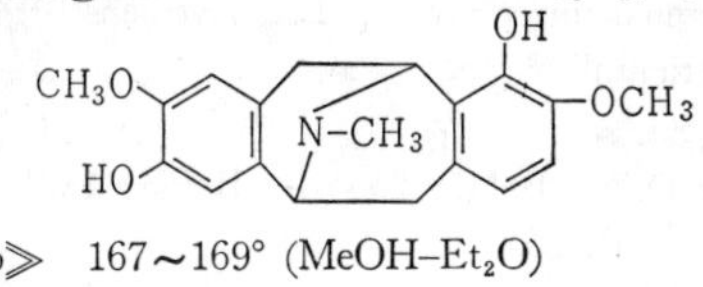

≪mp≫ 167∼169° (MeOH–Et_2O)
≪[α]$_D$≫ −239°
≪Ph.≫ CR[6] UV[6] IR[6] NMR[6] Mass[6] Rf[6]
≪OP.≫ Argemone hispida[6], A. munita[6]
≪C. & So.≫ Fine colourless needles
≪Sa. & D.≫ B–O, O–DiBz: 239∼241°
≪SD.≫ 6)
≪S. & B.≫ BG: 6)

Bisnorargemonine $C_{19}H_{21}O_4N \equiv 327.37$

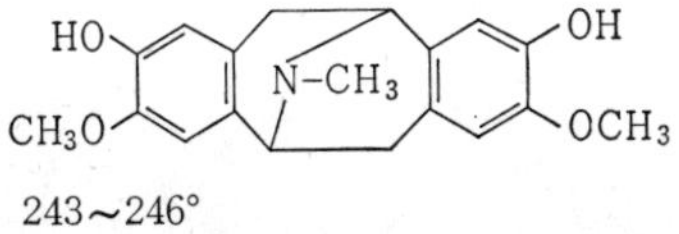

≪mp≫ 243∼246°
≪[α]$_D$≫ −265.8° (MeOH)
≪Ph.≫ UV[6)10)] IR[6)10)] NMR[6)8)] Mass[6] Rf[6]
≪OP.≫ Argemone hispida[6], A. munita[6)7)]
≪C. & So.≫
≪Sa. & D.≫
≪SD.≫ 8), 9), 10)
≪S. & B.≫ BG: 6)

Eschscholtzidine* $C_{20}H_{21}O_4N$≡339.38

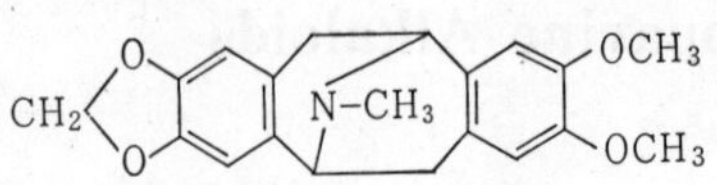

≪mp≫

≪$[\alpha]_D$≫ +194.2° (MeOH)

≪Ph.≫ UV[11] NMR[11] Rf[11]
≪OP.≫ Eschscholtzia californica[11]
≪C. & So.≫ Oil
≪Sa. & D.≫
≪SD.≫ 11)
≪S. & B.≫

d-O*-Methylcaryathine $C_{20}H_{21}O_4N$≡339.38

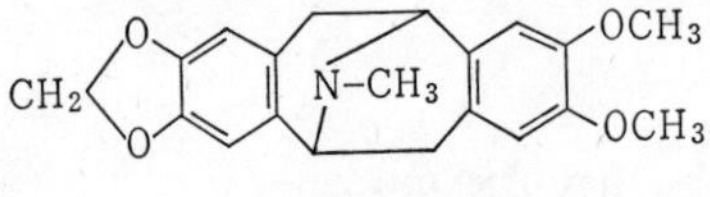

≪mp≫

≪$[\alpha]_D$≫ +195.1° (EtOH)

≪Ph.≫ IR[2] NMR[2] Mass[3] Rf[2]
≪OP.≫ Cryptocarya chinensis[2]
≪C. & So.≫ Oil
≪Sa. & D.≫ B–HCl: 178~180°
≪SD.≫ 2)
≪S. & B.≫

Norargemonine $C_{20}H_{23}O_4N$≡341.39

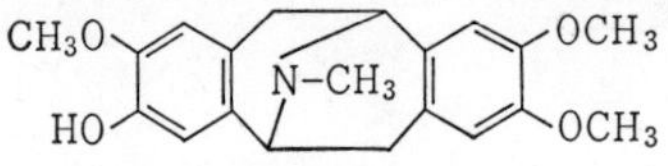

≪mp≫ 238°

≪$[\alpha]_D$≫ −153.7° ($CHCl_3$)

≪Ph.≫ IR[17] NMR[6,17] Rf[17]
≪OP.≫ Argemone hispida[6,12,13], A. munita[7,12,13,14]
≪C. & So.≫
≪Sa. & D.≫
≪SD.≫ 9), 13), 14), 15), 16), 17), 18)
≪S. & B.≫ 17), BG: 6)

Argemonine $C_{21}H_{25}O_4N$≡343.41

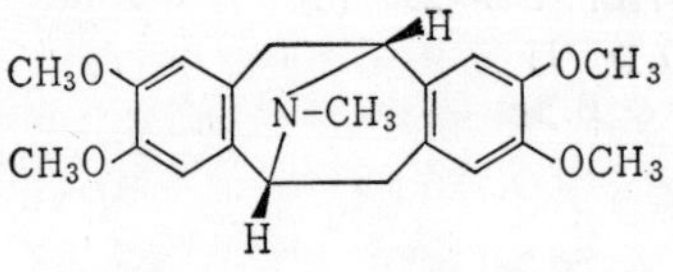

≪mp≫ 244~250°[6], 152.5~153°[12,13]

≪$[\alpha]_D$≫ −187.93° ($CHCl_3$), −214.22° (EtOH)

≪Ph.≫ UV[6,14,18,19] IR[6,14,18] NMR[5,6,14] Mass[14] ORD[19] Rf[6] CR[6]
≪OP.≫ Argemone hispida[6,7,12,13,14], A. munita[7,12,13,14]
≪C. & So.≫ Colourless cubes
≪Sa. & D.≫ B–Pic: 242~245° B–Styph: 247~249° B–MeI: 273~274°
≪SD.≫ 9), 12), 14), 15), 16), 17), 18), 19)
≪S. & B.≫ 14), BS: 24), BG: 6)

Amurensine (Xanthopetaline) $C_{19}H_{19}O_4N$≡325.33

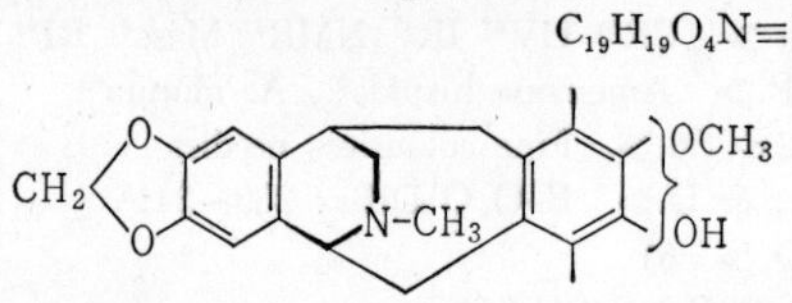

≪mp≫ 221~223°

≪$[\alpha]_D$≫ −178±4° (MeOH)

≪Ph.≫ UV[25] IR[25] NMR[24,25]
≪OP.≫ Papaver alpinum[21,22,23,24], P. nudicaule[23], P. pyrenaicum[21,22,23,24], P. suaveolens[21,22,23,24], P. tatricum[21,22,23,24]
≪C. & So.≫
≪Sa. & D.≫ B–HI: 255°(d) B–Pic: 132°
≪SD.≫ 24), 25)
≪S. & B.≫

* Eschscholtzidine may be identical with *d–O*–methylcaryachine

Amurensinine $C_{20}H_{21}O_4N \equiv 339.38$

≪mp≫ 162～164°

≪$[\alpha]_D$≫ $-162\pm4°$ ($CHCl_3$), $-175\pm4°$ (MeOH)

≪Ph.≫ UV[24)25)] IR[24)25)] NMR[24)25)] ORD[24)]

≪OP.≫ Papaver alpinum[21)22)23)24)], P. nudicaule[23)], P. pyrenaicum[21)22)23)24)], P. suaveolens[21)22)23)24)], P. tatricum[21)22)23)24)]

≪C. & So.≫

≪Sa. & D.≫

≪SD.≫ 24), 25)

≪S. & B.≫

Remrefine

≪mp≫ 241～242°

≪$[\alpha]_D$≫ $-147°$ (H_2O)

≪Ph.≫

≪OP.≫ Roemeria refracta[26)]

≪C. & So.≫

≪Sa. & D.≫

≪SD.≫ 26)

≪S. & B.≫

1) S.-T. Lu, P.-K. Lan: J. Pharm. Soc. Japan, **86**, 177 (1966).
2) S.-T. Lu: J. Pharm. Soc. Japan, **86**, 296 (1966).
3) M. Tomita, P.-K. Lan, T. Ibuka: J. Pharm. Soc. Japan, **86**, 414 (1966).
4) R.H.F. Manske, K.H. Shin: Can. J. Chem., **43**, 2180 (1965).
5) R.H.F. Manske, A.R. Battersby, D.F. Shaw: Can. J. Chem., **43**, 2183 (1965).
6) F.R. Stermitz, J.N. Seiber: J. Am. Chem. Soc., **88**, 2925 (1966).
7) L.B. Kier, T.O. Soine: J. Pharm. Sci., **49**, 187 (1960).
8) T.O. Soine, L.B. Kier: J. Pharm. Sci., **52**, 1013 (1963).
9) L.B. Kier, T.O. Soine: J. Pharm. Sci., **50**, 321 (1961).
10) A.D. Cross, L. Dolejs, V. Hanus, M. Maturova, F. Santavy: Coll. Czech. Chem. Comm., **30**, 1335 (1965).
11) R.H.F. Manske, K.H. Shin: Can. J. Chem., **44**, 1259 (1966).
12) T.O. Soine, O. Gisvold: J. Am. Pharm. Assoc. (Sci. Ed.), **33**, 185 (1944).
13) J.W. Shermerhorn, T.O. Soine: J. Am. Pharm. Assoc. (Sci. Ed.), **40**, 19 (1951).
14) M.J. Martell, Jr., T.O. Soine, L.B. Kier: J. Am. Chem. Soc., **85**, 1022 (1963).
15) T.O. Soine, L.B. Kier: J. Pharm. Sci., **51**, 1196 (1962).
16) M. Shamma: Experientia, **18**, 64 (1962).
17) F.R. Stermitz, J.N. Seiber: Tetrahedron Letters, **1966**, 1177.
18) F.R. Stermitz, S.-Y. Lus, G. Kallos: J. Am. Chem. Soc., **85**, 1551 (1963).
19) O. Červinka, A. Fábryová, V. Novák: Tetrahedron Letters, **1966**, 5375.
20) D.H.R. Barton, R.H. Hesse, G.W. Kirby: J. Chem. Soc., **1965**, 6379.
21) M. Maturová, B.K. Moza, J. Sitař, F. Santavý: Planta medica, **10**, 255 (1962).
22) H.-G. Boit, H. Flentze: Naturwiss., **46**, 514 (1959).
23) H.-G. Boit, H. Flentze: Naturwiss., **47**, 180 (1960).
24) F. Santavý, M. Maturová, L. Hruban: Chem. Comm., **1966**, 36.
25) F. Santavy, L. Hruban, M. Matsurova: Coll. Czech. Chem. Comm., **31**, 4286 (1966).
26) M.S. Yunusov, S.T. Akramov, S. Yu. Yunusov: Dokl. Akad. Nauk Uz. SSR., **23**, 38 (1966). [C.A., **65**, 13781 (1966)].

Chapter 5 *N*-Benzylisoquinoline Alkaloids

Sendaverine (Alkaloid F 28) $C_{18}H_{21}O_3N \equiv 299.36$

CH_3O
HO
N
OCH_3

≪mp≫ 138° (MeOH)
≪[α_D]≫ ±0°[6]

≪Ph.≫ UV[2,3,4] IR[3,4] NMR[3,4,5] Mass[3,4] Rf[2,3,4]
≪OP.≫ Corydalis aurea[1]
≪C. & So.≫ Colourless fine needles
≪Sa. & D.≫ B-HCl : 202∼207°, cf. 225∼220° (d)
≪SD.≫ 2), 3), 4)
≪S. & B≫ 3), 4)

1) R. H. F. Manske : Can. J. Res., B **16**, 81 (1936).
2) T. Kametani, K. Ohkubo, I. Noguchi : J. Chem. Soc. (C), **1966**, 715.
3) T. Kametani, K. Ohkubo : Tetrahedron Letters, **1965**, 4317.
4) T. Kametani, K. Ohkubo : Chem. Pharm. Bull., **15**, 608 (1967).
5) T. Kametani, K. Ohkubo, S. Takano : J. Pharm. Soc. Japan, **87**, 563 (1967).
6) T. Kametani, K. Ohkubo, R. H. F. Manske : Tetrahedron Letters, **1966**, 985.

Chapter 6 Bisbenzylisoquinoline Alkaloids

Magnoline $C_{36}H_{40}O_6N_2$≡596.70

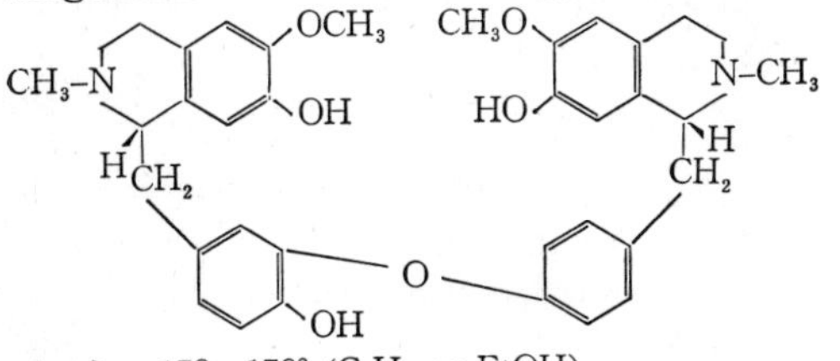

≪mp≫ 178~179° (C_6H_6 or EtOH)

≪[α]D≫ −9.6° (Py)
≪Ph.≫ IR[5]
≪OP.≫ Magnolia (Michelia) fuscata[1]
≪C. & So.≫ Sol. in C_6H_6, alkali. Spar. sol. in most org. solv.
≪Sa. & D.≫ B–2 Pic: 160~162° (d)
B–2 Picrol: 190° (d)
≪SD.≫ 2), 3)
≪S. & B.≫ 4), 5)

Berbamunine $C_{36}H_{40}O_6N_2$≡596.70

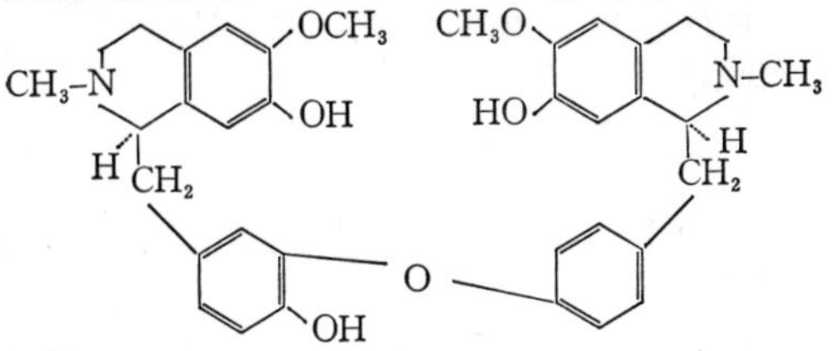

≪mp≫ 190~191° (Me_2CO)
≪[α]D≫ +87° (MeOH), +55.1° (Me_2CO)
+25.6° (Py)

≪Ph.≫ CR[6] UV[6] Rf[6]
≪OP.≫ Berberis amurense var. japonica
≪C. & So.≫ Needles, Sol. in Me_2CO, THF. Spar. sol. in Et_2O, C_6H_6, EtOH, $CHCl_3$, dioxane, MeOH
≪Sa. & D.≫
B–O, O, O–TriMe: 52~58°
B–O, O, O–TriMe·MeI: 82~84° (d)
≪SD.≫ 3), 6), 7)
≪S. & B.≫

Thalisopine $C_{38}H_{44}O_5N_2$≡608.75

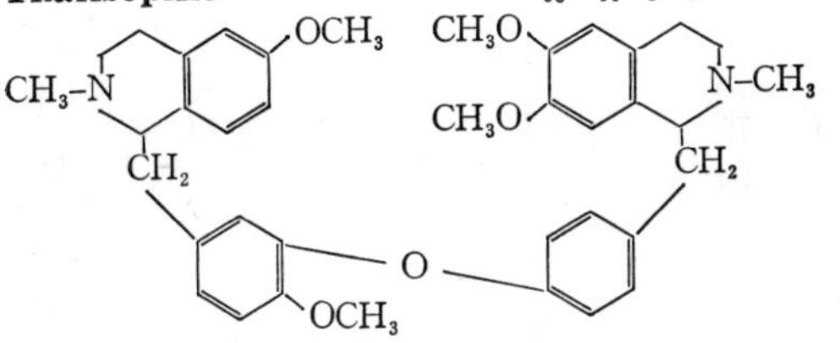

≪mp≫
≪[α]D≫
≪Ph.≫
≪OP.≫ Thalictrum isopyroides[8]
≪C. & So.≫
≪Sa. & D.≫
≪SD.≫ 8)
≪S. & B.≫

Daulinoline $C_{37}H_{42}O_6N_2$≡610.72

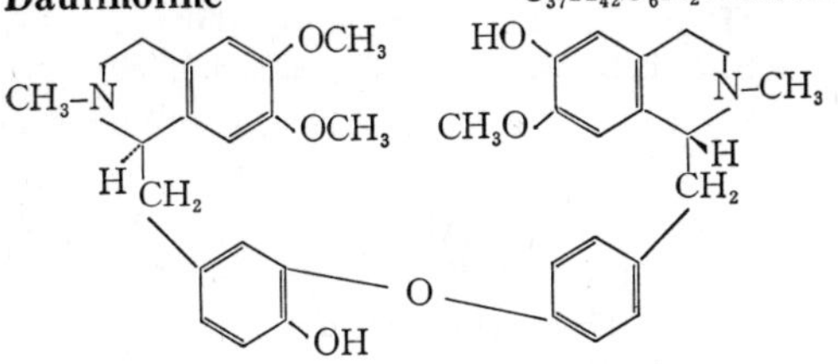

≪mp≫ 100~103° ($CHCl_3$–C_6H_{14})
≪[α]D≫ −108° (MeOH)
≪Ph.≫ CR[9] IR[9] NMR[9]
≪OP.≫ Menispermum dauricum[9]
≪C. & So.≫ Prisms
≪Sa. & D.≫
≪SD.≫ 9)
≪S. & B.≫

Cuspidaline $C_{37}H_{42}O_6N_2$≡610.72

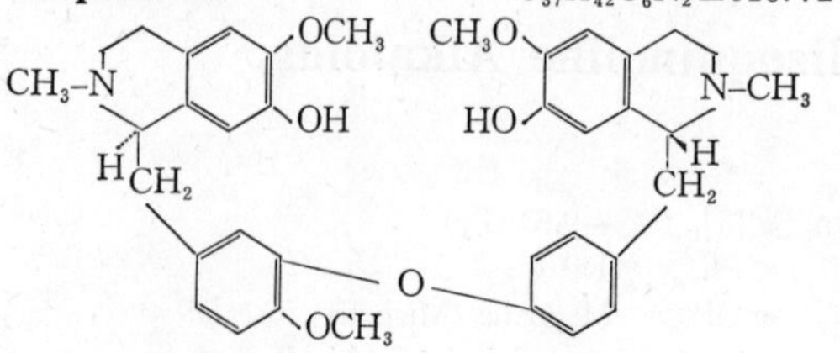

OCH_3 CH_3O
CH_3–N N–CH_3
OH HO
H CH_2 H CH_2
O
OCH_3

≪mp≫
≪[α_D]≫ −48° ($CHCl_3$)
≪Ph.≫ IR[10] NMR[10]
≪OP.≫ Limacia cuspidata[10]
≪C. & So.≫ Oil
≪Sa. & D.≫ B–O, O–DiMe·2–MeI: 181∼184°
≪SD.≫ 10)
≪S. & B.≫

Dauricine $C_{38}H_{44}O_6N_2$≡624.75

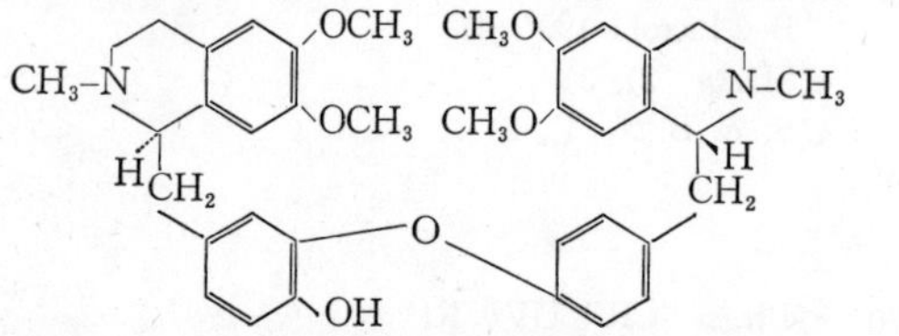

OCH_3 CH_3O
CH_3–N N–CH_3
OCH_3 CH_3O
H CH_2 H CH_2
O
OH

≪mp≫ 100∼103° [B+$CHCl_3$] ($CHCl_3$–C_6H_{14})
≪[α_D]≫ −139° (MeOH)
≪Ph.≫ UV[27,33] IR[20,23,26,27,30] NMR[20,23,34] Mass[35,36] ORD[37] Rf[20,26,27]
≪OP.≫ Menispermum canadense[11,12], M. dauricum[13]
≪C. & So.≫ Sol. in MeOH, EtOH, Me_2CO, $CHCl_3$, C_6H_6
≪Sa. & D.≫ B–2$HClO_4$: 160∼163°
B–2MeI: 196∼198°
B–O, O–DiMe·2MeI: 181∼182°
≪SD.≫ 3), 12), 13), 14)∼23)
≪S. & B.≫ 18), 20), 24)∼32)

Aztequine $C_{36}H_{40}O_7N_2$≡612.70

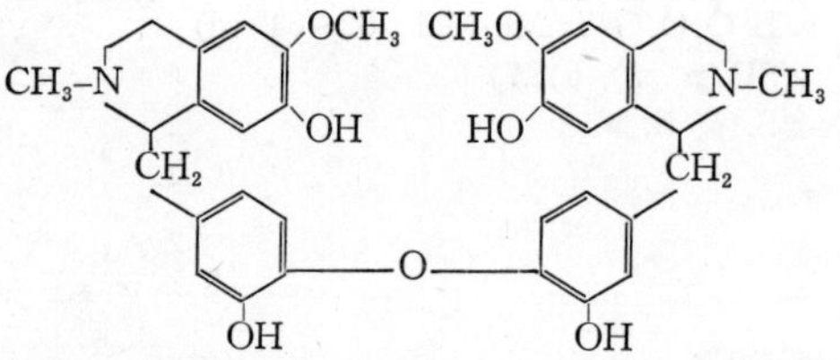

OCH_3 CH_3O
CH_3–N N–CH_3
OH HO
CH_2 CH_2
O
OH OH

≪mp≫ 176°
≪[α_D]≫
≪Ph.≫
≪OP.≫ Yoloxochitl (Talauma mexicana)[38]
≪C. & So.≫ Sol. in CH_2Cl_2. Spar. sol. in $CHCl_3$. Insol. in Et_2O, EtOH, H_2O
≪Sa. & D.≫
≪SD.≫ 38)
≪S. & B.≫ 39), cf. 40)

Magnolamine $C_{36}H_{40}O_7N_2$≡612.70

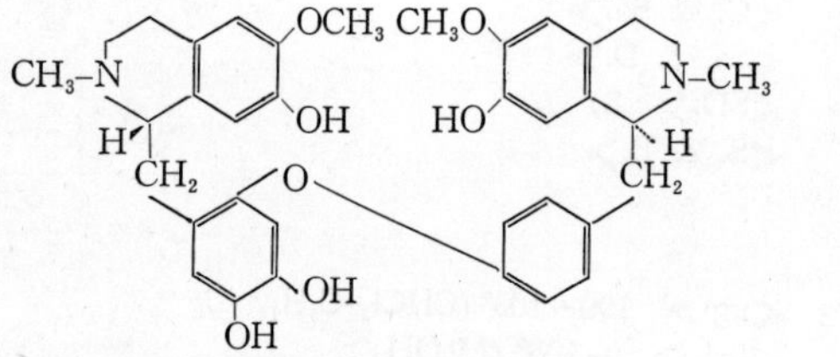

OCH_3 CH_3O
CH_3–N N–CH_3
OH HO
H CH_2 H CH_2
O
OH
OH

≪mp≫ 117∼119° (C_6H_6)
≪[α_D]≫ +116° (EtOH)
≪Ph.≫ UV[48,49] IR[48,49] NMR[49] Mass[51] Rf[48,49]
≪OP.≫ Magnolia (Michelia) fuscata[1]
≪C. & So.≫ Needles, Sol. in EtOH, $CHCl_3$. Insol. in C_6H_6
≪Sa. & D.≫ B–2 Pic: 142∼145° (d)
B–2 Picrol: 163∼164° (d)
B–O, O, O, O–TetraMe: 151∼152°
≪SD.≫ 3), 41), 42), 43), 44), 45)
≪S. & B.≫ 45), 46), 47), 48), 49), 50)

Fetidine $C_{41}H_{50}O_8N_2$≡698.83

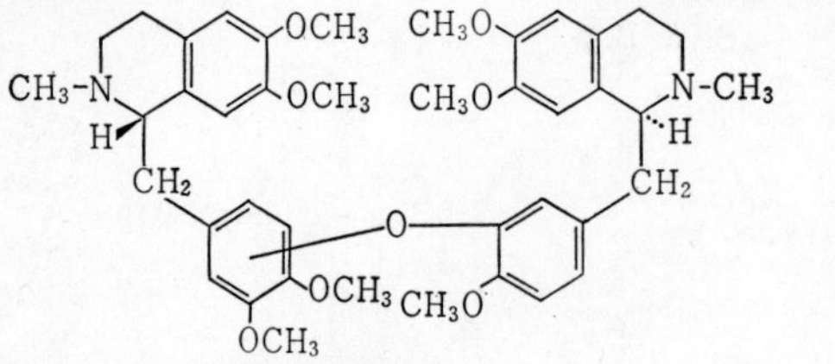

OCH_3 CH_3O
CH_3-N N-CH_3
OCH_3 CH_3O
H H
CH_2 CH_2
O
OCH_3 CH_3O
OCH_3

≪mp≫ 132∼135° [B+H_2O]
≪[α_D]≫ +121.4° (MeOH)
≪Ph.≫ UV[52] Rf[52]
≪OP.≫ Thalictrum fetidum[52]
≪C. & So.≫
≪Sa. & D.≫ B–2 HCl: 228∼230° (d)
B–2 HBr: 225∼230° B–2HNO_3: 200°
B–H_2SO_4: 215∼218° (d) B–2 MeI: 205∼210°
≪SD.≫ 52)
≪S. & B.≫

Liensinine $C_{37}H_{42}O_6N_2 \equiv 610.72$

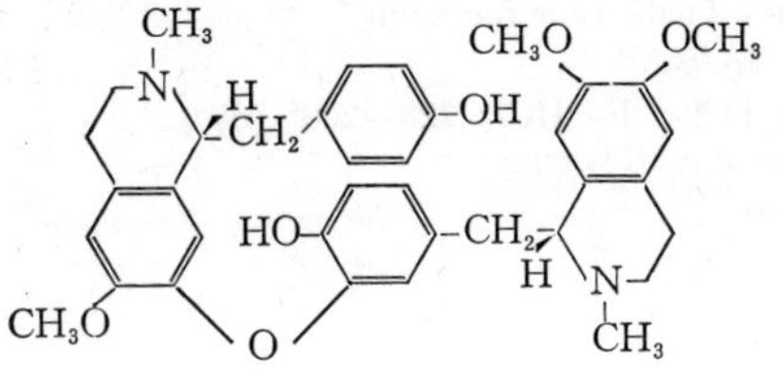

≪mp≫ 95～99°
≪[α_D]≫ +15.85° (Me_2CO)

≪Ph.≫ UV[59] IR[59] NMR[59,60] Rf[59]
≪OP.≫ Nelumbo nucifera[53,54,55]
≪C. & So.≫
≪Sa. & D.≫ B-2 $HClO_4$: 212～214°
B-O, O-DiMe: 105～110°,
B-O, O-DiMe·2MeBr: 203°
B-O, O-DiAc: 124°
≪SD.≫ 56), 57)
≪S. & B.≫ 58), 59)

Isoliensinine $C_{37}H_{42}O_6N_6 \equiv 610.72$

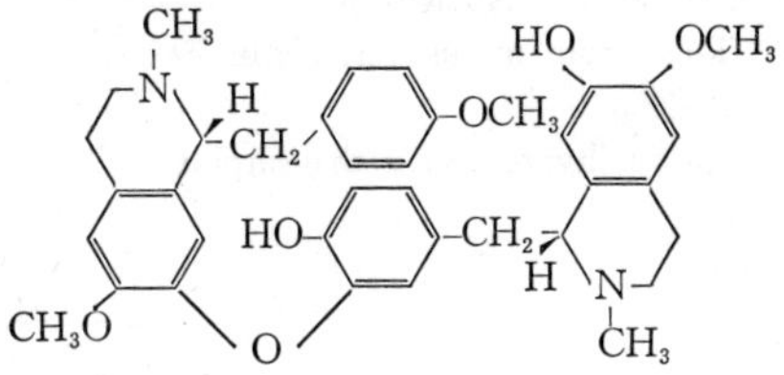

≪mp≫
≪[α_D]≫ −43.3° ($CHCl_3$), +49.3° (Me_2CO)

≪Ph.≫ UV[61] IR[61] NMR[60,61]
≪OP.≫ Nelumbo nucifera[61,62,63]
≪C. & So.≫ Oil
≪Sa. & D.≫ B-2HCl: 185～186°
B-2 $HClO_4$: 200～203°
B-O, O-DiMe: 133～135°
B-O, O-DiMe·2 Styph: 133～135°
≪SD.≫ 61)
≪S. & B.≫ 61)

Neferine $C_{38}H_{44}O_6N_2 \equiv 624.75$

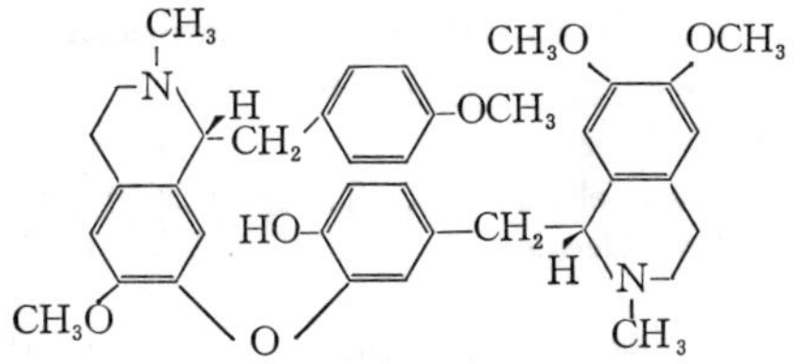

≪mp≫
≪[α_D]≫ −37.8° ($CHCl_3$)

≪Ph.≫ IR[62] NMR[60,62]
≪OP.≫ Nelumbo nucifera[54,55,62]
≪C. & So.≫ Oil
≪Sa. & D.≫ B-2MeI: 180～182°
B-OMe·2 Styph: 139～140°
B-OEt·2 Styph: 138～140°
≪SD.≫ 62)
≪S. & B.≫ 62)

Thalmelatine $C_{40}OH_{46}O_8N_2 \equiv 682.78$

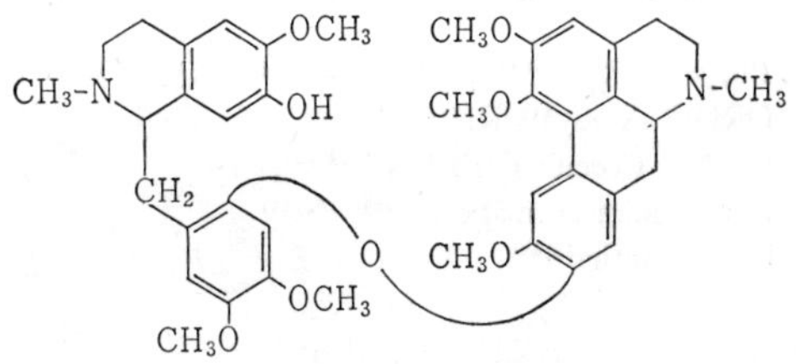

≪mp≫ 130～135° (120～123°)(EtOH)
≪[α_D]≫ +110° (EtOH)

≪Ph.≫ IR[66]
≪OP.≫ Thalictrum minus var. elatum[65]
≪C. & So.≫
≪Sa. & D.≫ B-OMe=thalicarpine
B-OEt: 133～135°
≪SD.≫ 66), 67), 68)
≪S. & B.≫ cf. 68)

Phetidine $C_{40}H_{46}O_8N_2 \equiv 682.78$

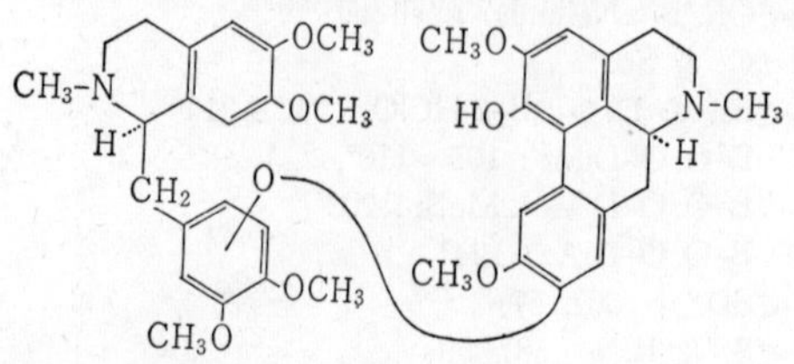

≪mp≫ 132～135° (AcOEt)
≪[α_D]≫ +121.4° (MeOH)

≪Ph.≫
≪OP.≫ Thalictrum foetidum[69]
≪C. & So.≫
≪Sa. & D.≫ B–2HCl: 228～230° (d)
≪SD.≫ 69)
≪S. & B.≫

Dehydrothalicarpine $C_{41}H_{46}O_8N_2 \equiv 694.79$

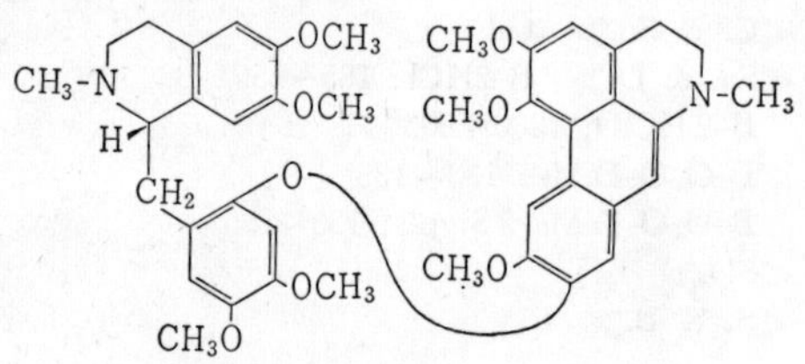

≪mp≫ 180～182° (MeOH–Et_2O)
≪[α_D]≫ +54° ($CHCl_3$)

≪Ph.≫ UV[70] IR[70] NMR[70] Rf[70]
≪OP.≫ Thalictrum minus var. elatum[70]
≪C. & So.≫
≪Sa. & D.≫ Dihydro–Der.=thalicarpine
≪SD.≫ 70)
≪S. & B.≫

Thalicarpine $C_{41}H_{48}O_8N_2 \equiv 696.81$

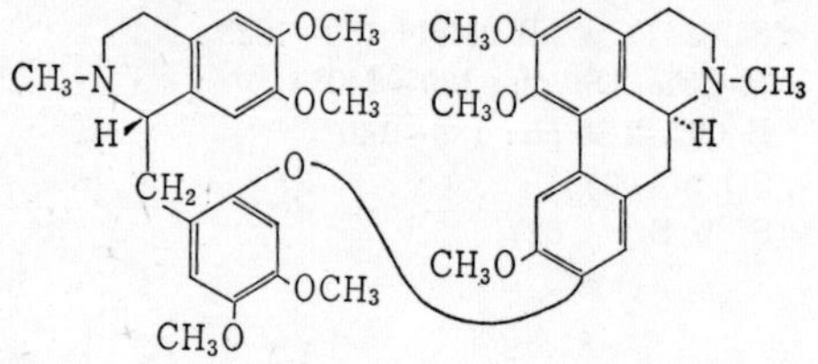

≪mp≫ 160～161° (153～155°) (Me_2CO–Et_2O)
≪[α_D]≫ +89° ($CHCl_3$), +133° (MeOH)

≪Ph.≫ UV[67,70,73,75,76,71] IR[67] NMR[67,75,76,363]
≪OP.≫ Hernandia ovigera[67,71,72],
Thalictrum dasycarpum[73],
T. minus var. elatum[65,68], T. resolutum[74]
≪C. & So.≫ Needles
≪Sa. & D.≫
≪SD.≫ 67), 68), 71), 75), 76)
≪S. & B.≫ 67), 71), 75), 76)

Daphnoline (Trilopamine) $C_{35}H_{36}O_6N_2 \equiv 580.65$

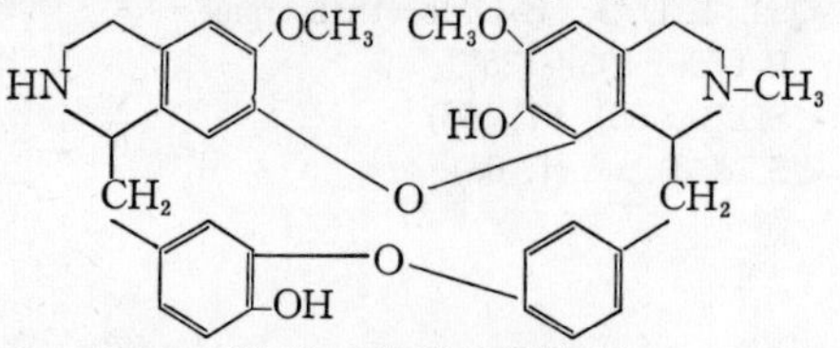

≪mp≫ 195～196° (B+$CHCl_3$) ($CHCl_3$)
≪[α_D]≫ +459° ($CHCl_3$), +356.6° (2%AcOH)

≪Ph.≫ CR[77,78,83] UV[81,86] IR[79,87] NMR[34]
ORD[37] X-ray[84] Rf[81]
≪OP.≫ Cocculus trilobus[77,78,79],
Daphnandra aromatica,[77,78,80,81,82]
D. micrantha[83,84]
≪C. & So.≫
≪Sa. & D.≫ B–2HCl: 290°(d)
B–2HI: 264°, 286°(d) B–2MeI: 266°(d)
B–O, O–DiMe=daphnandrine
≪SD.≫ 3), 77), 78), 79), 81), 82), 83), 84), 85)
≪S. & B.≫

Aromoline (Thalicrine) $C_{36}H_{38}O_6N_2$≡594.68

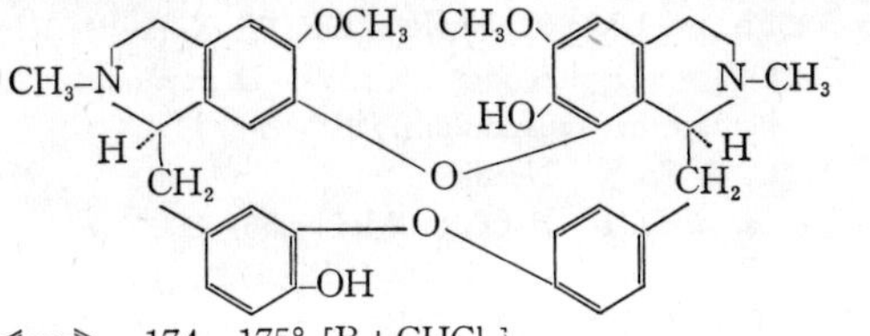

≪mp≫ 174～175° [B+$CHCl_3$]
≪[α_D]≫ +327°

≪Ph.≫ CR[80] UV[81,86] NMR[34] Rf[81]
≪OP.≫ Daphnandra aromatica[80,81], D. tenuipes[84]
≪C. & So.≫
≪Sa. & D.≫ B–O, O–DiMe=oxyacanthine–Me-ether
≪SD.≫ 81), 82)
≪S. & B.≫

Daphnandrine $C_{36}H_{38}O_6N_2$≡594.68

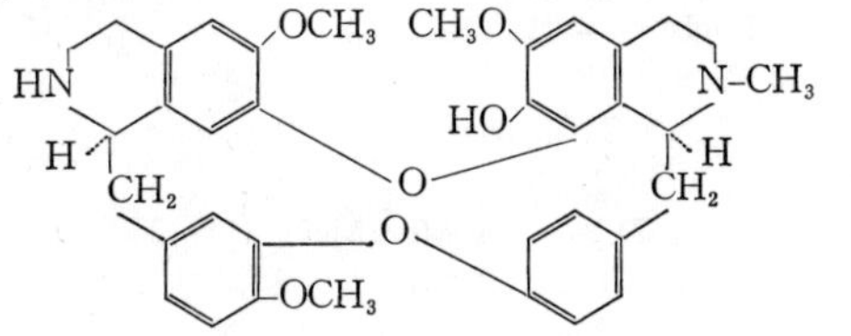

≪mp≫ 280° ($CHCl_3$ or $CHCl_3$–MeOH)
≪[α_D]≫ +475°

≪Ph.≫ CR[83] UV[81,86] NMR[34] Mass[35] ORD[37] Rf[81] X-ray[81]
≪OP.≫ Daphnandra micrantha[81,83,84]
≪C. & So.≫
≪Sa. & D.≫ B–2HCl : 282° B–2HBr : 291°
B–O, N–DiMe·2MeI : 255～256° (d)
≪SD.≫ 3), 81), 82), 83), 84)
≪S. & B.≫

Sepeerine $C_{36}H_{38}O_6N_2$≡594.68

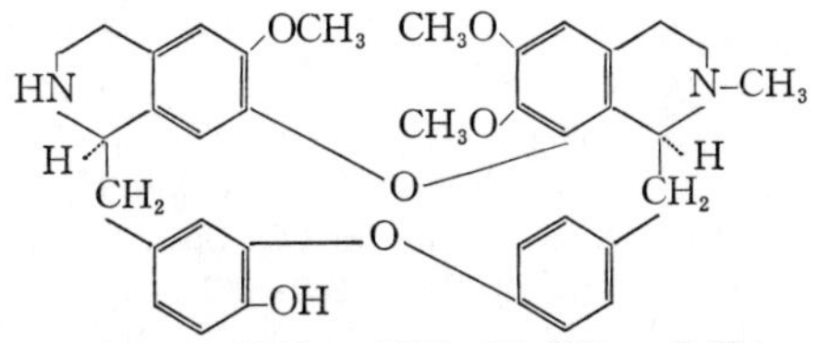

≪mp≫ 197～199°(164～166°) (MeOH or C_6H_6)
≪[α_D]≫ B–2HCl : +250° (H_2O)

≪Ph.≫ CR[88] UV[86,88] IR[87,88,89]
≪OP.≫ Ocotea (Nectandra) rodiei[87,88]
≪C. & So.≫ Prisms or rods
≪Sa. & D.≫ B–2 HCl : 254～256° (d)
B–H_2SO_4 : >300° B–2Pic : 178～180°
B–2MeI : 249～255° (d) B–OMe : 118～122°
B–NMe·2 MeI : 253～258° (d)
B–OMe·2 MeI : 230～235° (d)
B–NAc : 174～176° B–O, N–DiAc : 156～158°
B–O, N–DiMe·MeI : 249～255°
≪SD.≫ 3), 87), 88), 89)
≪S. & B.≫

Oxyacanthine $C_{37}H_{40}O_6N_2$≡608.71

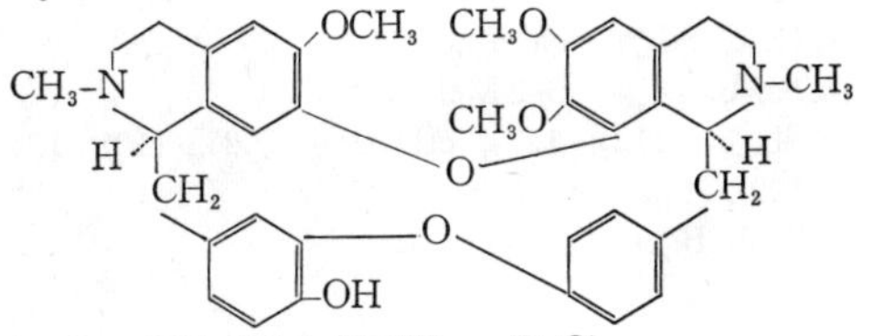

≪mp≫ 216～217° (EtOH or Et_2O)
≪[α]$_D$≫ +279° ($CHCl_3$)

≪Ph.≫ CR[114] UV[37,81,86,114,115,116] IR[117] NMR[34] Mass[35,36,74] ORD[37] X-ray[81] Rf[81]
≪OP.≫ Berberius equifolium,[90,91] B. floribunda (aristata)[92], B. (Mahonia) fortunei[93], B. heteropoda (vulgaris),[94] B. lamberti,[95] B. thunbergi var. maximowiczii[96,97]*
≪C. & So.≫ Needles, Sol. in $CHCl_3$, C_6H_6. Insol. in ligroin
≪Sa. & D.≫ B–2HCl : 270～271°
B–2HBr : 273～275° B–2MeI : 261～263°
B–OMe·MeI : 255～260°
≪SD.≫ 3), 21), 92), 95), 96), 101), 107)～112) cf. 113)
≪S. & B.≫

---

* B. tschonoskyana[98], B. vulgaris[9,99,100,101], Coccululas eba[102], Mahonia acanthifolia[103], M. borealis[104], M. griffithii[105], M. leschenaultii[106], M. manipurensis[104,106], M. sikkimensis[106], M. simonsii[104,105]

Repandine $C_{37}H_{40}O_6N_2 \equiv 608.71$

≪mp≫ 255° (Me_2CO–EtOH)

≪[α]$_D$≫ −106° ($CHCl_3$)
≪Ph.≫ UV[81][86] NMR[34] ORD[34] X-ray[119] Rf[81]
≪OP.≫ Daphnandra dielsii[84], D. repandula (Atherosperma repandulum)[118][119][120]
≪C. & So.≫ Needles
≪Sa. & D.≫ B–OMe·2MeI: 255∼260°
≪SD.≫ 3), 112), 119), 121), 122), 123) cf. 84), 108), 109), 110)
≪S. & B.≫ cf. 122), 123)

Thalmine $C_{37}H_{40}O_6N_2 \equiv 608.71$

≪mp≫

≪[α]$_D$≫ B–OMe: −68.5° ($CHCl_3$), B–OAc: −219° ($CHCl_3$)
≪Ph.≫
≪OP.≫ Thalictrum minus[124]
≪C. & So.≫
≪Sa. & D.≫ B–OMe·2MeI: 244∼250°(d)
B–OEt: 190∼193°(d)
B–OEt·2MeI: 234∼235°(d)
≪SD.≫ 124), 125)
≪S. & B.≫

Obaberine $C_{38}H_{42}O_6N_2 \equiv 622.73$

≪mp≫ 139∼140° (Et_2O–Petr. ether)

≪[α]$_D$≫ +302°
≪Ph.≫ UV[126] NMR[34]
≪OP.≫ Berberis tschonoskiana[126]
≪C. & So.≫ Needles, Sol. in MeOH, EtOH, $CHCl_3$. Mod. Sol. in Me_2CO. Spar. Sol. in Et_2O
≪Sa. & D.≫ B–2HCl: 260∼261°
B–2Pic: 178∼180°(d) B–2MeI: 256∼258°(d)
≪SD.≫ 3), 126)
≪S. & B.≫

***O*-Methylrepandine** $C_{38}H_{42}O_6N_2 \equiv 622.73$

≪mp≫ 211° (MeOH)

≪[α]$_D$≫ −73°($CHCl_3$), −108°(0.1 N-HCl)
≪Ph.≫ UV[37] NMR[34] Mass[51] ORD[37]
≪OP.≫ Daphnandra dielsii[84], D. repandula[84]
≪C. & So.≫ Needles
≪Sa. & D.≫ B–2MeI: 255∼260°
≪SD.≫ 112), 127), cf. 108), 109), 110), 121), 122), 123)
≪S. & B.≫

Hypoepistephanine(Pseudoepistephanine) $C_{36}H_{36}O_6N_2 \equiv 592.66$

≪mp≫ 256∼257° (MeOH)
≪[α]$_D$≫ +187° ($CHCl_3$)
≪Ph.≫ UV[130] CR[117][130]
≪OP.≫ Stephania japonica[128][129]
≪C. & So.≫ Prisms
≪Sa. & D.≫ B–OMe=epistephanine
≪SD.≫ 3), 124), 130), 131), cf. 117), 128), 132), 133)
≪S. & B.≫

Epistephanine $C_{37}H_{38}O_6N_2$≡606.69

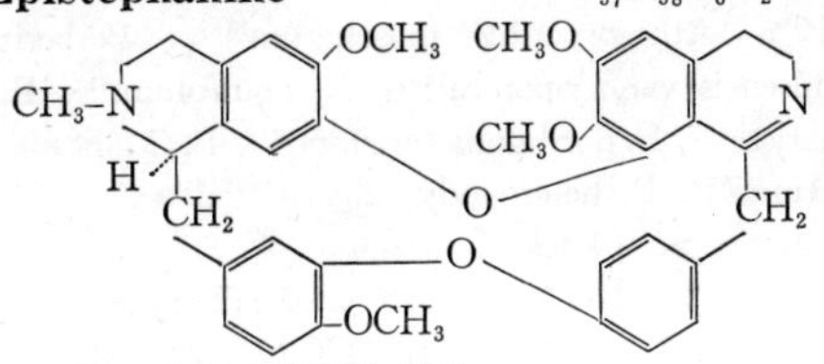

≪mp≫ 202～204°(MeOH)
≪[α]$_D$≫ +180°($CHCl_3$)

≪Ph.≫ UV[33,115,130,131,137,140] IR[117,130] NMR[139,142] Mass[51,142]
≪OP.≫ Stephania japonica[129,134,135]
≪C. & So.≫ Prisms
≪Sa. & D.≫ B–2HBr: 144°(d)
B–2MeI: 242～245°
≪SD.≫ 3), 115), 117), 128), 130), 131), 132), 133), 135), 136)～140)
≪S. & B.≫ BS: 141), 142)

Stebisimine $C_{36}H_{34}O_6N_2$≡590.65

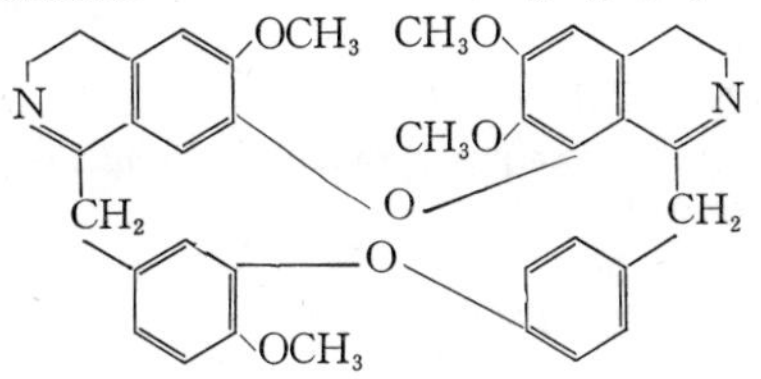

≪mp≫ 233～235°
≪[α]$_D$≫ 0°

≪Ph.≫ UV[141,142] IR[141,142] NMR[141,142] Mass[141,142]
≪OP.≫ Stephania japonica[141,142]
≪C. & So.≫
≪Sa. & D.≫ B–2HCl: >300° (d)
B–2Pic: 254～256° B–2MeI: >290°(d)
≪SD.≫ 141), 142)
≪S. & B.≫ BS: 142)

Atherospermoline $C_{35}H_{36}O_6N_2$≡580.65

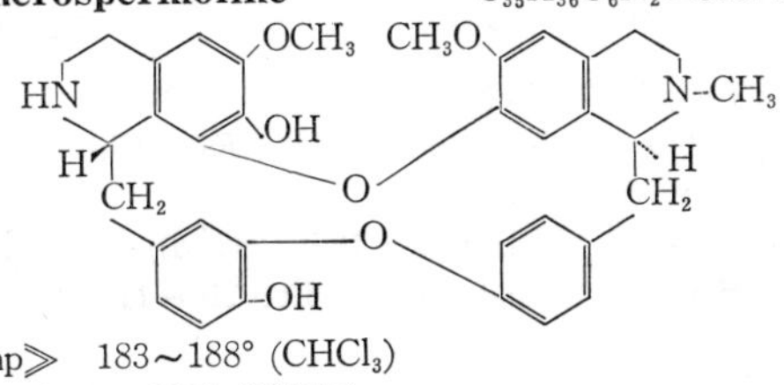

≪mp≫ 183～188° ($CHCl_3$)
≪[α]$_D$≫ +202° ($CHCl_3$)

≪Ph.≫ UV[143] NMR[143] ORD[143]
≪OP.≫ Atherosperma moschatum[143], Tasmanian sassafras[143]
≪C. & So.≫ Prisms
≪Sa. & D.≫
≪SD.≫ 143)
≪S. & B.≫

Obamegine (Stepholine)[144] $C_{36}H_{38}O_6N_2$≡594.68

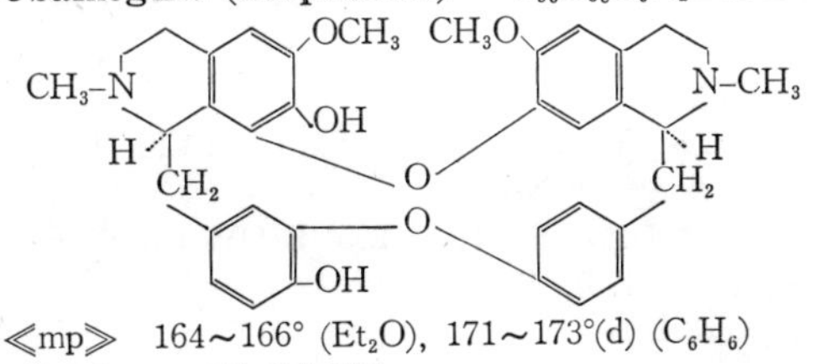

≪mp≫ 164～166° (Et_2O), 171～173°(d) (C_6H_6)
≪[α]$_D$≫ +99° (MeOH)

≪Ph.≫ CR[145,147] UV[146,147] IR[146] NMR[146] Mass[36,51] Rf[145]
≪OP.≫ Berberis tschonoskyana[98], Stephania japonica[129], Thalictrum rogosum[146]
≪C. & So.≫ Needles
≪Sa. & D.≫ B–2MeI: 271～273°(d)
B–O,O–DiMe=tetrandrine
B–O,O–DiEt: 196～198°
≪SD.≫ 3), 144), 145), 146)
≪S. & B.≫

Berbamine $C_{37}H_{40}O_6N_2$≡608.71

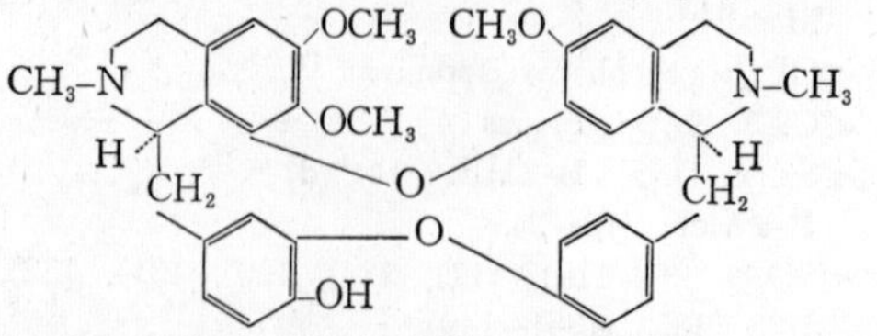

≪mp≫ 172°, cf. 156° (EtOH), 197～210° (Petr.-ether)
≪[α]$_D$≫ +109°

≪Ph.≫ CR[114] UV[114] NMR[34] Mass[51] ORD[37]
≪OP.≫ Atherosperma moschatum[148], Berberis amurensis var. japonica[149], B. aquifolium[91], B. asiatica[150], B. floribunda (aristata)[92], B. (Mahonia) fortunei[93], B. heteropoda(vulgaris)[91][194]*
≪C. & So.≫ Plates
≪Sa. & D.≫ B-2 HCl: 258° B-2 HBr: 283° (d) B-2 HI: 260～264°(d) B-2 $HClO_4$: 278°(d) B-2 MeI: 289° B-OMe=isotetrandrine
≪SD.≫ 3), 92), 95), 110), 113), 114), 145), 148), 158)～162), cf. 163)～168)
≪S. & B.≫

Pycnamine $C_{37}H_{40}O_6N_2$≡608.71

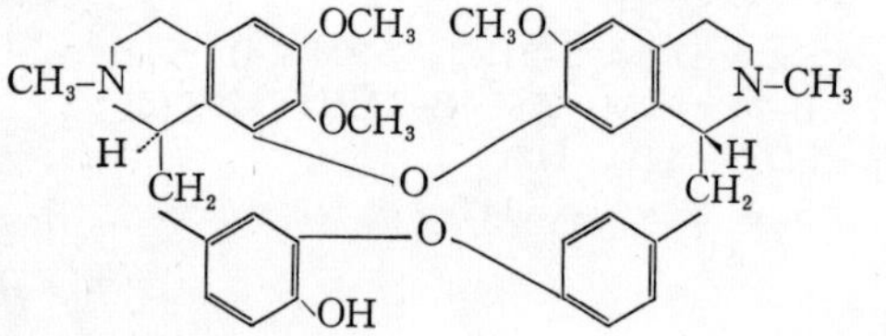

≪mp≫ 183° (EtOH)
≪[α]$_D$≫ −283° ($CHCl_3$)

≪Ph.≫ NMR[34][170] ORD[37]
≪OP.≫ Gyrocarpus jacquini[169], Pycnarrhena manillensis[155]
≪C. & So.≫
≪Sa. & D.≫ B-OMe: 218°
≪SD.≫ 3), 155), 170)
≪S. & B.≫

Fangchinoline $C_{37}H_{40}O_6N_2$≡608.71

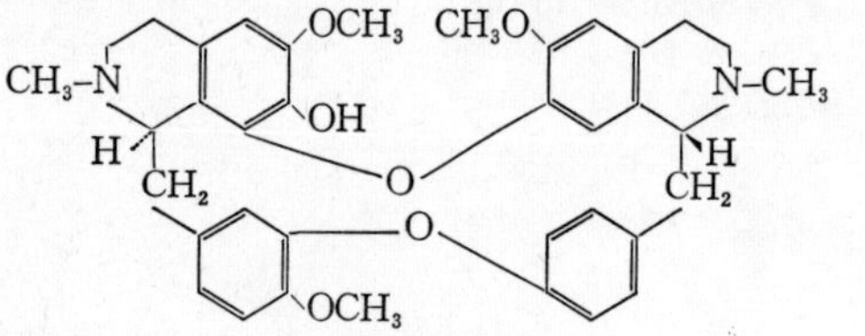

≪mp≫ 237～238° (EtOH or Me_2CO)
≪[α]$_D$≫ +255.1° ($CHCl_3$)

≪Ph.≫ CR[171] NMR[34]
≪OP.≫ Cyclea peltata[86], Han Fang Chi[171]
≪C. & So.≫
≪Sa. & D.≫ B-2 Pic: 186° (242°(d)) B-OMe=tetrandrine B-OEt: 116～117°
≪SD.≫ 3), 171), 172)
≪S. & B.≫

Menisidine $C_{37}H_{40}O_6N_2$≡608.71

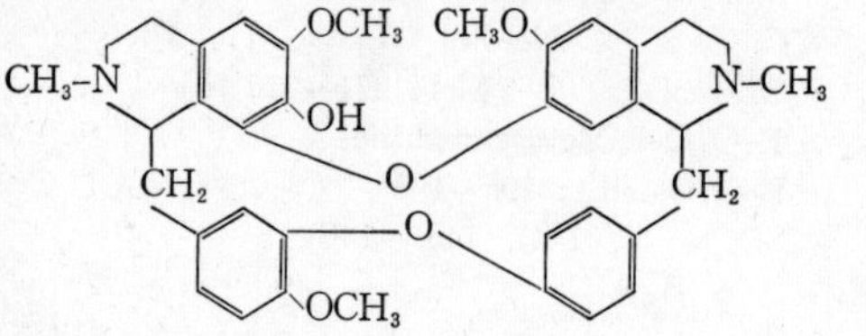

≪mp≫ 176°(EtOH)
≪[α]$_D$≫ +260°

≪Ph.≫
≪OP.≫ Mu Fang Chi[173], Shi Chan Chu[174]
≪C. & So.≫ Needles
≪Sa. & D.≫ B-2 HCl: 265° B-OMe=menisine
≪SD.≫ 173), 175)
≪S. & B.≫

* B. japonica[151][152][153], B. lamberti[95], B. swaseyi[154], B. thunbergii var. maximowiczii[96][97], B. tinctoria (aristata)[114], B. vulgaris[100], Mahonia griffithii[105], Pycnarrhena manillensis[155], Stephania cepharantha[156][157], S. sasakii[157]

Limacusine $C_{37}H_{40}O_6N_2 \equiv 608.71$

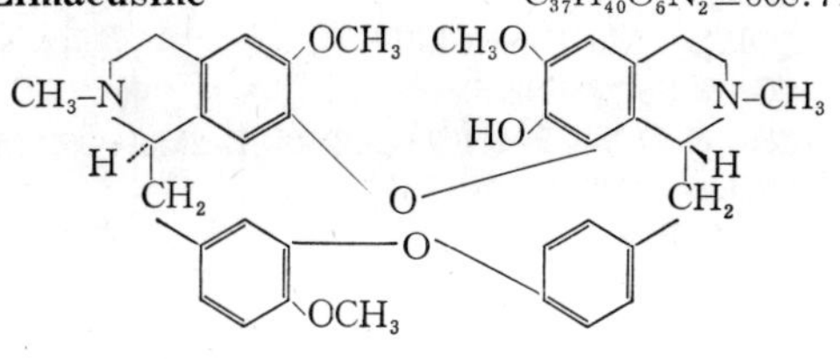

≪mp≫ 235～237° (MeOH–Me_2CO)
≪$[\alpha]_D$≫ +110° ($CHCl_3$)
≪Ph.≫ UV[10] IR[10] NMR[10] Mass[10]
≪OP.≫ Limacia cuspidata[10]
≪C. & So.≫ Needles
≪Sa. & D.≫ B–OMe: 210～212°
≪SD.≫ 10)
≪S. & B.≫ BG: 10)

Limacine $C_{37}H_{40}O_6N_2 \equiv 608.71$

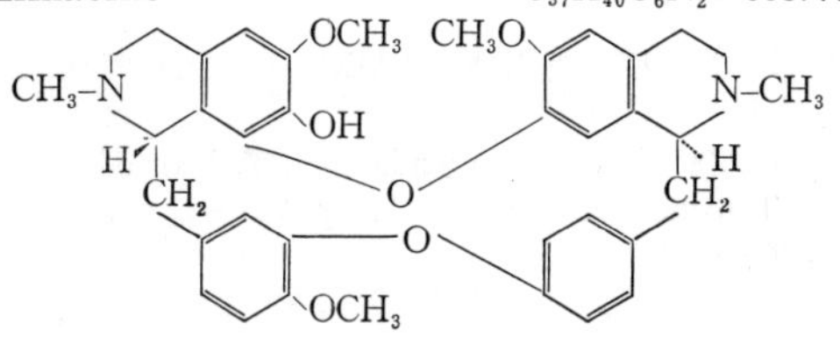

≪mp≫ 154～156° (C_6H_6–Me_2CO)
≪$[\alpha]_D$≫ −212° ($CHCl_3$)
≪Ph.≫ UV[10] IR[10] NMR[10]
≪OP.≫ Limacia cuspidata[10]
≪C. & So.≫ Needles
≪Sa. & D.≫ B–2 Pic: 184～186°
≪SD.≫ 10)
≪S. & B.≫ 10)

Tetrandrine $C_{38}H_{42}O_6N_2 \equiv 622.73$

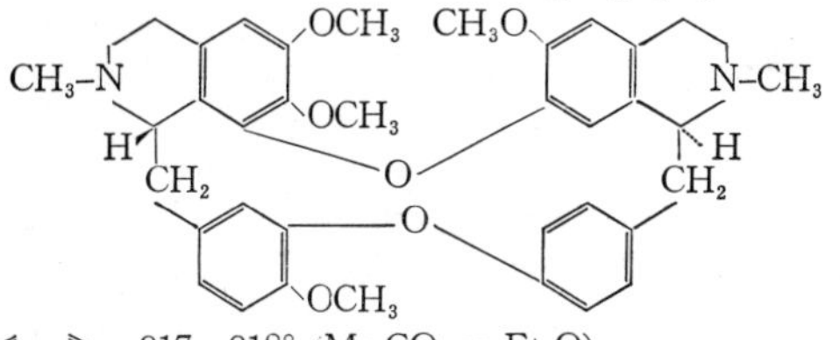

≪mp≫ 217～218° (Me_2CO or Et_2O)
≪$[\alpha]_D$≫ +285° ($CHCl_3$)
≪Ph.≫ CR[171][178] UV[33][115][188] IR[191] NMR[34] Mass[35][51] ORD[37]
≪OP.≫ Cyclea burmanni[176], C. peltata[86], Han Fang Chi[171][177], Menispermum canadense[11], Shi Chan Chu[174], Stephania tetrandra[178]
≪C. & So.≫ Needles, Sol. in Et_2O, Me_2CO, $CHCl_3$. Mod. sol. in EtOH. Insol. in Petr. ether
≪Sa. & D.≫ B–2 HCl: 266～268° (d) B–2 HBr: 270° (d) B–2 HI: 258° B–2 Pic: 235～245° (d) B–2 Oxal: 165～167° (d) B–2 MeI: 265～269°
≪SD.≫ 3), 158), 163), 165), 171), 176), 178), 182)～186)
≪S. & B.≫ 190)

Phaeanthine (*l*-Tetrandrine)

≪mp≫ 216～217° (Et_2O)
≪$[\alpha]_D$≫ −285° ($CHCl_3$)
≪Ph.≫ IR[191] NMR[34] Mass[51] ORD[37] Rf[169]
≪OP.≫ Cyclea peltata[86], Gyrocarpus americanus[179], G. jacquini[169], Phaeanthus ebracteolatus[179][180], Triclisia Patens Oliver[181]
≪C. & So.≫ Prisms, Sol. in Et_2O, $CHCl_3$. Mod. sol. in MeOH, Me_2CO. Spar. sol. in EtOH
≪Sa. & D.≫ B–2 HCl: 268° B–2 HI: 268° (d) B–2 MeI: 265° B–2 Pic: 245°
≪SD.≫ 37), 180), 185), 186), 187), 188), 189)
≪S. & B.≫

Isotetrandrine $C_{38}H_{42}O_6N_2 \equiv 622.73$

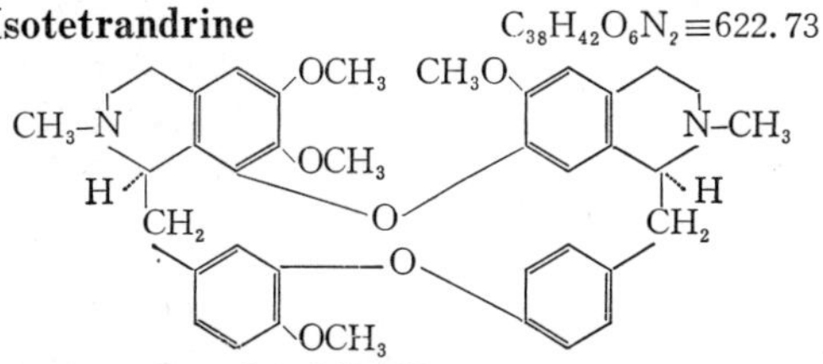

≪mp≫ 181～182° (Me_2CO)
≪$[\alpha]_D$≫ +142° ($CHCl_3$)
≪Ph.≫ CR[192] UV[37] NMR[34] IR[191][200] Mass[51] ORD[37]
≪OP.≫ Atherosperma moschatum[148], Berberis (Mahonia) japonica[151][152][153], Pycnarrhena manillensis[155], Stephania cepharantha[156][157][192]
≪C. & So.≫ Prisms
≪Sa. & D.≫ B–2MeI: 242°
≪SD.≫ 3), 113), 156), 160), 163), 165), 166), 167), 168), 185), 189), 192), 193), 194), 195)
≪S. & B.≫ cf. 196), 197), 198), 199)

Menisine $C_{38}H_{42}O_6N_2 \equiv 622.73$

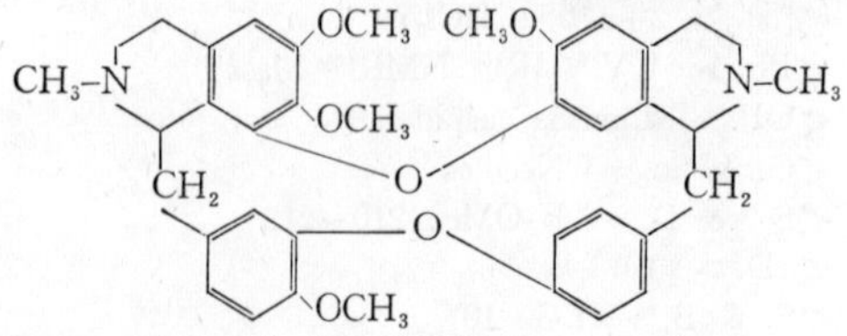

≪mp≫ 152°, cf. 127° [B+H_2O]
≪[α]D≫ +290° ($CHCl_3$)

≪Ph.≫
≪OP.≫ Mu Fang Chi[173]
≪C. & So.≫ Needles
≪Sa. & D.≫ B–2HCl : ca. 260° B–2MeI : 263°(d)
≪SD.≫ 173), 175)
≪S. & B.≫

Hernandezine $C_{39}H_{44}O_7N_2 \equiv 652.76$

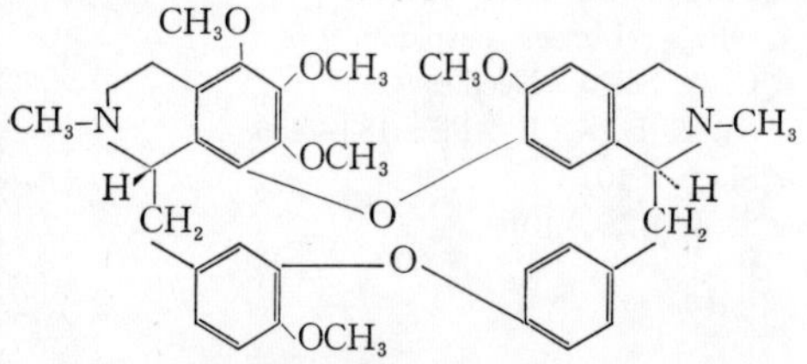

≪mp≫ 192~193° ($CHCl_3$), cf. 122~124° (Me_2CO), 157~158° (MeOH), 158~159° (Et_2O)
≪[α]D≫ +250° ($CHCl_3$)

≪Ph.≫ UV[200] IR[200] NMR[200][201] Mass[35][201] pKa[200] ORD[37]
≪OP.≫ Thalictrum fendleri[201], T. hernandezii[200], T. rochebrunianum[201], T. simplex[202]
≪C. & So.≫
≪Sa. & D.≫ B–2MeI : 250~260°(d)
≪SD.≫ 200), 201), 202), 203), 204)
≪S. & B.≫

Thalsimine $C_{38}H_{40}O_7N_2 \equiv 636.72$

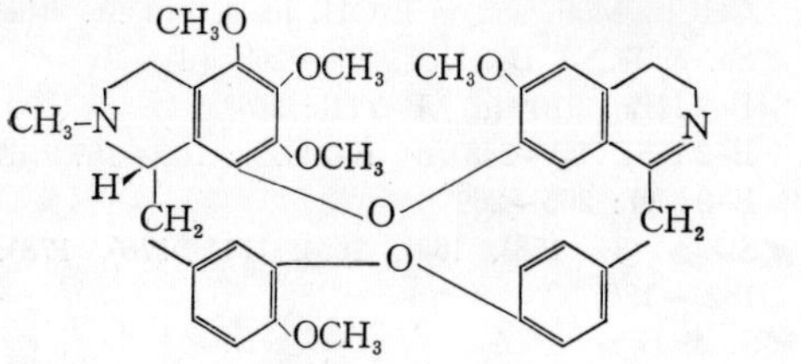

≪mp≫ 128~130°, cf. 140~142°
≪[α]D≫ +20.5° ($CHCl_3$)

≪Ph.≫ IR[202][206] Mass[201]
≪OP.≫ Thalictrum simplex[202][205][206]
≪C. & So.≫
≪Sa. & D.≫ B–2HNO_3 : 198~218°(d)
Dihydro–Der. 2HCl : 237~241° 2HBr : 250~251° 2HI : 236~237°(d)
NAc : 171° NMe=hernandezine
≪SD.≫ 201), 202), 203), 206), 207)
≪S. & B.≫

Phaeantharine $C_{38}H_{36}O_6N_2^{\oplus\oplus} \equiv 616.68$

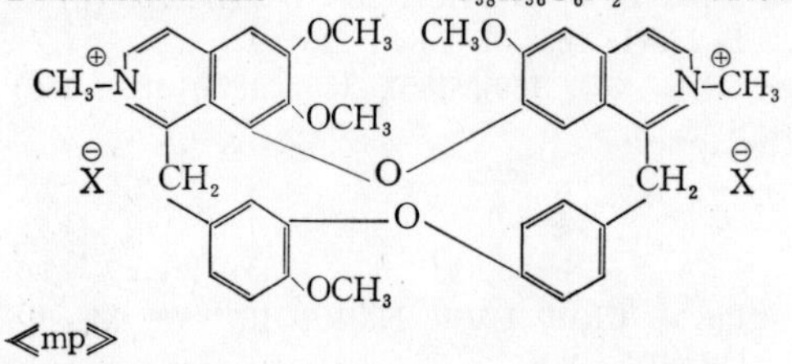

≪mp≫
≪[α]D≫ 0°

≪Ph.≫
≪OP.≫ Phaeanthus ebracteolatus[208][209]
≪C. & So.≫
≪Sa. & D.≫ B⊕⊕–2Cl⊖ : 215~220°
B⊕⊕–2I⊖ : 203~207°
B⊕⊕–2ClO_4⊖ : 180~184°
≪SD.≫ 209), 210)
≪S. & B.≫

Cepharanthine $C_{37}H_{38}O_6N_2 \equiv 606.69$

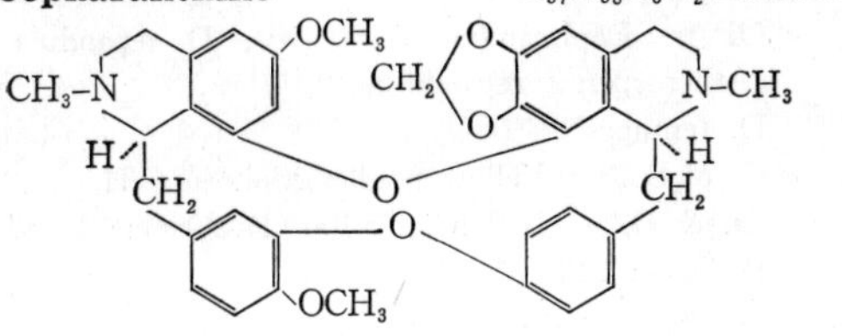

≪mp≫ 103° [B+C_6H_6], cf. anhyd. 144～155° (Me_2CO–C_6H_6)
≪$[\alpha]_D$≫ +277° ($CHCl_3$)
≪Ph.≫ CR[211][221] NMR[34] Mass[35][51] ORD[37]
≪OP.≫ Stephania cepharantha[156][157][211][212][213], S. sasaki[214][215]
≪C. & So.≫ Sol. in most org. solv. Spar. sol. in Petr. ether
≪Sa. & D.≫ B–2MeI: ca. 268°
≪SD.≫ 3), 211), 212), 214), 216), 217), 218), 219), 220), 221)
≪S. & B.≫ 222～228)

Nortenuipine $C_{37}H_{38}O_7N_2 \equiv 622.69$

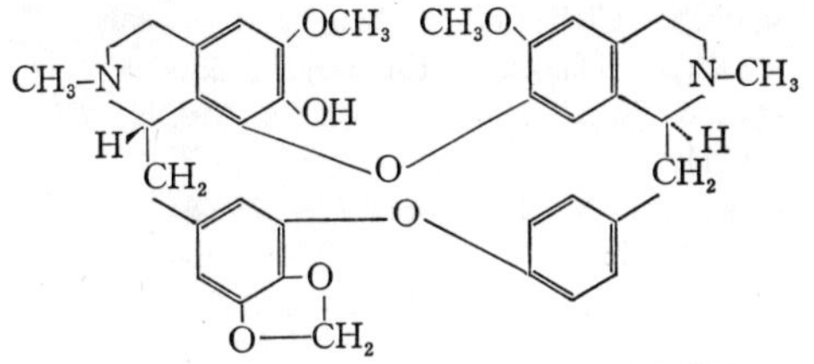

≪mp≫ 211°(d) (95% EtOH or $CHCl_3$–MeOH)
≪$[\alpha]_D$≫ −218° ($CHCl_3$)
≪Ph.≫ CR[84] UV[84] NMR[229][230] Mass[35] ORD[37][229] X-ray[84]
≪OP.≫ Daphnandra tenuipes[84][120][229]
≪C. & So.≫ Needles
≪Sa. & D.≫ B–2MeI: 261～265°(d)
B–OMe=tenuipine
≪SD.≫ 229), 230), cf. 84), 120)
≪S. & B.≫

Tenuipine $C_{38}H_{40}O_7N_2 \equiv 636.72$

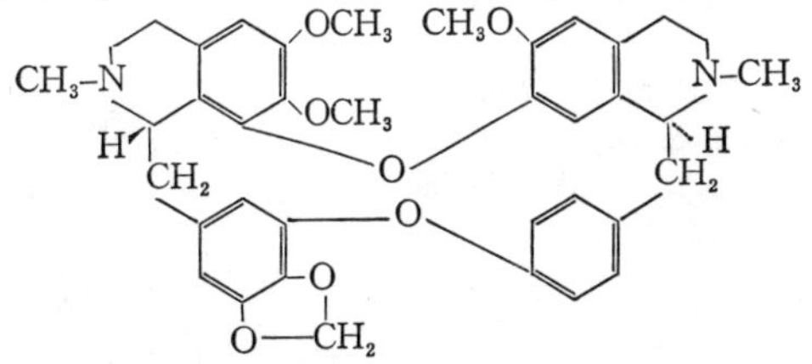

≪mp≫ 140～145° and 170°
≪$[\alpha]_D$≫ −258° ($CHCl_3$)
≪Ph.≫ UV[84][86] IR[84] NMR[229][230] ORD[37][229] X-ray[84] CR[84]
≪OP.≫ Daphnandra dielsii[84], D. tenuipes[84][229]
≪C. & So.≫ Prisms
≪Sa. & D.≫ B–2MeI: 267～272°
≪SD.≫ 84), 120), 229), 230)
≪S. & B.≫

Repandinine (*dl*-Tenuipine)
≪mp≫ 243° (Me_2CO–MeOH)
≪$[\alpha]_D$≫ 0°
≪Ph.≫ → tenuipine
≪OP.≫ Daphnandra dielsii[84], D. repandura[84], D. tenuipes[84]
≪C. & So.≫ Needles
≪Sa. & D.≫ B–2MeI: 275°
≪SD.≫ → tenuipine
≪S. & B.≫

Repanduline $C_{37}H_{34}O_7N_2 \equiv 618.66$

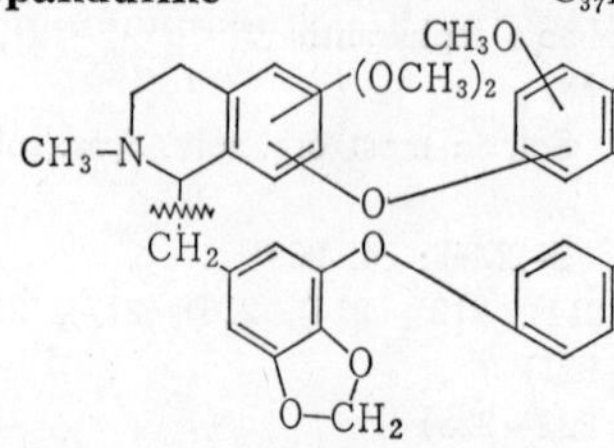

≪mp≫ 215~232° (darkens at 180~185°) (MeOH—Petr. ether)
≪[α]$_D$≫ +434° (MeOH), +473° ($CHCl_3$)

≪Ph.≫ CR[84] UV[120]
≪OP.≫ Daphnandra dielsii[84,231], D. repandula (Atherosperma repandulum)[84,118,120], D. tenuipes[84,120]
≪C. & So.≫ Yellow needles, Sol. in C_6H_6
≪Sa. & D.≫ B–2HCl: >100° B–2Oxal: 214°(d) B–2MeI: 240°(d)
≪SD.≫ 84), 120)
≪S. & B.≫

Thalicberine $C_{37}H_{40}O_6N_2 \equiv 608.71$

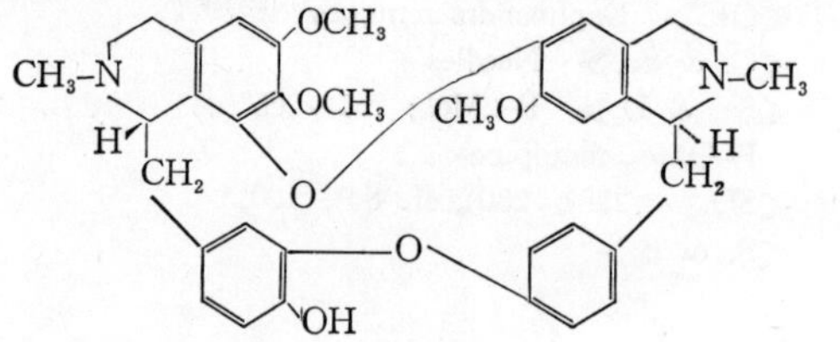

≪mp≫ 161° (sint. 155°) (Et_2O)
≪[α]$_D$≫ +231.2° ($CHCl_3$)

≪Ph.≫ CR[232] UV[232] Mass[36]
≪OP.≫ Thalictrum thunbergii (minus)[232]
≪C. & So.≫ Needles, Sol. in MeOH, $CHCl_3$, C_6H_6
≪Sa. & D.≫ B–2HBr: 272~273°(d) B–2Oxal: 239~240° B–OMe: 186~187°
≪SD.≫ 3), 233), 234), 235), 236), 237), 238)
≪S. & B.≫

***O*-Methylthalicberine** $C_{38}H_{42}O_6N_2 \equiv 620.72$

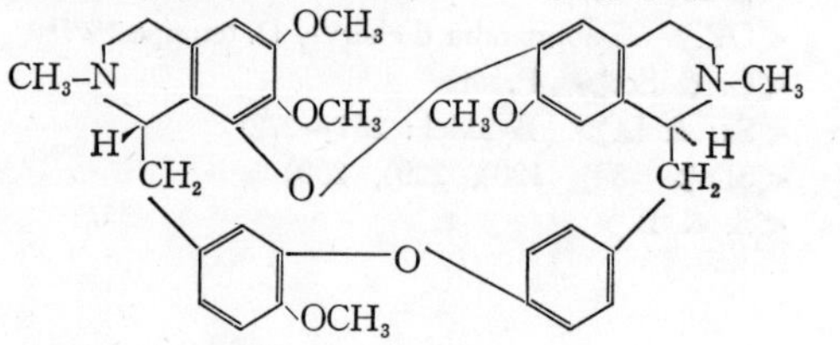

≪mp≫ 186~187°
≪[α]$_D$≫ +266° ($CHCl_3$)

≪Ph.≫ → thalicberine
≪OP.≫ Thalictrum thunbergii[232]
≪C. & So.≫ Needles
≪Sa. & D.≫
≪SD.≫ → thalicberine
≪S. & B.≫

Thalfoetidine $C_{38}H_{42}O_7N_2 \equiv 636.72$

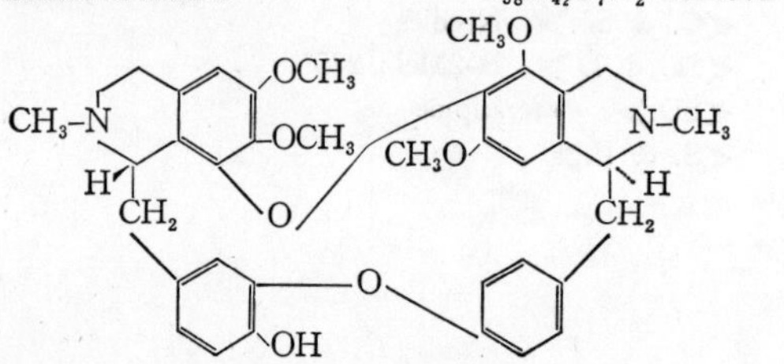

≪mp≫ 168~170°
≪[α]$_D$≫ −88.6° ($CHCl_3$)

≪Ph.≫ UV[239] IR[239] NMR[239]
≪OP.≫ Thalictrum foetidum[239]
≪C. & So.≫ Prisms, Sol. in Et_2O
≪Sa. & D.≫ B–OMe: 108~109°
≪SD.≫ 239)
≪S. & B.≫

Thalmetine $C_{36}H_{36}O_6N_2$≡592.66

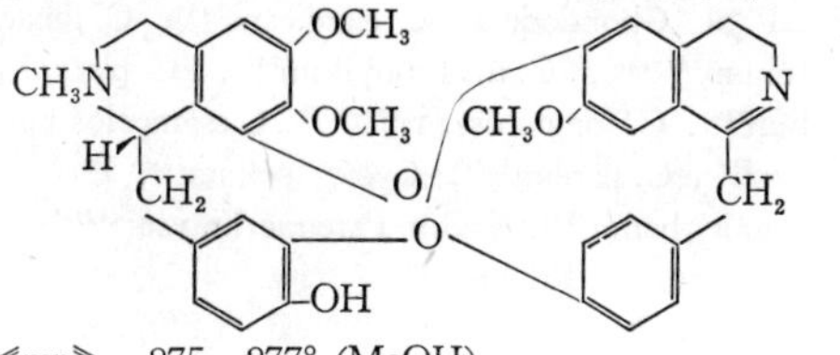

≪mp≫ 275～277° (MeOH)
≪$[\alpha]_D$≫ +200° ($CHCl_3$)

≪Ph.≫ UV[240] IR[240] NMR[240]
≪OP.≫ Thalictrum minus[240]
≪C. & So.≫
≪Sa. & D.≫ B–OMe: 245～246°
Dihydro–Der.: 196～197° Dihydro–Der.–NMe= thalicberine
≪SD.≫ 240)
≪S. & B.≫

Thalmidine (*O*-Methylthalmethine) $C_{37}H_{38}O_6N_2$≡606.69

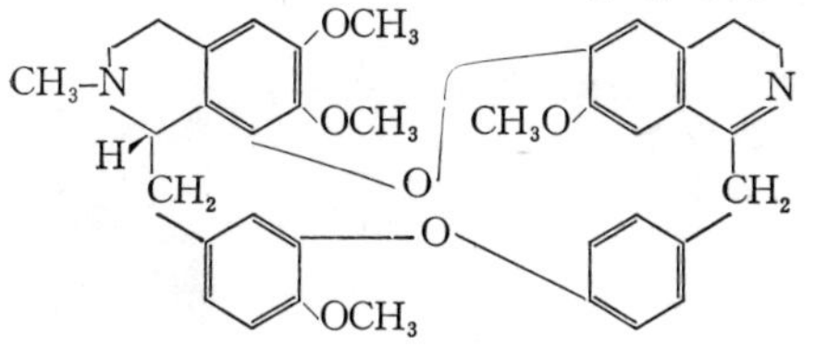

≪mp≫ 245～246° (C_6H_6)
≪$[\alpha]_D$≫ +237° ($CHCl_3$)

≪Ph.≫ UV[240] IR[240] NMR[240]
≪OP.≫ Thalictrum minus[124][240]
≪C. & So.≫ Needles
≪Sa. & D.≫ B–2HCl: 198～200°
Dihydro–Der.: 278～280°
≪SD.≫ 124), 125), 240)
≪S. & B.≫

Thalicrine $C_{36}H_{38}O_6N_2$≡594.68

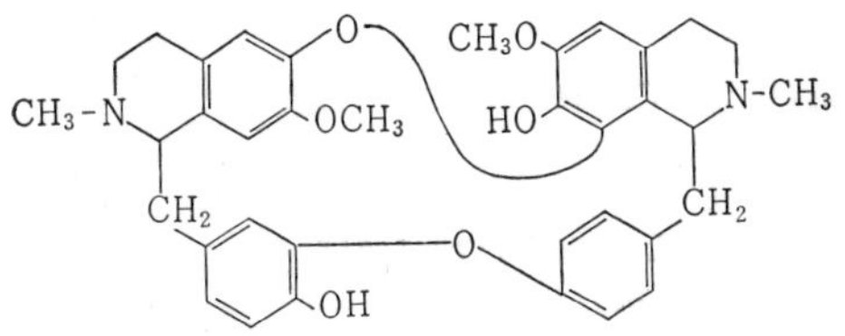

≪mp≫ 221～222° (MeOH)
≪$[\alpha]_D$≫ +341.2° ($CHCl_3$)

≪Ph.≫ CR[241] UV[241]
≪OP.≫ Thalictrum thunbergii[241]
≪C. & So.≫ Needles
≪Sa. & D.≫ B–2Oxal: 206～207°
B–OMe: 235～236° (d)
B–O, O–DiMe·2HBr: 265°
≪SD.≫ 3), 241), 242), 243)
≪S. & B.≫

Homothalicrine $C_{37}H_{40}O_6N_2$≡608.71

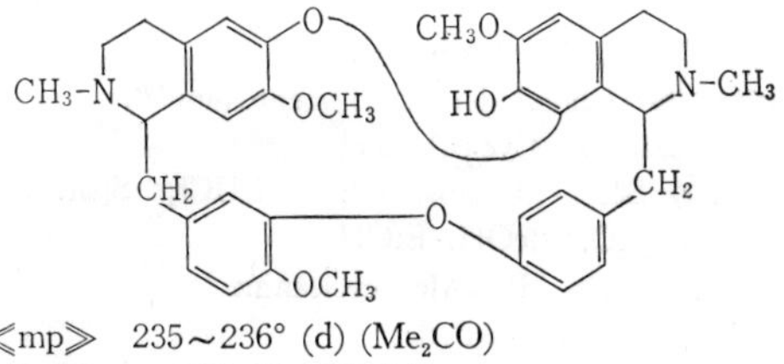

≪mp≫ 235～236° (d) (Me_2CO)
≪$[\alpha]_D$≫ +425.3° ($CHCl_3$)

≪Ph.≫ → thalicrine
≪OP.≫ Thalictrum thunbergii[241]
≪C. & So.≫ Prisms, Sol. in C_6H_6, Me_2CO. Spar. sol. in EtOH, Et_2O
≪Sa. & D.≫ B–2HBr: 263.5° (d)
B–2MeI: 247° (d)
≪SD.≫ → thalicrine
≪S. & B.≫

Isochondodendrine (Isobebeerine) $C_{36}H_{38}O_6N_2 \equiv 594.68$

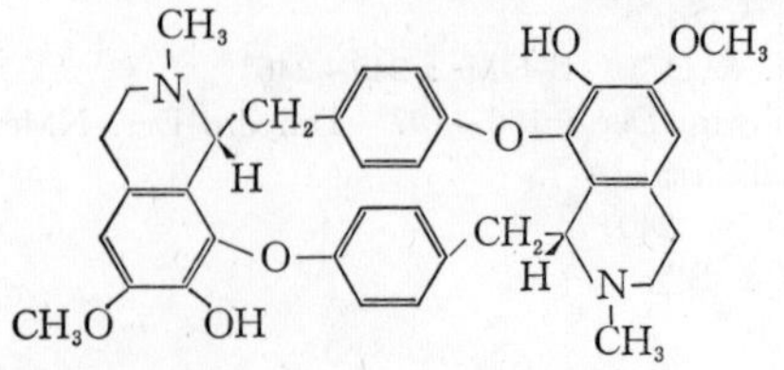

≪mp≫ 316° (MeOH)
≪$[\alpha]_D$≫ +50° (Py), −290° ($CHCl_3$)

≪Ph.≫ Mass[308] NMR[34] ORD[47] Rf[251]
≪OP.≫ Chondodendron candicans[244], C. limacii-folium[245][246], C. microphyllum[244], C. platyphyllum[244], C. tomentosum[247]~[250], Cissampelos insularis[251], C. pareira[252], Cyclea peltata[86], Pleogyne cunninghamii[253], Radix Pareirae bravae[244][254]
≪C. & So.≫ Prisms
≪Sa. & D.≫ B–2HCl: 333° B–H_2SO_4: 291° B–2MeI: 287°(d) B–O,O–DiMe=cycleanine
≪SD.≫ 3), 16), 186), 187), 244), 245), 246), 250), 255)~262)
≪S. & B.≫ 25), 263)

Protocuridine $C_{36}H_{38}O_6N_2 \equiv 594.68$

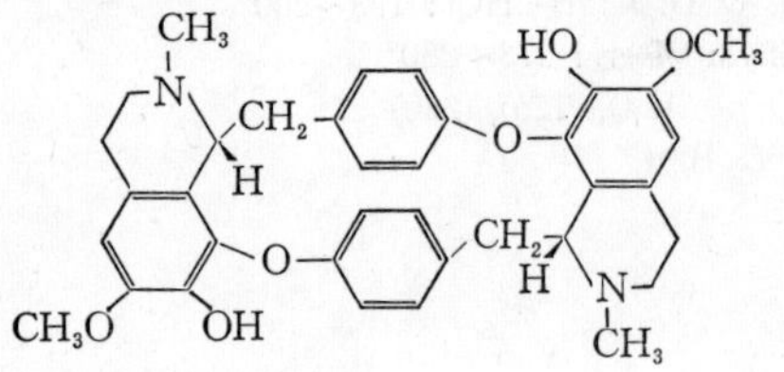

≪mp≫ 295° [B+Py]
≪$[\alpha]_D$≫ +8° (HCl)

≪Ph.≫
≪OP.≫ Curare[264][265]
≪C. & So.≫
≪Sa. & D.≫ B–2HCl: 295°
≪SD.≫ 244), 265)
≪S. & B.≫

Neoprotocuridine $C_{36}H_{38}O_6N_2 \equiv 594.68$

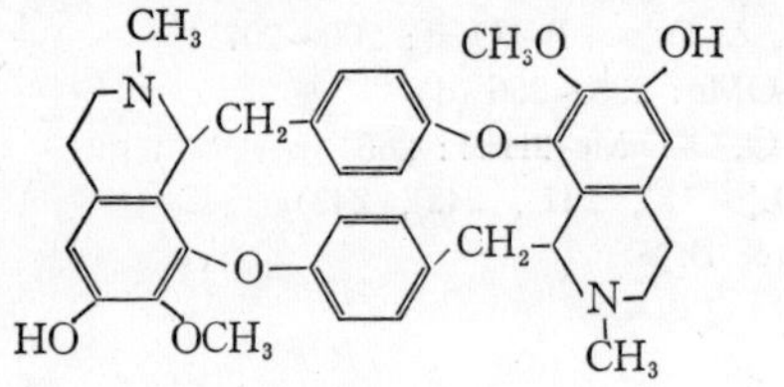

≪mp≫ 232°
≪$[\alpha]_D$≫ 0°

≪Ph.≫
≪OP.≫ Curare[265]
≪C. & So.≫
≪Sa. & D.≫ B–2HCl: >300° B–O,O–DiMe·2MeI: >300°
≪SD.≫ 265)
≪S. & B.≫

Norcycleanine $C_{37}H_{40}O_6N_2 \equiv 608.71$

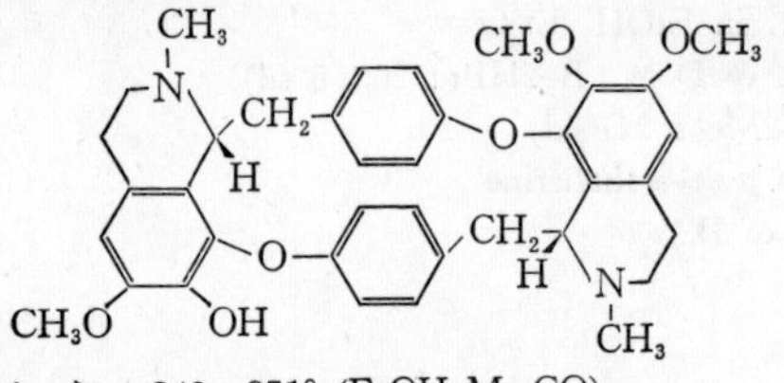

≪mp≫ 249~251° (EtOH–Me_2CO)
≪$[\alpha]_D$≫ −26.5° (MeOH)

≪Ph.≫ CR[266] UV[266] Rf[266]
≪OP.≫ Chondodendron tomentosum[250], Cissampelos (Cyclea) insularis[266]
≪C. & So.≫ Prisms, Sol. in $CHCl_3$. Spar. sol. in Me_2CO, MeOH, EtOH
≪Sa. & D.≫ B–OMe=cycleanine B–OEt: 231~233°(d)
≪SD.≫ 3), 250), 266), 267)
≪S. & B.≫

Cycleanine (Methylisochondodendrine) $C_{38}H_{42}O_6N_2 \equiv 622.73$

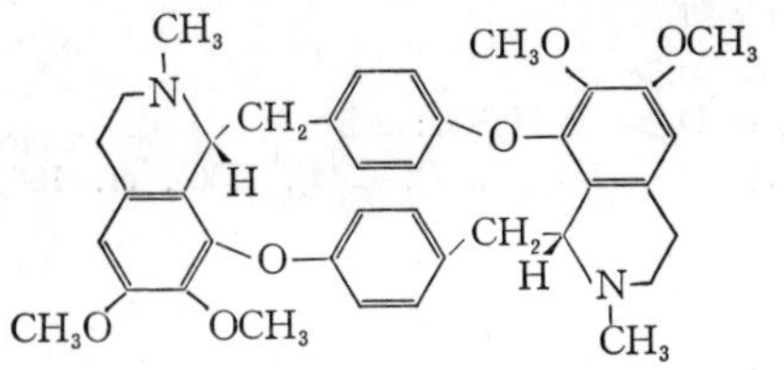

≪mp≫ 272∼273° (Me_2CO)
≪$[\alpha]_D$≫ −15° ($CHCl_3$), −30° (MeOH)

≪Ph.≫ NMR[34)272)] Mass[51)] Rf[251)]
≪OP.≫ Chondodendron tomentosum[250)], Cissampelos insularis[251)266)269)], Curare[249)], Stephania capitata[268)], S. cepharantha[156)269)], S. glabra[270)]
≪C. & So.≫ Needles
≪Sa. & D.≫ B-2Pic: 181° B-2MeI: 312°
≪SD.≫ 3), 163), 250), 268), 269), 271), cf. 261)
≪S. & B.≫ 272)

Chondrofoline $C_{35}H_{36}O_6N_2 \equiv 580.65$

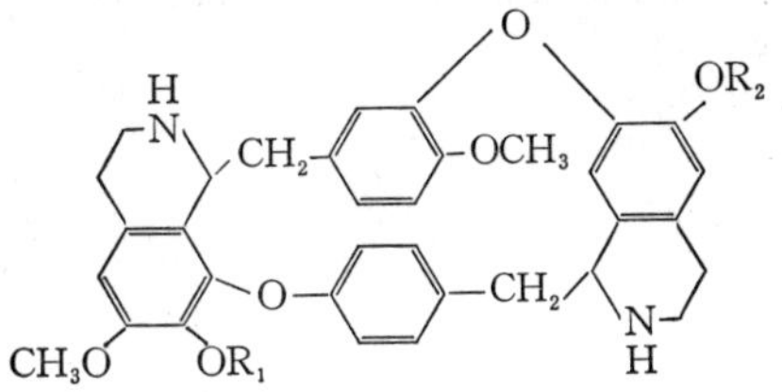

R_1=H, R_2=CH_3 or
R_1=CH_3, R_2=H

≪mp≫ 135° [B+H_2O] (MeOH)
≪$[\alpha]_D$≫ −281° (HCl)

≪Ph.≫ CR[244)] NMR[230)]
≪OP.≫ Chondodendron platyphyllum[244)]
≪C. & So.≫ Plates
≪Sa. & D.≫ B-2HNO_3: 225°
≪SD.≫ 244) → curine
≪S. & B.≫

***d*-Curine(*d*-Bebeerine, *d*-Chondodendrine)** $C_{36}H_{38}O_6N_2 \equiv 594.68$

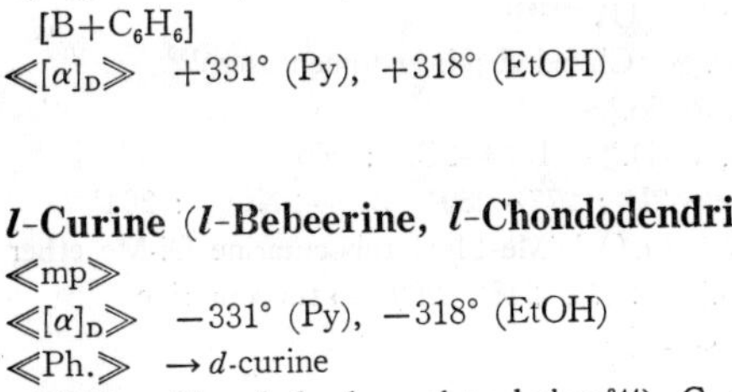

≪mp≫ 213° (220°) ($CHCl_3$-MeOH), cf. 166° [B+C_6H_6]
≪$[\alpha]_D$≫ +331° (Py), +318° (EtOH)

≪Ph.≫ CR[297)] UV[86)292)] NMR[34)] Mass[35)36)308)] ORD[37)]
≪OP.≫ Chondodendron candicans[244)], C. microphyllum[244)], Domerara (Ocotea or Nectandra) rodiei[273)], Radix Pareirae bravae[244)274)275)], Tubocurare[276)]
≪C. & So.≫ Needles
≪Sa. & D.≫ B-2HCl: 271° B-2MeI: 250°
B-O,O-DiAc: 147∼148°
B-O,O-DiBz: 139∼140°
≪SD.≫ 3), 186), 187), 189), 250), 253), 254), 271), 275), 276), 277), 278), 280), 282∼290)
≪S. & B.≫ 291), 292), cf. 293∼296), 362)

***l*-Curine (*l*-Bebeerine, *l*-Chondodendrine)**
≪mp≫
≪$[\alpha]_D$≫ −331° (Py), −318° (EtOH)
≪Ph.≫ → *d*-curine
≪OP.≫ Chondodendron platyphylum[244)], C. tomentosum[247)∼250)277)], Cissampelos pareira[252)278)], Pleogyne cunninghamii[253)], Radix Pareirae bravae[274)279)280)], Curare[249)276)281)]
≪C. & So.≫
≪Sa. & D.≫
≪SD.≫ → *d*-curine
≪S. & B.≫

Chondocurine $C_{36}H_{38}O_6N_2 \equiv 594.68$

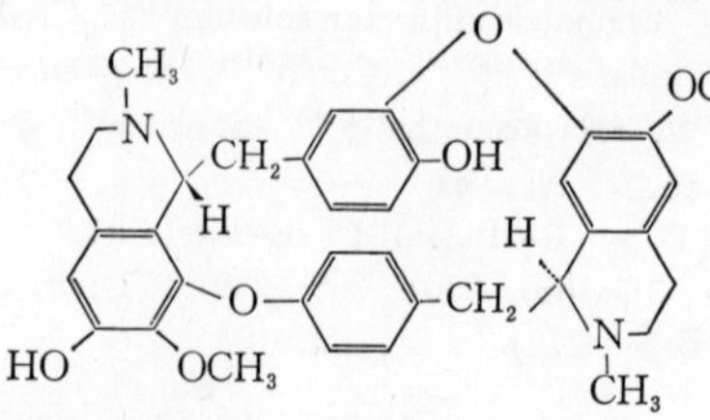

≪mp≫ 232~234° (MeOH)
≪[α]$_D$≫ +173° ($CHCl_3$), +200° (HCl)

≪Ph.≫ NMR[34)] ORD[37)]
≪OP.≫ Chondodendron tomentosum[247)~250)], Curare[249)]
≪C. & So.≫
≪Sa. & D.≫ B–H_2SO_4 : 265°
≪SD.≫ 3), 189), 247), 248), 250), cf. 186), 187)
≪S. & B.≫

Tomentocurine $C_{36}H_{38}O_6N_2 \equiv 594.68$

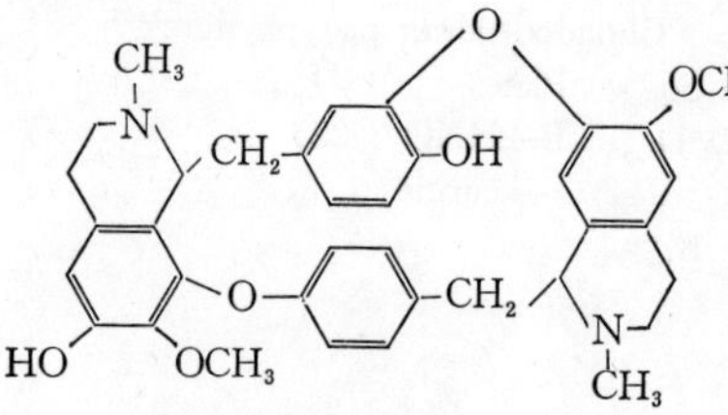

≪mp≫ 260° (MeOH–$CHCl_3$)
≪[α]$_D$≫ +202° (0.1 N–HCl)

≪Ph.≫
≪OP.≫ Chondodendron tomentosum[249)250)]
≪C. & So.≫
≪Sa. & D.≫
≪SD.≫ 250)
≪S. & B.≫

Protochondocurarine $C_{37}H_{41}O_6N_2^{\oplus} \equiv 609.71$

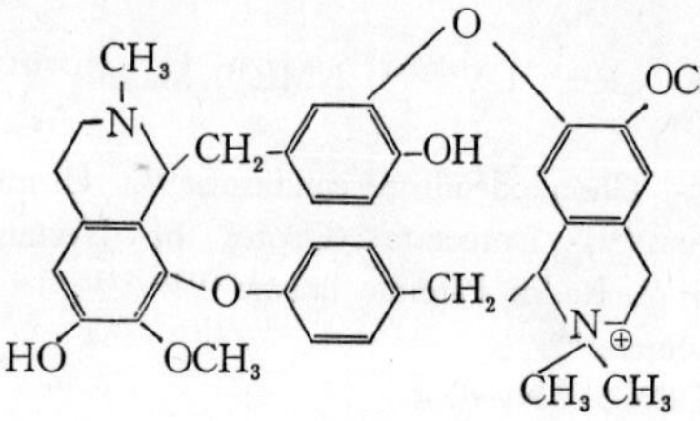

≪mp≫
≪[α]$_D$≫ B⊕–NO_3⊖ : +175°

≪Ph.≫
≪OP.≫ Curare[298)]
≪C. & So.≫
≪Sa. & D.≫ B⊕–I⊖·2MeI=chondocurarine diiodide B⊕–NO_3⊖ : 265°
≪SD.≫ 298) → tubocuraine
≪S. & B.≫

Chondocurarine $C_{38}H_{44}O_6N_2^{\oplus\oplus} \equiv 624.75$

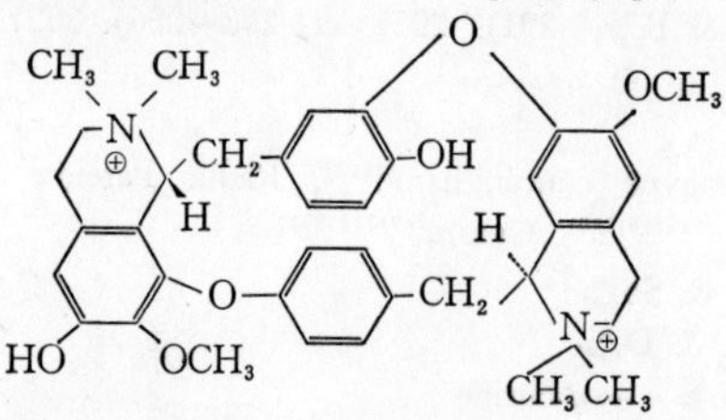

≪mp≫
≪[α]$_D$≫ B⊕⊕–2I⊖ : +150° (H_2O)

≪Ph.≫ UV[86)299)]
≪OP.≫ Chondodendron tomentosum[299)]
≪C. & So.≫
≪Sa. & D.≫ B⊕⊕–2Cl⊖ : 265°
B⊕⊕–2I⊖ : 277~280° B⊕⊕–2NO_3⊖ : 264°
B⊕⊕–O, O–DiMe–2I⊖=tubocurarine Di-Me ether
≪SD.≫ 247), 248), 299) → tubocurarine
≪S. & B.≫

Neochondocurarine $C_{38}H_{44}O_6N_2$⊕⊕≡624.75

O
CH_3 CH_3 OR_3
N ⊕ CH_2 OR_2
O CH_2 ⊕ N
R_1O OR_4 CH_3 CH_3

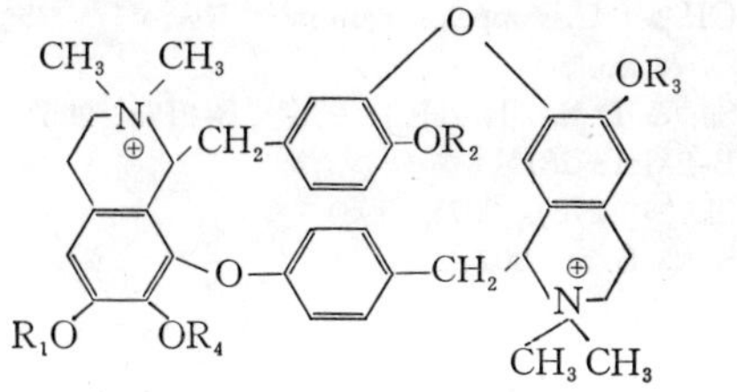

R_1, R_2, R_3, R_4=2OCH_3, 2OH

≪mp≫
≪[α]$_D$≫ B⊕⊕-2Cl⊖ +179°

≪Ph.≫
≪OP.≫ Curare[298]
≪C. & So.≫
≪Sa. & D.≫ B⊕-2Cl⊖ : 268°
B-O, O-DiMe=tubocurarine DiMe ether
≪SD.≫ 298) →tubocurarine
≪S. & B.≫

***d*-Tubocurarine** $C_{38}H_{44}O_6N_2$⊕⊕≡624.75

O
CH_3 CH_3 OCH_3
N ⊕ CH_2 OH
H H
O CH_2 ⊕ N
CH_3O OH CH_3 CH_3

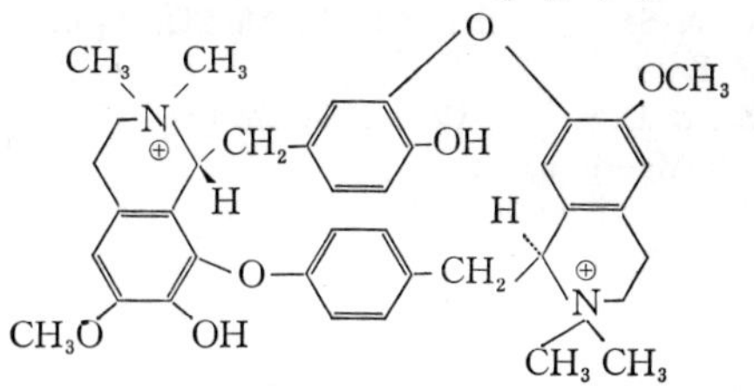

≪mp≫
≪[α]$_D$≫ B⊕⊕-2Cl⊖ +215° (H_2O)

≪Ph.≫ UV[86,299,301,303] Mass[308]
≪OP.≫ Chondodendron tomentosum[247,248,250], Tubocurare[249,264,276,281]
≪C. & So.≫
≪Sa. & D.≫ B⊕⊕-2Cl⊖ : 274~275°(d)
B_2⊕-2I⊖ : 263~265°(d)
B_2⊕-O, O-DiAc-2NO_3⊖ : 212°
B_2⊕-O, O-DiMe-2I⊖ : 267°(d)
≪SD.≫ 3), 189), 247), 248), 249), 276), 281), 287), 298), 299), 300)
≪S. & B.≫ 291), 293), 301~306)

***l*-Tubocurarine**

≪mp≫
≪[α]$_D$≫ B⊕⊕-2Cl⊖ : −215° (H_2O)
≪Ph.≫
≪OP.≫ Chondodendron tomentosum[277]
≪C. & So.≫
≪Sa. & D.≫
≪SD.≫ →*d*-tubocurarine
≪S. & B.≫ →*d*-tubocurarine

Isochondocurarine $C_{38}H_{44}O_6N_2$⊕⊕≡624.75

O
CH_3 CH_3 OCH_3
N ⊕ CH_2 OH
O CH_2 ⊕ N
HO OCH_3 CH_3 CH_3

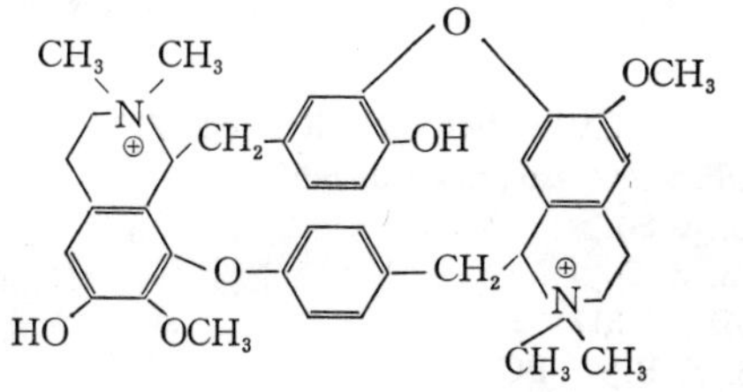

≪mp≫
≪[α]$_D$≫ B⊕⊕-2Cl⊖ : −150°

≪Ph.≫
≪OP.≫ Curare[298]
≪C. & So.≫
≪Sa. & D.≫ B⊕⊕-2Cl⊖ : 278°
B⊕⊕-O, O-DiMe-2I⊖ : 246°
≪SD.≫ →curine or tubocurarine
≪S. & B.≫

Hayatine $C_{36}H_{38}O_6N_2 \equiv 596.70$

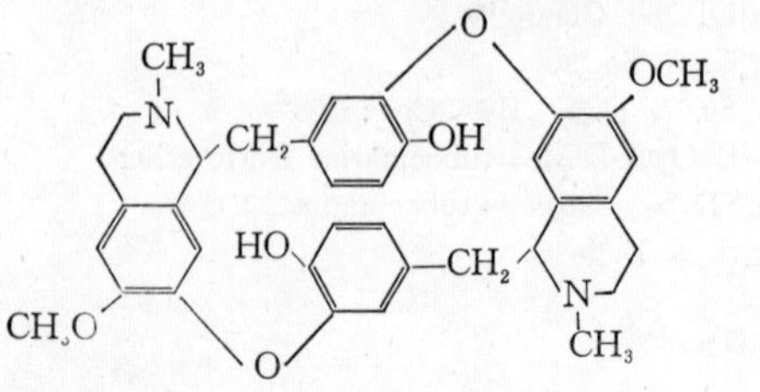

O
CH_3 OCH_3
N CH_2 OH
HO CH_2
CH_3O N
O CH_3

≪mp≫ 303°
≪[α]D≫ 0°

≪Ph.≫ UV[307] Mass[308]
≪OP.≫ Cissampelos pareira[252][278]
≪C. & So.≫
≪Sa. & D.≫ B–2HCl: 286° B–2Pic: 235° B–2MeI: 281°
≪SD.≫ 278), 307), 308)
≪S. & B.≫ 307)

Insulanoline $C_{37}H_{38}O_6N_2 \equiv 606.69$

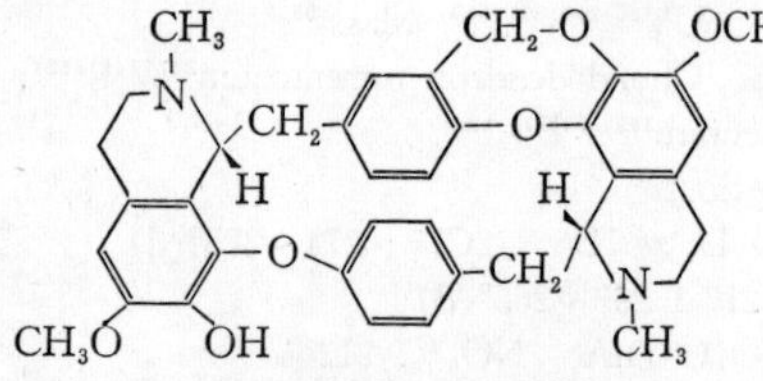

CH_3 CH_2–O OCH_3
N CH_2 O
H H
O CH_2 N
CH_3O OH CH_3

≪mp≫ 195° [B+H_2O] (Me_2CO)
≪[α]D≫ +99° (MeOH)

≪Ph.≫ CR[309] UV[309] Rf[309]
≪OP.≫ Cissampelos insularis[309]
≪C. & So.≫ Sol. in $CHCl_3$, Me_2CO. Spar. sol. in Et_2O
≪Sa. & D.≫ B–2MeI: 275～278°(d) B–OMe=insularine
≪SD.≫ 3), 309), 310)
≪S. & B.≫

Insularine $C_{38}H_{40}O_6N_2 \equiv 620.72$

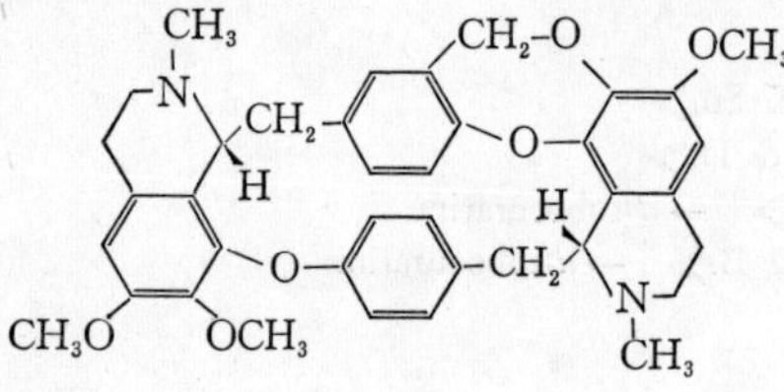

CH_3 CH_2–O OCH_3
N CH_2 O
H H
O CH_2 N
CH_3O OCH_3 CH_3

≪mp≫ 160° (aq. MeOH)
≪[α]D≫ +28° (EtOH)

≪Ph.≫ CR[311] UV[33][312][313] IR[312] NMR[34] ORD[37] Rf[251]
≪OP.≫ Cissampelos insularis[266][269][311], Stephania japonica[312]
≪C. & So.≫ Amorph, Sol. in most org. solv. except Petr.–ether
≪Sa. & D.≫ B–2HCl: 280°(d) B–2Pic: 232° B–2Oxal: 188～190° B–2MeI: 292°
≪SD.≫ 3), 269), 311), 312), 313), 314), 315), 316)
≪S. & B.≫ cf. 314)

Cissampareine $C_{37}H_{38}O_6N_2 \equiv 606.69$

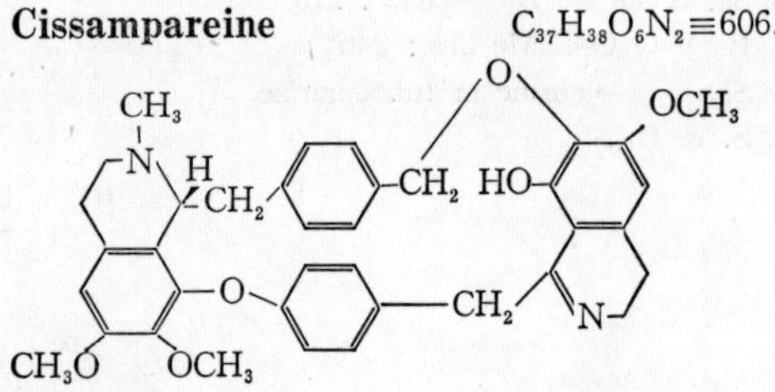

O
CH_3 OCH_3
N H CH_2 CH_2 HO
O CH_2 N
CH_3O OCH_3

≪mp≫
≪[α]D≫

≪Ph.≫ UV[317] IR[317] NMR[317] Mass[317]
≪OP.≫ Cissampelos pareira[317]
≪C. & So.≫
≪Sa. & D.≫
≪SD.≫ 317)
≪S. & B.≫ BG: 317)

Trilobine $C_{35}H_{34}O_5N_2 \equiv 562.64$

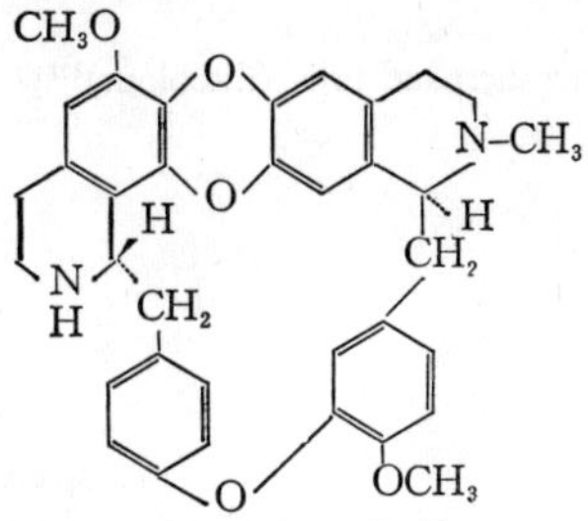

≪mp≫ 235~238° (Et_2O)
≪$[\alpha]_D$≫ +281.4° ($CHCl_3$)

≪Ph.≫ CR[320,321] UV[33,86,116,164,343] IR[332] Mass[35,51,343]
≪OP.≫ Cocculus hirsutus[318], C. laurifolius[319], C. sarmentosus[321], C. trilobus[77,78,320]
≪C. & So.≫ Needles
≪Sa. & D.≫ B-2HCl: 302~306°
B-2MeI: 256~257° B-NMe=isotrilobine
B-NAc: 149~161°(d)
≪SD.≫ 97), 289), 320), 321), 322~339)
≪S. & B.≫ cf. 97), 164), 331), 338), 340)
BG: 341), 342)

Isotrilobine $C_{36}H_{36}O_5N_2 \equiv 576.66$

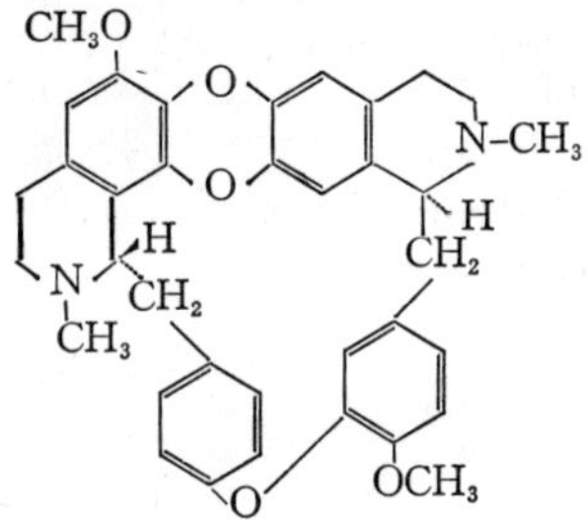

≪mp≫ 213~215° (Me_2CO)
≪$[\alpha]_D$≫ +343° ($CHCl_3$)

≪Ph.≫ UV[33,164,333] Mass[35,51] Rf[333] CR[322]
≪OP.≫ Cocculus laurifolius[319], C. sarmentosus[321,323], C. trilobus[322], Stephania hernandifolia (discolor)[333]
≪C. & So.≫ Pale yellow prisms, Sol. in C_6H_6, $CHCl_3$. Spar. sol. in EtOH, Et_2O
≪Sa. & D.≫ B-2HCl: 210°(d)
B-2MeI: 262°(d)
≪SD.≫ →trilobine
≪S. & B.≫

Micranthine $C_{34}H_{32}O_6N_2 \equiv 552.60$

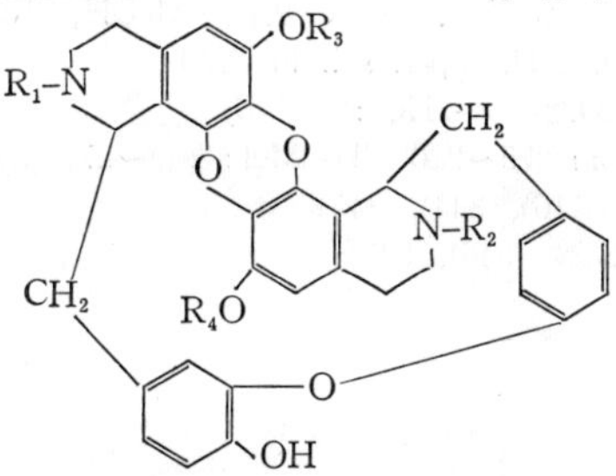

R_1, R_2=H and CH_3
R_3, R_4=H and CH_3

≪mp≫ 194~196° (AcOEt or Me_2CO)
≪$[\alpha]_D$≫ −231°($CHCl_3$)
≪Ph.≫ CR[83] UV[86,116,343]
≪OP.≫ Daphnandra micrantha[83,84,116]
≪C. & So.≫ Needles, sol. in EtOH, $CHCl_3$. Spar. sol. in MeOH, AcOEt, C_6H_6. Very Spar. sol. in Et_2O
≪Sa. & D.≫ B-$2H_2SO_4$: 310°(d)
B-O,O,N-TriMe-2MeI: 255~260°
≪SD.≫ 83), 84), 116)
≪S. & B.≫ BG: 341)

No Name $C_{35}H_{34}O_6N_2 \equiv 578.64$

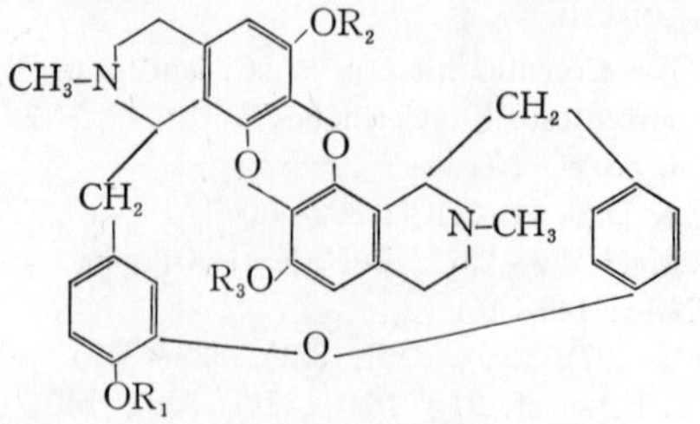

R_1, R_2, R_3 = $1CH_3$, $2H$

≪mp≫ 205~206°

≪[α]D≫ +268°
≪Ph.≫ UV[333]
≪OP.≫ Stephania hernandifolia (discolour)[333]
≪C. & So.≫
≪Sa. & D.≫
≪SD.≫ 333)
≪S. & B.≫

Normenisarine $C_{35}H_{32}O_6N_2 \equiv 576.62$

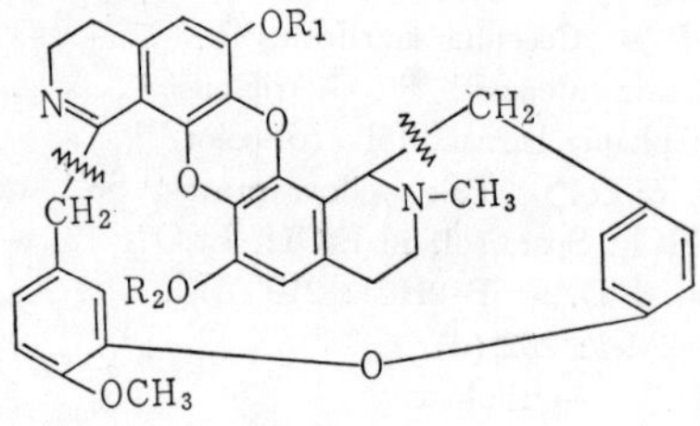

R_1 or R_2 = H or CH_3

≪mp≫ 223°

≪[α]D≫ +190°
≪Ph.≫
≪OP.≫ Cocculus sarmentosus[344], C. trilobus[77)78)344]
≪C. & So.≫
≪Sa. & D.≫ B–OMe=menisarine
≪SD.≫ 344) → menisarine
≪S. & B.≫

Menisarine $C_{36}H_{34}O_6N_2 \equiv 590.65$

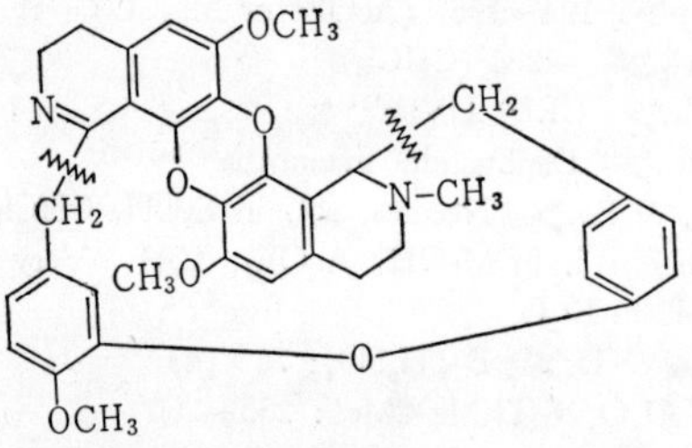

≪mp≫ 203°
≪[α]D≫ +149° ($CHCl_3$)

≪Ph.≫ CR[344] UV[343)347]
≪OP.≫ Cocculus sarmentosus[344], C. trilobus[77)78]
≪C. & So.≫ Sol. in $CHCl_3$, MeOH, EtOH, Me_2CO, C_6H_6. Spar. sol. in Et_2O
≪Sa. & D.≫ B–2HCl: 279° (d)
B–2HBr: 275~280° B–2MeI: 269~270° (d)
≪SD.≫ 116), 344), 345)
≪S. & B.≫ 346), 347)

Tiliarine $C_{36}H_{36}O_6N_2 \equiv 592.67$

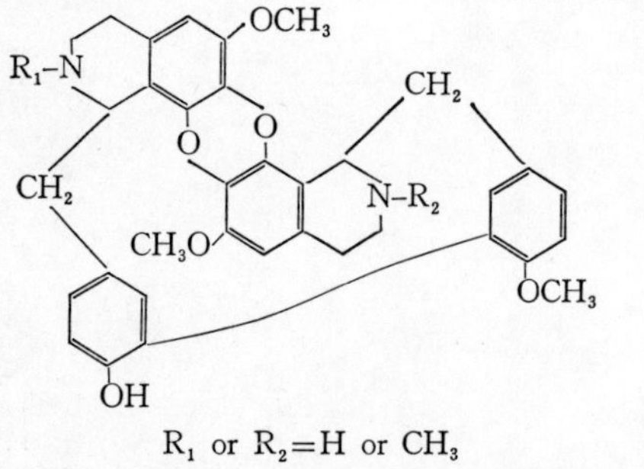

R_1 or R_2 = H or CH_3

≪mp≫ 203~207° (d) (Me_2CO or C_6H_6–Petr. ether)

≪[α]D≫ +283.4° ($CHCl_3$)
≪Ph.≫ UV[343]
≪OP.≫ Tiliacora racemosa[343)348)349]
≪C. & So.≫ Prisms
≪Sa. & D.≫ B–2HCl: 302~305° (d)
B–H_2SO_4: 297~300° (d)
B–NMe–2MeI: 279~281° (d)
≪SD.≫ 343), 349), 350)
≪S. & B.≫

Tiliacorine $C_{37}H_{38}O_6N_2$≡606.69

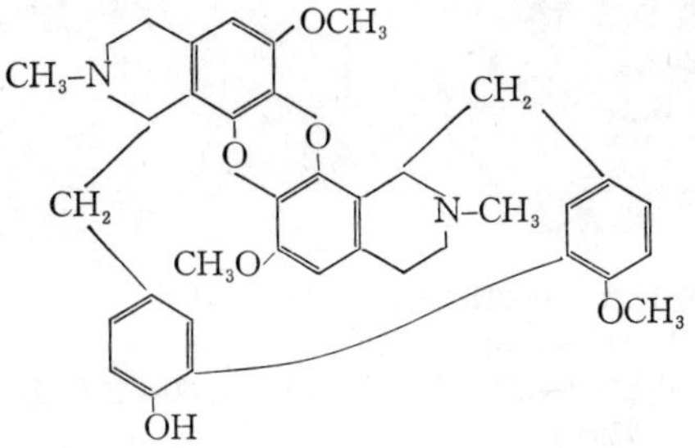

≪mp≫ 271～272°
≪[α]$_D$≫ +133° (HCl)
≪Ph.≫ CR[343)350)] UV[343)352)] NMR[354)]
≪OP.≫ Tiliacora acuminata (racemosa)[348)351)352)]
≪C. & So.≫ Sol. in Claisen alkali
≪Sa. & D.≫ B-2MeI: 273～275°
B-OMe: 210～212° B-OEt: 192～194°
≪SD.≫ 343), 348), 349), 350), 351), 352), 353), 354)
≪S. & B.≫

Rodiasine $C_{38}H_{42}O_6N_2$≡602.70

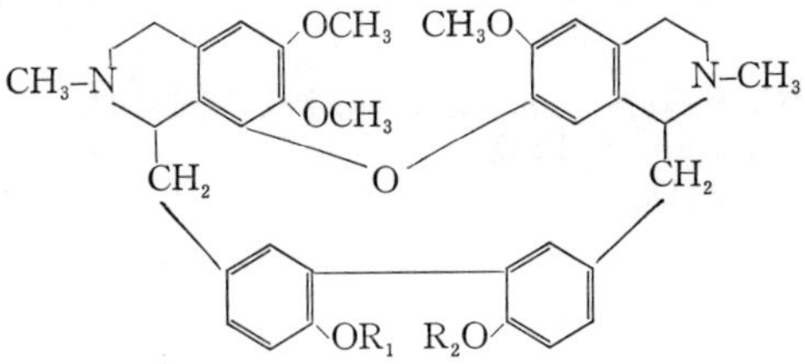

R_1, R_2=unknown

≪mp≫ 195° (EtOH)
≪[α]$_D$≫ +134° ($CHCl_3$)
≪Ph.≫ UV[86)88)273)] NMR[356)] Mass[359)] X-ray[273)]
≪OP.≫ Ocotea (Noctandra) rodiei[88)273)355)]
≪C. & So.≫ Prisms
≪Sa. & D.≫ B-2HCl: 255～259°(d)
B-2MeI: 291～295°(d) B-OMe: 172～173°
B-OMe-2HCl: 232～236°(d)
B-OMe-2MeI: 294～298°(d)
≪SD.≫ 356)
≪S. & B.≫ BG: 356)

Ocotine $C_{35}H_{38}O_6N_2$≡582.67

≪mp≫ 162～164° (EtOH)
≪[α]$_D$≫ +32°
≪Ph.≫ UV[86)88)] IR[88)]
≪OP.≫ Ocotea (Nectandra) rodiei[88)89)]
≪C. & So.≫ Needles
≪Sa. & D.≫ B-2HCl: 240°
B-2Pic: 178～180° B-2MeI: 250°(d)
≪SD.≫ 89)
≪S. & B.≫

Base B $C_{35}H_{36}O_8N_2$=614.67

≪mp≫ ～230° (d)
≪[α]$_D$≫ +31° ($CHCl_3$)
≪Ph.≫
≪OP.≫ Chondodendron limaciifolium[245)]
≪C. & So.≫
≪Sa. & D.≫
≪SD.≫ 245)
≪S. & B.≫

Base A $C_{36}H_{38}O_6N_2$≡596.70

≪mp≫ 279～300° (MeOH)
≪[α]$_D$≫ ～0°
≪Ph.≫
≪OP.≫ Chondodendron limaciifolium[245)]
≪C. & So.≫ Needles, Sol. in $CHCl_3$
≪Sa. & D.≫ B-H_2SO_4: 289°
≪SD.≫
≪S. & B.≫

Dinklageine $C_{36}H_{38}O_6N_2$≡596.70

≪mp≫ 285°
≪[α]$_D$≫ −24.4°
≪Ph.≫ UV[357)] Rf[357)]
≪OP.≫ Stephania dinklagei[357)]
≪C. & So.≫
≪Sa. & D.≫
≪SD.≫
≪S. & B.≫

Funiferine $C_{36}H_{40}O_6N_2$≡596.70

≪mp≫ 232° (EtOH)
≪[α]$_D$≫ +184.3 ($CHCl_3$), +171.4° (MeOH)
≪Ph.≫ CR[358)] UV[358)]
≪OP.≫ Tiliacora funifera[358)]
≪C. & So.≫ Sol. in Me_2CO, $CHCl_3$
≪Sa. & D.≫
≪SD.≫ 358)
≪S. & B.≫

Himanthine $C_{37}H_{40}O_6N_2 \equiv 608.71$
≪mp≫ 206~207° (MeOH)
≪[α]D≫ −202°
≪Ph.≫ CR[114] UV[114]
≪OP.≫ Berberis himalaica[114]
≪C. & So.≫ Needles
≪Sa. & D.≫
≪SD.≫
≪S. & B.≫

Hayatinine $C_{37}H_{40}O_6N_2 \equiv 608.71$
≪mp≫ 160° and 231~232° ($CHCl_3$, C_6H_6, Me_2CO)
≪[α]D≫ −3.2° (EtOH), −5.1° ($CHCl_3$)
≪Ph.≫ CR[359] UV[359] IR[359] Rf[359]
≪OP.≫ Cissampelos pareira[252][359]
≪C. & So.≫
≪Sa. & D.≫ B–2HCl: 252~253° B–2HI: 262~263° B–2$HClO_4$ 245°(d) B–2MeI: 262~263° B–OMe–2MeI: 250~251°(d)
≪SD.≫ 359)
≪S. & B.≫

Thalictrinine $C_{38}H_{46}O_7N_2 \equiv 642.76$
≪mp≫ 170°
≪[α]D≫ −80.9° ($CHCl_3$)
≪Ph.≫
≪OP.≫ Thalictrum simplex[360]
≪C. & So.≫ Sol. in $CHCl_3$, Me_2CO, EtOH. Spar. Sol. in Et_2O
≪Sa. & D.≫
≪SD.≫
≪S. & B.≫

Thalibrunine $C_{39}H_{46}O_8N_2 \equiv 670.77$
≪mp≫ 172~173°
≪[α]D≫ +160.0°
≪Ph.≫
≪OP.≫ Thalictrum rochebrunianum[361]
≪C. & So.≫
≪Sa. & D.≫
≪SD.≫
≪S. & B.≫

Norrodiasine
≪mp≫
≪[α]D≫
≪Ph.≫
≪OP.≫ Ocotea rodiei[355]
≪C. & So.≫
≪Sa. & D.≫
≪SD.≫
≪S. & B.≫

Dirosine
≪mp≫
≪[α]D≫
≪Ph.≫
≪OP.≫ Ocotea rodiei[355]
≪C. & So.≫
≪Sa. & D.≫
≪SD.≫
≪S. & B.≫

Ocoteamine
≪mp≫
≪[α]D≫
≪Ph.≫
≪OP.≫ Ocotea rodiei[355]
≪C. & So.≫
≪Sa. & D.≫
≪SD.≫
≪S. & B.≫

Ococamine
≪mp≫
≪[α]D≫
≪Ph.≫
≪OP.≫ Ocotea rodiei[355]
≪C. & So.≫
≪Sa. & D.≫
≪SD.≫
≪S. & B.≫

Demerarine

≪mp≫
≪$[\alpha]_D$≫
≪Ph.≫
≪OP.≫ Ocotea rodiei[355]
≪C. & So.≫
≪Sa. & D.≫
≪SD.≫
≪S. & B.≫

Ocodemerine

≪mp≫
≪$[\alpha]_D$≫
≪Ph.≫
≪OP.≫ Ocotea rodiei[355]
≪C. & So.≫
≪Sa. & D.≫
≪SD.≫
≪S. & B.≫

---

1) N. F. Proskurnina, A. P. Orekhov: Bull. Soc. Chim. France, **5**, 1357 (1938). [C. A., **33**, 1439 (1939)].
2) N. F. Proskurnina, A. P. Orekhov: Zhur. Obshchei Khim., **10**, 707 (1940). [C. A., **35**, 2520 (1941)].
3) M. Tomita, J. Kunitomo: J. Pharm. Soc. Japan, **82**, 741 (1962).
4) I. N. Gorbachev, L. P. Varnakova, E. M. Kleiner, I. I. Chernova, N. A. Preobrazhenskii: Zhur. Obshchei Khim., **28**, 167 (1958). [C. A., **52**, 12879 (1958)].
5) T. Kametani, R. Yanase, R. Kano, K. Sakurai: J. Heterocyclic Chem., **3**, 239 (1966).
6) M. Tomita, T. Kugo: J. Pharm. Soc. Japan, **77**, 1075 (1957).
7) M. Tomita, T. Kugo: J. Pharm. Soc. Japan, **77**, 1079 (1957).
8) Z. F. Ismailov, A. U. Rakhmatkaviev, S. Yu. Yunusov: Dokl. Akad. Nauk Uz. SSR., **20**, 21 (1963). [C. A., **61**, 4407 (1964)].
9) M. Tomita, Y. Okamoto: J. Pharm. Soc. Japan, **85**, 456 (1965).
10) M. Tomita, H. Furukawa: Tetrahedron Letters, **1966**, 4293.
11) H. Kondo, Z. Narita, M. Murakami: J. Pharm. Soc. Japan, **61**, 375 (1941).
12) R. H. F. Manske: Can. J. Res., **21**B, 17 (1943).
13) H. Kondo, Z. Narita: J. Pharm. Soc. Japan, **47**, 40, 279 (1927).
14) H. Kondo, Z. Narita: J. Pharm. Soc. Japan, **49**, 688 (1929).
15) H. Kondo, Z. Narita: J. Pharm. Soc. Japan, **50**, 589 (1930).
16) F. Faltis, H. Frauendorfer: Ber., **63**, 806 (1930).
17) H. Kondo, Z. Narita: Ber., **63**, 2420 (1930).
18) H. Kondo, Z. Narita, S. Uyeo: Ber., **68**, 519 (1935).
19) Y. Inubushi, H. Niwa: J. Pharm. Soc. Japan, **72**, 762 (1952).
20) M. Tomita, K. Ito, H. Yamaguchi: Pharm. Bull., **3**, 449 (1955).
21) J. Kunitomo: J. Pharm. Soc. Japan, **82**, 1577 (1962).
22) M. Tomita, Y. Okamoto: J. Pharm. Soc. Japan, **84**, 1030 (1964).
23) R. H. F. Manske, M. Tomita, K. Fujitani, Y. Okamoto: Chem. Pharm. Bull., **13**, 1476 (1965).
24) I. N. Gorbatscheva, G. V. Buschbek, L. P. Warnakova, N. A. Preobrazhenskii: Zhur. Obshchei Khim., **27**, 2297 (1957). [C. A., **52**, 6375 (1958)].
25) E. N. Tsvetkov, I. N. Gorbachev, N. A. Preobrazhenskii: Zhur. Obshchei Khim., **27**, 3370 (1957). [C. A., **52**, 9170 (1958)].
26) T. Kametani, K. Fukumoto: Tetrahedron Letters, **1964**, 2771.
27) T. Kametani, K. Fukumoto: J. Chem. Soc., **1964**, 6141.
28) K. Fujitani, Y. Aoyagi, Y. Makoshi: J. Pharm. Soc. Japan, **84**, 1234 (1964).
29) T. Kametani, S. Takano, R. Yanase, C. Kibayashi, H. Iida, S. Kano: Chem. Pharm. Bull., **14**, 73 (1966).
30) F. D. Popp, H. W. Gibson, A. C. Noble: J. Org. Chem., **31**, 2296 (1966).
31) K. Fujitani, Y. Aoyagi, Y. Makoshi: J. Pharm. Soc. Japan, **86**, 654 (1966).
32) K. Fujitani, Y. Makoshi: J. Pharm. Soc. Japan, **86**, 660 (1966).
33) E. Ochiai: J. Pharm. Soc. Japan, **49**, 425 (1929).
34) I. R. C. Bick, J. Harley-Mason, N. Sheppard, M. J. Vernengo: J. Chem. Soc., **1961**, 1896.
35) J. Baldas, Q. N. Porter, I. R. C. Bick, M. J. Vernengo: Tetrahedron Letters, **1966**, 2059.
36) D. C. De. Jongh, S. R. Shrader, M. P.

Cava: J. Am. Chem. Soc., 88, 1052 (1966).
37) A. R. Battersby, I. R. C. Bick, W. Klyne, J. P. Jennings, P. H. Scopes, M. J. Vernengo: J. Chem. Soc., **1965**, 2239.
38) E. S. Pallares, H. M. Garza: Arch. Biochem., **16**, 275 (1948).
39) T. Kametani, M. Ro, Y. Iwabuchi: J. Pharm. Soc. Japan, **85**, 355 (1965).
40) T. Kametani, K. Fukumoto, M. Ro: J. Pharm. Soc. Japan, **84**, 532 (1964).
41) N. F. Proskurnina: Zhur. Obshchei Khim., **16**, 129 (1946). [C. A., **41**, 460 (1947)].
42) M. Tomita, E. Fujita: J. Pharm. Soc. Japan, **70**, 411 (1950).
43) M. Tomita, E. Fujita, T. Nakamura: J. Pharm. Soc. Japan, **71**, 1075 (1951).
44) M. Tomita: Pharm. Bull., **4**, 411 (1956).
45) M. Tomita, H. Ito: J. Pharm. Soc. Japan, **78**, 103 (1958).
46) I. N. Gorbacheva, M. I. Lerner, G. G. Zapesochnaya, L. P. Varnakova, N. P. Preobrazhenskii: Zhur. Obshchei Khim., **27**, 3353 (1957). [C. A., **52**, 9129 (1958)].
47) M. Tomita, K. Ito: J. Pharm. Soc. Japan, **78**, 605 (1958).
48) T. Kametani, H. Yagi: Tetrahedron Letters, **1965**, 953.
49) T. Kametani, H. Yagi: Chem. Pharm. Bull., **14**, 78 (1966).
50) T. Kametani, H. Yagi, S. Kaneda: Chem. Pharm. Bull., **14**, 974 (1966).
51) M. Tomita, K. Fujitani, A. Kato, H. Furukawa, Y. Aoyagi, M. Kitano, T. Ibuka: Tetrahedron Letters, **1966**, 857.
52) D. Sargabakov, Z. F. Ismailov, S. Yu. Yunusov: Dokl. Akad. Nauk Uz. SSR., **20**, 28 (1963). [C. A., **59**, 15336 (1963)].
53) Y. C. Chao, Y.-L. Chou, P.-C. Yang, C.-K. Chao: Sci. Sinica, **11**, 215 (1962). [C. A., **57**, 7383 (1962)].
54) H. Furukawa: J. Pharm. Soc. Japan, **85**, 353 (1965).
55) H. Furukawa: J. Pharm. Soc. Japan, **86**, 75 (1966).
56) P. C. Pan, Y.-L. Chou, T.-C. San, I. C. Kao: Sci. Sinica, **11**, 321 (1962). [C. A., **58**, 3467 (1963)].
57) Y.-Y. Hsieh, W.-C. Chen, Y.-S. Kao: Sci. Sinica, **12**, 2018 (1964). [C. A., **62**, 9183 (1965)].
58) Y.-Y. Hsieh, P.-C. Pan, W.-C. Chen, Y.-S. Kao: Sci. Sinica, **12**, 2020 (1964). [C. A., **62**, 9184 (1965)]. cf. Yao Hsueh Hsueh Pao, **13**, 166 (1966). [C. A., **65**, 8979 (1966)].
59) T. Kametani, S. Takano, K. Masuko, F. Sasaki: Chem. Pharm. Bull., **14**, 67 (1966).
60) H. Furukawa: J. Pharm. Soc. Japan, **86**, 883 (1966).
61) M. Tomita, H. Furukawa, T.-H. Yang, T.-J. Lin: Tetrahedron Letters, **1964**, 2637.
62) H. Furukawa: J. Pharm. Soc. Japan, **85**, 335 (1965).
63) H. Furukawa, T.-H. Yang, T.-J. Lin: J. Pharm. Soc. Japan, **85**, 472 (1965).
64) M. Tomita, N. Furukawa, T.-H. Yang T.-J. Lin: Chem. Pharm. Bull., **13**, 39 (1965).
65) N. Mollov, H. Putschewska, D. Panov: Compt. rend. Acad. Bulg. Sci., **17**, 709 (1964).
66) N. M. Mollov, H. B. Dutschewska: Tetrahedron Letters, **1964**, 2219.
67) M. Tomita, H. Furukawa, S.-T. Lu, S. M. Kupchan: Tetrahedron Letters, **1965**, 3409.
68) N. M. Mollov, H. B. Dutschewska: Tetrahedron Letters, **1966**, 853.
69) Z. F. Ismailov, S. Yu. Yunusov: Khim. Pridn. Soedin. Akad. Nauk Uz. SSR., **2**, 43, (1966). [C. A., **65**, 2320 (1966)].
70) H. B. Dutschevska, N. M. Mollov: Chem. & Ind., **1966**, 770.
71) M. Tomita, S.-T. Lu, Y.-Y. Chen: J. Pharm. Soc. Japan, **86**, 763, 1148 (1966).
72) M. Tomita, H. Furukawa, S.-T. Lu, S. M. Kupchan: Tetrahedron Letters, **1965**, 4309.
73) S. M. Kupchan, K. K. Chakravarti, N. Yokoyama: J. Pharm. Sci., **50**, 985 (1963).
74) T. Tomimatsu, M. P. Cava: J. Pharm. Soc., **54**, 1389 (1965).
75) S. M. Kupchan, N. Yokoyama: J. Am. Chem. Soc., **85**, 1361 (1963).
76) S. M. Kupchan, N. Yokoyama: J. Am. Chem. Soc., **86**, 2177 (1964).
77) H. Kondo, M. Tomita: Arch. Pharm., **269**, 433 (1931).
78) H. Kondo, M. Tomita: J. Pharm. Soc. Japan, **51**, 452 (1931).
79) Y. Inubushi: Pharm. Bull., **3**, 384 (1955).
80) I. R. C. Bick, T. G. Whalley: Univ. Queensland Papers, Dept. Chem., **1**, No. 33, p. 7 (1948). [C. A., **43**, 6787 (1949)].
81) I. R. C. Bick, E. S. Ewen, A. R. Todd: J. Chem. Soc., **1949**, 2767.
82) I. R. C. Bick, P. S. Clezy, M. J. Vernengo: J. Chem. Soc., **1960**, 4928.
83) F. E. Pyman: J. Chem. Soc., **105**, 1679 (1914).
84) I. R. C. Bick, E. S. Ewen, A. R. Todd: J. Chem. Soc., **1953**, 695.
85) H. Kondo, M. Tomita: J. Pharm. Soc. Japan, **55**, 646 (1935).
86) A. W. Sangster, K. L. Stuart: Chem. Revs., **65**, 69 (1965).

87) M. F. Grundon, J. F. B. McGarvey: J. Chem. Soc., **1962**, 2077.
88) M. F. Grundon, J. F. B. McGarvey: J. Chem. Soc., **1960**, 2739.
89) M. F. Grundon: Chem. & Ind., **1955**, 1772.
90) Parsons: Pharm. J., [3], **13**, 46 (1882).
91) C. Rüdel: Arch. Pharm., **229**, 631 (1891).
92) R. Chatterjee: J. Indian Chem. Soc., **28**, 225 (1951).
93) M. Tomita, T. Abe: J. Pharm. Soc. Japan, **72**, 773 (1952).
94) A. Orechov: Arch. Pharm., **271**, 323 (1933).
95) R. Chatterjee, A. Banerjee: J. Indian Chem. Soc., **30**, 705 (1953).
96) H. Kondo, M. Tomita: Arch. Pharm., **268**, 549 (1930).
97) Y. Inubushi, M. Kozuka: Pharm. Bull., **2**, 215 (1954).
98) M. Tomita, T. Kugo: J. Pharm. Soc. Japan, **79**, 317 (1959).
99) Polex: Arch. Pharm., **6**, 265 (1836).
100) O. Hesse: Ber., **19**, 3190 (1886).
101) E. Späth, A. Kolbe: Ber., **58**, 2280 (1925).
102) L. Beauquense: Bull. Sci. Pharmacol., **45**, 7 (1938). [C. A., **32**, 3089 (1938)].
103) R. Chatterjee, M. P. Guha: J. Am. Pharm. Assoc., **39**, 577 (1950).
104) R. Chatterjee, M. P. Guha, S. K. Sen: J. Am. Pharm. Assoc., **40**, 36 (1951).
105) R. Chatterjee, M. P. Guha: J. Am. Pharm. Assoc., **39**, 181, (1950).
106) R. Chatterjee, M. P. Guha: J. Am. Pharm. Assoc., **40**, 229 (1951).
107) F. von Bruchhausen, H. Schultze: Arch. Pharm., **267**, 617 (1929).
108) E. Späth, J. Pikl: Ber., **62**, 2251 (1929).
109) F. von Bruchhausen, P. H. Gericke: Arch. Pharm., **269**, 115 (1931).
110) F. von Bruchhausen, H. Oberembt, A. Feldhaus: Ann., **507**, 144 (1933).
111) E. Fujita: J. Pharm. Soc. Japan, **72**, 213, 217 (1952).
112) M. Tomita, E. Fujita: Pharm. Bull., **1**, 101 (1953).
113) M. Tomita, E. Fujita, F. Murai: J. Pharm. Soc. Japan, **71**, 226 (1951).
114) R. Chatterjee, M. P. Guha, A. K. Das-Gapta: J. Indian Chem. Soc., **29**, 921 (1952).
115) H. Kondo, K. Tanaka: J. Pharm. Soc. Japan, **63**, 267, 273 (1943).
116) I. R. C. Bick, A. R. Todd: J. Chem. Soc., **1950**, 1606.
117) M. Tomita, Y. Watanabe: Pharm. Bull., **4**, 126 (1956).
118) I. R. C. Bick, W. H. Whaley: Univ. Queensland Papers, Dept. Chem., **1**, No. 28, p. 8 (1946). [C. A., **41**, 1390 (1947)].
119) I. R. C. Bick, A. R. Todd: J. Chem. Soc., **1948**, 2170.
120) I. R. C. Bick, K. Doebel, W. I. Taylor, A. R. Todd: J. Chem. Soc., **1953**, 692.
121) F. von Bruchhausen: Arch. Pharm., **283**, 44 (1950).
122) E. Fujita, T. Saijoh: J. Pharm. Soc. Japan, **72**, 1232 (1952).
123) M. Tomita, Y. Inubushi, E. Fujita: Pharm. Bull., **3**, 97 (1955).
124) M. V. Telezhenetskaya, S. Yu. Yunusov: Dokl. Akad. Nauk SSSR., **162**, 254 (1965). [C. A., **63**, 5689 (1965)].
125) M. V. Telezhenetskaya, Z. F. Ismailov, S. Yu. Yunusov: Khim. Prirodn. Soedin., Akad. Nauk Uz. SSR., **2**, 107 (1966). [C. A., **65**, 10629 (1966)].
126) T. Kugo, M. Tanaka, T. Sanae: J. Pharm. Soc. Japan, **80**, 1425 (1960).
127) I. R. C. Bick, P. S. Clezy, M. J. Vernengo: J. Chem. Soc., **1961**, 4928.
128) H. Kondo, T. Sanada: J. Pharm. Soc. Japan, **48**, 163, 1141 (1928).
129) M. Tomita, T. Ibuka: J. Pharm. Soc. Japan, **83**, 996 (1963).
130) H. Kondo, T. Nozoe: J. Pharm. Soc. Japan, **63**, 333 (1943).
131) T. Nakano: J. Pharm. Soc. Japan, **64**, 27 (1944).
132) H. Kondo, T. Sanada: J. Pharm. Soc. Japan, **51**, 509 (1931).
133) Y. Watanabe: J. Pharm. Soc. Japan, **80**, 166 (1960).
134) H. Kondo, T. Sanada: J. Pharm. Soc. Japan, **517**, 1034 (1924).
135) H. Kondo, T. Sanada: J. Pharm. Soc. Japan, **47**, 31, 126, 177, 930 (1927).
136) H. Kondo, Y. Watanabe: J. Pharm. Soc. Japan, **58**, 268 (1938).
137) M. Tomita, S. Uyeo, S. Sawa, K. Dai, T. Miwa: J. Pharm. Soc. Japan, **69**, 22 (1949).
138) M. Tomita, E. Fujita: Pharm. Bull., **2**, 378 (1954).
139) H. Furukawa: J. Pharm. Soc. Japan, **86**, 253 (1966).
140) Y. Watanabe, H. Furukawa, M. Kurita: J. Pharm. Soc. Japan, **86**, 257 (1966).
141) D. H. R. Barton, G. W. Kirby, A. Wiechers: Chem. Comm., **1966**, 266.
142) D. H. R. Barton, G. W. Kirby, A. Wiechers: J. Chem. Soc. (C), **1966**, 2313.
143) I. R. C. Bick, G. K. Douglas: Chem. & Ind., **1965**, 694.
144) M. Tomita, T. Ibuka: J. Pharm. Soc. Japan,

85, 557 (1965).
145) M. Tomita, T. Ibuka: J. Pharm. Soc. Japan, **83**, 940 (1963).
146) T. Tomimatsu, J. L. Beal: J. Pharm. Sci., **55**, 208 (1965).
147) T. Kugo: J. Pharm. Soc. Japan, **79**, 322 (1959).
148) I. R. C. Bick, P. S. Clezy, W. D. Crow: Austral. J. Chem., **9**, 111 (1956).
149) M. Tomita, T. Kugo: J. Pharm. Soc. Japan, **75**, 753 (1965).
150) R. Chatterjee, A. Banerjee, A. K. Barua, A. K. DasGupta: J. Indian Chem. Soc., **31**, 83 (1954).
151) M. Tomita, T. Abe: J. Pharm. Soc. Japan, **72**, 735 (1952).
152) M. Tomita, Y. Inubushi, N. Mizoguchi: J. Pharm. Soc. Japan, **73**, 776 (1953).
153) M. Tomita, Y. Inubushi, N. Mizoguchi: Pharm. Bull., **1**, 53 (1953).
154) G. A. Graethouse, N. E. Rigler: Plant Physiol., **15**, 563 (1940). [C.A., **36**, 6205 (1942)].
155) F. von Bruchhausen, A. C. Santos, C.Schäfer: Arch. Pharm., **293**, 454 (1960).
156) H. Kondo, M. Tomita, M. Satomi, T. Ikeda: J. Pharm. Soc. Japan, **58**, 276 (1938).
157) M. Tomita, H. Kitagishi: J. Pharm. Soc. Japan, **64**, 240 (1944).
158) M. Tomita, E. Fujita, M. Murai: J. Pharm. Soc. Japan, **71**, 1034 (1951).
159) Y. Inubushi: J. Pharm. Soc. Japan, **72**, 220 (1952).
160) M. Tomita, Y. Inubushi: J. Pharm. Soc. Japan, **72**, 221 (1952).
161) F. von Bruchhausen, J. Knabe: Arch. Pharm., **287**, 601 (1954).
162) F. von Bruchhausen, C. Schäfer: Arch. Pharm., **290**, 357 (1957).
163) M. Tomita, E. Fujita, F. Murai: J. Pharm. Soc. Japan, **71**, 301 (1951).
164) M. Tomita, Y. Inubushi, M. Kozuka: Pharm. Bull., **1**, 360, 368 (1953).
165) M. Tomita, Y. Inubushi, K. Ito: Pharm. Bull., **2**, 372 (1954).
166) M. Tomita, H. Ishii: J. Pharm. Soc. Japan, **76**, 1322 (1956).
167) M. Tomita, Y. Kondo: J. Pharm. Soc. Japan, **77**, 1019 (1957).
168) M. Tomita, Y. Sato: J. Pharm. Soc. Japan, **78**, 543 (1958).
169) L. R. Row, A. S. R. Anjaneyulu: J. Sci. Ind. Res., **21**B, 581 (1962).
170) F. von Bruchhausen, A. C. Santos, C. Schäfer: Arch. Pharm., **293**, 785 (1960).
171) C.-K. Chuang, C.-Y. Hsing, Y.-S. Kao, K.-J. Chang: Ber., **72**, 519 (1939).
172) C.-Y. Hsing, K.-J. Chang: Sci. Sinica, **7**, 59 (1958). [C. A., **52**, 18494 (1958)].
173) J.-H. Chou: Chin. J. Physiol., **9**, 267 (1935). [C. A., **30**, 471 (1936)].
174) J.-H. Chou: Chin. J. Physiol., **14**, 315 (1939). [C. A., **34**, 4739 (1940)].
175) J.-H. Chou: Science Record, **3**, 107 (1950). [C., **1951**, II, 2184].
176) G. R. Chaudhury, M. L. Dhar: J. Sci. Ind. Res., **17** B, 163 (1958).
177) S. Kubota: Folia Pharmacol. Japan, **12**, No. 2. 328 (1921). [C. A., **25**, 5736 (1931). cf. C. A., **28**, 7422 (1934)].
178) H. Kondo, K. Yano: J. Pharm. Soc. Japan, **48**, 15, 107 (1928).
179) A. W. McKenzie, J. R. Price: Austral. J. Chem., **6**, 180 (1953).
180) A. C. Santos: Rev. Filippina. Med. Farm., **22**, No. 9. (1931). [C. A., **26**, 729 (1932)].
181) J. R. Boissier, A. Bouquet, G. Combes, C. Dumont, M. Debray: Ann. Pharm. Franc., **21**, 767 (1963).
182) H. Kondo, K. Yano: J. Pharm. Soc. Japan, **49**, 315 (1929).
183) H. Kondo, K. Yano: J. Pharm. Soc. Japan, **50**, 224 (1930).
184) E. Fujita, F. Murai: J. Pharm. Soc. Japan, **71**, 1039 (1951).
185) D. A. A. Kidd, J. Walker: Chem. & Ind., **1953**, 243.
186) D. A. A. Kidd, J. Walker: J. Chem. Soc., **1954**, 669.
187) A. C. Santos: Ber., **65**B, 472 (1932).
188) H. Kondo, I. Keimatsu: Ber., **68**B, 1503 (1935).
189) I. R. C. Bick, P. S. Clezy: J. Chem. Soc., **1953**, 3893.
190) M. Tomita, K. Fujitani, T. Kishimoto: J. Pharm. Soc. Japan, **82**, 1148 (1962).
191) M. Tomita, Y. Inubushi: Pharm. Bull., **4**, 413 (1956).
192) H. Kondo, I. Keimatsu: J. Pharm. Soc. Japan, **55**, 63, 234 (1935).
193) M. Tomita, Y. Inubushi, H. Niwa: J. Pharm. Soc. Japan, **72**, 211 (1952).
194) F. von Bruchhausen, C. Schäfer: Arch. Pharm., **294**, 112 (1961).
195) M. Tomita, T. Ibuka: J. Pharm. Soc. Japan, **82** 1656 (1962).
196) W.M. Whaley, L.N. Starker, M. Meadow: J. Org. Chem., **18**, 833 (1953).
197) W.M. Whaley, L.N. Staker, W.L. Dean, M. Meadow: J. Org. Chem., **19**, 1018, 1020, 1022 (1954).
198) H. Kondo, H. Kataoka, Y. Baba: Ann. Rept. ITSUU Lab., **5**, 59 (1954).

199) H. Kondo, H. Kataoka, K. Kigasawa: Ann. Rept. ITSUU Lab., **6**, 46 (1955).
200) J. Padilla, J. Herran: Tetrahedron, **18**, 427 (1962).
201) M. Shamma, B.S. Dudock, M.P. Cava, K.V. Rao, D.R. Dalton. D.C. DeJongh, S.R. Shrader: Chem. Comm., **1966**, 7.
202) N.M. Mollov, V. St. Georgiev, D. Jordanov, P. Panov: Compt. rend. Acad. Bulgare Sci., **19**., 491 (1966). [C.A., **65**, 13780 (1966)].
203) S. Kh. Maekh, S. Yu. Yunusov: Khim. Prirodn. Soedin., Akad. Nauk Uz. SSR., **1965**, 188. [C.A., **63**, 14929 (1965)].
204) S. Kh. Maekh, S. Yu. Yunusov: Khim. Prirodn. Soedin., Akad. Nauk Uz. SSR., **1965**, 294. [C.A., **63**, 18183 (1965)].
205) Z.F. Ismailov, S. Kh. Moekh, S. Yu. Yunusov: Dokl. Akad. Nauk Uz. SSR., **1960**, 22. [C.A., **56**, 11646 (1962)].
206) S. Kh. Maekh, S. Yu Yunusov: Izv. Akad. Nauk SSR., Ser. Khim., **1966**, 112. [C.A., **64**, 14233 (1966)].
207) S. Kh. Maekh, S. Yu. Yunusov: Dokl. Akad. Nauk Uz. SSR., **21**., 27 (1964). [C. A., **62**, 13191 (1965)].
208) A.C. Santos: Arch. Pharm., **284**, 360 (1951).
209) F. von Bruchhausen, A.C. Santos, J. Knabe, G.A. Santos: Arch. Pharm., **290**, 232 (1957).
210) J. Knabe: Ber., **91**, 1612 (1958).
211) H. Kondo, Y. Yamashita, I. Keimatsu: J. Pharm. Soc. Japan, **54**, 108, 620 (1934).
212) H. Kondo, I. Keimatsu: J. Pharm. Soc. Japan, **55**, 121 (1935).
213) H. Kondo, S. Hasegawa, M. Tomita: U.S. Pat. 2,206,407 [C.A., **34**, 7542 (1940)].
214) M. Tomita: J. Pharm. Soc. Japan, **59**, 207 (1939).
215) H. Kondo, M. Tomita: J. Pharm. Soc. Japan, **59**, 542 (1939).
216) M. Tomita, Y. Sasaki: Pharm. Bull., **1**, 105 (1953).
217) M. Tomita, Y. Sasaki: Pharm. Bull., **2**, 89 (1954).
218) M. Tomita, Y. Sasaki: Pharm. Bull., **2**, 375 (1954).
219) Y. Sasaki, H. Ohnishi, N. Satoh: Pharm. Bull., **3**, 178 (1955).
220) Y. Sasaki: Pharm. Bull., **3**, 250 (1955).
221) H. Kondo, I. Keimatsu: Ber., **71 B**, 2553 (1938).
222) H. Kondo, H. Kataoka, G. Ito, K. Kigasawa, M. Ikechi: Ann. Rept. Itsuu Lab., **1**, 15 (1950).
223) H. Kondo: Ann. Rept. Itsuu Lab., **3**, 27, 43, 48, 50 (1951).
224) H. Kondo, H. Kataoka, K. Nakagawa: Ann. Rept. Itsuu Lab., **3**, 49 (1952).
225) H. Kondo, H. Kataoka, Y. Baba: Ann. Rept. ITSUU Lab., **4**, 70 (1953).
226) W.M. Whalley, C.N. Robinson: J. Org. Chem., **19**, 1029 (1954).
227) H. Kondo, H. Kataoka, Y. Baba: Ann. Rept. ITSUU Lab., **5**, 55 (1954).
228) H. Kondo, H. Kataoka, Y. Baba: Ann. Rept. ITSUU Lab., **6**, 41 (1955).
229) I.R.C. Bick, J. Harley–Mason, M.J. Vernengo: Anales Assoc. Quim. Argo., **51**, 135 (1963). [C.A., **60**, 1811 (1964)].
230) I.R.C. Bick, J. Harley–Mason, N Sheppard. M.J. Vernengo: J. Chem. Soc., **1961**, 1896.
231) I.R.C. Bick, W.M. Whaley: Univ. Queensland Papers, Dept. Chem., **1**, No. 30 p. 4 (1947). [C.A., **42**, 3909 (1948)].
232) E. Fujita, T. Tomimatsu: J. Pharm. Soc. Japan, **79**, 1256 (1959).
233) E. Fujita, T. Tomimatsu: J. Pharm. Soc. Japan, **79**, 1260 (1959).
234) T. Tomimatsu: J. Pharm. Soc. Japan, **79**, 1386 (1959).
235) E. Fujita, T. Tomimatsu, Y. Kano: J. Pharm. Soc. Japan, **80**, 1137 (1960).
236) T. Tomimatsu, Y. Kano: J. Pharm. Soc. Japan, **83**, 153 (1963).
237) T. Tomimatsu, Y. Kano: J. Pharm. Soc. Japan, **83**, 159 (1963).
238) E. Fujita, K. Fuji, T. Suzuki: Bull. Inst. Chem. Res., Kyoto Univ., **43**, 449 (1965). [C.A., **65**, 7229 (1966)].
239) N.M. Mollov, V. St. Georgiev: Chem. & Ind., **1966**, 1178.
240) N.M. Mollov, Kh. B. Duchevska, Khr. G. Kiryakov: Chem. & Ind., **1965**, 1595.
241) E. Fujita, T. Tomimatsu, Y. Kano: J. Pharm. Soc. Japan, **82**, 311 (1962).
242) E. Fujita, T. Tomimatsu, Y. Kano: J. Pharm. Soc. Japan, **82**, 315 (1962).
243) E. Fujita, T. Tomimatsu, Y. Kano: J. Pharm. Soc. Japan, **82**, 320 (1962).
244) H. King: J. Chem. Soc., **1940**, 737.
245) J.A. Barltrop. J.A.D. Jeffreys: J. Chem. Soc., **1954**, 159.
246) J.A.D. Jeffreys: J. Chem. Soc., **1956**, 4451.
247) O. Wintersteiner, J.D. Dutcher: Science, **97**, 467 (1943).
248) J.D. Dutcher: J. Am. Chem. Soc., **68**, 419 (1946).
249) H. King: J. Chem. Soc., **1948**, 1945.
250) I.R.C. Bick, P.S. Clezy: J. Chem. Soc., **1960**, 2402.
251) M. Tomita, T. Kikuchi: J. Pharm. Soc. Japan, **77**, 69 (1957).
252) R.M. Srivastava, M.P. Khara: Current Sci.,

32, 114 (1963). [C. A., **59**, 27578 (1963)].
253) F. A. L. Anet, G. K. Hughes, E. Ritchie: Austral, J. Sci. Res., **A 3**, 346 (1956).
254) F. Faltis: Monatsh., **33**, 873 (1912).
255) F. Faltis, F. Neumann: Monatsh., **42**, 331 (1923).
256) F. Faltis, T. Heczko: Monatsh., **43**, 377 (1924).
257) F. Faltis, A. Troller: Ber., **61 B**, 345 (1928).
258) F. Faltis, K. Zwerina: Ber., **62 B**, 1034 (1929).
259) F. Faltis, S. Wrann, E. Kühas: Ann., **497**, 69 (1932).
260) F. Faltis, H. Dietreich: Ber., **67B**, 231 (1934).
261) M. Tomita, E. Fujita, F. Murai: J. Pharm. Soc. Japan, **71** 1043 (1951).
262) M. Tomita, T. Kikuchi: J. Pharm. Soc. Japan, **77**, 238 (1955).
263) X. A. Dominguez, B. Gomez, E. Homberg, J. Slim: J. Am. Chem. Soc., **77**, 1288 (1955).
264) R. Boehm: Arch. Pharm., **235**, 660 (1897).
265) H. King: J. Chem. Soc., **1937**, 1472.
266) T. Kikuchi, K. Bessho: J. Pharm. Soc. Japan, **78**, 1408 (1958).
267) T. Kikuchi, K. Bessho: J.Pharm. Soc. Japan, **79**, 192, 262 (1959).
268) M. Tomita, H. Shirai: J. Pharm. Soc. Japan, **62**, 381 (1942).
269) H. Kondo, M. Tomita, S. Uyeo: Ber., **70**, 1890 (1937).
270) I. I. Shchelchkova, T. N. Ilinskaya, A. D. Kuzovkov: Khim. Prirodn. Soedin., Akad. Nauk Uz. SSR., **1965**, 271. [C. A., **64**, 6709 (1966)].
271) M. Tomita, T. Sasaki, S. Matsumura: J. Pharm. Soc. Japan, **79**, 1120 (1959).
272) M. Tomita, K. Fujitani, Y. Aoyagi: Tetrahedron Letters, **1966**, 4243.
273) H. Mckennis, Jr., P. H. Hearst, R. W. Drisko, T. Roe, Jr., R. L. Alumbaugh: J. Am. Chem. Soc., **78**, 245 (1956).
274) M. Scholtz: Arch. Pharm., **237**, 199 (1899); **244**, 555 (1906).
275) H. Hildebrandt: Arch. exptl. Pathol. Pharmacol., **57**, 279 (1907).
276) H. King: J. Chem. Soc., **1935**, 1381.
277) H. King: Nature, **158**, 515 (1946); J. Chem. Soc., **1947**, 936.
278) S. Bhattacharji, V. N. Sharma, M. L. Dhar: Bull. Nat. Inst. Sci. India, **4**, 39 (1955); J. Sci. Ind. Res., **15 B**, 363 (1956).
279) Wiggers: Ann., **33**, 81 (1840).
280) M. Scholtz: Ber., **29**, 2054 (1896).
281) H. King: Chem. & Ind., **1935**, 739.
282) J. Herzig, H. Meyer: Monatsh., **18**, 379 (1890).
283) M. Scholtz: Arch. Pharm., **236**, 530 (1898).
284) M. Scholtz: Arch. Pharm., **250**, 684 (1912).
285) E. Späth, W. Leithe, F. Ladek: Ber., **61 B**, 1698 (1928).
286) E. Späth, F. Kuffner: Ber., **67B**, 55 (1934).
287) H. King: J. Chem. Soc., **1936**, 1276.
288) F. Faltis, K. Kadiera, F. Doblhammer: Ber., **69**, 1029 (1936).
289) F. Faltis, L. Holtzinger, P. Ita, R. Schwarz: Ber., **74B**, 79 (1941).
290) H. King: J. Chem. Soc., **1939**, 1157.
291) M. F. Grundon, H. J. H. Perry: J. Chem. Soc., **1954**, 3531.
292) V. I. Shvets, L. V. Volkova, O. N. Tolkachev: Izv. Vysshikh. Uchebn. Zavedenii. Khim. i Khim. Technol., **5**, 445 (1962). [C. A., **59**, 2876 (1963)].
293) H. Kondo, H. Kataoka, K. Kigasawa: Ann. Rept. ITSUU Lab., **7**, 37 (1956).
294) J. R. Crowder, M. F. Grundon, J. R. Lewis: J. Chem. Soc., **1958**, 2142.
295) M. F. Grundon: J. Chem. Soc., **1959**, 3010.
296) J. R. Crowder, E. E. Glover, M. F. Grundon, H. X. Kaempfen: J. Chem. Soc., **1963**, 4578.
297) L. P. J. Palet: Anales soc. quim. Argentina, **6**, 156 (1918). [C. A., **13**, 216 (1919)].
298) K. Bodendorf, W. Schreibe: Arch. Pharm., **287**, 555 (1954).
299) J. A. D. Dutcher: J. Am. Chem, Soc., **74**, 2221 (1952).
300) H. King: J. Chem. Soc., **1948**, 265.
301) O. N. Tolkachev, V. G. Voronin, N. A. Preobrazhenskii: Zhur. Obshchei Khim., **29**, 1192 (1959). [C. A., **54**, 1578 (1960)].
302) H. Hellmann, W. Elser: Ann., **639**, 77 (1961).
303) V. G. Voronin, O. N. Tolkachev, N. A. Preobrazhenskii: Izv. Vysshikh. Uchebn. Zavedenii. Khim. i. Khim. Technol., **5**, 449 (1962). [C. A., **59**, 2877 (1963)].
304) E. Hultin: Acta Chem. Scand., **16**, 559 (1962); **17**, 753 (1963).
305) L. V. Volkova, O. N. Tolkachev, N. A. Preobrazhenskii: Dokl. Akad. Nauk USSR., **102**, 521 (1955). [C., **1956**, 1024].
306) V. G. Voronin, O. N. Tolkachev, N. A. Preobrazhenskii: Dokl. Akad. Nauk USSR., **122**, 77 (1958). [C. A., **53**, 1345 (1959)].
307) K. P. Agarwal, S. Rakhit, S. Bhattercharji, M. M. Dhar: J. Sci. Ind. Res., **19 B**, 479 (1960). [C. A., **55**, 16585 (1961)].
308) G. W. Milne, J. R. Plimmer: J. Chem. Soc. (C), **1966**, 1966.
309) T. Kikuchi, K. Bessho: J. Pharm, Soc. Japan, **78**, 1408 (1958).
310) T. Kikuchi, K. Bessho: J. Pharm. Soc. Japan, **78**, 1413 (1958).
311) H. Kondo, K. Yano: J. Pharm. Soc. Japan,

47, 107, 815 (1927).
312) M. Satomi: Ann. Rept. Itsuu Lab., **6**, 31 (1955).
313) M. Tomita, S. Uyeo: J. Chem. Soc. Japan, **64**, 64, 70, 77, 142, 147 (1943).
314) Y. Inubushi: J. Pharm. Soc. Japan, **72**, 1223 (1952).
315) M. Tomita, T. Kikuchi: J. Pharm. Soc. Japan, **77**, 997 (1957).
316) T. Kikuchi: J. Pharm. Soc. Japan, **78**, 576 (1958).
317) S.M. Kupchan, S. Kubota, E. Fujita, S. Kobayashi, J.H. Block, S.A. Telang: J. Am. Chem. Soc., **88**, 4212 (1966).
318) K.V.J. Rao, L.R. Row: J. Sci. Ind. Res., **20B**, 125 (1961). [C.A., **55**, 21153 (1961)].
319) M. Tomita, F. Kusuda: Pharm. Bull., **1**, 1 (1953).
320) H. Kondo, T. Nakazato: J. Pharm. Soc. Japan, **44**, 691 (1924).
321) H. Kondo, M. Tomita: J. Pharm. Soc. Japan, **47**, 39, 265 (1927).
322) H. Kondo, T. Nakazato: J. Pharm. Soc. Japan, **46**, 461, 465 (1926).
323) H. Kondo, M. Tomita: J. Pharm. Soc. Japan, **48**, 83, 659 (1928).
324) H. Kondo, T. Tomita: J. Pharm, Soc. Japan, **50**, 1035 (1930).
325) H. Kondo, T. Tomita: Ann., **497**, 104 (1932).
326) F. Faltis: Ann., **499**, 301 (1932).
327) M. Tomita, C. Tani: J. Pharm. Soc. Japan, **62**, 468 (1942).
328) M. Tomita, C. Tani: J. Pharm. Soc. Japan, **62**, 481 (1942).
329) M. Tomita, Y. Inubushi: Pharm. Bull., **2**, 6 (1954).
330) M. Tomita, Y. Inubushi, H. Niwa: J. Pharm. Soc. Japan, **72**, 206 (1952).
331) Y. Inubushi: Pharm. Bull., **2**, 1 (1954).
332) M. Tomita, Y. Inubushi: Pharm. Bull., **3**, 7 (1955).
333) M. Tomita, S. Ueda: J. Pharm. Soc. Japan, **79**, 977 (1959).
334) Y. Inubushi: Pharm. Bull., **2**, 11 (1954).
335) Y. Inubushi, K. Nomura: Tetrahedron Letters, **1963**, 1133.
336) Y. Inubushi, K. Nomura, M. Miyasaki: J. Pharm. Soc. Japan, **83**, 282 (1963).
337) Y. Inubushi, K. Nomura: J. Pharm. Soc. Japan, **83**, 288 (1963).
338) M. Tomita, H. Furukawa: J. Pharm. Soc. Japan, **83**, 676 (1963).
339) M. Tomita, H. Furukawa: J. Pharm. Soc. Japan, **83**, 760 (1963).
340) M. Tomita, H. Furukawa: J. Pharm. Soc. Japan, **84**, 1027 (1964).
341) D.H.R. Barton, T. Cohen: "Festshrift Arthur Stoll", p. 117. Birkhauser, Basel. (1957).
342) M.F. Grundon: Progress of Organic Chemistry, Vol. **6**, p. 38 (1964).
343) K.V.J. Rao, L. Row: J. Sci. Ind. Res., **18B**, 247 (1959). [C.A., **54**, 18577 (1960)].
344) H. Kondo, M. Tomita: J. Pharm. Soc. Japan, **55**, 633, 637, 911 (1935).
345) M. Tomita: J. Pharm. Soc. Japan, **55**, 100 (1935).
346) M. Tomita, S. Ueda, A. Teraoka: Tetrahedron Letters, **1962**, 635.
347) M. Tomita, S. Teraoka: J. Pharm. Soc. Japan, **83**, 87 (1963).
348) K.V.J. Rao, L.R. Row: J. Sci. Ind. Res., **16B**, 156 (1957).
349) K.V.J. Rao, L.R. Row: J. Org. Chem., **25**, 981 (1960).
350) K.V.J. Rao, L.R. Row: Chem. & Ind., **1960**, 406.
351) L. van Itallie, A.S. Steenhauer: Pharm. Weekbl., **59**, 1381 (1922). [C.A., **17**, 611 (1923)].
352) B. Anjaneyulu, K.W. Gopinath, T.R. Govindachari: Chem. & Ind., **1959**, 702.
353) B. Anjaneyulu, K.W. Gopinath, T.R. Govindachari: Chem. & Ind., **1959**, 1119.
354) B. Anjaneyulu, K.W. Gopinath, T.R. Govindachari, B.R. Pai: J. Sci. Ind. Res., **21 B**, 602 (1962).
355) P.J. Hearst: J. Org. Chem., **29**, 466 (1964).
356) M.F. Grundon, J.E.B. McGarvey: J. Chem. Soc. (C), **1966**, 1082.
357) P.A. Paris, J. Lemen: Ann. Pharm. franc., **13**, 200 (1955). [C.A., **49**, 11959 (1955)].
358) A.N. Tackie, A. Thomas: Ghana J. Sci., **5**, 11 (1965). [C.A., **65**, 3922 (1966)].
359) S. Bhattachacharji, A.C. Roy, M.L. Dhar: J. Sci. Ind. Res., **21 B**, 428 (1962).
360) S.S. Norkina, N.A. Pakhareva: Zhur. Obshchei Khim., **20**, 1720 (1950). [C.A., **45**, 1306 (1951)].
361) H.H.S. Fong, J.L. Beal, M.P. Cava: Lloydia, **29**, 94 (1966). [C.A., **65**, 7229 (1966)].
362) O.N. Tolkachev, L.P. Kvashnina, N.A. Preobrazhenskii: Zhur. Obshchei Khim., **30**, 1704 (1966).
363) N.M. Mollov, H.B. Dutschewska: Dokl. Bolgar. Akad. Nauk, **19**, 495 (1966).

Chapter 7 Cularine and Related Alkaloids

Cularicine

$C_{18}H_{17}O_4N \equiv 311.32$

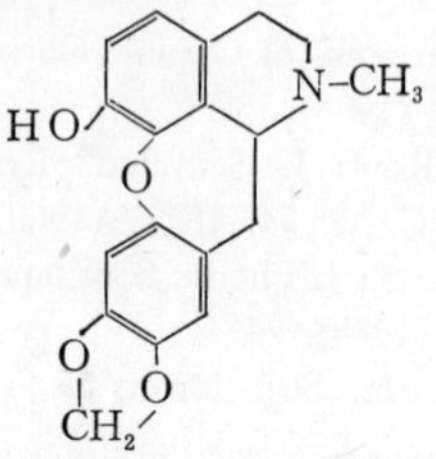

≪mp≫ 185° (MeOH)
≪$[\alpha]_D$≫ +295° ($CHCl_3$)

≪Ph.≫ NMR[3] Mass[3]
≪OP.≫ Corydalis claviculata[1,2]
≪C. & So.≫ Prisms
≪Sa. & D.≫ B–OMe-HCl: 267° B–OEt: 114° B–OEt–HCl: 275°
≪SD.≫ 2), 3)
≪S. & B.≫

Cularidine

$C_{19}H_{21}O_4N \equiv 327.37$

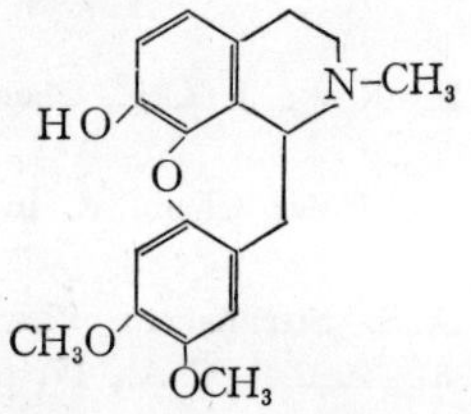

≪mp≫ 157° (MeOH)
≪$[\alpha]_D$≫ +292° (MeOH)

≪Ph.≫ NMR[3] Mass[3]
≪OP.≫ Corydalis claviculata[1,2], Dicentra cucullaria[4]
≪C. & So.≫ Fine needles
≪Sa. & D.≫ B–$HClO_4$: 297° B–OMe=cularine
≪SD.≫ 3), 5)
≪S. & B.≫

Cularimine

$C_{19}H_{21}O_4N \equiv 327.37$

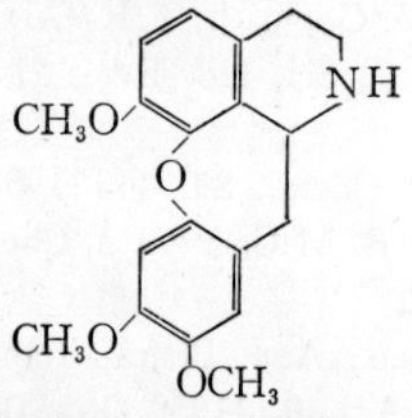

≪mp≫ 102° (Et_2O)
≪$[\alpha]_D$≫

≪Ph.≫ UV[11] IR[10] NMR[3] Mass[3]
≪OP.≫ Dicentra eximia[4]
≪C. & So.≫ Fine needles or rhombic plates
≪Sa. & D.≫ B–NBz: 174° B–NMe=cularine
≪SD.≫ 6)
≪S. & B.≫ 7), 8), 9), 10), 11)

Cularine

$C_{20}H_{23}O_4N \equiv 341.39$

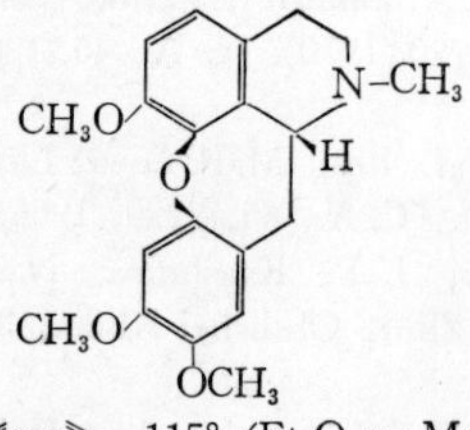

≪mp≫ 115° (Et_2O or MeOH)
≪$[\alpha]_D$≫ +285° (MeOH)

≪Ph.≫ UV[13,15] IR[10,15] NMR[3,13] Mass[3,12] Mass[3,12] ORD[13] CD[13] Rf[15]
≪OP.≫ Corydalis claviculata[1], Dicentra cucullaria[4], D. eximia[4], D. formasa[4], D. oregana[4]
≪C. & So.≫ Large stout prisms, Sol. in most org. solv.
≪Sa. & D.≫ B–HCl: 207° B–Oxal: 245° (d) B–MeI: 205°
≪SD.≫ 6), 12), 13)
≪S. & B.≫ 7), 8), 9), 10), 14), 15), 16)
BG: 6), 17)

---

1) R. H. F. Manske: Can. J. Res., **18 B**, 97 (1940).
2) R. H. F. Manske: Can. J. Chem., **43**, 989 (1965).
3) T. Kametani, S. Shibuya, S. Sasaki: Tetrahedron Letters, **1966**, 3215.
4) R. H. F. Manske: Can. J. Res., **16 B**, 81 (1938).
5) R. H. F. Manske: Can. J. Chem., **44**, 561 (1966).
6) R. H. F. Manske: J. Am. Chem. Soc., **72**, 55 (1950).
7) T. Kametani, S. Shibuya, S. Seino, K. Fukumoto: Tetrahedron Letters, **1964**, 25.
8) T. Kametani, S. Shibuya, S. Seino, K. Fukumoto: J. Chem. Soc., **1964**, 4146.
9) T. Kametani, S. Shibuya: Tetrahedron Letters, **1965**, 1897.
10) T. Kametani, S. Shibuya: J. Chem. Soc., **1965**, 5565.
11) T. Kametani, S. Shibuya, I. Noguchi: J. Pharm. Soc. Japan, **85**, 667 (1965).
12) M. Ohashi, J. M. Wilson, H. Budzikiewicz, M. Shamma, W.A. Slusarchyk, C. Djerassi: J. Am. Chem. Soc., **85**, 2807 (1963).
13) N. S. Bhacca, J. C. Craig, R. H. F. Manske, S. K. Roy, M. Shamma, W. A. Slusarchyk: Tetrahedron, **22**, 1467 (1966).
14) T. Kametani, K. Fukumoto: Chem. & Ind., **1963**, 291.
15) T. Kametani, K. Fukumoto: J. Chem. Soc., **1963**, 4289.
16) T. Kametani, K. Ogasawara: J. Chem. Soc., **1964**, 4142.
17) K. W. Bentley: The Isoquinoline Alkaloids, p. 250, Pergamon Press, 1965.

Chapter 8 Proaporphine Alkaloids

Crotonosine $C_{17}H_{17}O_3N \equiv 283.31$

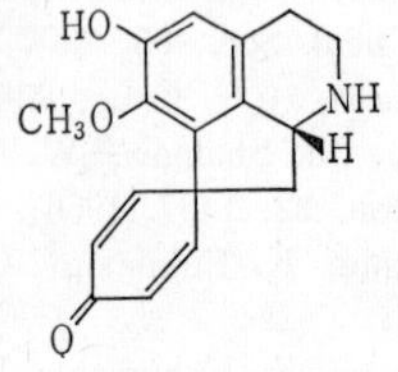

≪mp≫ 197° (soft), >300° (d) (i-PrOH)
≪$[\alpha]_D$≫ +180° (MeOH)

≪Ph.≫ CR[2,8] UV[2,3] IR[2,3] NMR[1,3,4,5] CD[6] Rf[1] pKa[2] Mass[27]
≪OP.≫ Croton discolor[1], C. linearis[2]
≪C. & So.≫
≪Sa. & D.≫ B–MeI: >250° (d) B–OMe=stepharine B–N,O–DiMe=pronuciferine B–NAc: 205° B–N,O–DiAc: 203~205°
≪SD.≫ 1), 3), 4), 5), 6)
≪S. & B.≫ BS: 7), BG: 8), 9)

Fugapavine (Mecambrine)[10,11] $C_{18}H_{17}O_3N \equiv 295.32$

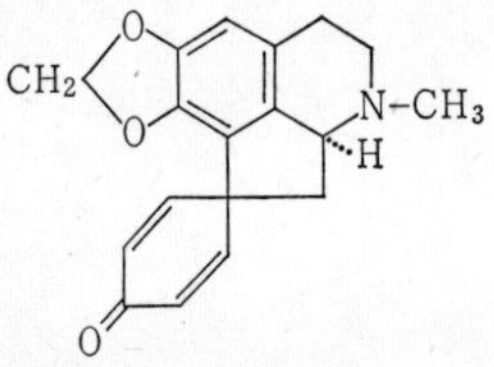

≪mp≫ 178.5~179.5° (Et_2O)
≪$[\alpha]_D$≫ −116° ($CHCl_3$), cf. −92±3° ($CHCl_3$)[15]

≪Ph.≫ UR[8] UV[10,13,15,17,18,19,21] IR[13,15,17,18,19,21] CD[6] Rf[10,12,13,16,18] pKa[13]
≪OP.≫ Meconopsis cambrica[12,13], Papaver armeniacum[10,14], P. caucasicum[10,14], P. dubium[15] P. dubium spp. arbiflorium[16], P. fugax[17,18,19,20], P. persicum[10,14], P. polycaetum[10,14], P. triniaefolium[10,14]
≪C. & So.≫ Colourless prisms, Sol. in $CHCl_3$, Spar. sol. in Et_2O, Me_2CO, EtOH
≪Sa. & D.≫ B–HCl: 269~270° (d) B–Pic: 165° B–Picrol: 159~163° Oxime: >285° Semicarb: 237° 2,4–D: >285° (d) Hexahydro-Der.: 267~269°
≪SD.≫ 6), 17), 18), 19), 20), 21), 22), 23), 48)
≪S. & B.≫

Stepharine $C_{18}H_{19}O_3N \equiv 297.34$

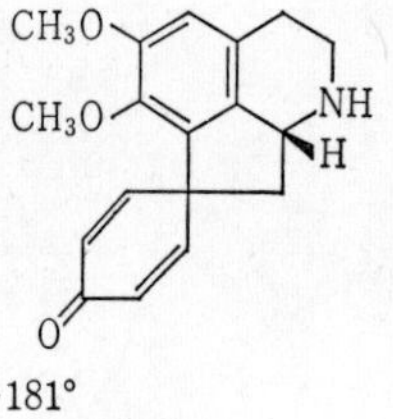

≪mp≫ 179~181°
≪$[\alpha]_D$≫ +143° ($CHCl_3$)

≪Ph.≫ Mass[27] CD[6]
≪OP.≫ Stephania glabra[24], S. rotunda[25]
≪C. & So.≫
≪Sa. & D.≫ B–NMe=pronuciferine B–NAc: 234~235°
≪SD.≫ 6), 24), 26)
≪S. & B.≫

Glaziovine $C_{18}H_{19}O_3N \equiv 297.34$

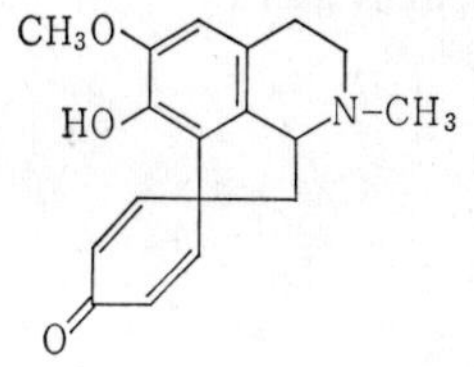

≪mp≫ 235~237° (d) (Et_2O, Et_2O-MeOH, or EtOAc)
≪$[\alpha]_D$≫ +7° ($CHCl_3$)

≪Ph.≫ UV[22] IR[22] NMR[22] Mass[22,27] Rf[1]
≪OP.≫ Ochotea glaziovii[22]
≪C. & So.≫ Colourless needles
≪Sa. & D.≫ B–Pic: 199~203°
Tetrahydro-Der.: 112~116°
≪SD.≫ 22)
≪S. & B.≫

Homolinearsine $C_{18}H_{19}O_3N \equiv 297.34$

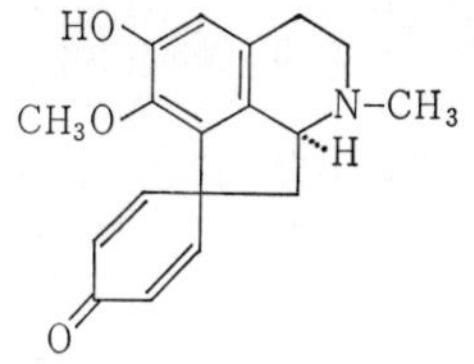

≪mp≫ 218~220° (d) (EtOH)
≪$[\alpha]_D$≫ −116.5° (MeOH)[1], +84.4 (MeOH)[2]

≪Ph.≫ CR[2,8] UV[2] IR[1,2] CD[6] Rf[2] pKa[2]
≪OP.≫ Croton linearis[2]
≪C. & So.≫ Plates (anisotropic)
≪Sa. & D.≫ B–HCl: >300°
B–$HClO_4$: 184~186°
≪SD.≫ 1), 4), 5), 6)
≪S. & B.≫

Pronuciferine (Base A)[1,5,28] $C_{19}H_{21}O_3N \equiv 311.37$

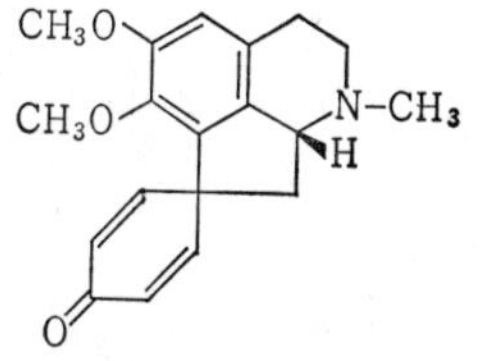

≪mp≫ 127~129°, cf. 148~151°[34]
≪$[\alpha]_D$≫ +99° ($CHCl_3$), +105.8° (MeOH), +111.2° (EtOH)

≪Ph.≫ CR[8] UV[2,28,34] IR[2,24,28,34] NMR[28,34] Mass[27] CD[6] Rf[1] pKa[2,28]
≪OP.≫ Croton linearis[2,25,29], Nelumbo nucifera[28,29,30,31,32], P. caucasicum[10,29], P. fugax[10,14], P. persicum[10,14], P. polycaetum[14], P. triniaefolium[14], Stephania glabra[24,29]
≪C. & So.≫
≪Sa. & D.≫ B–HCl: 201~204° (d) B–HBr: 224~227° (d) B–HI: 225~227° (d) B–Oxal: 170~172° B–MeI: 244~245° (d) Oxime·HCl: >210° (d)
Hexahydro-Der.HCl: I 253~254°, II 205~208°
≪SD.≫ 1), 3), 6), 24)
≪S. & B.≫ 34), 35)

Orientalinone $C_{19}H_{21}O_4N \equiv 327.37$

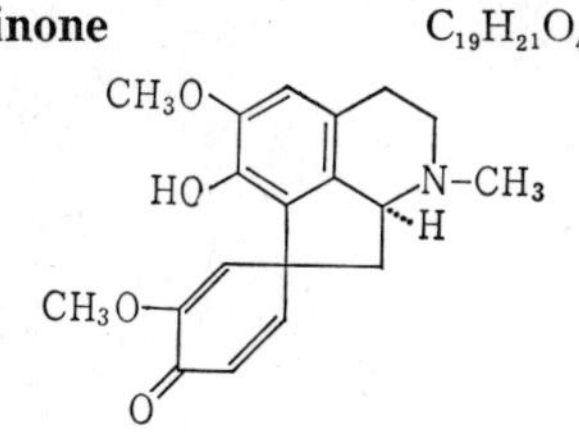

≪mp≫ 230~232° (d) (Et_2O)[36], cf. 183~184° (d) ($CHCl_3$–C_6H_6)[37]
≪$[\alpha]_D$≫ −76±10° ($CHCl_3$) or −62.6° ($CHCl_3$)[39]

≪Ph.≫ UV[36] IR[37,38,39] NMR[36,37,38,39] Mass[36,37,39] Rf[36]
≪OP.≫ Papaver bracteatum[36], P. orientale[37]
≪C. & So.≫ Sol. in MeOH, Spar. sol. in Et_2O
≪Sa. & D.≫
≪SD.≫ 36), 37), 38), 39)
≪S. & B.≫ 38), 39), BS: 37), 38), 39), BG: 36)

Acetylstepharine $C_{20}H_{21}O_4N \equiv 339.38$

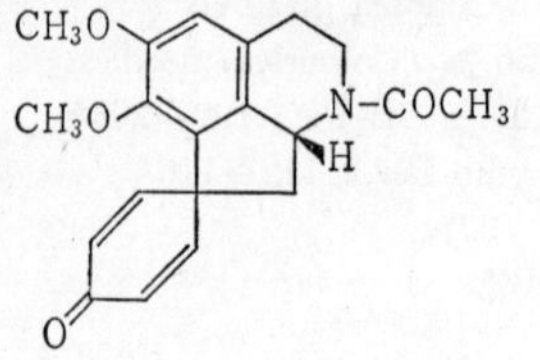

≪Ph.≫
≪OP.≫ Stephania glabra[24]
≪C. & So.≫
≪Sa. & D.≫
≪SD.≫ 24)
≪S. & B.≫

≪mp≫ 234～235°
≪[α]D≫ −80° ($CHCl_3$)

Linearsine $C_{18}H_{21}O_3N \equiv 323.38$

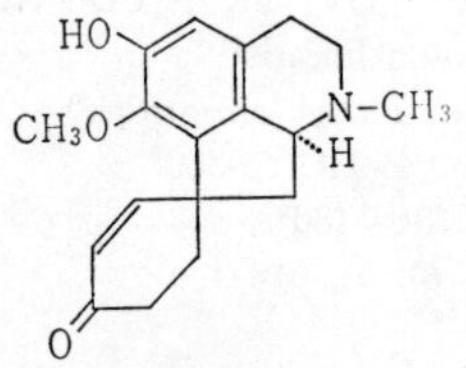

≪Ph.≫ CR[2] UV[1,2] IR[1,2] NMR[1] Rf[2] pKa[2]
≪OP.≫ Croton linearis[2]
≪C. & So.≫ Rods (anisotropic)
≪Sa. & D.≫ B–HCl: ＞300°(d) Dihydro-Der.: $[\alpha]_D$ −60.6°
≪SD.≫ 1), 3), cf. 4), 5)
≪S. & B.≫

≪mp≫ 219～220° (EtOH)
≪[α]D≫ +116° (MeOH)

Dihydroorientalinone $C_{19}H_{23}O_4N \equiv 329.38$

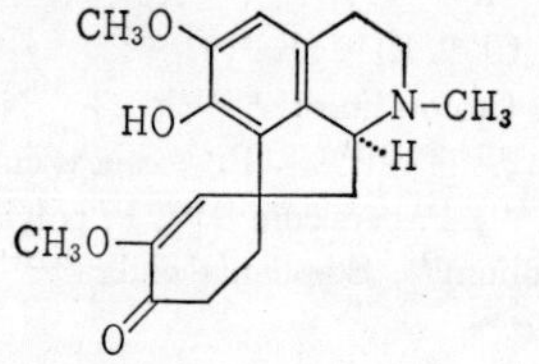

≪Ph.≫ UV[37] UMR[37] Mass[37]
≪OP.≫ Papaver orientale[37]
≪C. & So.≫
≪Sa. & D.≫
≪SD.≫ 37)
≪S. & B.≫

≪mp≫
≪[α]D≫ +50° ($CHCl_3$)

No Name $C_{18}H_{23}O_3N \equiv 301.37$

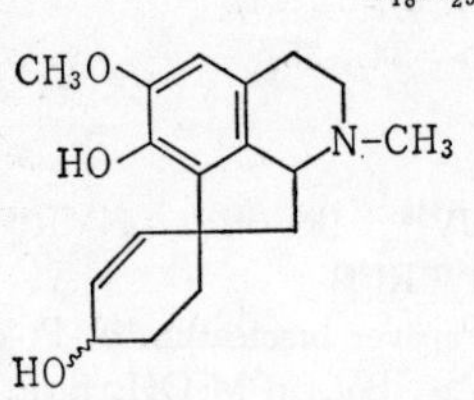

≪Ph.≫ UV[40] IR[40] NMR[40] Rf[40]
≪OP.≫ Croton linearis[40]
≪C. & So.≫
≪Sa. & D.≫ B–O,O–DiAc: 190～193°
≪SD.≫ 40)
≪S. & B.≫

≪mp≫
≪[α]D≫ B–O,O–DiAc: −18° (MeOH)

Litsericine $C_{17}H_{21}O_3N \equiv 287.35$

≪mp≫ 157～158° (C_6H_6)

≪$[\alpha]_D$≫ +67° (EtOH)
≪Ph.≫ UV[42)43)] IR[41)42)43)] NMR[43)] ORD[45)] Rf[41)]
≪OP.≫ Neolitsea sericea[41)42)]
≪C. & So.≫ Colourless needles
≪Sa. & D.≫ B–HCl: 265～266° B–HBr: 267° B–NMe: 185° B–NMe·MeI: 173～175° B–NO–DiAc: 157°
≪SD.≫ 43), 45)
≪S. & B.≫

Amurine[33),44)] → Chap. 15 Morphine and its related alkaloids.
≪Ph.≫
≪OP.≫
≪C. & So.≫
≪Sa. & D.≫
≪SD.≫
≪S. & B.≫

Amuronine $C_{19}H_{23}O_3N \equiv 313.38$

≪mp≫ 119～120° (ligroin) or 131～132°

≪$[\alpha]_D$≫ +124° ($CHCl_3$)
≪Ph.≫ IR[46)] CD[47)] pKa[47)]
≪OP.≫ Papaver nudicaule var. amurense[46)49)]
≪C. & So.≫
≪Sa. & D.≫ B–HI: 249～250° B–$HClO_4$: 252～254° B–MeI: 235～237° Dihydro-Der.: 125～126°
≪SD.≫ 46), 47)
≪S. & B.≫

Amuroline $C_{19}H_{25}O_3N \equiv 315.40$

≪mp≫ 169～170° (Me_2CO)

≪$[\alpha]_D$≫ +106° ($CHCl_3$)
≪Ph.≫ pKa[47)]
≪OP.≫ Papaver nudicaule var. amurense[46)49)]
≪C. & So.≫
≪Sa. & D.≫ B–$HClO_4$: 151～153° B–OAc: 153～154°
≪SD.≫ 46), 47)
≪S. & B.≫

---

1) L. J. Haynes, K. L. Stuart, D. H. R. Barton, G. W. Kirby: J. Chem. Soc. (C), **1966**, 1676.
2) L. J. Haynes, K. L. Stuart: J. Chem. Soc., **1963**, 1784.
3) L. J. Haynes, K. L. Stuart: J. Chem. Soc., **1963**, 1789.
4) L. J. Haynes, K. L. Stuart, D. H. R. Barton, G. W. Kirby: Proc. Chem. Soc., **1964**, 261.
5) L. J. Haynes, K. L. Stuart, D. H. R. Barton, G. W. Kirby: Proc. Chem. Soc., **1964**, 280.
6) G. Snatzke, G. Wollenberg: J. Chem. Soc. (C), **1966**, 1681.
7) L. J. Haynes, K. L. Stuart, D. H. R. Barton, D. S. Bhakuni, G. W. Kirby: Chem. Comm., **1965**, 141.
8) L. Kuhn, S. Pfeifer: Pharmazie, **20**, 559 (1965).
9) D. H. R. Barton, T. Cohen: "Festshrift Arthur Stoll", p. 117, Birkhauser, Basel (1957).
10) L. Kuhn, S. Pfeifer, J. Slavík, J. Appelt: Naturwiss., **51**, 556 (1964).
11) A. W. Sangster, K. L. Stuart: Chem. Revs., **65**, 69 (1965).
12) J. Slavík: Coll. Czech. Chem. Comm., **25**, 1663 (1960).
13) J. Slavík, L. Slaviková: Coll. Czech. Chem. Comm., **28**, 1720 (1963).
14) L. Kuhn, S. Pfeifer: Pharmazie, **20**, 520 (1965).

15) J. Slavík: Coll. Czech. Chem. Comm., **28**, 1738 (1963).
16) J. Slavík: Coll. Czech. Chem. Comm., **29**, 1314 (1964).
17) M. Shamma, W. A. Slusarchyk: Chem. Revs., **64**, 59 (1964).
18) S. Yu. Yunusov, V. A. Mnatsakanyan, S. T. Akramow: Dokl. Akad. Nauk Uz. SSR., **1961**, 43.
19) V. A. Mnatsakanyan, S. Yu. Yunusov: Dokl. Akad. Nauk Uz. SSR., **1961**, 36.
20) S. Yu. Yunusov, V. A. Mnatsakanyan, S. T. Akramow: Dokl. Akad. Nauk Uz. SSR., **1965**, 502.
21) J. Slavíc: Coll. Czech. Chem. Comm., **30**, 914 (1965).
22) B. Gilbert, M. E. A. Gilbert, M. M. DeOriveira, O. Ribeiro, E. Wenkert, B. Wickberg, U. Hollstein, H. Rapoport: J. Am. Chem. Soc., **86**, 694 (1964).
23) I. R. C. Bick: Experientia, **20**, 362 (1964).
24) M. P. Cava, K. Nomura, R. H. Schlessinger, K. T. Buck, B. Douglas, R. F. Raffauf, J. Weisback: Chem. & Ind., **1964**, 282.
25) M. Tomita, M. Kozuka, S. Uyeno: J. Pharm. Soc. Japan, **86**, 460 (1966).
26) M. Tomita, M. Kozuka: J. Pharm. Soc. Japan, **86**, 871 (1966).
27) M. Tomita, A. Kato, T. Ibuka, H. Furukawa, M. Kozuka: Tetrahedron Letters, **1965**, 2825.
28) K. Bernauer: Helv. Chim. Acta, **46**, 1783 (1963).
29) S. Pfeifer, L. Kuhn: Pharmazie, **20**, 394 (1965).
30) K. Bernauer: Helv. Chim. Acta, **47**, 2119 (1964).
31) K. Bernauer: Helv. Chim. Acta, **47**, 2122 (1964).
32) K. Bernauer: Chimia, **17**, 392 (1963).
33) M. Maturova, L. Hruban, F. Santavy, W. Wiegrebe: Arch. Pharm., **298**, 209 (1965).
34) K. Bernauer: Experientia, **20**, 380 (1964).
35) K. Bernauer: Neth. Appl., 6,507,700 [C.A., **64**, 19707 (1966)].
36) K. Heydenreich, S. Pfeifer: Pharmazie, **21**, 12 (1966).
37) A. R. Battersby, T. H. Brown: Chem. Comm., **1966**, 170.
38) A. R. Battersby, T. H. Brown: Proc. Chem. Soc., **1964**, 85.
39) A. R. Battersby, J. H. Brown, J. H. Clements: J. Chem. Soc., **1965**, 4550.
40) L. J. Haynes, G. E. M. Husbans, K. L. Stuart: J. Chem. Soc. (C), **1966**, 1680.
41) T. Nakasato, S. Nomura: J. Pharm. Soc. Japan, **79**, 1267 (1959).
42) T. Nakasato, S. Asada, Y. Koezuka: J. Pharm. Soc. Japan, **86**, 129 (1966).
43) T. Nakasato, S. Asada: J. Pharm. Soc. Japan, **86**, 134 (1966).
44) H. Flentje, W. Doepke, P. W. Jeffs: Naturwiss., **52**, 259 (1965).
45) T. Nakasato, S. Asada: J. Pharm. Soc. Japan, **86**, 1205 (1966).
46) H. G. Boit, H. Flentje: Naturwiss., **46**, 514 (1959).
47) H. Flentje, W. Doepke, P. W. Jeffs: Pharmazie, **21**, 379 (1966).
48) J. Slavik: Coll. Czech. Chem. Comm., **31**, 4184 (1966).
49) H. G. Boit, H. Flentje: Naturwiss., **47**, 180 (1960).

Chapter 9 Aporphine Alkaloids

Anonaine $C_{17}H_{15}O_2N \equiv 265.30$

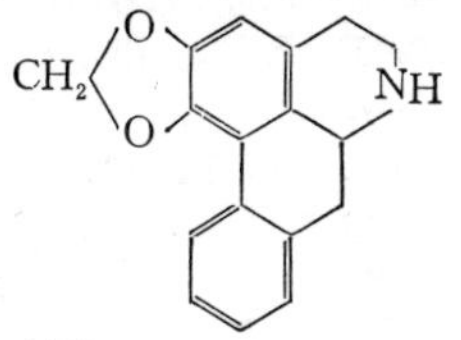

≪mp≫ 122～123°
≪[α]$_D$≫ −52° ($CHCl_3$), −73.8±1° (EtOH)

≪Ph.≫
≪OP.≫ Anona reticulata[1)2)3)], A. squamosa[1)2)4)], Nelumbo nucifera[5)], Roemeria refracta[6)]
≪C. & So.≫
≪Sa. & D.≫ B–HCl: 277° (287°～288°)
B–H_2PtCl_6: 248.5° B–NMe=roemerine
B–NAc: 230°
≪SD.≫ 1), 3), 7)
≪S. & B.≫ 7), 8), 9), BS: 10), BG: 11), 12), 13)

Caaverine $C_{17}H_{17}O_2N \equiv 267.31$

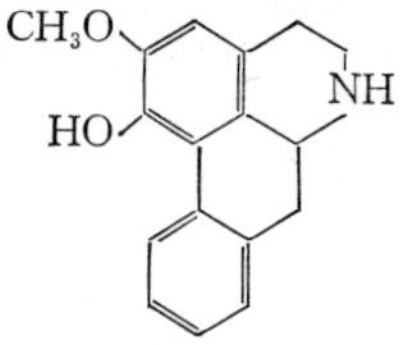

≪mp≫ 208～210° (d) (C_6H_6)

≪[α]$_D$≫ −89° (MeOH)
≪Ph.≫ CR[14)] UV[14)15)] IR[14)] NMR[14)] ORD[14)] Rf[14)]
≪OP.≫ Symplocos celastrinea[14)]
≪C. & So.≫ Sensitive to light
≪Sa. & D.≫ B–urethan: 245～246°
B–O,N–DiAc: 236～238°
≪SD.≫ 14)
≪S. & B.≫ 14)

Asimilobine $C_{17}H_{17}O_2N \equiv 267.31$

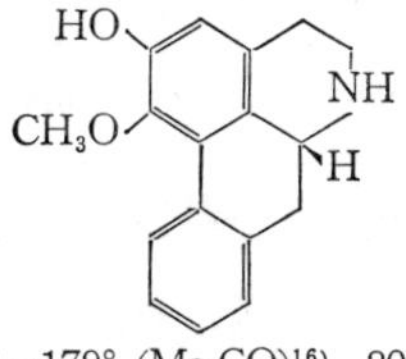

≪mp≫ 177～179° (Me_2CO)[16)], 201～202° (EtOH)[5)]

≪[α]$_D$≫ −213° ($CHCl_3$)
≪Ph.≫ UV[16)] NMR[16)]
≪OP.≫ Asimia triloba[16)], Nelumbo nucifera[5)]
≪C. & So.≫
≪Sa. & D.≫ B–NMe: 191～193°=nor-nuciferine B–O, N–DiAc: 146°
≪SD.≫ 5), 16)
≪S. & B.≫

***l-N*-Nornuciferine** $C_{18}H_{19}O_2N \equiv 281.34$

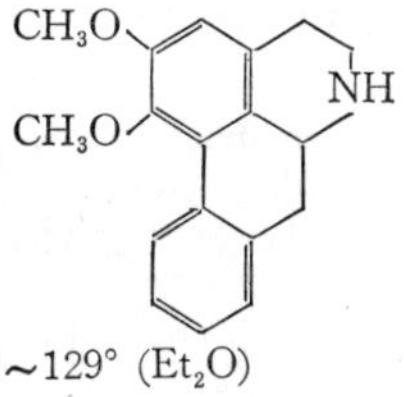

≪mp≫ 128～129° (Et_2O)

≪[α]$_D$≫ −145° (EtOH)
≪Ph.≫ UV[17)20)] Mass[22)]
≪OP.≫ Nelumbo lutea[17)18)]
≪C. & So.≫ Small needles
≪Sa. & D.≫ B–NMe=nuciferine
≪SD.≫ 19)
≪S. & B.≫ 20), 21)

Aporpheine $C_{18}H_{17}O_2N \equiv 279.32$

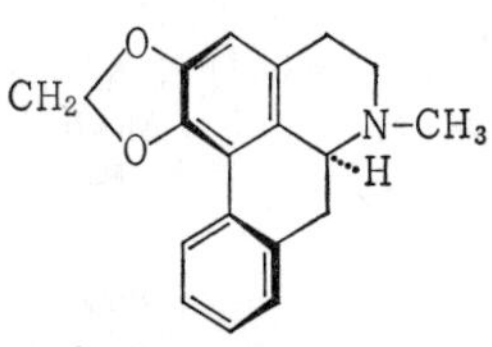

≪mp≫ 102～103° (Et_2O)
≪[α]$_D$≫ +80±3° (EtOH)

≪Ph.≫ CR[23)] UV[23)] Rf[23)] pKa[23)]
≪OP.≫ Papaver dubium[23)], P. fugax[24)]
≪C. & So.≫
≪Sa. & D.≫ B–HCl: 266～267°
B–HBr: 275～276° B–HI: 264～265°
B–MeI: 225～226°
≪SD.≫ 23)
≪S. & B.≫

Roemerine (Remerine) $C_{18}H_{17}O_2N \equiv 279.32$

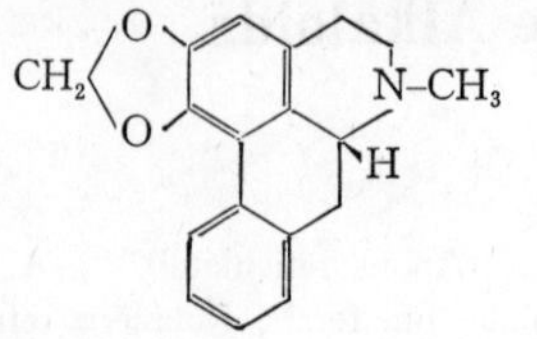

CH2 O O N-CH3 H

≪mp≫ 102~103°
≪$[\alpha]_D$≫ −72.5° (EtOH)

≪Ph.≫ UV[28)33)34)61)] IR[18)] NMR[35)] ORD[61)]
≪OP.≫ Cryptocarya angulata[25)], C. triplinervis[25)], Nelumbo nucifera[5)8)26)], Neolitsea (Laurus) sericea[27)28)], Roemeria refracta[29)30)]
≪C. & So.≫ Colourless needles
≪Sa. & D.≫ B–HCl: 271~272° B–HI: 245° B–$HClO_4$: 245° B–Pic: 195~196° B–MeI: 215~216°
≪SD.≫ 7), 25), 28), 30), 31)
≪S. & B.≫ 7), 8), 32)

1-Methoxy-2-hydroxyaporphine
$C_{18}H_{19}O_2N \equiv 281.34$

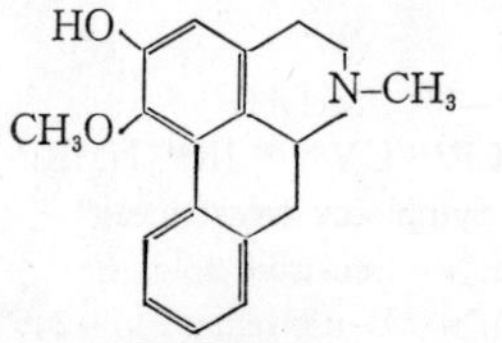

HO CH3O N-CH3

≪mp≫ 195~196° (EtOH)
≪$[\alpha]_D$≫ −265° ($CHCl_3$)

≪Ph.≫ IR[38)] Mass[22)]
≪OP.≫ Nelumbo nucifera[18)26)36)37)38)], N. nucifera var. prolifera[39)]
≪C. & So.≫ Colourless prisms
≪Sa. & D.≫ B–Oxal: 190~191° B–OMe=nuciferine B–OEt: 156~156.5° B–OEt·MeI: 192~193°
≪SD.≫ 19), 26), 36)
≪S. & B.≫ 26)

Nuciferine $C_{19}H_{21}O_2N \equiv 295.37$

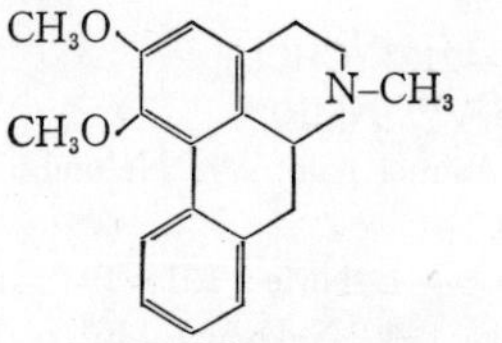

CH3O CH3O N-CH3

≪mp≫ 166~168° (EtOH)
≪$[\alpha]_D$≫ −157.5° (EtOH), cf. −203.2° (EtOH)[18)]

≪Ph.≫ CR[41)] UV[17)41)46)61)360)] IR[17)18)41)] ORD[46)61)] NMR[355)]
≪OP.≫ Nelumbo lutea[17)40)], N. nucifera (speciosum)[5)18)26)41)], N. nucifera var. prolifera[39)], Lysichiton camtschatcense var. japonicum[360)]
≪C. & So.≫
≪Sa. & D.≫ B–HCl: 241° (vacuum) B–MeI: 177~178°
≪SD.≫ 19), 41), 42), cf. 30)
≪S. & B.≫ 20), 21), 43), 44), 45), BG: 13)

Stephanine $C_{19}H_{19}O_3N \equiv 309.35$

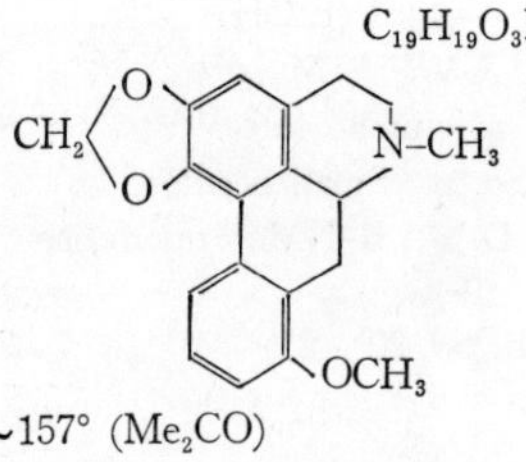

CH2 O O N-CH3 OCH3

≪mp≫ 155~157° (Me_2CO)

≪$[\alpha]_D$≫ −92.5° ($CHCl_3$)
≪Ph.≫ CR[40)] UV[34)47)48)52)54)55)] IR[52)] Mass[22)] Rf[52)]
≪OP.≫ Stephania capitata[47)48)], S. japonica[49)50)]
≪C. & So.≫ Colourless needles
≪Sa. & D.≫ B–HCl: 280° B–MeI: 212~214°
≪SD.≫ 47), 48), 51)
≪S. & B.≫ 52), 53), cf. 54), BG: 12)

Anolobine (Analobine) $C_{17}H_{15}O_3N \equiv 281.30$

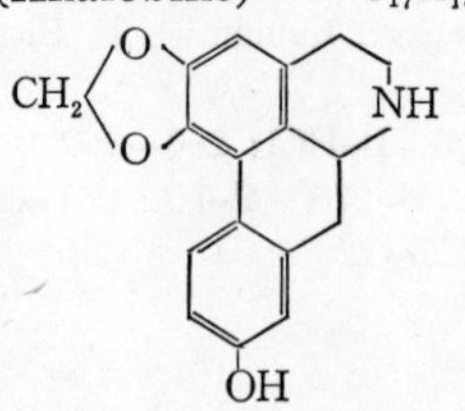

CH2 O O NH OH

≪mp≫ 262° ($CHCl_3$–MeOH)

≪$[\alpha]_D$≫ −22.5° ($CHCl_3$–MeOH)
≪Ph.≫ UV[46)61)] Mass[22)] ORD[46)62)]
≪OP.≫ Asimia triloba[16)56)], Xylopia discreta[57)]
≪C. & So.≫ Spar. sol. in MeOH
≪Sa. & D.≫ B–HCl: 262° B–OMe=xylopine B–N,O–DiMe: 266° (d)
≪SD.≫ 42), 56), 58), 59), 60)
≪S. & B.≫ cf. 58), 60)

Xylopine $C_{18}H_{17}O_3N \equiv 295.32$

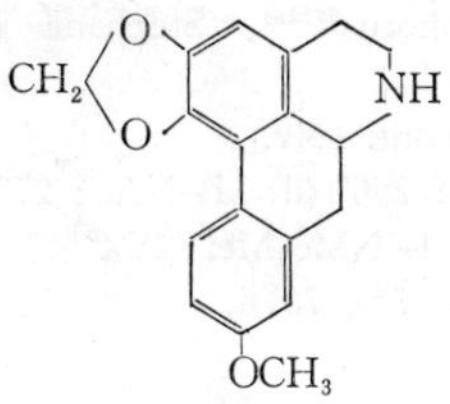

≪mp≫ 78～102° (Me_2CO–Et_2O)
≪$[\alpha]_D$≫ −23.4° (MeOH)

≪Ph.≫ UV[57] CR[57] ORD[62]
≪OP.≫ Xylopia (Unona) discreta[57][63]
≪C. & So.≫
≪Sa. & D.≫ B–HCl: >250°
B–Tart: 190～192° B–MeI: 232°
B–NMe= isolaureline B–NAc: 213～214°
≪SD.≫ 57)
≪S. & B.≫ cf. 64)

Mecambroline (Isofungipavine) $C_{18}H_{17}O_3N \equiv 295.32$

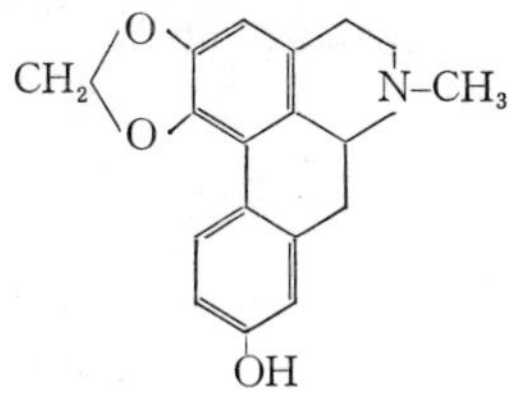

≪mp≫ 145° (Et_2O), 225～253° ($CHCl_3$)
≪$[\alpha]_D$≫ +76±3° ($CHCl_3$)

≪Ph.≫ CR[65] UV[65] IR[65] pKa[65]
≪OP.≫ Meconopsis cambrica[65]
≪C. & So.≫ Colourless needles or prisms
≪Sa. & D.≫ B–HCl: 264～266°
B–Pic: 179～180° B–OMe: 112°
≪SD.≫ 65)
≪S. & B.≫

Isolaureline $C_{19}H_{19}O_3N \equiv 309.34$

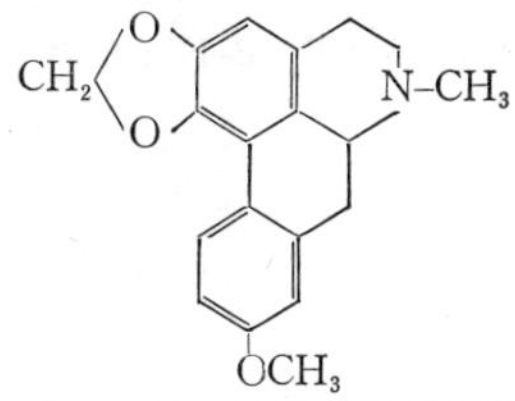

≪mp≫ 108～110° (Me_2CO–Et_2O)
≪$[\alpha]_D$≫ −36.7° (EtOH)

≪Ph.≫
≪OP.≫ Xylopia discreta[57]
≪C. & So.≫
≪Sa. & D.≫ B–HCl: 243～246°
B–MeI: 229～232°
≪SD.≫ 57)
≪S. & B.≫ 60), 64), 66)

Sparsiflorine $C_{17}H_{17}O_3N \equiv 283.31$

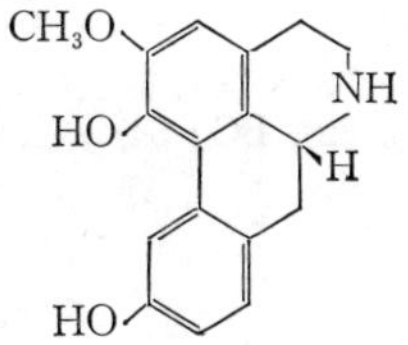

≪mp≫ 230～232° (d)
≪$[\alpha]_D$≫ B–HCl +43° (H_2O)

≪Ph.≫ CR[68] UV[68] IR[68] NMR[68] Mass[68]
≪OP.≫ Croton sparsiflorus[67][68]
≪C. & So.≫ Silky needles
≪Sa. & D.≫ B–MeI: 218° (d)
B–NMe·MeI: 236～238° B–O,O,N–TriMe= N,O–DiMe-tuduranine B–O,N–DiAc: 245°
B–O,O,N–TriAc: 196～197°
≪SD.≫ 68)
≪S. & B.≫

Tuduranine $C_{18}H_{19}O_3N \equiv 297.34$

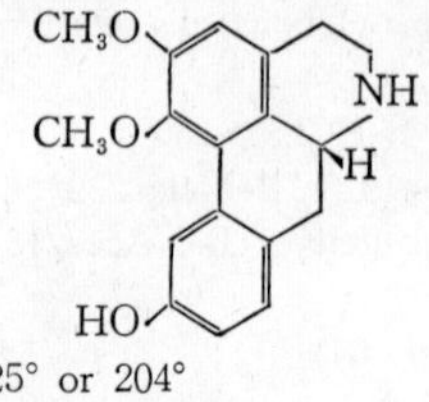

CH_3O NH CH_3O H HO

≪mp≫ 125° or 204°

≪$[\alpha]_D$≫ −127.5° (EtOH)

≪Ph.≫ UV[70] IR[70] NMR[68,70,74,75] Rf[70]

≪OP.≫ Croton sparsiflorus[67,68], Stephania acutum[69], S. japonica[70]

≪C. & So.≫ Sol. in org. solv.

≪Sa. & D.≫ B–HCl: 286° (d) B–NAc: 277° B–O,N–DiAc: 170° B–NMe·MeI: 224°

≪SD.≫ 68), 69), 71), 72), 73)

≪S. & B.≫ 71)

Isoroemerine $C_{18}H_{17}O_3N \equiv 295.32$

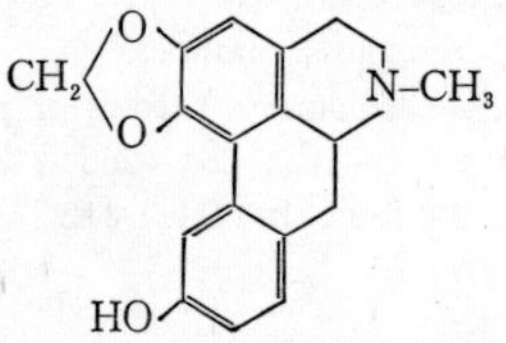

O CH_2 O N–CH_3 HO

≪mp≫ 87∼88°

≪$[\alpha]_D$≫ +69.9° (EtOH)

≪Ph.≫ UV[24,76] Rf[24]

≪OP.≫ Papaver fugax[24,76]

≪C. & So.≫

≪Sa. & D.≫ B–HCl: 263∼265° (d) B–MeI 221∼222° (d)

≪SD.≫ 24), 76)

≪S. & B.≫

No Name $C_{18}H_{19}O_3N \equiv 297.34$

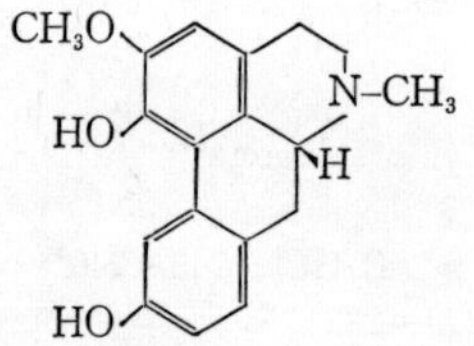

CH_3O N–CH_3 HO H HO

≪mp≫ 149∼152° (d)

≪$[\alpha]_D$≫ −35° ($CHCl_3$)

≪Ph.≫ UV[15] IR[15] NMR[15]

≪OP.≫ Ocotea glaziovii[15,66]

≪C. & So.≫ Colourless cryst, Suffers readily air oxidn.

≪Sa. & D.≫ B–HCl: >300° B–HI: 277° (d) B–MeI: 251∼253°

≪SD.≫ 15), 66)

≪S. & B.≫ 77)

Laureline $C_{19}H_{19}O_3N \equiv 309.35$

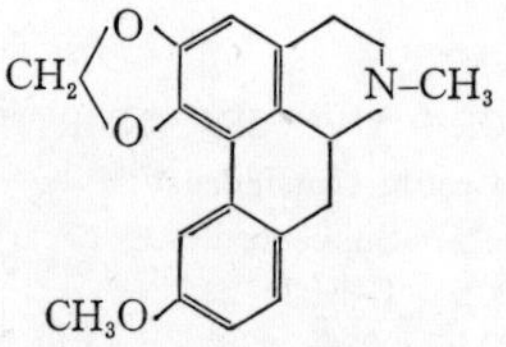

O CH_2 O N–CH_3 CH_3O

≪mp≫ 97° (Et_2O) or 114° (C_6H_{14})

≪$[\alpha]_D$≫ −98.5° (EtOH)

≪Ph.≫ CR[64,80] UV[61] ORD[61]

≪OP.≫ Laurelia novae-zelandiae[78,79]

≪C. & So.≫ Tablets

≪Sa. & D.≫ B–HCl: 280° B–HNO_3: 238∼240° B_2–H_2SO_4: 105° B–MeI: 223°

≪SD.≫ 79), 80)

≪S. & B.≫ 64), 66), 81), 82), BG; 64)

Michepressine $C_{19}H_{20}O_3N^{\oplus} \equiv 310.35$

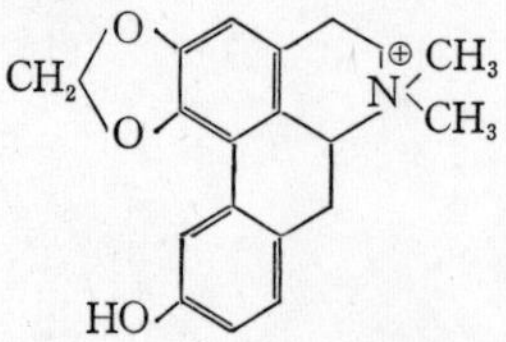

O CH_2 O $N^{\oplus}$ CH_3 CH_3 HO

≪mp≫

≪$[\alpha]_D$≫ $B^{\oplus}$–$I^{\ominus}$ −130.8° (MeOH)

≪Ph.≫ CR[83] UV[34,83] IR[83] Rf[83]

≪OP.≫ Michelia compressa[83]

≪C. & So.≫

≪Sa. & D.≫ $B^{\oplus}$–$I^{\ominus}$: 235∼236° (d)

≪SD.≫ 83)

≪S. & B.≫

Pukateine $C_{18}H_{17}O_3N \equiv 295.32$

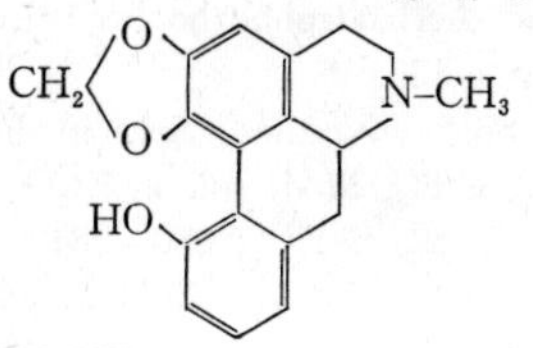

≪mp≫ 200° (EtOH)
≪$[\alpha]_D$≫ −220° (EtOH)

≪Ph.≫ CR[78)79)] UV[85)61)] ORD[61)]
≪OP.≫ Laurelia novae-zelandiae[78)]
≪C. & So.≫ Sol. in $CHCl_3$, Mod. sol. in EtOH, MeOH, Insol. in Et_2O
≪Sa. & D.≫ B-OMe: 137°
≪SD.≫ 78), 79)
≪S. & B.≫ cf. 84), BG: 64)

Isothebaine $C_{19}H_{21}O_3N \equiv 311.37$

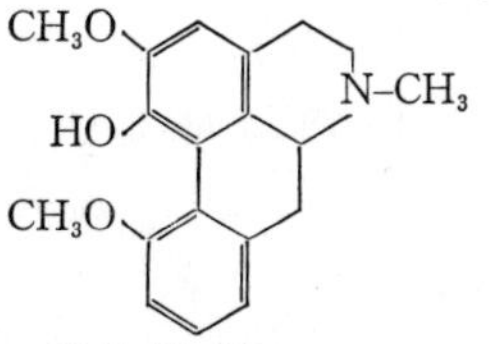

≪mp≫ 203～204° (Et_2O)
≪$[\alpha]_D$≫ +285.1° (EtOH)

≪Ph.≫ CR[91)] UV[61)95)96)101)103)] IR[95)96)98)101)] NMR[100)101)] Mass[101)] ORD[61)]
≪OP.≫ Papaver bracteatum[86)87)88)], P. lateritium[90)], P. orientale[89)90)91)92)]
≪C. & So.≫ Prisms, Sol. in $CHCl_3$, MeOH, EtOH
≪Sa. & D.≫ B-HCl: 213～214°
B-$HClO_4$: 255° B-MeI: 237°
B-OMe·MeI: 256° B-OAc: 143～145°
≪SD.≫ 90), 91), 92), 93), 94), 95), 96), 97)
≪S. & B.≫ 95), 96), 98), 99), 100), 101)
cf. 92), 93), 102)～104), BG: 12), 13)
BS: 89), 105), 106)

Artabotrinine $C_{18}H_{17}O_3N \equiv 295.32$

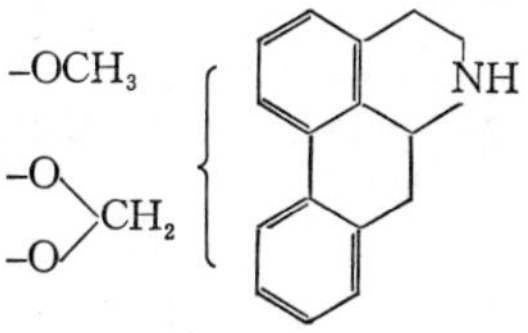

≪mp≫
≪$[\alpha]_D$≫ −19° ($CHCl_3$)

≪Ph.≫ CR[107)]
≪OP.≫ Artabotrys suaveolens[107)]
≪C. & So.≫ Amorph
≪Sa. & D.≫ B-HCl: 274° B-NMe: 133°
B-NMe·MeI: 224°
≪SD.≫ 57), 107)
≪S. & B.≫

Crebanine $C_{20}H_{21}O_4N \equiv 339.38$

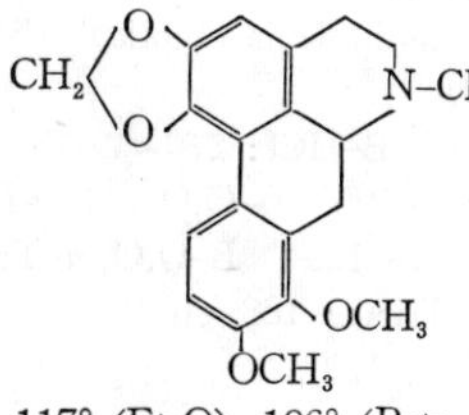

≪mp≫ 115～117° (Et_2O), 126° (Petr. ether)
≪$[\alpha]_D$≫ −57.5° ($CHCl_3$)

≪Ph.≫ CR[111)] UV[54)55)113)114)] IR[114)] Mass[22)] Rf[114)]
≪OP.≫ Stephania capitata[47)108)109)], S. sasakii[109)110)111)]
≪C. & So.≫ Colourless needles
≪Sa. & D.≫ B-HCl: 225° B-HBr: 220°
B-HI: 250° B-$HClO_4$: 265～266°
B-MeI: 220°
≪SD.≫ 47), 108), 112)
≪S. & B.≫ 54), 113), 114)

Actinodaphnine

$C_{18}H_{17}O_4N \equiv 311.32$

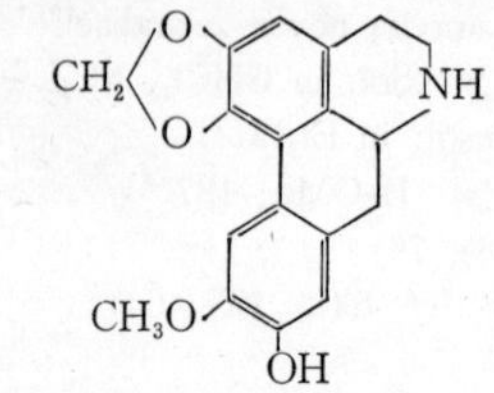

≪mp≫ 210~211° (EtOH), cf. 280°[116]
≪[α]$_D$≫ +32.8° (EtOH)

≪Ph.≫ CR[117] IR[116] UV[61] ORD[61]
≪OP.≫ Actinodaphne hookeri[115], Laurus nobilis[116]
≪C. & So.≫ Needles, Sol. in EtOH, Me_2CO, $CHCl_3$, C_6H_6, Mod. sol. in Et_2O, Insol. in H_2O
≪Sa. & D.≫ B–HCl: 280~281°
B–HI: 264~265° (d) B–MeI: 249~250° (d)
B–Pic: 220~222° (d) B–Ac: 229~230° (d)
B–NMe: 210~211°
≪SD.≫ 31), 115), 117)
≪S. & B.≫ 116), 118), 119)

No Name

$C_{18}H_{19}O_4N \equiv 313.34$

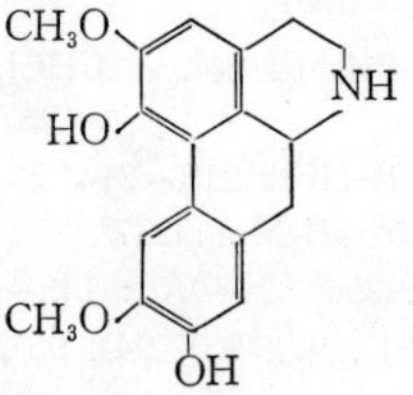

≪mp≫
≪[α]$_D$≫

≪Ph.≫
≪OP.≫ Beilschmiedia elliptica[120]
≪C. & So.≫
≪Sa. & D.≫
≪SD.≫ 120)
≪S. & B.≫

Laurolitsine

$C_{18}H_{19}O_4N \equiv 313.34$

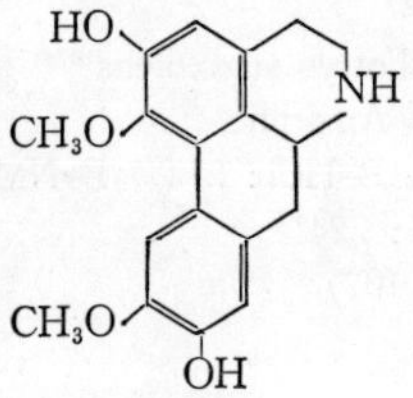

≪mp≫ 138~140° (C_6H_6)
≪[α]$_D$≫ +102.5° (EtOH)

≪Ph.≫ CR[28]
≪OP.≫ Hernandia catalpifolia[121], Litsea japonica[122], Neolitsea (Laurus) sericea[28][33]
≪C. & So.≫ Amorph
≪Sa. & D.≫ B–Pic: 212° (d) B–Picrol: 239° (d)
B–NMe=boldine
≪SD.≫ 28), 33)
≪S. & B.≫

Laurelliptine

$C_{18}H_{19}O_4N \equiv 313.34$

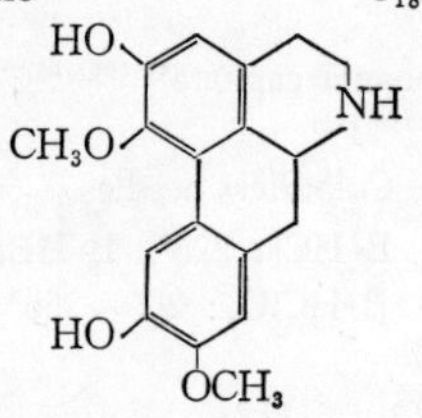

≪mp≫ 190~192°
≪[α]$_D$≫ +47° (EtOH)

≪Ph.≫ CR[123] UV[120][123] IR[123] NMR[120] Rf[123]
≪OP.≫ Beilschmiedia elliptica[120][123]
≪C. & So.≫
≪Sa. & D.≫ B–HCl: 270~280°
B–Pic: 194~195° B–O,O–DiMe·Pic: 135~137°
B–NMe: 121~123° B–O,O,N–TriMe=glaucine
B–O,O,N–TriAc: 188~189°
≪SD.≫ 120), 123)
≪S. & B.≫

Laurotetanine $C_{19}H_{21}O_4N \equiv 327.37$

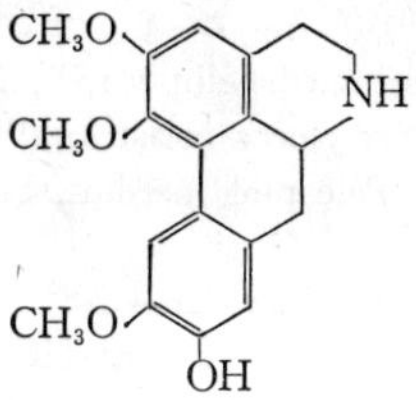

≪mp≫ 125° (Me_2CO) or 134° (Et_2O)
≪[α]D≫ +98.5° (EtOH)

≪Ph.≫ CR[124),135)] UV[135)] IR[135)]
≪OP.≫ Actinodaphne procera[124)], Hernandia ovigera[121)], Illigera pulchra[125),126)], Litsea cubeta[127)], L. crysocoma[125)], L. citrata[124),125)], Papaver somniferum[128)], Haasia squarrosa[124)], Litsea amara[124)], L. javanica[124)]
≪C. & So.≫
≪Sa. & D.≫ B–HCl: 245° B–Oxal: 233° B–Pic: 148～150° B–O,N–DiMe=glaucine
≪SD.≫ 31), 124), 127), 129), 130), 131), 132), 133)
≪S. & B.≫ 134), 135)

Neolitsine $C_{19}H_{17}O_4N \equiv 323.33$

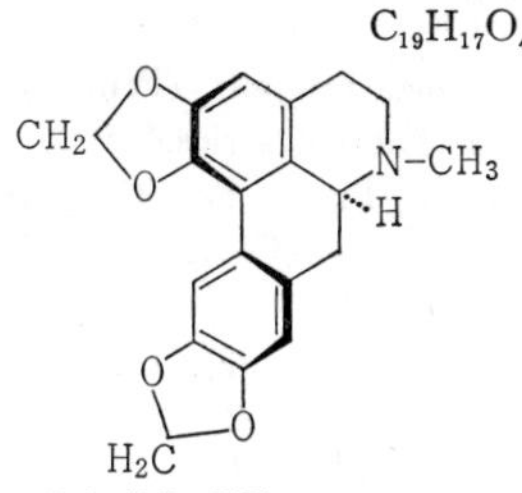

≪mp≫ 149～150° (Me_2CO)
≪[α]D≫ +56.5° ($CHCl_3$)

≪Ph.≫ CR[136)] UV[136)] IR[136)] NMR[136)] Mass[136)]
≪OP.≫ Neolitsea pulchella[136)]
≪C. & So.≫
≪Sa. & D.≫ B–HCl: 219° B–Pic: 186° (d) B–MeI: 221～224°
≪SD.≫ 136)
≪S. & B.≫

Phanostenine $C_{19}H_{19}O_4N \equiv 325.35$

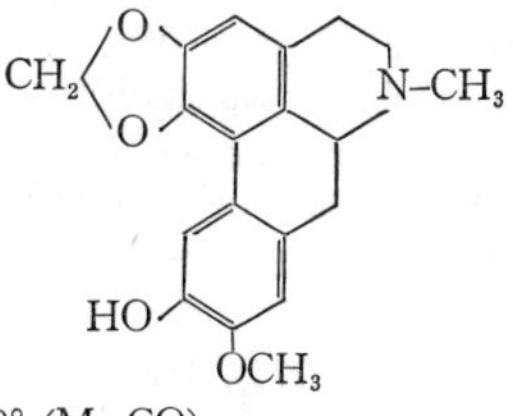

≪mp≫ 210° (Me_2CO)
≪[α]D≫ −36.7° ($CHCl_3$)

≪Ph.≫ CR[110),111)]
≪OP.≫ Stephania sasakii[109),110),111)], S. capitata[110),137)]
≪C. & So.≫ Colourless column
≪Sa. & D.≫ B–HCl: 264° B–OMe=*l*-dicentrine
≪SD.≫ 109), 110), 138)
≪S. & B.≫ 109), 118), 119), 137)

Domesticine $C_{19}H_{19}O_4N \equiv 325.35$

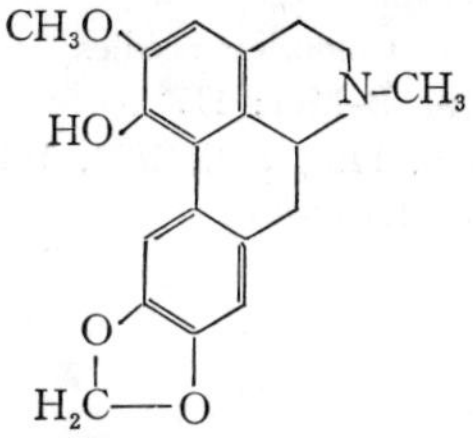

≪mp≫ 115～117° (Et_2O or aq. MeOH), 84～85° (MeOH–C_6H_6)

≪[α]D≫ +64 ±3° ($CHCl_3$)
≪Ph.≫ CR[139),149)] UV[139),140),141)] ORD[62)] Rf[139)]
≪OP.≫ Glaucium oxylobum[139)], Nandina domestica[140)～143)]
≪C. & So.≫
≪Sa. & D.≫ B–OMe=nantenine B–OEt: 131°
≪SD.≫ 140), 141), 144), 145), 146), 147), 148)
≪S. & B.≫ 141), 144), 146)

Isoboldine $C_{19}H_{21}O_4N \equiv 327.37$

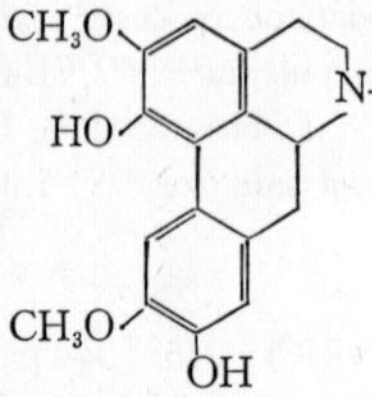

≪mp≫ 97～98° (C_6H_6), 121～123° (EtOH), 118～120° ($CHCl_3$)
≪[α]$_D$≫ +65.3° ($CHCl_3$)

≪Ph.≫ CR[143] UV[152] IR[14)152] NMR[151)152)357] Mass[356] Rf[14]
≪OP.≫ Beilschmiedia elliptica[120], Nandina domestica[143)150], Symplocos celastrinea[14]
≪C. & So.≫ Pale pink needles, Sol. in $CHCl_3$, AcOEt, Py
≪Sa. & D.≫ B–MeI: 226～228°
B–OMe·HI: 242° (d) B–OEt·$HClO_4$: 196～198°
≪SD.≫ 14), 120), 123), 143), 150)
≪S. & B.≫ 151), 152), 153)

Boldine $C_{19}H_{21}O_4N \equiv 327.37$

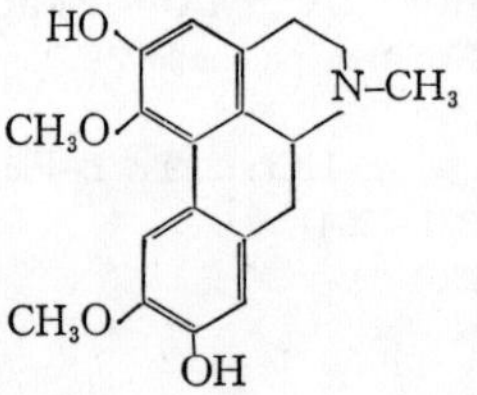

≪mp≫ 161～163° (C_6H_6)
≪[α]$_D$≫ +73° ($CHCl_3$–EtOH), +112° (EtOH)

≪Ph.≫ UV[46)61)156)160] NMR[355)357] Mass[22)356] ORD[46)61)62]
≪OP.≫ Bolda fragrans (Peumus boldus)[154)～156], Neolitsea (Laurus) sericea[27)28]
≪C. & So.≫ Plates, Sensitive to light
≪Sa. & D.≫ B–O,O–DiMe=*d*-glaucine
B–O,O–DiBz: 127°
≪SD.≫ 31), 126), 133), 157), 158), 159)
≪S. & B.≫ 152), 159)

Dicentrine (Eximine)[161] $C_{20}H_{21}O_4N \equiv 339.38$

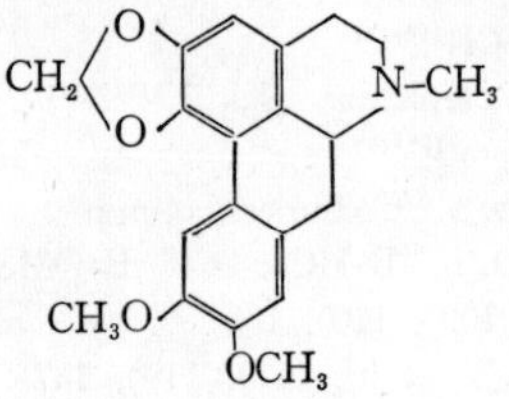

≪mp≫ 169° (EtOH or Et_2O)
≪[α]$_D$≫ +52° (EtOH), +62° ($CHCl_3$)

≪Ph.≫ CR[47)166)174] UV[61] NMR[74)175)355] ORD[61)62)141)214] Mass[356]
≪OP.≫ Dicentra eximia[80)161)162], D. formosa[163], D. oregana[164], D. pusilla[165)166], Ocotea leucoxylon[167], Stephania capitata[47)48]
≪C. & So.≫ Colourless needles
≪Sa. & D.≫ B–Pic: 180° B–MeI: 224°
≪SD.≫ 31), 133), 162), 168), 169), 170), 171), 172)
≪S. & B.≫ 170), 171), 173)

Nantenine (Domestine) $C_{20}H_{21}O_4N \equiv 339.38$

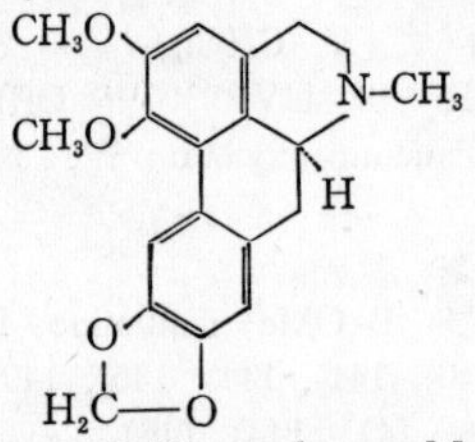

≪mp≫ 138～139° (Petr. ether or MeOH)
≪[α]$_D$≫ +111° ($CHCl_3$)

≪Ph.≫ CR[149)176)177] UV[28)34)46] Mass[22] ORD[46)62] CD[359]
≪OP.≫ Nandina domestica[140)149)176)～178]
≪C. & So.≫ Colourless needles
≪Sa. & D.≫ B–MeI: 117°
≪SD.≫ 140), 172), 179), 180), 181), 182), 183)
≪S. & B.≫ 180)

Rogersine $C_{20}H_{23}O_4N \equiv 341.39$

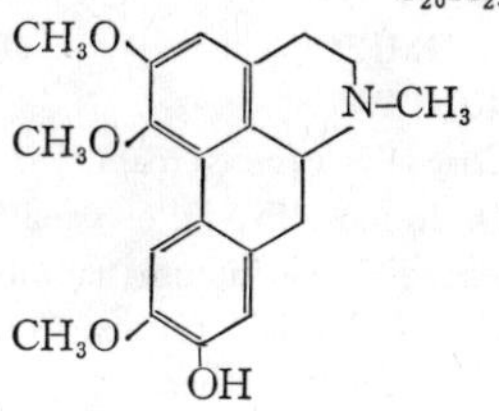

CH_3O CH_3O N–CH_3 CH_3O OH

≪mp≫ 100～105°
≪[α]$_D$≫ +111° (MeOH)

≪Ph.≫ UV[61,137,160,185] IR[137,185] NMR[137,185,355] ORD[61,62] X-ray[185] Rf[137]
≪OP.≫ Hernandia catalpifolia[121], Litsea citrata[184], Peumus boldus[156,160], Phylica(Papaver) rogersii[137], Symplocos celastrinea[14]
≪C. & So.≫ cf. bp 205～215°/0.01
2 forms { amorph. cryst.
≪Sa. & D.≫ B–HBr: 223～224° B–HI: 245°
B–MeI: 188° or 199～200°
B–OMe=glaucine
≪SD.≫ 132), 133), 137), 156), 160), 184), 185), 186)
≪S. & B.≫ 134), 135)

Glaucentrine $C_{20}H_{23}O_4N \equiv 341.39$

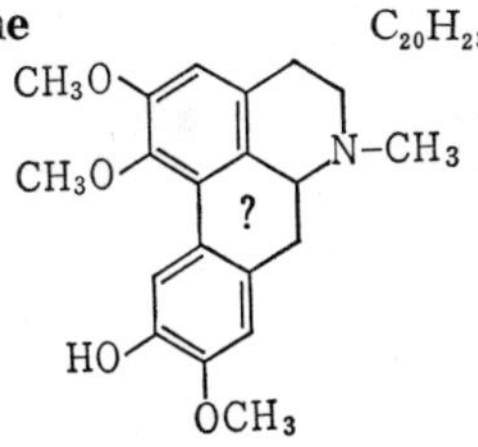

CH_3O CH_3O N–CH_3 ? HO OCH_3

≪mp≫ 148°
≪[α]$_D$≫

≪Ph.≫ CR[187] UV[189,190] IR[189] NMR[185,189,191] Mass[189]
≪OP.≫ Dicentra eximia[162], D. formosa[163], D. oregana[164]
≪C. & So.≫
≪Sa. & D.≫ B–HCl: 236～237° (d)
B–HBr: 239° B–MeI: 225°
B–OMe=glaucine B–OEt·MeI: 205～207°
≪SD.≫ 123), 133), 187), 188), 189)
≪S. & B.≫ cf. 187), 189)

No Name $C_{20}H_{23}O_4N \equiv 341.39$

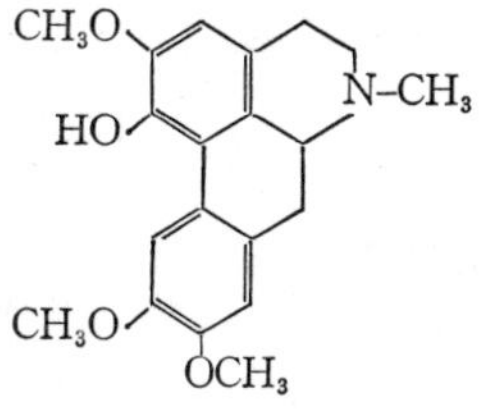

CH_3O HO N–CH_3 CH_3O OCH_3

≪mp≫
≪[α]$_D$≫

≪Ph.≫
≪OP.≫ Fagara tuiguassoiba[189]
≪C. & So.≫
≪Sa. & D.≫
≪SD.≫ 189), cf. 14)
≪S. & B.≫ 189)

Thalicmidine $C_{20}H_{23}O_4N \equiv 341.39$

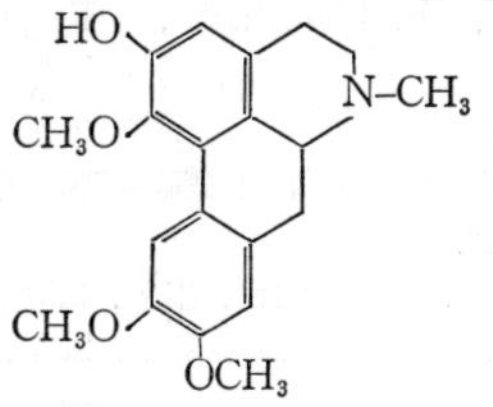

HO CH_3O N–CH_3 CH_3O OCH_3

≪mp≫ 192～193°
≪[α]$_D$≫ −84° (EtOH)

≪Ph.≫ CR[192] UV[61,136] ORD[61,195]
≪OP.≫ Thalictrum minus[192]
≪C. & So.≫
≪Sa. & D.≫ B–HI: 222～226° (d)
B–Tart: 240° B–MeI: 217～217.5°
≪SD.≫ 14), 123), 136), 192), 193), 194), 195), 196)
≪S. & B.≫ 136)

d-Glaucine

$C_{21}H_{25}O_4N \equiv 355.42$

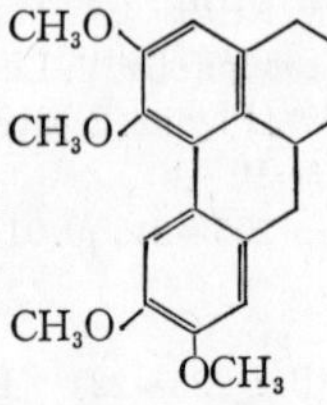

≪mp≫ 119~120° (AcOEt or Et_2O)
≪$[\alpha]_D$≫ +115.4° (aq. MeOH), +120° ($CHCl_3$)

≪Ph.≫ CR[168,169,198,199] UV[46,61,152,205,213,214] IR[152,213] NMR[74,152,210,355,357] ORD[46,61,62,214] Mass[356] Rf[152,213]
≪OP.≫ Corydalis cava (ternata)[168], Dicentra eximia[162], D. formosa[163], D. oregana[164], Glaucium fimbrilligerund[197], G. luteum (flavum)[193,198~202]*
≪C. & So.≫ Colourless prisms, Sol. in org. solv, Insol. in C_6H_6, Petr. ether
≪Sa. & D.≫ B–HCl: 243~246°
B–HBr: 241° (d) B–HI: 243°
B–MeI: 224~225°
≪SD.≫ 31), 129), 133), 157), 158), 162), 169), 183)
≪S. & B.≫ 151), 152), 153), 157), 169), 209), 211), 212), 213)

l-Glaucine

≪mp≫ 124~125°
≪$[\alpha]_D$≫ −114.5° ($CHCl_3$)
≪Ph.≫
≪OP.≫ Corydalis ternata[208]

≪C. & So.≫
≪Sa. & D.≫
≪SD.≫
≪S. & B.≫

Hernandaline

$C_{29}H_{37}O_7N \equiv 511.59$

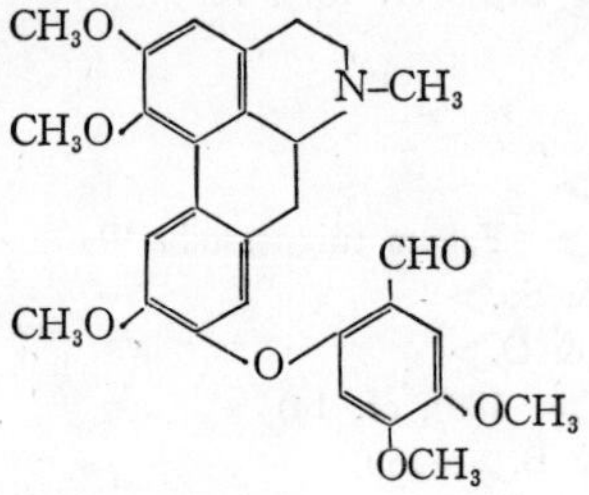

≪mp≫ 170~171.5° (EtOH)
≪$[\alpha]_D$≫ +35.6° ($CHCl_3$)

≪Ph.≫ UV[121] IR[121] NMR[121,362] Mass[121]
≪OP.≫ Hernandia ovigera[121]
≪C. & So.≫ Needles
≪Sa. & D.≫
≪SD.≫ 121)
≪S. & B.≫ 121)

Laurifoline

$C_{20}H_{24}O_4N^{\oplus} \equiv 324.39$

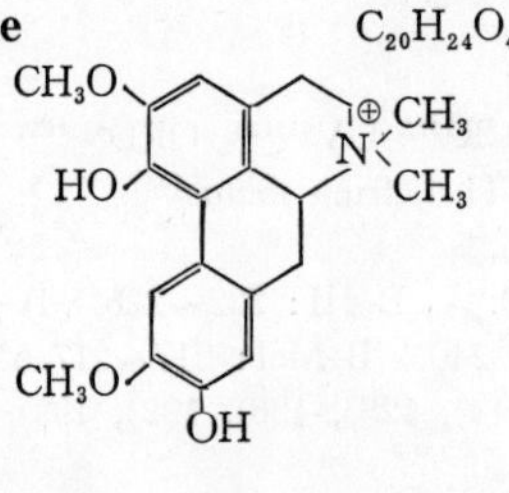

≪mp≫
≪$[\alpha]_D$≫ B⊕–Cl⊖ +26.3° (H_2O)

≪Ph.≫ CR[218] UV[61,214,222] IR[225] ORD[61,62,214]
≪OP.≫ Cocculus laurifolius[215], Fagara (Xanthoxylum) ailanthoides[216], F. chilopcrone var. angustifolia[217], F. hiemalis[217], F. pterota[217], F. shoifolia[179]
≪C. & So.≫
≪Sa. & D.≫ B⊕–Cl⊖: 253~257° (d)
B–Pic: 222°
≪SD.≫ 24), 183), 218), 219), 220)
≪S. & B.≫ 218), 221), 222), 223), 224)

---

* G. serpieri[203,204], G. squamigerum[201], Liliodendron tulipifera[205,206], Magnolia kachirachirai[207], Neolitsea sericea[28], Papaver somniferum[128], Thalictrum minus[193]

Xanthoplanine $C_{21}H_{26}O_4N^{\oplus} \equiv 356.42$

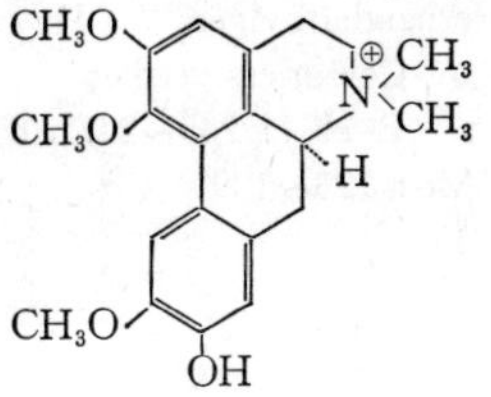

CH_3O CH_3 N CH_3 CH_3O H CH_3O OH

≪mp≫
≪$[\alpha]_D$≫ B⊕–I⊖ +71° (EtOH)

≪Ph.≫ UV[214,226,228] IR[226,228,230] ORD[214]
≪OP.≫ Fagara hyemalis[217], F. nigerescens[217], Hernandia ovigera[226], Xanthoxylum planispium[227,228]
≪C. & So.≫
≪Sa. & D.≫ B⊕–Cl⊖ : 218～220° (d)
B⊕–I⊖ : 148～149° (d) B–Pic : 228° (d)
cf. 213～214° B⊕–OMe–I⊖ : 225～227° (d)
≪SD.≫ 227), 228), 229), 230)
≪S. & B.≫

Cocsarmine $C_{21}H_{26}O_4N^{\oplus} \equiv 356.42$

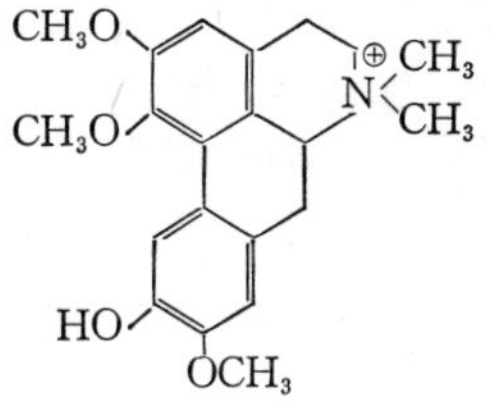

CH_3O CH_3 N CH_3 CH_3O HO OCH_3

≪mp≫
≪$[\alpha]_D$≫ B⊕–I⊖ +27.9° (EtOH)

≪Ph.≫ UV[24,196] IR[196] Rf[196]
≪OP.≫ Cocculus sarmentosus[196]
≪C. & So.≫
≪Sa. & D.≫ B⊕–I⊖ : 205～207°
B–Pic : 226～227° (d)
≪SD.≫ 196)
≪S. & B.≫

No Name $C_{21}H_{26}O_4N^{\oplus} \equiv 356.42$

CH_3O CH_3 N CH_3 HO CH_3O OCH_3

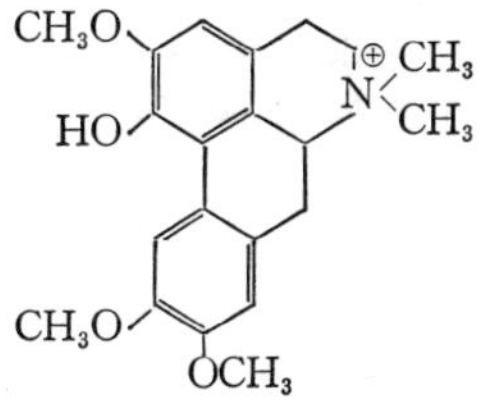

≪mp≫
≪$[\alpha]_D$≫

≪Ph.≫ NMR[284]
≪OP.≫ Fagara tinguassoiba[26,36,37]
≪C. & So.≫
≪Sa. & D.≫ B⊕–I⊖ : 226～229°
≪SD.≫ 26), 36), 37)
≪S. & B.≫ 189)

No Name $C_{21}H_{26}O_4N^{\oplus} \equiv 356.42$

HO CH_3 N CH_3 CH_3O CH_3O OCH_3

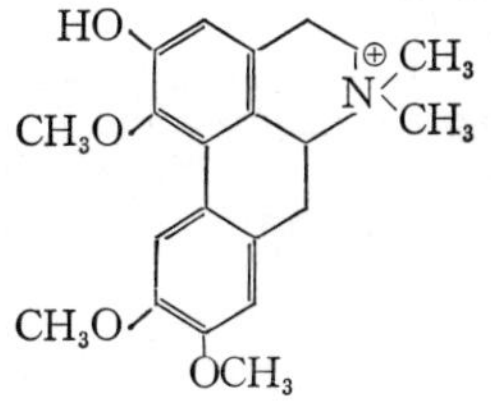

≪mp≫
≪$[\alpha]_D$≫ B⊕–Cl⊖ + 29.6°

≪Ph.≫ UV[75,190,326] IR[75,326]
≪OP.≫ Fagara rhoifolia[75], F. tinguassoiba[75,326]
≪C. & So.≫
≪Sa. & D.≫ B⊕–Cl⊖ : 218～220°
B⊕–I⊖ : 228～229° B–Pic : 147～149°
B⊕–OMe–I⊖ : 216～218°
B⊕–OAc–I⊖ : 234～236°
≪SD.≫ 75), 326)
≪S. & B.≫

Nandigerine $C_{18}H_{17}O_4N \equiv 311.32$

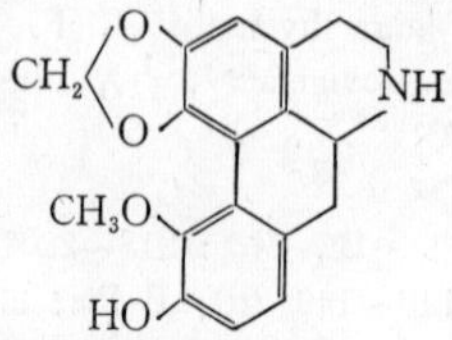

≪mp≫ 176～177° (MeOH)
≪[α]D≫ +248° (EtOH)

≪Ph.≫ UV[231] NMR[231] Rf[231]
≪OP.≫ Hernandia ovigera[231], H. catalpifolia[121]
≪C. & So.≫ Colourless needles
≪Sa. & D.≫ B–HCl: 245～247° (d)
B–O,N–DiMe: 129～130°
≪SD.≫ 231)
≪S. & B.≫

Launobine $C_{18}H_{17}O_4N \equiv 311.32$

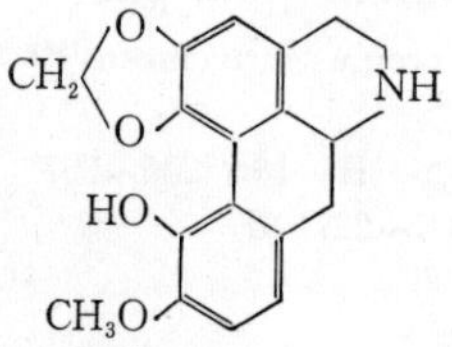

≪mp≫ 214～215° (d)
≪[α]D≫ +192.7° ($CHCl_3$)

≪Ph.≫
≪OP.≫ Laurus nobilis[116]
≪C. & So.≫
≪Sa. & D.≫ B–HCl: 275° (d)
B–NMe=bulbocapnine
≪SD.≫ 116)
≪S. & B.≫

Ovigerine (Hernovine)[226] $C_{18}H_{15}O_4N \equiv 309.31$

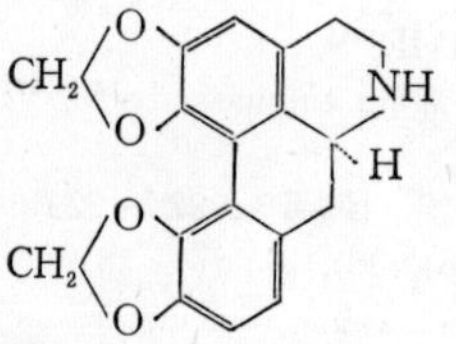

≪mp≫ 143～145° (Me_2CO)
≪[α]D≫ +217° ($CHCl_3$)

≪Ph.≫ UV[226][231] NMR[226][231]
≪OP.≫ Hernandia catalpifolia[121], H. ovigera[121][226][231]
≪C. & So.≫ Yellowish-grey prisms
≪Sa. & D.≫ B–HCl: 300° (d)
≪SD.≫ 226), 231)
≪S. & B.≫

Hernovine $C_{18}H_{19}O_4N \equiv 313.34$

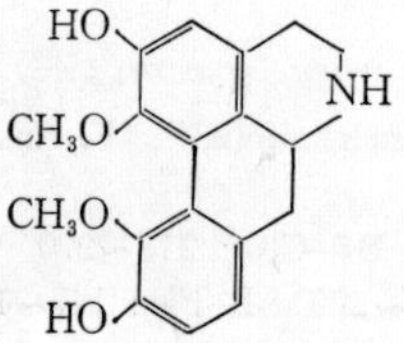

≪mp≫ 236～240° (d) (MeOH)
≪[α]D≫

≪Ph.≫ UV[231] IR[231] NMR[231]
≪OP.≫ Hernandia catalpifolia[121], H. ovigera[121][226][231]
≪C. & So.≫
≪Sa. & D.≫ B–NMe·HCl: 245～247°
B–O,O–DiMe: 174～175°
≪SD.≫ 231)
≪S. & B.≫

Norisocorydine $C_{19}H_{21}O_4N \equiv 327.37$

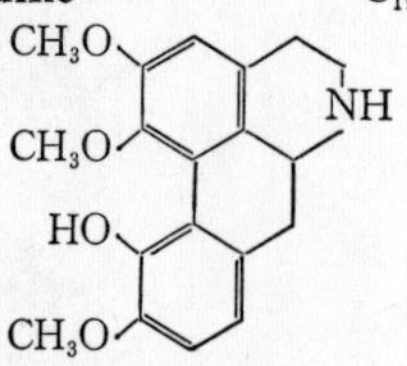

≪mp≫ 203～205°
≪[α]D≫ +203～206° ($CHCl_3$)

≪Ph.≫ UV[156] ORD[62]
≪OP.≫ Hernandia catalpifolia[121], Peumus boldus[156][160]
≪C. & So.≫ Unstable
≪Sa. & D.≫ B–HBr: 203～205°
B–NMe=isocorydine
≪SD.≫ 156)
≪S. & B.≫

Laurepukine $C_{18}H_{17}O_4N \equiv 311.32$

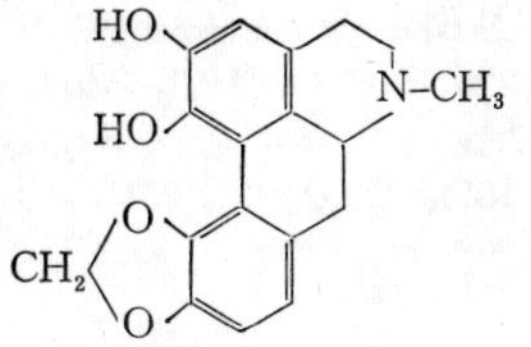

≪mp≫ 230～231° (Et_2O–$CHCl_3$)
≪$[\alpha]_D$≫ −222° ($CHCl_3$), −252° (EtOH)

≪Ph.≫ CR[232] UV[61,233,235] ORD[61]
≪OP.≫ Laurelia novae-zelandiae[79,232,233,234]
≪C. & So.≫ Needles
≪Sa. & D.≫ B_2–H_2SO_4: 99～100°
B–MeI: 249～250° B–O,O–DiMe: 134°
≪SD.≫ 79), 232), 233), 235)
≪S. & B.≫

***N*-Methylovigerine** $C_{19}H_{17}O_4N \equiv 323.33$

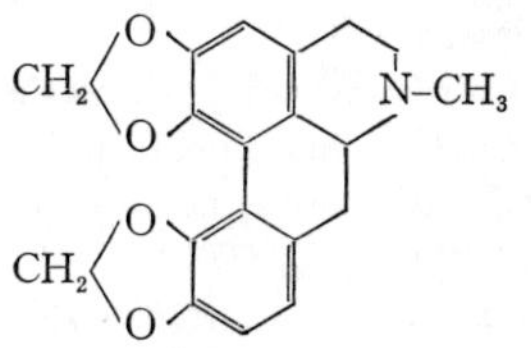

≪mp≫
≪$[\alpha]_D$≫

≪Ph.≫ CR[231]
≪OP.≫ Hernandia catalpifolia[121],
H. ovigera[121,231]
≪C. & So.≫
≪Sa. & D.≫ B–HBr: 243～245°
B–MeI: 252～253° (d)
≪SD.≫ 231)
≪S. & B.≫

***N*-Methylnandigerine**
(Hernangerine)[236] $C_{19}H_{19}O_4N \equiv 325.35$

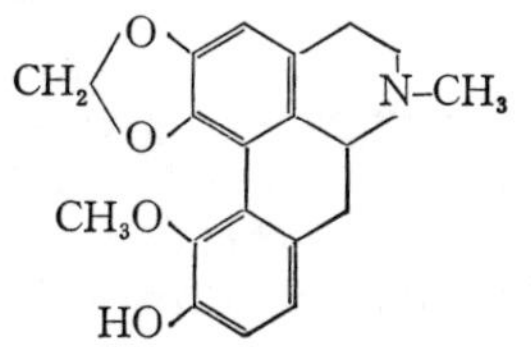

≪mp≫
≪$[\alpha]_D$≫ B–HBr +170° (H_2O)

≪Ph.≫ UV[236] IR[231,236] NMR[231,236] Rf[231]
≪OP.≫ Hernandia catalpifolia[121],
H. ovigera[231,236]
≪C. & So.≫ Amorph
≪Sa. & D.≫ B–HBr: 243～245° (d)
B–OMe·Pic: 180～182° (d)
≪SD.≫ 231), 236)
≪S. & B.≫

Bulbocapnine $C_{19}H_{19}O_4N \equiv 325.35$

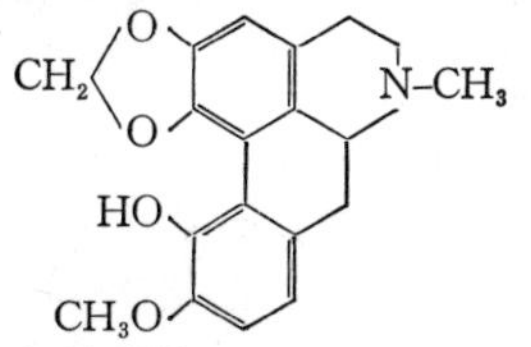

≪mp≫ 199° (EtOH)
≪$[\alpha]_D$≫ +237° ($CHCl_3$)

≪Ph.≫ CR[240,246] UV[54,98,121] NMR[61,74,175,210,355,358] Mass[22,356] ORD[54,61,62,240]
≪OP.≫ Bulbocapnos cavus[363], Coryda liscava (tuberosa)[237,238,239], C. decumbens[240], C. solida (bulbosa)[239], Dicentra canadensis[241]
≪C. & So.≫ Prisms, Sol. in EtOH
≪Sa. & D.≫ B–MeI: 258° B–OMe: 128～129°
≪SD.≫ 50), 210), 242), 243), 244), 245), 246), 247)
≪S. & B.≫ 246), 248)

Suaveoline $C_{19}H_{21}O_4N \equiv 327.37$

CH_3O / HO (substituent positions on aporphine skeleton, N–CH_3)

≪mp≫ 232° (MeOH), cf. 182°[250]
≪[α]D≫ +164° ($CHCl_3$)

≪Ph.≫ CR[107]
≪OP.≫ Artabotrys suaveolens[107,250]
≪C. & So.≫ Sol. in $CHCl_3$, Mod. sol. in MeOH
≪Sa. & D.≫
≪SD.≫ 107), 220), 247), 251)
≪S. & B.≫

Corytuberine $C_{19}H_{21}O_4N \equiv 327.37$

≪mp≫ 240° (EtOH)
≪[α]D≫ +282.7° (EtOH)

≪Ph.≫ CR[124,163,203] UV[61] IR[259] NMR[151,357] Rf[259] Mass[356] ORD[61]
≪OP.≫ Corydalis cava[203], C. nobilis[252], Dicentra formosa[163], Glaucium vitellinum[253]
≪C. & So.≫ Colourless needles, Sol. in hot H_2O, Insol. in Et_2O, C_6H_6, $CHCl_3$
≪Sa. & D.≫ B–$HAuCl_4$: 195~196°
B–OBz: 214° B–O,O–DiBz: 140°
B–O,O–DiAc: 72° B–OMe=corydine
B–O,O–DiMe–MeI: 248°
≪SD.≫ 220), 242), 247), 254), 255), 256), 257), 258)
≪S. & B.≫ 151), 255), 256), 259), 321), 357)

Isocorydine[254] $C_{20}H_{23}O_4N \equiv 341.39$
(Artabotrine[107,251]**, Luteanine**[260]**)**

≪mp≫ 185~186° (MeOH)
≪[α]D≫ +195.3° ($CHCl_3$)

≪Ph.≫ CR[261,267,274,278] UV[61,137,156,160,263,214] IR[137] NMR[74,355] Mass[22,356] ORD[61,62,214]
≪OP.≫ Asimina triloba[16], Artabotrys suaveolens[107,250,251,261], Atherosperma moschatum[262], Corydalis govaniana[262], C. knolle[264], C. lutea[260,265,266], C. platycarpa[266], C. ternata[208,267]*
≪C. & So.≫ Prisms, Sol. in $CHCl_3$, Spar. sol. in MeOH
≪Sa. & D.≫ B–HCl: 227° B–HBr: 215°
B–HI: 225° B–MeI: 225° B–OMe–MeI: 253°
≪SD.≫ 220), 244), 247), 250), 251), 254), 255), 257), 258)
≪S. & B.≫ 151), 264), 274), 275), 276), 277)

* Dicentra canadensis[241], Dicranostigma franchetianum[268], D. lactucoides[80], Glaucium corniculatum[269,270,271], G. flavum[80,200,202,272], G. serpieri[203,204], G. vitellinum[253], Hernandia ovigera[121], H. catalpifolia[121], Legnephora moorei[273], Peumus boldus[156,160], Phylica (Papaver) rogersii[137], Xanthoxylum brachyacanthum[273]

_d_-Corydine $C_{20}H_{23}O_4N \equiv 341.39$

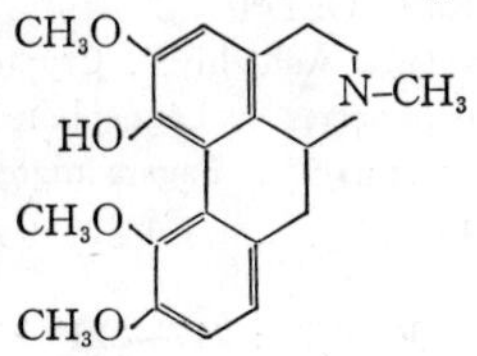

CH_3O / HO / CH_3O / CH_3O / N–CH_3

≪mp≫ 149° (Et_2O)
≪[α]D≫ +206° (EtOH)

≪Ph.≫ CR[279] UV[61,151,213,214,283] IR[151,213] NMR[74,151,284,355,357] Mass[151,356] ORD[61,62,214,249] Rf[213,270]
≪OP.≫ Corydalis cava (tuberosa)[238,279], C. ternata[280], Dicentra canadensis[241], D. eximia[162,163], D. formosa[163], D. oregana[164], Glaucium corniculatum[269,270]*
≪C. & So.≫ Colourless needles, Sol. in EtOH, $CHCl_3$, AcOEt
≪Sa. & D.≫ B–HCl: 258° B–HI: 248°
B–Pic: 195~196° B–MeI: 230°
B–OMe·HCl: 250~252° (d) B–OMe–MeI: 253°
≪SD.≫ 220), 242), 244), 247), 254), 255), 257), 258)
≪S. & B.≫ 151), 213), 233), 274), 282), 283), 357), BG: 357)

_l_-Corydine→ _d_-Corydine

≪mp≫ 148~150°
≪[α]D≫ −206° (EtOH)
≪Ph.≫
≪OP.≫

≪C. & So.≫
≪Sa. & D.≫
≪SD.≫
≪S. & B.≫

Magnoflorine $C_{20}H_{24}O_4N^{\oplus} \equiv 324.39$
(Thalictrine)[285]

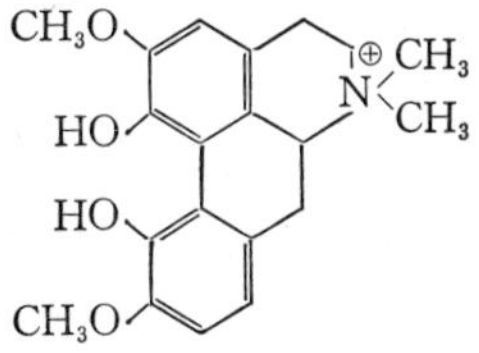

CH_3O / HO / HO / CH_3O / $N^{\oplus}(CH_3)_2$

≪mp≫
≪[α]D≫ B⊕–I⊖ +200° (MeOH)

≪Ph.≫ UV[214,285,292,308,318] IR[259,285,286,287,289,290,291,300,301,304,306,307,313,314,316,318] ORD[214] Rf[292,293,297,315]
≪OP.≫ Aristochia clematitis[286], A. debilis[287,288], A. kaempferi[287], Berberis amurensis var. japonica[289], B. (Mahonia) fortunei[290], B. japonica[291], B. thunbergii (vulgaris)[292]**
≪C. & So.≫
≪Sa. & D.≫ B⊕–I⊖: 248~249° (d)
B–Pic: 206° B–Styph: 233°
≪SD.≫ 183), 220), 247), 259), 285), 308), 320)
≪S. & B.≫ 259), 308), 321)

---

* G. fimbrilligerum[197,281], G. flavum[202], G. flavum var. fulvum[202], G. oxylobum[139], G. elegans[271], G. squamigerum[201], G. vitellinum[253]

** B. tschonoskyana[293], B. kawakami[295], B. mingestensis[296], Caulophyllum robastum[297], Cissampelos insularis[298], Cocculus laurifolius[299], C. trilobus[300], Coptis japonica (orientalis)[301], Epimedium cremeum[302], Fagara chiloperone var. angustifolia[217], F. coco[210], F. hyemalis[217], F. naranjillo var. paraquarensis[217], F. pterota[217], F. rhoifolia[217], F. rhoifolia var. petiolutatum[217], Epimedium grandiflorum var. thunbergianum[303], E. rugosum[304], Magnolia coco[305], M. denunata (obovata)[306], M. kobus var. borealis[307], M. grandiflora[308], M. kachirachirai[305], M. parviflora[309], Mahonia compressa var. formosana[310], M. champaka[311], Michelia compressa[294], M. fuscata[312], Nandina domestica[313], Papaver somniferum[128], Phellodendron amurense[314], P. amarense var. sachalinense[315], Sinomenium acutum[316], Thalictrum fendleri[317], T. foliolosum[285], T. thunbergi (minus)[318], Xanthoxylum ailanthoides[216], X.piperitum[319], X. piperitum var. inerme[319], X. planispium[227,228], X. planispium (alatum)[228]

Menisperine $C_{21}H_{26}O_4N^{\oplus} \equiv 356.42$
(Chakramine)[322]

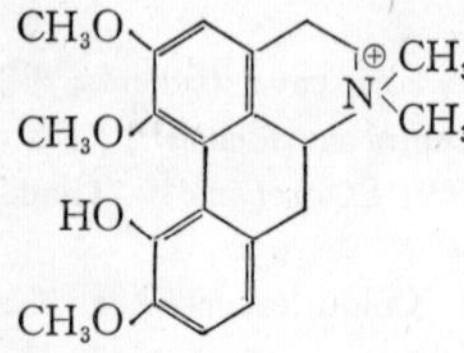

≪mp≫
≪$[\alpha]_D$≫ $B^{\oplus}$–$I^{\ominus}$ +136° (H_2O), +139° (MeOH)
$B^{\oplus}$–$Cl^{\ominus}$ +168.6° (H_2O)

≪Ph.≫ CR[325] UV[61)83)322)323)325] IR[313)322)323]
NMR[322] pKa[322] ORD[61]
≪OP.≫ Bragantic wallichi[322], Cryptocarya angulata[25], C. triplinervis[25], Legnephora moorei (Pericampylus incanus)[273], Fagara nigrescens[217], F. rhoifolia[217]*
≪C. & So.≫
≪Sa. & D.≫ $B^{\oplus}$–$Cl^{\ominus}$: 217～221°
$B^{\oplus}$–$I^{\ominus}$: 219° (238～240°) B–Pic : 204°
B–Styph : 202°
≪SD.≫ 220), 247), 320), 322), 323), 325)
≪S. & B.≫ 275)

***d*-N-Methylcorydine** $C_{21}H_{26}O_4N^{\oplus} \equiv 356.42$

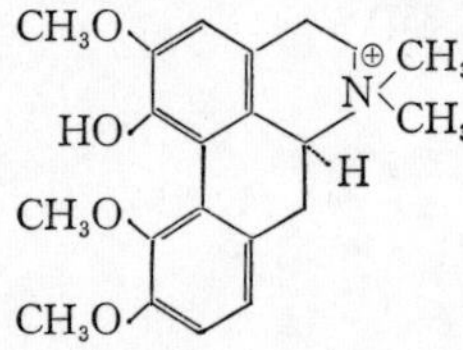

≪mp≫
≪$[\alpha]_D$≫ $B^{\oplus}$–$I^{\ominus}$ +154.2° (50% EtOH)

≪Ph.≫ UV[234] IR[234] NMR[234)284] Rf[234]
≪OP.≫ Fagara nigrescens[217)234]
≪C. & So.≫
≪Sa. & D.≫ $B^{\oplus}$–$I^{\ominus}$: 190～200° (d)
B–Pic : 186～187° B–Styph : 206～207°
≪SD.≫ 234), 284)
≪S. & B.≫

Hernandine $C_{19}H_{19}O_5N \equiv 341.35$

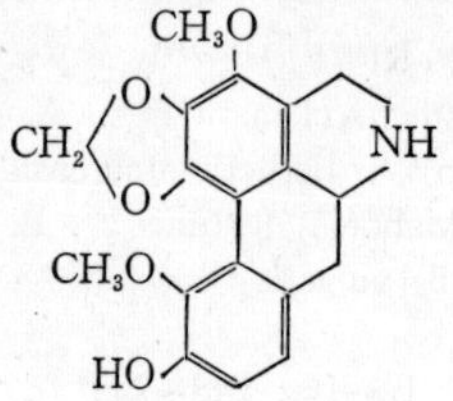

≪mp≫
≪$[\alpha]_D$≫

≪Ph.≫ NMR[358]
≪OP.≫ Hernandia bivalvis[327]
≪C. & So.≫
≪Sa. & D.≫
≪SD.≫ 327), 358)
≪S. & B.≫

Cassythidine $C_{19}H_{17}O_5N \equiv 339.33$

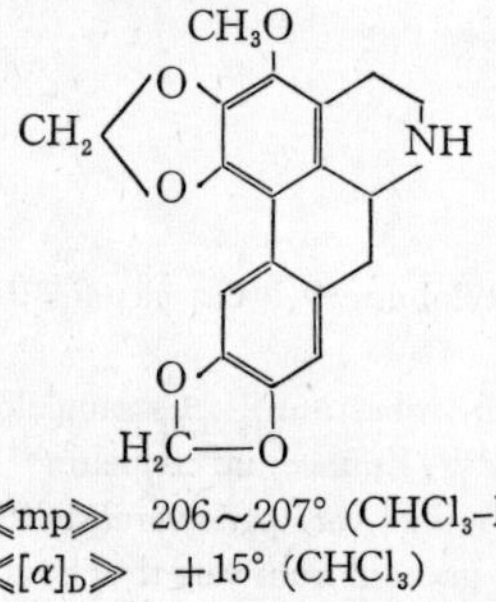

≪mp≫ 206～207° ($CHCl_3$–MeOH)
≪$[\alpha]_D$≫ +15° ($CHCl_3$)

≪Ph.≫ UV[328]
≪OP.≫ Cassytha filiformis[328]
≪C. & So.≫ Colourless prisms
≪Sa. & D.≫
≪SD.≫ 328)
≪S. & B.≫

* Menispermam dauricum[323], Nandina domestica[313], Xanthoxylum brachyacanthum[324], X (Fagara) coco[217)322)325], X. veneticum[324]

Cassyfiline $C_{19}H_{19}O_5N \equiv 341.35$

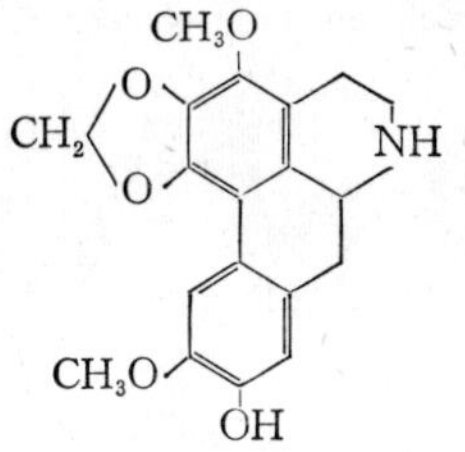

≪mp≫ 217° (d) (MeOH)
≪$[\alpha]_D$≫ −89.6° ($CHCl_3$)

≪Ph.≫ UV[329)] NMR[329)]
≪OP.≫ Cassytha filiformis[329)]
≪C. & So.≫ Microgranules
≪Sa. & D.≫ B–NMe: 210∼211°
B–OMe: 150∼151° B–OMe·HCl: 260∼261°
B–O,N–DiMe: 139∼140°=ocoteine
≪SD.≫ 329)
≪S. & B.≫

Cassythine $C_{19}H_{19}O_5N \equiv 341.35$

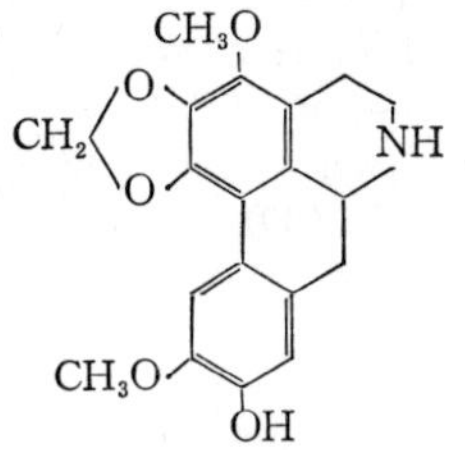

≪mp≫ 217∼219° ($CHCl_3$–EtOH)
≪$[\alpha]_D$≫ +24° ($CHCl_3$)

≪Ph.≫ UV[329)] NMR[329)]
≪OP.≫ Cassytha filiformis[328)]
≪C. & So.≫ Colourless needles
≪Sa. & D.≫ B–NMe: 208°
B–O, N–DiMe: 138∼139°=ocoteine
≪SD.≫ 328)
≪S. & B.≫

Ocoteine (Thalicmine) $C_{21}H_{23}O_5N \equiv 369.40$

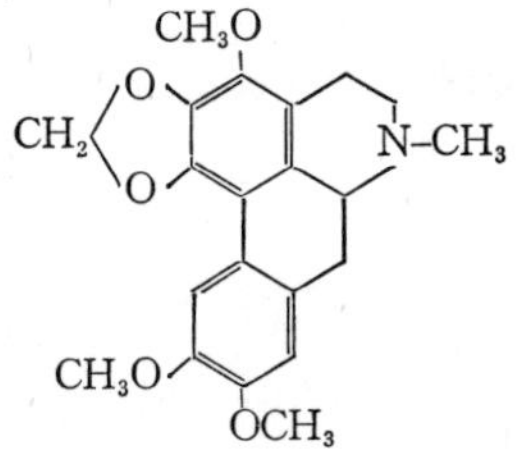

≪mp≫ 137∼138° (50% EtOH)
≪$[\alpha]_D$≫ +33.3° ($CHCl_3$), +37° (EtOH)
cf. +255.3° ($CHCl_3$)[192)]

≪Ph.≫ CR[330)] UV[61)141)205)214)330)334)]
NMR[141)334)355)358)] ORD[61)141)214)334)]
≪OP.≫ Ocotea puberula[330)331)], Thalictrum minus[192)]
≪C. & So.≫
≪Sa. & D.≫ B–HCl: 268∼270°
B–HBr: 258∼260° B–HI: 223∼224° (d)
B–Pic: 178° B–MeI: 222∼224°
≪SD.≫ 55), 141), 193), 331), 332), 333), 334), 335), cf. 192)
≪S. & B.≫ 52), 332), 333), 335)

Liriodenine $C_{17}H_9O_3N \equiv 275.31$
(Oxoushinsunine, Spermatheridine[336)], Mickeline B[337)])

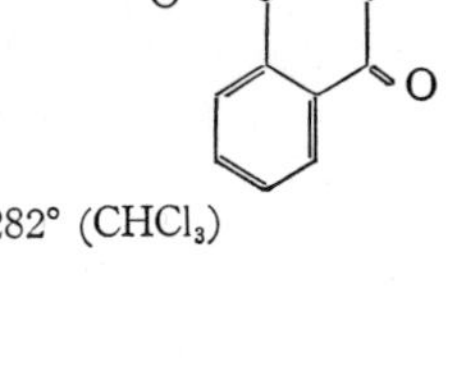

≪mp≫ 282° ($CHCl_3$)
≪$[\alpha]_D$≫

≪Ph.≫ UV[336)339)346)] IR[336)341)344)345)] NMR[336)]
Rf[262)345)]
≪OP.≫ Asimia triloba[16)], Atherosperma moschatum[262)336)], Doryphora sassafras[338)], Liriodendron tulipifera[206)339)∼341)], Lysichiton camtschatcense var. japonicum[360)], Magnolia coco[305)], Michelia alba[305)342)], M. champaca[311)317)343)], M. compressa[310)344)], M. compressa var. formosana[278)345)]
≪C. & So.≫ Yellow needles
≪Sa. & D.≫ B–HCl: 289∼292° (d)
B–$HClO_4$: 250∼253° (d) B–Pic: 232∼234°
B–Oxal: 250∼252° (d) B–Oxime: 270∼271° (d)
≪SD.≫ 35), 305), 310), 337), 339), 340), 345)
≪S. & B.≫ 9), 339), BG: 348)

Lysicamine $C_{18}H_{13}O_3N \equiv 291.26$

≪Ph.≫ UV[360] IR[360] NMR[360]
≪OP.≫ Lysichiton camtschatcense var. japonicum[360]
≪C. & So.≫
≪Sa. & D.≫
≪SD.≫ 360)
≪S. & B.≫ 360), 361)

≪mp≫ 210~211° (d) (EtOH)
≪$[\alpha]_D$≫

Atherospermidine
(Psilopine[336]) $C_{18}H_{11}O_4N \equiv 305.28$

≪Ph.≫ UV[336,347] IR[336,349] NMR[336] Rf[262]
≪OP.≫ Atherosperma moschatum[142,262], Guatteria psilopus[347]
≪C. & So.≫ Yellow needles
≪Sa. & D.≫ B-HCl: 256~258° (d)
B-Oxime: 245~250°
≪SD.≫ 262), 336), 347), 348), 349)
≪S. & B.≫ 348), 349)

≪mp≫ 276~278° (d) ($CHCl_3$ or Py)
≪$[\alpha]_D$≫

Moschatoline
(Atheroline[350]) $C_{18}H_{13}O_4N \equiv 307.29$

≪Ph.≫ CR[351] UV[351] IR[351] NMR[351]
≪OP.≫ Atherosperma moschatum[350,351]
≪C. & So.≫ Oil or yellow amorph, Sol. in $CO_3^{2\ominus}$, Insol. in $HCO_3^{\ominus}$
≪Sa. & D.≫ B-OAc: 190~200°
≪SD.≫ 351)
≪S. & B.≫ 351)

≪mp≫
≪$[\alpha]_D$≫

Atheroline $C_{19}H_{15}O_5N \equiv 337.32$

≪Ph.≫ UV[350] IR[350,351] NMR[350,351]
≪OP.≫ Atherosperma moschatum[350]
≪C. & So.≫
≪Sa. & D.≫ B-OMe: 235~236°
B-OAc: 190~195°
≪SD.≫ 350), 351)
≪S. & B.≫ 350), 351)

≪mp≫ 250~260° (d)
≪$[\alpha]_D$≫

No Name $C_{20}H_{17}O_5N$≡351.34

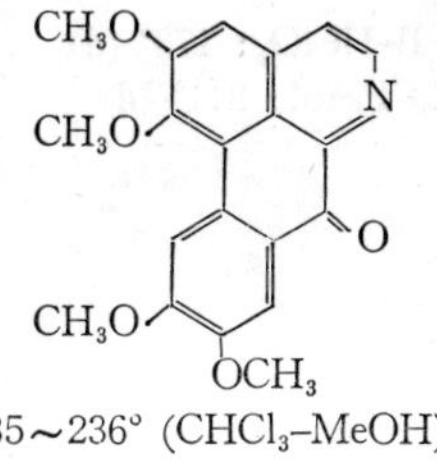

CH_3O, CH_3O, N, O, CH_3O, OCH_3

≪mp≫ 235～236° ($CHCl_3$–MeOH)
≪[α]D≫

≪Ph.≫ UV[205)340)] IR[205)]
≪OP.≫ Liriodendron tulipifera[339)340)]
≪C. & So.≫
≪Sa. & D.≫
≪SD.≫ 205), 339)
≪S. & B.≫ 205), 339), 352), BG: 205)

Norushinsunine $C_{17}H_{15}O_3N$≡281.30
(Michelalbine, Normicheline A)

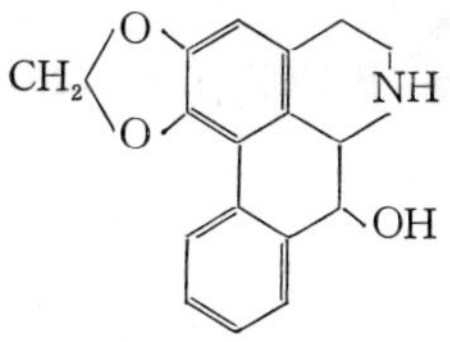

CH_2, O, O, NH, OH

≪mp≫ 205～207°
≪[α]D≫ −105.2° ($CHCl_3$)

≪Ph.≫ CR[342)] UV[34)342)] IR[342)]
≪OP.≫ Asimia triloba[16)], Michelia alba[342)]
≪C. & So.≫ Colourless prisms
≪Sa. & D.≫ B-HCl: 265° (d)
B-Pic: 203～205° or 220° B-NMe=ushinsunine
≪SD.≫ 305), 342)
≪S. & B.≫

Ushinsunine $C_{18}H_{17}O_3N$≡295.32
(Micheline A)[345)]

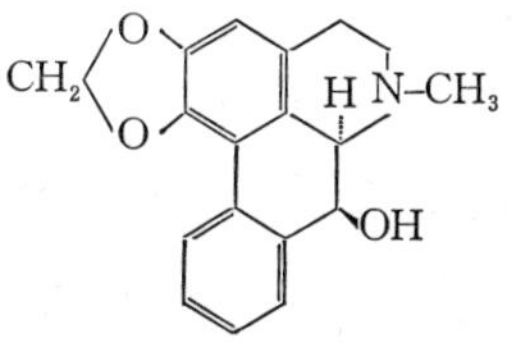

CH_2, O, O, H, N–CH_3, OH

≪mp≫ 180～181° (Me_2CO), 122～123° (EtOH)
≪[α]D≫ −117° ($CHCl_3$), −121.8° (MeOH)

≪Ph.≫ CR[310)345)] UV[34)310)] IR[310)] NMR[35)] Rf[345)]
≪OP.≫ Michelia alba[342)], M. compressa[344)], M. compressana var. formosana[310)345)], M. champaca[311)]
≪C. & So.≫ Colourless prisms
≪Sa. & D.≫ B-HCl: 261° (d) B-Pic: 134°
B-Oxal: 155° B-OAc: 202～203°
B-MeI: 234～235° (d)
≪SD.≫ 35), 305), 310), 337), 345), 347)
≪S. & B.≫

Guatterine $C_{19}H_{19}O_4N$≡325.36

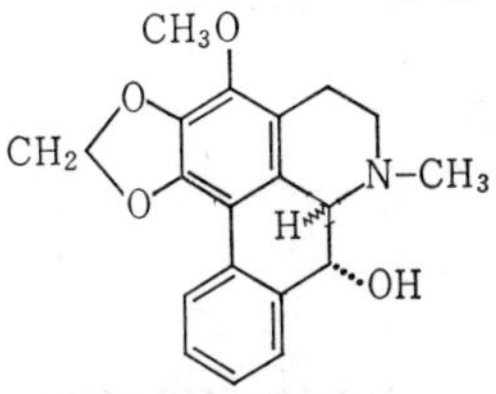

CH_3O, CH_2, O, O, N–CH_3, H, OH

≪mp≫ 146～148° (Me_2CO–H_2O)
≪[α]D≫ −57.1° ($CHCl_3$)

≪Ph.≫ CR[347)] UV[347)] IR[347)] NMR[347)] Rf[347)]
≪OP.≫ Guatteria psilopus[347)351)]
≪C. & So.≫ Colourless needles
≪Sa. & D.≫
≪SD.≫ 347)
≪S. & B.≫

Vitricine $C_{17}H_{15}O_3N \equiv 281.30$

≪mp≫ 237° (d)
≪$[\alpha]_D$≫ +108° ($CHCl_3$)
≪Ph.≫ UV[353] IR[353]
≪OP.≫ Vitex trifolia[353]
≪C. & So.≫
≪Sa. & D.≫ B–$HClO_4$: 178° (d)
B–Pic: 228° B–Picrol: 211° (d)
≪SD.≫
≪S. & B.≫

Michelavine $C_{17}H_{15}O_3N \equiv 281.30$

≪mp≫ 205~207°
≪$[\alpha]_D$≫ −105.2°
≪Ph.≫ UV[342] IR[342]
≪OP.≫ Michelia alba[342]
≪C. & So.≫
≪Sa. & D.≫
≪SD.≫ 342)
≪S. & B.≫

Muricinine $C_{18}H_{19}O_4N \equiv 313.34$

≪mp≫
≪$[\alpha]_D$≫
≪Ph.≫ CR[354]
≪OP.≫ Anona nobilis[354]
≪C. & So.≫
≪Sa. & D.≫ B–$HClO_4$: 206~208°
≪SD.≫ 354)
≪S. & B.≫

Pulchelline $C_{18}H_{23}O_3N \equiv 301.37$

≪mp≫ 189° (Me_2CO)
≪$[\alpha]_D$≫ +80.1°
≪Ph.≫
≪OP.≫ Neolitsea pulchella[136]
≪C. & So.≫
≪Sa. & D.≫
≪SD.≫
≪S. & B.≫

Neolitsinine $C_{18}H_{23}O_3N \equiv 301.37$

≪mp≫ 214~215° (Me_2CO)
≪$[\alpha]_D$≫ +94.7°
≪Ph.≫ Mass[136]
≪OP.≫ Neolitsca pulchella[136]
≪C. & So.≫
≪Sa. & D.≫
≪SD.≫ 136)
≪S. & B.≫

Muricine $C_{19}H_{21}O_4N \equiv 327.37$

≪mp≫
≪$[\alpha]_D$≫
≪Ph.≫ CR[354]
≪OP.≫ Anona nobilis[354]
≪C. & So.≫
≪Sa. & D.≫ B–HBr: 242~243°
≪SD.≫ 354)
≪S. & B.≫

Leucoxine $C_{20}H_{21}O_5N \equiv 355.38$

≪mp≫ 213~217°
≪$[\alpha]_D$≫ +81° (EtOH)
≪Ph.≫ UV[167]
≪OP.≫ Ocotea leucoxylon[167]
≪C. & So.≫
≪Sa. & D.≫ B–OMe: 116~117°
B–OMe·MeI: 214~217°
≪SD.≫
≪S. & B.≫

Eximine $C_{20}H_{23}O_4N \equiv 341.39$

≪mp≫ 142° (MeOH–Et_2O)
≪$[\alpha]_D$≫
≪Ph.≫
≪OP.≫ Dicentra eximia[162]
≪C. & So.≫
≪Sa. & D.≫ B–HBr: 270° (d) B–MeI: 212°
≪SD.≫
≪S. & B.≫

Eximidine $C_{20}H_{23}O_4N \equiv 341.39$

≪mp≫ 133° (Et_2O)
≪[α]D≫
≪Ph.≫
≪OP.≫ Dicentra eximia[162]
≪C. & So.≫
≪Sa. & D.≫ B–MeI: 218°
≪SD.≫
≪S. & B.≫

Platycerine $C_{20}H_{23}O_4N \equiv 341.39$

≪mp≫ 130～132° (Et_2O)
≪[α]D≫ −267±3° ($CHCl_3$)
≪Ph.≫ CR[65] UV[65] pKa[65]
≪OP.≫ Argemone platyceras[65]
≪C. & So.≫
≪Sa. & D.≫ B–Pic: 155～160°
B–O, O–DiMe·1/2 H_2SO_4: 145°
≪SD.≫
≪S. & B.≫

Catalpifoline $C_{20}H_{23}O_4N \equiv 341.39$

≪mp≫ 174～175°
≪[α]D≫ +220° (EtOH)
≪Ph.≫
≪OP.≫ Hernandia catalpifolia[121]
≪C. & So.≫
≪Sa. & D.≫
≪SD.≫ 121)
≪S. & B.≫

Leucoxylone $C_{22}H_{25}O_6N \equiv 399.43$

≪mp≫ 54～58°
≪[α]D≫ +52° (EtOH)
≪Ph.≫ UV[167] NMR[167]
≪OP.≫ Ocotea leucoxylon[167]
≪C. & So.≫
≪Sa. & D.≫ B–MeI: 227～230°
≪SD.≫
≪S. & B.≫

Spermatherine

≪mp≫ 124～125° (Me_2CO–Petr. ether)
≪[α]D≫
≪Ph.≫ Rf[262]
≪OP.≫ Atherosperma moschatum[262]
≪C. & So.≫
≪Sa. & D.≫
≪SD.≫
≪S. & B.≫

Atherosperminine $C_{20}H_{23}O_2N \equiv 341.39$

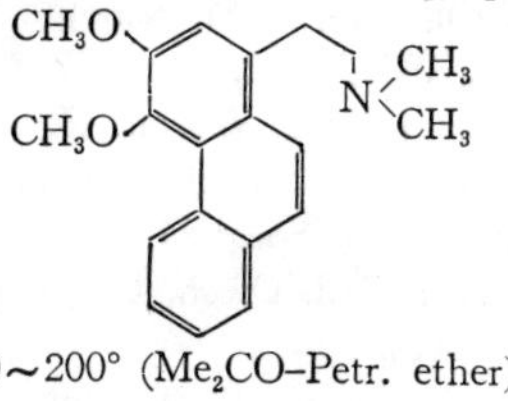

≪mp≫ 199～200° (Me_2CO–Petr. ether)
≪[α]D≫
≪Ph.≫ Rf[262]
≪OP.≫ Atherosperma moschatum[262][349], Cryptocarya angulata[25], C. triplinervis[25]
≪C. & So.≫
≪Sa. & D.≫ B–HI: 234～235°
B–Pic: 186～188° B–MeI: 270.5～276.5°
≪SD.≫ 262), 349)
≪S. & B.≫

Methoxyatherosperminine $C_{21}H_{25}O_3N \equiv 339.42$

CH_3O CH_3O CH_3O N CH_3 CH_3

≪mp≫
≪[α]D≫
≪Ph.≫
≪OP.≫ Atherosperma moschatum[349]
≪C. & So.≫ Oil
≪Sa. & D.≫ B–Pic: 161~162°
B–MeI: 243~245°
≪SD.≫ 349)
≪S. & B.≫ 349)

Thalicthuberine $C_{21}H_{23}O_4N \equiv 353.40$

CH_3O CH_3O N CH_3 CH_3 O H_2C—O

≪mp≫ 126~127°
≪[α]D≫
≪Ph.≫ UV[179] IR[179]
≪OP.≫ Thalictrum thunbergii[179]
≪C. & So.≫ Colourless silky needles
≪Sa. & D.≫ B–HCl: 209~210°
B–Oxal: 206~207°
≪SD.≫ 179)
≪S. & B.≫

1) A. C. Santos: Philippine J. Sci., **43**, 561 (1930); **47**, 357 (1932).
2) F. R. Reyes, A. C. Santos: Philippine J. Sci., **44**, 409 (1931).
3) K. P. Gopinath, T. R. Govindachari, B. R. Pai, N. Viswanathan: Ber., **92**, 776 (1959).
4) N. Trimurti: J. Indian Inst. Sci., **7**, 232 (1924). [C., **1925**, I, 679].
5) K. Bernauer: Helv. Chim. Acta, **47**, 2119 (1964).
6) M. S. Yunusov, S. T. Akramov, S. Yu. Yunusov: Dokl. Akad. Nauk Uz. SSR., **23**, 38 (1966). [C. A., **65**, 13781 (1966)].
7) G. Barger, G. Weitnauer: Helv. Chim. Acta, **22**, 1036 (1939).
8) L. Marion, L. Lemay, R. Ayotte: Can. J. Res., **28 B**, 211 (1950).
9) M. P. Cava, D. R. Dolton: J. Org. Chem., **31**, 1281 (1966).
10) D. H. R. Barton, D. S. Bhakuni, G. M. Chapman, G. W. Kirby: Chem. Comm., **1966**, 259.
11) R. Robinson: "The Structural Relations of Natural Products", Clarendon Press, Oxford (1955).
12) D. H. R. Barton, T. Cohen: "Festshrift Arthur Stoll," p. 117. Birkhauser, Basel (1957).
13) D. H. R. Barton: Proc. Chem. Soc., **1963**, 293.
14) R. Tschesche, P. Welzel, R. Moll, G. Legler: Tetrahedron, **20**, 1435 (1964).
15) B. Gilbert, M. E. A. Gilbert, M. M. De-Oliveira, O. Ribeiro, E. Wenkert, B. Wickberg, U. Hollstein, H. Rapoport: J. Am. Chem. Soc., **86**, 694 (1964).
16) M. Tomita, M. Kozuka: J. Pharm. Soc. Japan, **85**, 77 (1965).
17) S. M. Kupchan, B. Dasgupta, E. Fujita, M. L. King: Tetrahedron, **19**, 227 (1963).
18) J. Kunitomo, C. Yamato, T. Otuki: J. Pharm. Soc. Japan, **84**, 1141 (1964).
19) J. Kunitomo, M. Kamimura: J. Pharm. Soc. Japan, **84**, 1100 (1964).
20) J. A. Weisbach, B. Douglas: J. Org. Chem., **27**, 3738 (1962).
21) S. M. Kupchan, R. M. Kanojia: Tetrahedron Letters, **1966**, 5353.
22) M. Ohashi, J. M. Wilson, H. Budzikiewicz, M. Shamma, W. A. Slusarchyk, C. Djerassi: J. Am. Chem. Soc., **85**, 2807 (1963).
23) J. Slavik: Coll. Czech. Chem. Comm., **28**, 1738 (1963).
24) S. Yu. Yunusov, V. A. Mnatsakanyan, S. T. Akramov: Dokl. Akad. Nauk Uz. SSR., **1961**, No. 8, p. 43. [C. A., **57**, 9900 (1962)].
25) R. G. Cooke, H. F. Haynes: Austral. J. Chem., **7**, 99 (1954).
26) M. Tomita, Y. Watanabe, H. Furukawa: J. Pharm. Soc. Japan, **81**, 469 (1961).
27) T. Nakasato, S. Nomura: J. Pharm. Soc. Japan, **77**, 816 (1957).
28) T. Nakasato, S. Nomura: J. Pharm. Soc. Japan, **79**, 1267 (1959); **78**, 540 (1958).
29) R. A. Konowalowa, S. Yu. Yunusov, A. P.

Orekhov: Bull. Soc. Chim. France, **6**, 811 (1939).

30) R. A. Konovalova, S. Yu. Yunusov, A. P. Orekhov: Zhur. Obshchei Khim., **9**, 1868 (1939).

31) K. W. Bentley, H. M. E. Cardwell: J. Chem. Soc., **1955**, 3253.

32) L. Marion, V. Grassie: J. Am. Chem. Soc., **66**, 1290 (1944).

33) T. Nakasato, S. Nomura: Chem. Pharm. Bull., **7**, 780 (1959).

34) M. Shamma, W. A. Slusarchyk: Chem. Revs., **64**, 74 (1964).

35) T.-H. Yang: J. Pharm. Soc. Japan, **82**, 804 (1962).

36) M. Tomita, Y. Watanabe, H. Furukawa: J. Pharm. Soc. Japan, **81**, 942 (1961).

37) M. Tomita, Y. Watanabe, H. Furukawa: J. Pharm. Soc. Japan, **81**, 1644 (1961).

38) M. Tomita, H. Furukawa: J. Pharm. Soc. Japan, **82**, 1458 (1962).

39) M. Tomita, Y. Watanabe, H. Furukawa: J. Pharm. Soc. Japan, **81**, 1202 (1961).

40) S. M. Kupchan, B. Dasgupta, E. Fujita, M. L. King: J. Pharm. Sci., **51**, 599 (1962).

41) H. R. Arthur. H. T. Cheung: J. Chem. Soc., **1959**, 2306.

42) S. Yu. Yunusov, R. Konowalowa, A.P.Orekhov: Bull. Soc. Chim. France, **7**, 70 (1940).

43) J. M. Gulland, R. D. Haworth: J. Chem. Soc., **1928**, 581.

44) M. P. Cava, S. C. Havlicek, A. Lindert, R. J. Spangler: Tetrahedron Letters, **1966**, 2937.

45) F. Hoffmann-La Roche & Co., A.-G. Neth. Appl. 6,507,700. [C. A., **64**, 19707 (1966)].

46) J. C. Craig, S. K. Roy: Tetrahedron, **21**, 395 (1964).

47) M. Tomita, H. Shirai: J. Pharm. Soc. Japan, **62**, 381 (1942).

48) H. Shirai: J. Pharm. Soc. Japan, **64 B**, 208 (1944).

49) H. Kondo, T. Sanada: J. Pharm. Soc. Japan, **44**, 1034 (1924).

50) H. Kondo, T. Sanada: J. Pharm. Soc. Japan, **48**, 163 (1928).

51) H. Shirai, N. Oda: J. Pharm. Soc. Japan, **76**, 1287 (1956).

52) M. Tomita, K. Hirai: J. Pharm. Soc. Japan, **77**, 290 (1957).

53) D. H. Hey, A. Husain: J. Chem. Soc., **1958**, 1876.

54) M. Tomita, K. Hirai: J. Pharm. Soc. Japan, **79**, 723 (1959).

55) K. Hirai, K. Harada: J. Pharm. Soc. Japan, **80**, 1429 (1960).

56) R. M. F. Manske: Can. J. Res., **16 B**, 76 (1938).

57) J. Schmutz: Helv. Chim. Acta, **42**, 335 (1959).

58) T. R. Govindachari: Current Sci. (India), **10**, 76 (1941).

59) T. R. Govindachari: Current Sci. (India), **11**, 238 (1942).

60) L. Marion: J. Am. Chem. Soc., **66**, 1125 (1944).

61) M. Shamma: Experientia, **16**, 484 (1960).

62) C. Djerassi, K. Mislow, M. Shamma: Experientia, **18**, 53 (1962).

63) E. N. Parabirsing: Dissertation, Leiden (1952).

64) F. Faltis, G. Wagner, E. Adler: Ber., **77**, 686 (1944).

65) J. Slavik, L. Slavikova: Coll. Czech. Chem. Comm., **28**, 1720, 1728 (1963).

66) M. Tomita, T. Kitamura: J. Pharm. Soc. Japan, **80**, 21 (1960).

67) S. K. Saha: Science & Culture, **24**, 572 (1959).

68) A. Chatterjee, P. L. Majumder, R. Mukherjee, S. K. Saha, S. K. Talapatra: Tetrahedron Letters, **1965**, 1539.

69) K. Goto: Ann., **521**, 175 (1936).

70) M. Tomita, M. Kozuka: J. Pharm. Soc. Japan, **86**, 871 (1966).

71) K. Goto, R. Inabe, H. Nozaki: Ann., **530**, 142 (1937).

72) K. Goto, H. Shishido: Proc. Imp. Acad. (Tokyo), **15**, 8 (1939).

73) K. Goto, H. Shishido: Ann., **539**, 262 (1939).

74) I. R. C. Bick, J. Harley-Mason, N. Sheppard, M. J. Vernengo: J. Chem. Soc., **1961**, 1896.

75) J. M. Calderwood, F. Fish: Chem. & Ind., **1966**, 237.

76) S. Yu. Yunusov, V. A. Mnatsakanyan, S. T. Akramov: Izu. Akad. Nauk SSSR., Jer. Khim., **1965**, 502. [C. A., **63**, 642 (1965)].

77) T. Kitamura: J. Pharm. Soc. Japan, **80**, 1104 (1960).

78) B. C. Aston: J. Chem. Soc., **97**, 1381 (1910).

79) G. Barger, A. Girardet: Helv. Chim. Acta, **14**, 481 (1931).

80) R. H. F. Manske: Can. J. Chem., **32**, 83 (1954).

81) E. Schlittler: Helv. Chim. Acta, **15**, 394 (1932).

82) M. S. Gibson, J. M. Walthew: Chem. & Ind., **1965**, 185.

83) K. Ito: J. Pharm. Soc. Japan, **81**, 703 (1961).

84) G. Barger, E. Schlittler: Helv. Chim .Acta, **15**, 381 (1932).

85) E. Schlittler, A. Landenmann: Helv. Chim. Acta, **32**, 1881 (1949).

86) B. L. Konson, P. P. Saksonov: Farmakol i. Toksikol, **9**, No. 4. 14 (1946).
87) V. V. Kislev, R. A. Konowalowa: Zhur. Obshchei Khim., **18**, 142 (1948). [C. A., **42**, 5037 (1948)].
88) K. Heydenreich, S. Pfeifer: Pharmazie, **21**, 121 (1966).
89) A. R. Battersby, T. H. Brown: Chem. Comm., **1966**, 170.
90) J. Gadamer, W. Klee: Arch. Pharm., **249**, 39 (1941).
91) W. Klee: Arch. Pharm., **252**, 211 (1914).
92) E. Schlittler, J. Müller: Helv. Chim. Acta, **31**, 1119 (1948).
93) R. K. Callow, J. M. Gulland, R. D. Haworth: J. Chem. Soc., **1929**, 1444.
94) V. V. Kiselev, R. A. Konowalowa: Zhur. Obshchei Khim., **19**, 148 (1949).
95) K. W. Bentley, S. F. Dyke: Experientia, **12**, 205 (1956).
96) K. W. Bentley, S. F. Dyke: J. Org. Chem., **22**, 429 (1957).
97) A. V. Bochanikova, V. V. Kiselev: Zhur. Obshchei Khim., **34**, 1984 (1964).
98) T. R. Govindachari, B. R. Pai: J. Org. Chem., **18**, 1352 (1953).
99) M. Tomita, T. Kitamura: J. Pharm. Soc. Japan, **79**, 997 (1959).
100) A. R. Battersby, T. H. Brown: Proc. Chem. Soc., **1964**, 85.
101) A. R. Battersby, T. H. Brown, J. H. Clements: J. Chem. Soc., **1965**, 4550.
102) J. Müller: Helv. Chim. Acta, **31**, 1111 (1948).
103) K. W. Bentley, E. T. Blues: J. Chem. Soc., **1956**, 1732.
104) K. W. Bentley, S. F. Dyke: Chem. & Ind., **1956**, 1054.
105) A. R. Battersby: Proc. Chem. Soc., **1963**, 189.
106) A. R. Battersby, R. T. Brown, J. H. Clements, G. G. Iverach: Chem. Comm., **1965**, 230.
107) G. Barger, L. J. Sargent: J. Chem. Soc., **1939**, 991.
108) M. Tomita, H. Shirai: J. Pharm. Soc. Japan, **62**, 109 (1942).
109) M. Tomita, H. Shirai: J. Pharm. Soc. Japan, **63**, 233 (1943).
110) M. Tomita: J. Pharm. Soc. Japan, **59**, 207 (1939).
111) H. Kondo, M. Tomita: J. Pharm. Soc. Japan, **59**, 542 (1939).
112) H. Shirai: J. Pharm. Soc. Japan, **63**, 517, 532 (1943).
113) T. R. Govindachari, K. Nagarajan, C. V. Ramadas: J. Chem. Soc., **1958**, 983.
114) M. Tomita, K. Hirai: J. Pharm. Soc. Japan, **78**, 738 (1958).
115) S. Krishna, T. P. Ghose: J. Indian Chem. Soc., **9**, 429 (1932).
116) M. Tomita, M. Kozuka, E. Nakagawa, Y. Mitsunori: J. Pharm. Soc. Japan, **83**, 763 (1963).
117) T. P. Ghose, S. Kirskna, E. Schlittler: Helv. Chim. Acta, **17**, 919 (1934).
118) D. H. Hey, L. Labo: J. Chem. Soc., **1954**, 2246.
119) M. Tomita, I. Kikkawa: J. Pharm. Soc. Japan, **77**, 1015 (1957).
120) P. S. Clezy, E. Gellert, D. Y. K. Lau, A. W. Nichol: Austral. J. Chem., **19**, 135 (1966).
121) M. P. Cava, K. Bessho, B. Douglas, S. Markey, J. A. Weisbach: Tetrahedron Letters, **1966**, 4279.
122) M. Kozuka: J. Pharm. Soc. Japan, **82**, 1567 (1962).
123) P. S. Clezy, A. W. Nichol, E. Gellert: Experientia, **19**, 1 (1963).
124) J. D. Filippo: Arch. Pharm., **236**, 601 (1898).
125) M. Greshoff: Ber., **23**, 3577 (1890).
126) E. Späth, K. Tharrer: Ber., **66**, 904 (1933).
127) K. Gorter: Bull. jard. bot. Buitenzorg, [iii], **3**, 180 (1921). [C., **1921**, III, 344].
128) S. Pfeifer: Pharmazie, **21**, 378 (1961).
129) G. Barger, R. Silberschmidt: J. Chem. Soc., **1928**, 2919.
130) E. Späth, F. Stauhal: Ber., **61**, 2395 (1928).
131) G. Barger, T. Eisenbrand, L. Eisenbrand, E. Schlittler: Ber., **66**, 450 (1933).
132) E. Späth, E. E. Suominen: Ber., **66**, 1344 (1933).
133) E. Faltis, E. Alder: Arch. Pharm., **284**, 281 (1951).
134) I. Kikkawa: J. Pharm. Soc. Japan, **79**, 83 (1959).
135) I. Kikkawa: J. Pharm. Soc. Japan, **79**, 425 (1959).
136) W. H. Hui, S. N. Loo, H. R. Arthur: J. Chem. Soc., **1965**, 2285.
137) R. R. Arndt, W. H. Baarschers: J. Chem. Soc., **1964**, 2244.
138) M. Tomita, I. Kikkawa: J. Pharm. Soc. Japan, **77**, 1011 (1957).
139) J. Slavik, L. Slavikova: Coll. Czech. Chem. Comm., **28**, 2530 (1963).
140) Z. Kitasato: J. Pharm. Soc. Japan, **535**, 71 (1926); **536**, 843 (1926).
141) Z. Kitasato: Acta Phytochim. Japan, **3**, 157 (1927).
142) M. Tomita, Y. Inubushi, S. Ishii, M. Yamagata: J. Pharm. Soc. Japan, **71**, 381 (1951).

143) H. Chikamatsu, M. Tomita, M. Kotake: J. Chem. Soc. Japan, **82**, 1708 (1961).
144) Z. Kitasato, H. Shishido: Ann., **527**, 176 (1937).
145) Z. Kitasato, H. Shishido: Acta Phytochim. Japan, **9**, 265 (1937).
146) H. Shishido: Bull. Chem. Soc. Japan, **12** 150 (1937).
147) T. Kitamura: J. Pharm. Soc. Japan, **80**, 1104 (1960).
148) T. Kitamura, M. Sugamoto: J. Pharm. Soc. Japan, **81**, 254 (1961).
149) M. Tomita, T. Kitamura: J. Pharm. Soc. Japan, **79**, 1092 (1959).
150) M. Tomita, M. Fujii: J. Pharm. Soc. Japan, **82**, 1457 (1962).
151) A. H. Jackson, J. A. Martin: Chem. Comm., **1965**, 142, 420.
152) W. W. C. Chan, P. Maitland: J. Chem. Soc. (C), **1966**, 753.
153) A. H. Jackson, J. A. Martin: J. Chem. Soc. (C), **1966**, 2061.
154) E. Bourgoin, C. Verne: Bull. Soc. Chim. France, [2] **18**, 481 (1872).
155) E. Merck: Jahresber., **36**, 110 (1922).
156) A. Rüegger: Helv. Chim. Acta, **42**, 754 (1959).
157) K. Warnat: Ber., **58**, 2768 (1925).
158) K. Warnat: Ber., **59**, 85 (1926).
159) E. Schlittler: Ber., **66**, 988 (1933).
160) A. Rüegger: Helv. Chim. Acta, **42**, 754 (1959).
161) W. W. Eggleston, O. F. Black, J. W. Kelly: J. Agric. Res., **39**, 477 (1929). [C., **1930**, I, 697].
162) R. H. F. Manske: Can. J. Res., **8**, 592 (1933).
163) R. H. F. Manske: Can. J. Res., **10**, 521 (1934).
164) R. H. F. Manske: Can. J. Res., **10**, 765 (1934).
165) Y. Asahina: Arch. Pharm., **247**, 201 (1909).
166) Y. Asahina: J. Pharm. Soc. Japan, **328**, 626 (1909).
167) S. Goodwin, A. F. Smith, E. C. Horning: Chem. & Ind., **1960**, 691.
168) J. Gadamer: Arch. Pharm., **249**, 224 (1911).
169) J. Gadamer: Arch. Pharm., **249**, 680 (1911).
170) R. D. Haworth, W. H. Perkin, Jr., J. Rankin: J. Chem. Soc., **127**, 2018 (1925).
171) R. D. Haworth, W. H. Perkin, Jr., J. Rankin: J. Chem. Soc., **1926**, 29.
172) S. Osada: J. Pharm. Soc. Japan, **48**, 423 (1928).
173) J. Haginiwa, I. Murakoshi: J. Pharm. Soc. Japan, **75**, 462 (1955).
174) G. Heyl: Arch. Pharm., **241**, 313 (1903).
175) N. S. Bhacca, L. F. Johnson, J. N. Shoolery: "NMR Spectra Catalog", Varian Associates, Palo Alto, Calif. (1962).
176) T. Takase, H. Ohashi: J. Pharm. Soc. Japan, **535**, 742 (1926).
177) Z. Kitasato: Acta Phytochim., **3**, 175 (1927).
178) H. Maniwa, R. Sakae, I. Kan: J. Pharm. Soc. Japan, **536**, 833 (1926).
179) E. Fujita, T. Tomimatsu: J. Pharm. Soc. Japan, **79**, 1252 (1959).
180) Z. Kitasato: J. Pharm. Soc. Japan, **536**, 843 (1926).
181) T. Kitamura: J. Pharm. Soc. Japan, **80**, 219 (1960).
182) T. Kitamura: J. Pharm. Soc. Japan, **80**, 613 (1960).
183) M. Tomita, K. Fukagawa: J. Pharm. Soc. Japan, **83**, 293 (1963).
184) E. Späth, E. E. Suominen: Ber., **66**, 1344 (1933).
185) W. H. Baarschers, R. R. Arndt: Tetrahedron, **21**, 2155 (1965).
186) K. G. R. Pachler, R. R. Arndt, W. H. Baarschers: Tetrahedron, **21**, 2159 (1965).
187) R. H. F. Manske, E. H. Chalesworth, W. R. Ashford: J. Am. Chem. Soc., **73**, 3751 (1951).
188) R. Tschesche, P. Welzel, G. Legler: Tetrahedron Letters, **1965**, 445.
189) M. Shamma, W. A. Slusarchyk: Tetrahedron Letters, **1965**, 1509.
190) N. V. Riggs, L. Antonaccio, L. Marion: Can. J. Chem., **39**, 1330 (1961).
191) R. Tschesche, P. Welzel, G. Legler: Tetrahedron Letters, **1965**, 3451.
192) S. Yu. Yunusov, N. N. Progressov: Zhur. Obshchei Khim., **20**, 1151 (1950). [C. A., **45**, 1608 (1951)].
193) S. Yu. Yunusov, N. N. Progressov: Zhur. Obshchei Khim., **22**, 1047 (1952). [C. A., **47**, 8084 (1953)].
194) S. Yu. Yunusov, Z. F. Ismailov: Zhur. Obshchei Khim., **30**, 1710, 1721 (1960). [C. A., **55**, 3631 (1961)].
195) M. Shamma: Experientia, **18**, 64 (1962).
196) M. Tomita, H. Furukawa: J. Pharm. Soc. Japan, **83**, 190 (1963).
197) R. A. Konowalowa, S. Yu. Yunusov, A. P. Orekhov: Zhur. Obshchei Khim., **9**, 1939 (1939). [C., **1941**, I, 2530].
198) D. Probst: Ann., **31**, 241 (1839).
199) R. Fischer: Arch. Pharm., **239**, 426 (1901).
200) R. H. F. Manske: Can. J. Res., **17**B, 399 (1939).
201) T. F. Platonova, P. S. Massagetov, A. D. Kusowkow, L. M. Utkin: Zhur. Obshchei Khim., **26**, 173 (1956).

202) J. Slavik, L. Slavikova: Coll. Czech. Chem. Comm., **24**, 3141 (1959).
203) J. J. Dobbie, A. Lauder: J. Chem. Soc., **63**, 485, 1131 (1893).
204) R. H. F. Manske: Can. J. Res., **20 B**, 53 (1942).
205) J. Cohen, W. von Rosenthal, W. I. Taylor: J. Org. Chem., **26**, 4143 (1961).
206) M. Tomita, J. Furukawa: J. Pharm. Soc. Japan, **82**, 1199 (1962).
207) T.-H. Yang, S.-T. Lu, C.-Y. Hsiao: J. Pharm. Soc. Japan, **82**, 816 (1962).
208) J. Go: J. Pharm. Soc. Japan, **49**, 801, 814 (1929); **50**, 122, 937, 940 (1930).
209) R. Pschorr: Ber., **37**, 1926 (1904).
210) S. Goodwin, J. N. Shoolery, L. F. Johnson: Proc. Chem. Soc., **1958**, 306.
211) B. Franck, G. Schlingloff: Ann., **659**, 123 (1963).
212) B. Franck, G. Blaschke, G. Schlingloff: Tetrahedron Letters, **1962**, No. 10, 439.
213) M. Shamma, W. A. Slusarchyk: Chem. Comm., **1965**, 528.
214) S. M. Albonico, J. Comin, A. M. Kuck, E. Sanchez, P. M. Scopes, R. J. Swan, M. J. Vernengo: J. Chem. Soc. (C), **1966**, 1340.
215) M. Tomita, F. Kusuda: Pharm. Bull., **1**, 1 (1953).
216) M. Tomita, H. Ishii: J. Pharm. Soc. Japan, **78**, 1441 (1958).
217) A. M. Kuck, S. M. Albonico, V. Deulofeu: Chem. & Ind., **1966**, 945.
218) M. Tomita, F. Kusuda: Pharm. Bull., **1**, 5 (1953).
219) F. Kusuda: Pharm. Bull., **1**, 55 (1953).
220) H. Corrodi, E. Hardegger: Helv. Chim. Acta, **39**, 889 (1956).
221) M. Tomita, I. Kikkawa: Pharm. Bull., **4**, 230 (1956).
222) Shionogi & Co. Ltd., Japan Pat., 7976 (1958). [C. A., **54**, 4651 (1960)].
223) S. M. Albonico, A. M. Kuck, V. Deulofeu: Chem. & Ind., **1964**, 1580.
224) S. M. Albonico, A. M. Kuck, V. Deulofeu: Ann., **685**, 200 (1965).
225) H. Ishii: J. Pharm. Soc. Japan, **86**, 631 (1966).
226) M. Tomita, S.-T. Lu, Y.-T. Chen: J. Pharm. Soc. Japan, **86**, 763 (1966).
227) M. Tomita, H. Ishii: J. Pharm. Soc. Japan, **79**, 1228 (1959).
228) H. Ishii, H. Harada: J. Pharm. Soc. Japan, **81**, 238 (1961).
229) H. Ishii: J. Pharm. Soc. Japan, **81**, 241 (1961).
230) H. Ishii: J. Pharm. Soc. Japan, **81**, 243 (1961).
231) M. P. Cava, K. Bessho, B. Douglas, S. Markey, R. F. Raffau, J. A. Weisbach: Tetrahedron Letters, **1966**, 1577.
232) A. Girardet: Helv. Chim. Acta, **14**, 504 (1931).
233) T. R. Govindachari, K. Nagarajan, S. Rajadurai, R. U. Ramadas: Ber., **91**, 40 (1958).
234) A. M. Kuck: Chem. & Ind., **1966**, 118.
235) A. Girardet: J. Chem. Soc., **1931**, 2630.
236) H. Furukawa, S.-T. Lu: J. Pharm. Soc. Japan, **86**, 1143 (1966).
237) M. Freund, W. Josephy: Ber., **25**, 2411 (1892); Ann., **277**, 1 (1893).
238) E. Merck: Arch. Pharm., **231**, 133 (1893).
239) O. Haars: Arch. Pharm., **243**, 154 (1905).
240) Y. Asahina, S. Motigase: J. Pharm. Soc. Japan, **463**, 766 (1920).
241) R. H. F. Manske: Can. J. Res., **7**, 258 (1932).
242) J. Gadamer, H. Ziegenbein: Arch. Pharm., **240**, 94 (1902).
243) J. Gadamer, F. Kuntze: Arch. Pharm., **249**, 598 (1911).
244) J. Gadamer, F. Knoch: Arch. Pharm., **259**, 135 (1921).
245) E. Späth, H. Holter, R. Posega: Ber., **61**, 322 (1928).
246) J. M. Gulland, R. D. Haworth: J. Chem. Soc., **1928**, 1132.
247) W. A. Ayer, W. I. Taylor: J. Chem. Soc., **1956**, 472.
248) I. Kikkawa: J. Pharm. Soc. Japan, **79**, 1244 (1959).
249) M. J. Vernengo: Experientia, **17**, 420 (1961).
250) A. C. Santos, F. R. Reyes: Univ. Philippines Natl. and Appl. Sci. Bull., **2**, 407 (1932).
251) E. Schlittler, H. U. Huber: Helv. Chim. Acta, **35**, 111 (1952).
252) R. H. F. Manske: Can. J. Res., **18** B, 288 (1940).
253) J. Slavik: Coll. Czech. Chem. Comm., **24**, 3999 (1959).
254) J. Gadamer: Arch. Pharm., **249**, 641, 661 (1911).
255) E. Späth, O. Hromatka: Ber., **61**, 1692 (1928).
256) J. M. Gulland, R. D. Haworth: J. Chem. Soc., **1928**, 1834.
257) J. Go: J. Pharm. Soc. Japan, **49**, 821 (1929).
258) E. Späth, F. Berger: Ber., **64**, 2038 (1931).
259) M. Tomita, I. Kikkawa: J. Pharm. Soc. Japan, **77**, 195 (1957).
260) R. H. F. Manske: Can. J. Res., **20** B, 57 57 (1942).
261) J. M. Maranõn: Philippine J. Sci., **38**, 259

(1929).
262) I. R. C. Bick, P. S. Clezy, W. D. Crow: Austral. J. Chem., **9**, 111 (1956).
263) O. E. Edwards, K. L. Handa: Can. J. Chem., **39**, 1801 (1961).
264) J. Go: J. Pharm. Soc. Japan, **49**, 814 (1929).
265) R. H. F. Manske: Can. J. Res., **17 B**, 89 (1939).
266) R. H. F. Manske: Can. J. Res., **21 B**, 13 (1943).
267) J. Go: J. Pharm. Soc. Japan, **49**, 126 (1929).
268) J. Slavik, L. Slavikova: Chem. Listy, **51**, 1923 (1957).
269) J. Slavik, L. Slavikova: Chem. Listy, **50**, 969 (1956).
270) J. Slavik, L. Slavikova: Coll. Czech. Chem. Comm., **22**, 279 (1957).
271) J. Slavik: Coll. Czech. Chem. Comm., **25**, 1698 (1960).
272) R. H. F. Manske: Can. J. Res., **16 B**, 438 (1938).
273) G. K. Hughes, F. P. Kaiser, N. Matheson, E. Ritchie: Austral. J. Res., **6**, 90 (1953).
274) D. H. Hey, A. L. Palluel: J. Chem. Soc., **1957**, 2926.
275) I. Kikkawa: J. Pharm. Soc. Japan, **78**, 1006 (1958).
276) A. M. Kuck, B. Frydman: J. Org. Chem., **26**, 5253 (1961).
277) Shionogi & Co. Ltd., Japan Pat., 7335 (1962).[C. A., **59**, 1698 (1963)].
278) J. Gadamer: Arch. Pharm., **249**, 669 (1911).
279) J. Gadamer, H. Ziegenbein, H. Wagner: Arch. Pharm., **240**, 19, 81 (1902).
280) J. Go: J. Pharm. Soc. Japan, **49**, 125 (1929).
281) R. A. Konowalowa, S. Yu. Yunusov, A. P. Orechov: Zhur. Obshchei Khim., **9**, 1939 (1939).
282) D. H. Hey, A. L. Palluel: J. Chem. Soc., **1957**, 2926.
283) N. Arumugam, T. R. Govindachari, K. Nagarajan, U. R. Rao: Ber., **91**, 40 (1958).
284) W. H. Baarschers, K. G. R. Pachler: Tetrahedron Letters, **1965**, 3451.
285) K. W. Gopinath, T. R. Govindachari, S. Rajappa, C. V. Ramadas: J. Sci. Ind. Res., **18**B, 444 (1959). [C. A., **54**, 14287 (1960)].
286) M. Pailer, G. Pruckmayr: Monatsh., **90**, 145 (1959).
287) M. Tomita, S. Kura: J. Pharm. Soc. Japan, **77**, 812 (1957).
288) M. Tomita, K. Fukagawa: J. Pharm. Soc. Japan, **82**, 1673 (1962).
289) M. Tomita, T. Kugo: Pharm. Bull., **4**, 121 (1956).
290) M. Tomita, T. Kugo: J. Pharm. Soc. Japan, **77**, 213 (1957).
291) M. Tomita, T. Kugo: J. Pharm. Soc. Japan, **76**, 599 (1956).
292) M. Tomita, T. Kugo: J. Pharm. Soc. Japan, **76**, 597 (1956).
293) M. Tomita, T. Kugo: J. Pharm. Soc. Japan, **79**, 317 (1959).
294) K. Ito: J. Pharm. Soc. Japan, **80**, 705 (1960).
295) T.-H. Yang, S.-T. Lu: J. Pharm. Soc. Japan, **80**, 847 (1960).
296) T.-H. Yang, S.-T. Lu: J. Pharm. Soc. Japan, **80**, 849 (1960).
297) M. Tomita, T. Takahashi: J. Pharm. Soc. Japan, **78**, 680 (1958).
298) M. Tomita, T. Kikuchi: J. Pharm. Soc. Japan, **77**, 69 (1957).
299) T. Nakano, M. Uchiyama: Pharm. Bull., **4**, 407 (1956).
300) T. Nakano: Pharm. Bull., **4**, 69 (1956).
301) M. Tomita, S. Kura: J. Pharm. Soc. Japan, **76**, 1425 (1956).
302) M. Tomita, H. Ishii: J. Pharm. Soc. Japan, **77**, 319 (1957).
303) M. Tomita, H. Ishii: J. Pharm. Soc. Japan, **77**, 212 (1957).
304) M. Tomita, H. Ishii: J. Pharm. Soc. Japan, **77**, 114 (1957).
305) T.-H. Yang, S.-T. Lu, C.-Y. Hsiao: J. Pharm. Soc. Japan, **82**, 816 (1962).
306) T. Nakano: Pharm. Bull., **4**, 67 (1956).
307) T. Nakano, M. Uchiyama: Pharm. Bull., **4**, 409 (1956).
308) T. Nakano: Pharm. Bull., **2**, 326, 329 (1954).
309) T. Nakano, M. Uchiyama: Pharm. Bull., **4**, 408 (1956).
310) T.-H. Yang: J. Pharm. Soc. Japan, **82**, 794 (1962).
311) T.-H. Yang, C.-Y. Hsiao: J. Pharm. Soc. Japan, **83**, 216 (1963).
312) K. Ito, I. Uchiyama: J. Pharm. Soc. Japan, **79**, 1108 (1959).
313) M. Tomita, T. Kugo: J. Pharm. Soc. Japan, **76**, 751 (1956).
314) M. Tomita, T. Nakano: Pharm. Bull., **5**, 10 (1957).
315) M. Tomita, J. Kunitomo: J. Pharm. Soc. Japan, **78**, 1444 (1958).
316) M. Tomita, T. Kugo: J. Pharm. Soc. Japan, **76**, 857 (1956).
317) M. Shamma, M. A. Greenberg, B. S. Dudock: Tetrahedron Letters, **1965**, 3595.
318) E. Fujita, T. Tomimatsu: Pharm. Bull., **4**, 489 (1956); **6**, 107 (1958).
319) M. Tomita, H. Ishii: J. Pharm. Soc. Japan, **77**, 810 (1957).
320) M. Tomita, Y. Takano: J. Pharm. Soc.

Japan, **80**, 1645 (1960).
321) Shionogi & Co. Ltd., Japan Pat., 6466 (1958).〔C. A., **54**, 1584 (1960)〕.
322) A. R. Katritzky, R. A. Y. Jones, S. S. Bhatnagar: J. Chem. Soc., **1960**, 1950.
323) M. Tomita, T. Kikuchi: Pharm. Bull., **3**, 100 (1955).
324) J. R. Cannon, G. K. Hughes, E. Ritchie, W. C. Taylor: Austral. J. Chem., **6**, 86 (1953).
325) J. Comin, V. Deulofeu: J. Org. Chem., **19**, 1774 (1954).
326) N. V. Riggs, L. Antonaccio, L. Marion: Can. J. Chem., **39**, 1330 (1961).
327) R. Greenhalgh, F. N. Lahay: "Heterocyclic Chemistry", Chemical Society (London), Butterworths, **1958**, 100.
328) S. R. Johns, J. A. Lamberton: Austral. J. Chem., **19**, 297 (1966).
329) M. Tomita, S.-T. Lu, S. Wang: J. Pharm. Soc. Japan, **85**, 827 (1965).
330) G. A. Iacobucci: Ciencia Invest. (Buenos Aires), **7**, 48 (1951). 〔C. A., **45**, 7129 (1951)〕. An. Assoc. Quim. Argent., **42**, 18 (1954). 〔C. A., **48**, 12376 (1954)〕.
331) M. J. Vernengo, A. S. Cerezo, G. A. Iacobucci, V. Deulofeu: Ann., **610**, 173 (1957).
332) T. R. Govindachari, S. Rajadurai, C. V. Ramadas, N. Viswanathan: Ber., **93**, 360 (1960).
333) K. Hirai: J. Pharm. Soc. Japan, **80**, 608 (1960).
334) M. J. Vernengo: Experientia, **19**, 294 (1963).
335) T. R. Govindachari, B. R. Pai, G. Shanmugasundasan: Tetrahedron, **20**, 2895 (1964).
336) I. R. C. Bick, G. K. Douglas: Tetrahedron Letters, **1964**, 1629.
337) T.-H. Yang: J. Pharm. Soc. Japan, **82**, 798 (1962).
338) S. A. Gharbo, J. L. Beal, R. H. Schlessinger, M. P. Cava, G. H. Svoboda: Lloydia, **28**, 237 (1965). 〔C. A., **64**, 2135 (1966)〕.
339) W. I. Taylor: Tetrahedron, **14**, 42 (1961).
340) M. A. Buchanan, E. E. Dickey: J. Org. Chem., **25**, 1388 (1960).
341) M. Tomita, K. Fukugawa: J. Pharm. Soc. Japan, **83**, 293 (1963).
342) T.-H. Yang: J. Pharm. Soc. Japan, **82**, 811 (1962).
343) P. S. Clezy, D. Y. K. Lau: Austral. J. Chem., **19**, 437 (1966).
344) M. Tomita, J. Furukawa: J. Pharm. Soc. Japan, **82**, 925 (1962).
345) S.-S. Yang, W.-Y. Huang, L.-C. Lin, P.-Y. Yeh: Chemistry, **1961**, 144. 〔C. A., **56**, 1489 (1962)〕.
346) M. A. Buchanan, E. E. Dickey: J. Org. Chem., **25**, 1389 (1960).
347) W. M. Harris, T. A. Geissman: J. Org. Chem., **30**, 432 (1965).
348) B. R. Pai, G. Shanmugasundram: Tetrahedron, **21**, 2579 (1965).
349) I. R. C. Bick, G. K. Douglas: Austral. J. Chem., **18**, 1997 (1965).
350) I. R. C. Bick, G. K. Douglas: Tetrahedron Letters, **1965**, 2399.
351) I. R. C. Bick, G. K. Douglas: Tetrahedron Letters, **1965**, 4655.
352) M. Tomita, T.-H. Yang, H. Furukawa, M.-H. Yang: J. Pharm. Soc. Japan, **82**, 1574 (1962).
353) W. Doepke: Naturwiss., **49**, 375 (1962).
354) T. M. Meyer: Ing. Ned.-Indië, 8, No. 6 VII, 64 (1941).
355) W. H. Baarschers, R. R. Arndt, K. Pachler, J. A. Weisbach, B. Douglas: J. Chem. Soc., **1964**, 4778.
356) A. H. Jackson, J. A. Martin: J. Chem. Soc. (C), **1966**, 2181.
357) A. H. Jackson, J. A. Martin: J. Chem. Soc. (C), **1966**, 2222.
358) K. S. Soh, F. N. Lahey, R. Greenhalgh: Tetrahedron Letters, **1966**, 5279.
359) K. Kotera, Y. Hamada, R. Mitsui: Tetrahedron Letters, **1966**, 6273.
360) N. Katsui, K. Sato, S. Tobinaga, N. Takeuchi: Tetrahedron Letters, **1966**, 6257.
361) M. Tomita, T.-H. Yang, H. Furukawa, H.-Y. Yang: J. Pharm. Soc. Japan, **82**, 1574 (1962).
362) N. M. Mollov, H. B. Dutschewska: Dokl. Bulgar. Akad. Nauk., **19**, 491 (1966).
363) "Dictionary of Organic Compounds", Vol. 1, p. 383, Edited by I. Heilbron, H. M. Bunbury. Oxford Univ. Press. (1953).

Chapter 10 Protoberberine Alkaloids

Coptisine

$C_{19}H_{14}O_4N^{\oplus}\equiv 320.30$

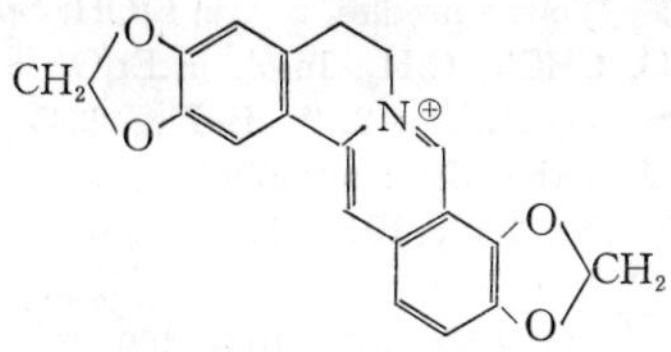

≪mp≫ 216~8° (EtOH)
≪[α]D≫ 0°

≪Ph.≫ CR[3] UV[274] Rf[3]
≪OP.≫ Argemone mexicana[1], Chelidonium majus[2], Chines Corydalis[3], Coptis japonica[4], C. teeta[5], Corydalis ambigua[6][7], C. incisa[8]*
≪C. & So.≫ Yellow needles
≪Sa. & D.≫ B⊕-I⊖ : >280°
≪SD.≫ 4), 20)
≪S. & B.≫ 21), cf. 20)

Berberrubine

$C_{19}H_{16}O_4N^{\oplus}\equiv 322.32$

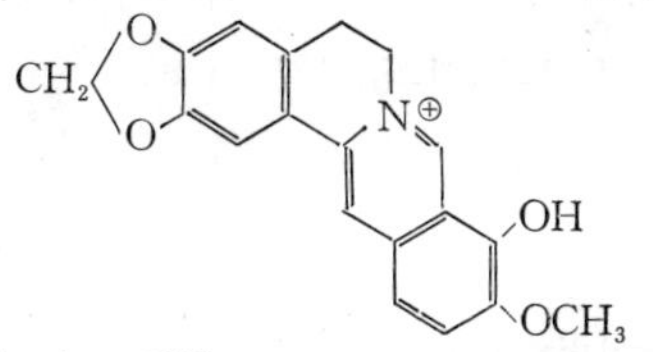

≪mp≫ 285°
≪[α]D≫ 0°

≪Ph.≫
≪OP.≫ Berberis vulgaris[22]
≪C. & So.≫
≪Sa. & D.≫
≪SD.≫ 23), 24), 25), 26)
≪S. & B.≫ 25), 27), cf. 23), 24)

Thalifendine

$C_{19}H_{16}O_4N^{\oplus}\equiv 322.32$

≪mp≫
≪[α]D≫ 0°

≪Ph.≫ CR[28] UV[28] NMR[28] Mass[28] Rf[28]
≪OP.≫ Thalictrum fendleri[28]
≪C. & So.≫
≪Sa. & D.≫ B⊕-Cl⊖ : >230° Tetrahydro-Der.: 209~211° B-OMe·tetrahydro-Der.: =tetrahydroberberine
≪SD.≫ 28)
≪S. & B.≫

---

* Dicentra spectabilis[9], Dicranostigma franchetianum[10], Glaucium corniculatum[11], G. oxylofum[12], Eschscholtzia glauca[13], Hunnemannia fumariaefolia[14], Macleaya (Bocconia) microcarpa[15], Papaver rhoeas[9], P. somniferum[16], Sanguinaria canadensis[17], Argemone alba[18], Meconopsis paniculata[19]

Berberine (Umbellatine) $C_{20}H_{18}O_4N^{\oplus} \equiv 336.34$

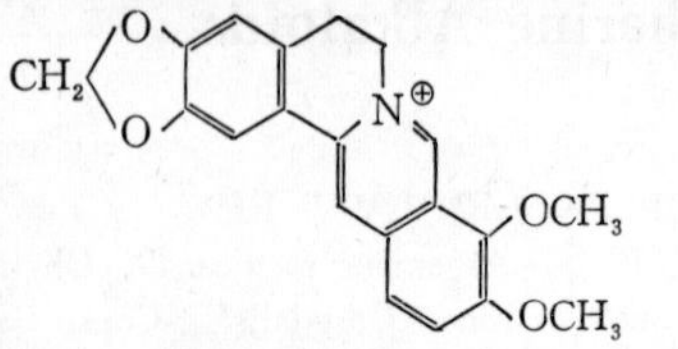

≪mp≫ 145° [B + $6H_2O$] (H_2O)
≪$[\alpha]_D$≫ 0°

≪Ph.≫ CR[67] UV[121)122)123)124] IR[123]
≪OP.≫ Archangelisia flava[29], Argemone alba[30], A. mexicana[31], Berberis aetnensis[32], B. amurensis var. japonica[33], B. aquifolium[34)35)36], B. aristata[37], B. asiatica[36]*
≪C. & So.≫ Yellow needles, Sol. in EtOH, Spar. sol. in H_2O, $CHCl_3$, C_6H_6, Insol. in Et_2O
≪Sa. & D.≫ B_2–H_2SO_4 : 274° B–Pic : 234° B–Ac : 168° Other der. : ref. 67)
≪SD.≫ 67), 99), 100), 101), 102), 103), 104), 105)
≪S. & B.≫ 101), 106), 107), 108), 109)
BS : 64), 110)～119), BG : 120)

Epiberberine $C_{20}H_{18}O_4N^{\oplus} \equiv 336.34$

≪mp≫ 162° (Me_2CO)
≪$[\alpha]_D$≫ 0°
≪Ph.≫
≪OP.≫ Berberis florbunda (aristata)[40]
≪C. & So.≫
≪Sa. & D.≫ $B^{\oplus}$–$I^{\ominus}$: 300° B–Pic : 222°
≪SD.≫ cf. 125)
≪S. & B.≫ cf. 125), 126), 127), 128)

Dehydrocorydalmine $C_{20}H_{20}O_4N^{\oplus} \equiv 338.36$

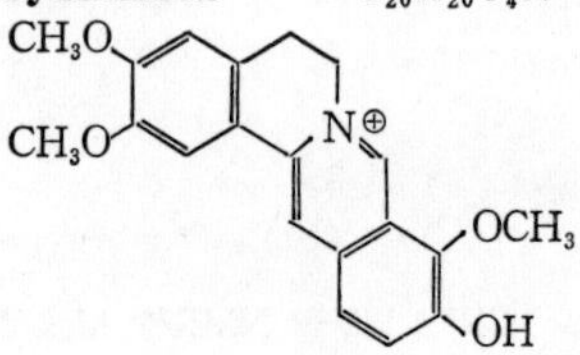

≪mp≫
≪$[\alpha]_D$≫

≪Ph.≫ CR[3] IR[3]
≪OP.≫ Corydalis spp[3]
≪C. & So.≫
≪Sa. & D.≫ $B^{\oplus}$–$I^{\ominus}$: 228～230° (d)
$B^{\oplus}$–OMe·$I^{\ominus}$ = palmatine iodide
≪SD.≫ 3)
≪S. & B.≫

---

* B. buxifolia[37)38], B. darwinii[37)39], B. floribunda (aristata)[40], B. (Mahonia) fortunei[41], B. heterobotnis[41], B. heteropoda (vulgaris)[42], B. himalaica[43], B. insignis[44], B. (Mahonia) japonia[45], B. kawakami[46], B. lambertii[47], B. laurine[48], B. lycium[49], B. mingetsensis[46], B. (Mahonia) nepalensis[50)51)52], B. nervosa[53], B. (Mahonia) swaseyi[54)55], B. thunbergi[56], B. thunbergi var. maximoviczii[57], B. tinctoria (aristata)[43)58], B. (Mahonia) trifoliolata[55], B. umbellata[59], B. tschonoskyana[60], B. vulgaris[61)62)63)64], B. wallichiana[49], Chelidonium majus[65)66], Cocculus palmatus[67], Coptis japonica (orientalis)[58], C. occidentalis[68], C. teeta[5)69)70], C. trifolia[68)71)72], Corydalis cheilanthifolia[73], C. ophiocarpa[74], Coscinum blumeanum[75], C. fenestratum[76)77)78)79], Dicranostigma franchetianum[10], Evodia meliaefolia[80], Glaucium corniculatum[11], Eschscholtzia glauca[13], Hunnemannia fumariaefolia[14], Hydrastis canadensis[69)72)81], Macleaya (Boconia) microcarpa[15], Mahonia acanthifolia[82], M. borealis[83], M. griffithii[84], M. leschenaultii[11], M. manipurensis[85], M. philippinensis[86], M. sikkimensis[85], M. simonsii[83], Nandina domestica[87], Papaver dubium var. lecogii[88], P. somniferum[16], Phellodendron amurense[58)89)90)91], Phellodendron amurense var. sachaliense[92], P. lavallei[93], P. wilsonii[94], Thalictrum fendleri[28], T. foliolosum[95], Toldalia aculeata[80], Xylopia (Coclocline) polycarpa[96], Xanthorhiza apiifolia[72], Xanthoxylum calva[67], X. calva var. herculis[69)97], Glaucium oxylobum[12], Michelia compressa[98], Sanguinaria canadensis[17]

Columbamine $C_{20}H_{20}O_4N^{\oplus} \equiv 338.36$

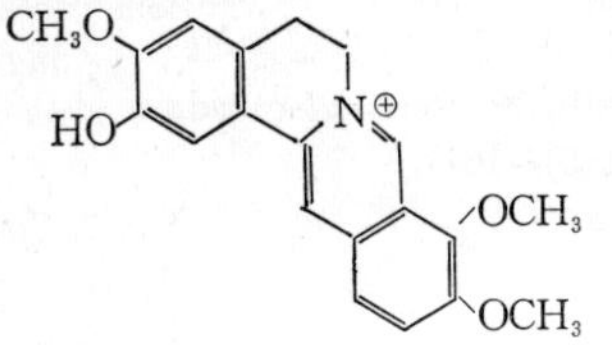

≪mp≫
≪$[\alpha]_D$≫ 0°

≪Ph.≫ UV[305] Rf[305]
≪OP.≫ Archangelisia flava[29], Berberis floribunda (aristata)[40], B. heteropoda (vulgaris)[47], B. thunbergii var. maximowiczii[57]*
≪C. & So.≫
≪Sa. & D.≫ Tetrahydro-Der. =*dl*-isocorypalmine
≪SD.≫ 25), 132), 133)
≪S. & B.≫ 133)

Jatrorrhizine (Jateorhizine, Neprotine) $C_{20}H_{20}O_4N^{\oplus} \equiv 338.36$

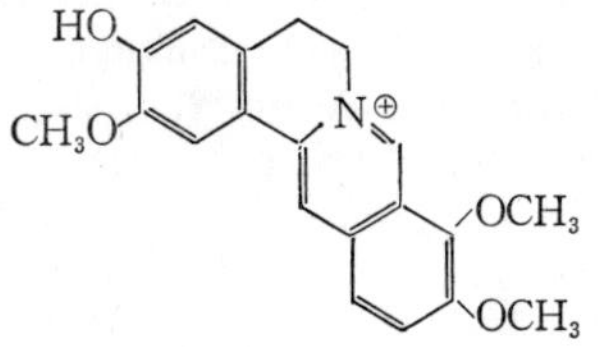

≪mp≫
≪$[\alpha]_D$≫ 0°

≪Ph.≫ UV[305] Rf[305]
≪OP.≫ Archangelisia flava[29], Berberis amurensis var. japonica[33], B. asiatica[36], B. floribunda (aristata)[40], B. (Mahonia) fortunei[41], B. heteropoda (vulgaris)[42], B. himalaica[43]**
≪O. & So.≫ Unknown in free state
≪Sa. & D.≫ $B^{\oplus}$–$Cl^{\ominus}$: 206° $B^{\oplus}$–$I^{\ominus}$: 208～210° $B^{\oplus}$–$NO_3^{\ominus}$: 225° B–Pic : 217～220° (d)
≪SD.≫ 25), 132), 133), 136), 139)
≪S. & B.≫ cf. 133), 140)

Palmatine $C_{21}H_{22}O_4N^{\oplus} \equiv 352.39$

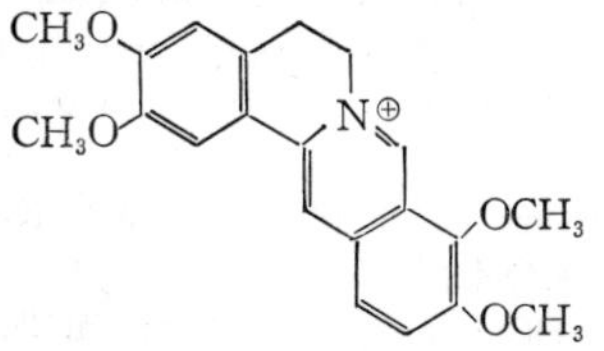

≪mp≫
≪$[\alpha]_D$≫ 0°

≪Ph.≫ UV[305] Rf[305]
≪OP.≫ Archangelisia flava[29], Berberis amurensis var. japonica[33], B. aristata[34], B. asiatica[36], B. floribunda (aristata)[40], B. (Mahonia) fortunei[41], B. heteropoda (vulgaris)[42], B. (Mahonia) japonica[45]***
≪C. & So.≫
≪Sa. & D.≫ $B^{\oplus}$–$Cl^{\ominus}$: 250° (d)
$B^{\oplus}$–$I^{\ominus}$: 241° (d) $B^{\oplus}$–$ClO_4^{\ominus}$: 262°
$B_2^{\oplus}$–$SO_4^{\ominus}$: 250°
≪SD.≫ 132)
≪S. & B.≫ 154), cf. 140)

---

* B. lamberti[47], B. vulgaris[22], Coptis japonica (orientalis)[58], Burasaia madagascariensis[129], Jateorhiza palmata (columba)[25][130][131], Fibraurea tinctoria[305]

** B. kawakami[134], B. mingetsensis[46], B. (Mahonia) nepalensis[50], B. thunbergii (vulgaris)[56], B. thunbergii var. maximowiczii[57], B. tinctoria (aristata)[43], B. tschonoskyana[60], B. vulgaris[22], Burasaia madagascariensis[129], Coptis japonica (orientalis)[58][136], C. teeta[5], Coscinium blumeanum[75], Fibraurea chloroleuca (tinctoria)[137][305], Jateorhiza palmata (columba)[131], Mahonia acanthifolia[82], M. borealis[83], M. griffithii[84], M. leschenaultii[85], M. manipurensis[85], M. philippinensis[86], M. shikkimensis[85], M. simonsii[83], Michelia compressa[98], Nandina domestica[87], Phellodendron amurense[91], P. amurense var. sachalinense[92], Thalictrum fendleri[28], T. foliolosum[138]

*** B. lamberti[47], B. thunbergii var. maximowiczii[57], B. tinctoria (aristata)[43], B. tschonoskyana[60], B. vulgaris[22], Burasaine madagascariensis[129][141], Calystegia hederacea[142][143], Cocculus leaeba[144], Coptis japonica (orientalis)[58][145], C. teeta[5], Coscinium blumeanum[75], Enantia chlorantha[146], E. polycarpa[147], Fibraurea chloroleuca (tinctoria)[137][148][305], Jateorhiza palmata (columba)[131][149], Fagara coco[150], F. hiemalis[150], Mahonia acanthifolia[82], M. borealis[83], M. griffithii[84], M. leschenaultii[85], M. phillippinensis[83][86], M. simonsii[83], Parabaena hirsuta[151], Phellodendron amurense[58], P. amurense var. sachalinense[92], Stephania glabra[152][153], Thalictrum foliolosum[138], Tinospora bakis[144]

Lambertine $C_{20}H_{19}O_4N \equiv 337.36$

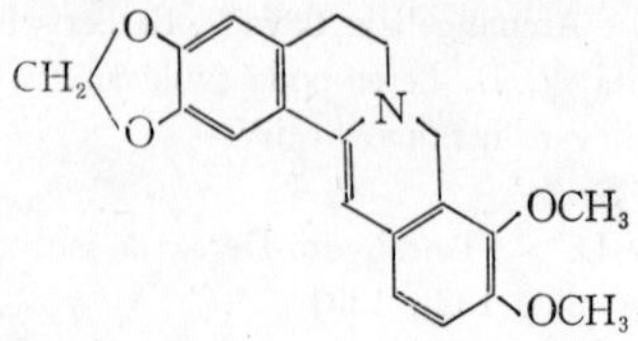

<mp> 164° or 172°
<$[\alpha]_D$> 0°

<Ph.> UV[155]
<OP.> Berberis lamberti[47)155]
<C. & So.>
<Sa. & D.> Dihydro-Der. = *dl*-canadine
<SD.> 123), 155)~161)
<S. & B.> 158)

Berlambine (Oxyberberine) $C_{20}H_{17}O_5N \equiv 351.34$

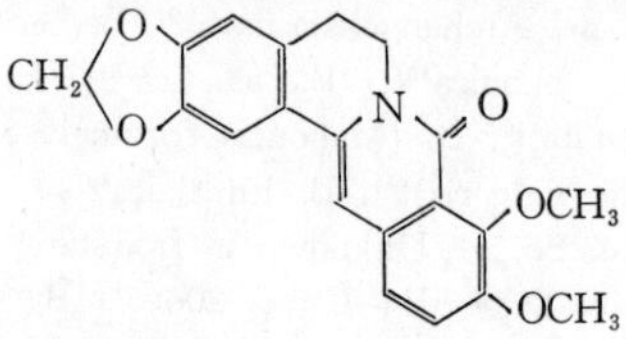

<mp> 200° (Me_2CO)
<$[\alpha]_D$> 0°

<Ph.> UV[155]
<OP.> Berberis amurensis var. japonica[33], B. lambertii[47)155], B. mingetsensis[46], B. morrisonensis[162], B. thunbergii[163], B. thunbergii var. maximowiczii[57], B. tschonoskayana[60], F. coco[150]
<C. & So.> Colourless plates
<Sa. & D.>
<SD.> 155), 164)
<S. & B.> 108), 154), 155), cf. 107), 140)

***l*-Stylopine (*l*-Tetrahydrocoptisine)**
$C_{19}H_{17}O_4N \equiv 323.33$

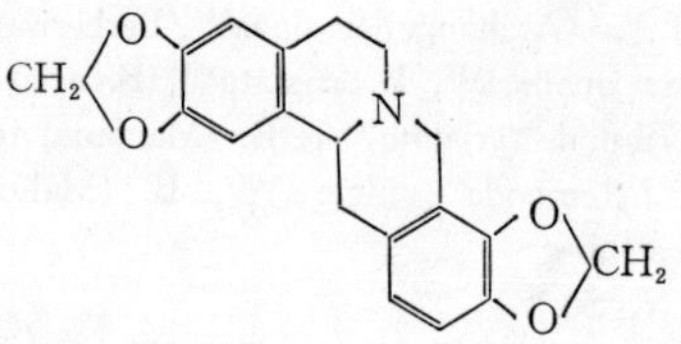

<mp> 203°
<$[\alpha]_D$> −310°

<Ph.> CR[3] UV[3)274] IR[3] Rf[3]
<OP.> Chelidonium majus[165], Corydalis ambigua[7)166], C. cheilantifolia[73], C. claviculata[167], C. lutea[168], C. nobilis[169], C. ophiocarpa[74], C. ternata[170], Dactylicapnos (Dicentra) macrocapnos[171]*
<C. & So.> Prisms or needles, Sol. in $CHCl_3$, Et_2O, Me_2CO, AcOH
<Sa. & D.> B-HCl: ca. 250° B-HBr: ca. 260° B-MeI: 248~252°
<SD.> 4), 20), 170), 180)
<S. & B.> 21), 128), cf. 20), BS: 181)

***d*-Stylopine**
<mp> 203°
<$[\alpha]_D$> +310°
<Ph.>
<OP.> Corydalis cava (ternata)[174], C. cornata[175], C. solida[176], C. thalictrifolia[177]
<C. & So.>
<Sa. & D.>
<SD.> → *l*-stylopine
<S. & B.> → *l*-stylopine

***dl*-Stylopine**
<dm> 221°
<$[\alpha]_D$> 0°
<Ph.>
<OP.> Chelidonium majus[165], Corydalis ambigua[7)166], C. cheilanthifolia[73]**
<C. & So.> bp 260°/0.01
<Sa. & D> B-HCl: 266~269° B-Pic: 138~141°
<SD.> → *l*-stylopine
<S. & B.> → *l*-stylopine

* Dicranostigma franchetianum[172], Fumaria officinalis[173], Stylophonum diphyllum[172], Corydalis tuberosa[174]

** C. claviculata[167], C. lutea[168], C. nobilis[169], C. platycarpa[179], C. thalictrifolia[177], Dactylicapnos (Dicentra) macrocapnos[171], Dicranostigma franchetianum[172], Fumaria officinalis[173], Corydalis tuberosa[174]

Nandinine $C_{19}H_{19}O_4N \equiv 325.35$

≪mp≫ 146°
≪$[\alpha]_D$≫ +63° (EtOH)
≪Ph.≫ CR[184]
≪OP.≫ Nandina domestica[182)183)184]
≪C. & So.≫
≪Sa. & D.≫ B–OMe: 139°
≪SD.≫ 25), 26), 184)
≪S. & B.≫ 25), 27)

Cheilanthifoline $C_{19}H_{19}O_4N \equiv 325.35$

≪mp≫ 184° (MeOH)
≪$[\alpha]_D$≫ −311° (MeOH)
≪Ph.≫
≪OP.≫ Corydalis cheilanthifolia[73)185)], C. scouleri[185)186)], C. sibirica[185)187)]
≪C. & So.≫ Plates, Sol. in most. org. solv.
≪Sa. & D.≫ B–OMe=*l*–sinactine B–OEt: 144°
≪SD.≫ 185)
≪S. & B.≫

l-Scoulerine $C_{19}H_{21}O_4N \equiv 327.37$

≪mp≫ 204° (MeOH)
≪$[\alpha]_D$≫ −70°
≪Ph.≫ UV[191)200)] IR[191)200)] NMR[191)] Mass[191)200)] ORD[200)] Rf[191)]
≪OP.≫ Corydalis caseana[188)], C. micrantha[189)], C. montana (aurea)[190)], C. scouleri[186)], C. sibirica[187)], C. pallida[194)], Fumaria officinalis[173)], Opium[191)]
≪C. & So.≫ Fine grey needles
≪Sa. & D.≫ B–HCl: 209° B–O,O–DiMe= tetrahydropalmatine B–O,O–DiEt: 155°
≪SD.≫ 193), 199), 200)
≪S. & B.≫ 200), BG: 191)

d-Scoulerine
≪mp≫ 204°
≪$[\alpha]_D$≫ +70°
≪Ph.≫
≪OP.≫ Corydalis cava (tuberosa)[192)], C. plotycarpa[179)]
≪C. & So.≫
≪Sa. & D.≫
≪SD.≫ → *l*–scoulerine
≪S. & B.≫ → *l*–scoulerine

dl-Scoulerine (Aurotensine)[193)]
≪mp≫ 126∼127° (MeOH)
≪$[\alpha]_D$≫ 0°
≪Ph.≫
≪OP.≫ Corydalis aurea[195)196)], C. ochotensis[197)], C. platycarpa[179)]*
≪C. & So.≫ Pink plates
≪Sa. & D.≫
≪SD.≫ → *l*–scoulerine
≪S. & B.≫ → *l*–scoulerine

* C. solida (bulbosa)[176)], Fumaria officinalis[173)], Glaucium flavum[198)], G. serpieri[172)]

l-Canadine $C_{20}H_{21}O_4N \equiv 339.38$

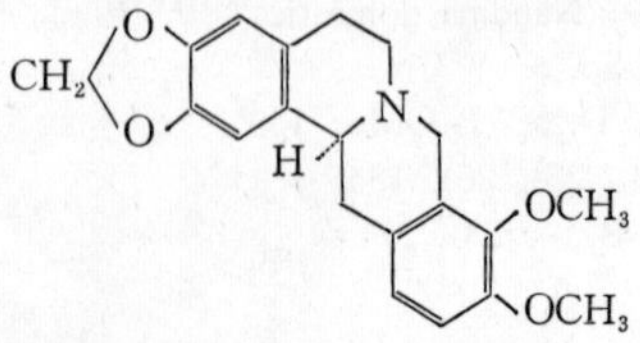

≪mp≫ 134~135° (MeOH)
≪[α]D≫ −298° ($CHCl_3$)
≪Ph.≫ CR[202] IR[210]
≪OP.≫ Corydalis cheilanthifolia[73][201], C. ophiocarpa[74], C. ternata[170][202], Hydrastis canadensis[203], Xanthoxylum brachyacanthum[204]
≪C. & So.≫ Needles, Sol. in Et_2O, $CHCl_3$, EtOH, Insol. in H_2O
≪Sa. & D.≫ B–MeI α: 220° β: 264°
≪SD.≫ 101), 105), 161), 180), 204), 205)
≪S. & B.≫ 101), 109), 206), BS: 118)

d-Canadine

≪mp≫ 134~135° (MeOH)
≪[α]D≫ +298° ($CHCl_3$)
≪Ph.≫
≪OP.≫ Corydalis cava (tuberosa)[129][174]
≪C. & So≫
≪Sa. & D.≫
≪SD.≫ → l–canadine
≪S. & B.≫ → l–canadine

l-Sinactine $C_{20}H_{21}O_4N \equiv 339.38$

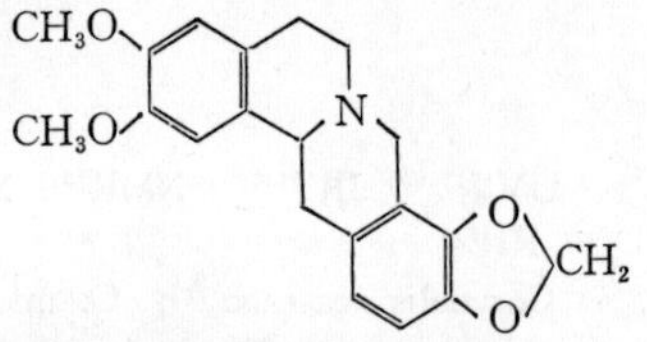

≪mp≫ 175° (EtOH)
≪[α]D≫ −312° ($CHCl_3$)
≪Ph.≫ CR[211] UV[213]
≪OP.≫ Coscinium blumeanum[211], Fumaria officinalis[173], Sinomenium acutum[212]
≪C. & So.≫ Prisms, Sol. in $CHCl_3$, Spar. sol. in MeOH, EtOH, Insol. in H_2O. Oxidised in air.
≪Sa. & D≫ B–HCl: 275° (d) B–H_2PtCl_6: 247°
≪SD.≫ 73), 180), 213), 214)
≪S. & B.≫ 21), 126), 128), 208), 214), 215)

dl-Sinactine

≪mp≫ 170° (EtOH)
≪[α]D≫ 0°
≪Ph.≫
≪OP.≫ Fumaria officinalis[173][185]
≪C. & So.≫ Needles
≪Sa. & D.≫ B–HCl: 286° (d)
≪SD.≫ → l–sinactine
≪S. & B.≫ → l–sinactine

Corydalmine $C_{20}H_{23}O_4N \equiv 341.39$

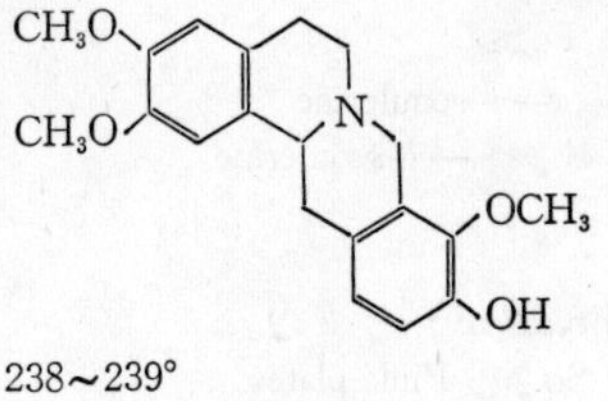

≪mp≫ 238~239°
≪[α]D≫ + 337.4° ($CHCl_3$)
≪Ph.≫ CR[3] UV[3] IR[3] Rf[3]
≪OP.≫ Corydalis spp[3]
≪C. & So.≫ Colourless needles
≪Sa. & D.≫
≪SD.≫ 3)
≪S. & B.≫

l-Isocorypalmine (Casealutine, Somniferine[216]) $C_{20}H_{23}O_4N \equiv 341.39$

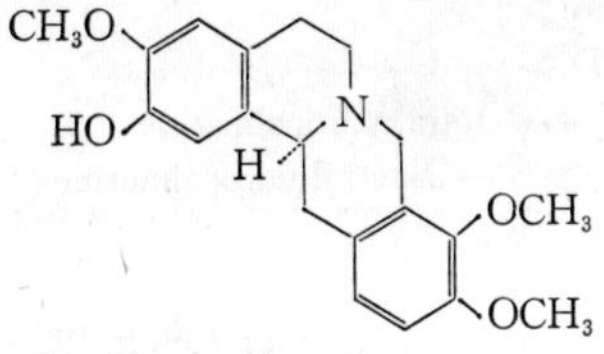

≪mp≫ 246° (in vacuo)
≪$[\alpha]_D$≫ −303° ($CHCl_3$)

≪Ph.≫ CR[188] UV[16] IR[16][218] Mass[16] Rf[16]
≪OP.≫ Corydalis caseana[188], C. lutea[168], C. ochroleuca[217], C. platycarpa[179], Papaver somniferum[16]
≪C. & So.≫ Colourless prisms
≪Sa. & D.≫ B–OMe=tetrahydropalmatine
B–Pic: 190～193°
≪SD.≫ 16), 133), 199)
≪S. & B.≫ 218), BG: 16)

d-Isocorypalmine
≪mp≫ 246° (in vacuo)
≪$[\alpha]_D$≫ +303° ($CHCl_3$)
≪Ph.≫
≪OP.≫ Corydalis cava (tuberosa)[192], C. nobilis[169]

≪C. & So.≫
≪Sa. & D.≫
≪SD.≫ → l–isocorypalmine
≪S. & B.≫ → l–isocorypalmine

l-Corypalmine $C_{20}H_{23}O_4N \equiv 341.39$

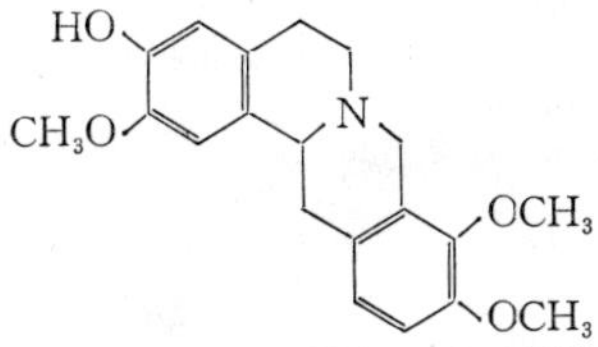

≪mp≫ 230° ($CHCl_3$–MeOH), cf. 266°[188]
≪$[\alpha]_D$≫ −274° ($CHCl_3$)

≪Ph.≫ CR[211] UV[211] IR[218][221]
≪OP.≫ Corydalis ambigua[166], C. caseana[188], C. cava[192], C. cheilanthifolia[73], C. insisa[138], C. ochroleuca[217], C. ophiocarpa[74], C. thalictrifolia[177], Coscinium blumeanum[211], Dicentra oregna[219]
≪C. & So.≫ Colourless prisms
≪Sa. & D.≫
≪SD.≫ 133), 139), 180), 199), 220)
≪S. & B.≫ 133), 218)

d-Corypalmine
≪mp≫ 230°
≪$[\alpha]_D$≫ +274° ($CHCl_3$)
≪Ph.≫
≪OP.≫ Corydalis cava[192], C. tuberosa[220]

≪C. & So.≫
≪Sa. & D.≫
≪SD.≫ → l–corypalmine
≪S. & B.≫ → l–corypalmine

d-Tetrahydropalmatine $C_{21}H_{25}O_4N \equiv 355.42$

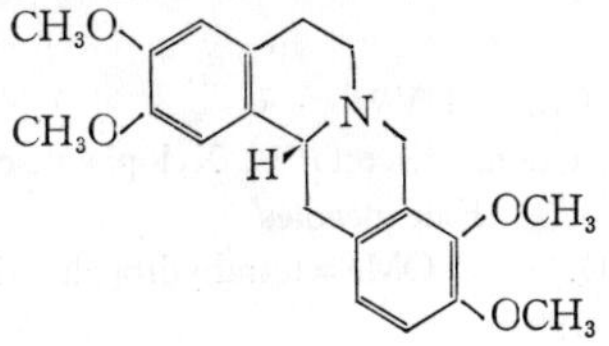

≪mp≫ 141～143°
≪$[\alpha]_D$≫ +291° (EtOH)

≪Ph.≫ CR[211] UV[3][211][227] IR[3][215][227] NMR[227][235] Mass[227] ORD[230][236] Rf[3]
≪OP.≫ Corydalis cava (tuberosa)[220], C. decumbens[222][223], C. nobilis[169], C. pallida[194], Stephania glabra[152][153]
≪C. & So.≫ Turns yellow in air
cf. bp 150°/5×10^{-5}
≪Sa. & D.≫ B–HCl: 266° B–Pic: 113°
≪SD.≫ 26), 180), 199), 220), 228), 229), 230)
≪S. & B.≫ 21), 154), 206), 209), 215), 218), 231), 232), 233), 234)

l-Tetrahydropalmatine

≪mp≫ 141～143°
≪[α]$_D$≫ −291° (EtOH)
≪Ph.≫
≪OP.≫ Corydalis aurea[195,196], C. cava (ternata)[224], C. casean[188], C. lutea[168], C. microcantha[189], C. ochroleuca[217], C. platycarpa[179], Stephania grabra[225]
≪C. & So.≫
≪Sa. & D.≫
≪SD.≫ → *d*-tetrahydropalmatine
≪S. & B.≫ → *d*-tetrahydropalmatine

dl-Tetrahydropalmatine

≪mp≫ 148°
≪[α]$_D$≫ 0°
≪Ph.≫
≪OP.≫ Corydalis ambigua[7,166], C. angustifolia[226], C. aurea[195,196], C. cava (ternata)[129], C. monteana[190], C. nobilis[169], C. pallida[194], C. solida[176], Coscinium blumeanum[211], Stephania glabra[152,153,225]
≪C. & So.≫
≪Sa. & D.≫ B–HCl: 215° B–HI: 241° B–$HAuCl_4$: 201° B–MeI: 230°&266° (d)
≪SD.≫ → *d*-tetrahydropalmatine
≪S. & B.≫ → *d*-tetrahydropalmatine

l-Bis-*O*,*O*′-bidemethyl-tetrahydropalmatine (Alkaloid HF) $C_{19}H_{21}O_4N$≡327.37

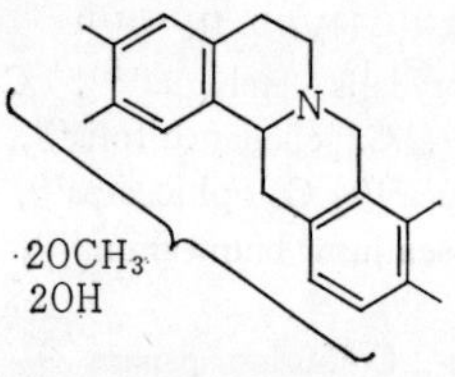

2OCH$_3$
2OH

≪mp≫ 201～202° (Et_2O or EtOH)
≪[α]$_D$≫ −356±3° ($CHCl_3$)
≪Ph.≫ CR[14] UV[14] IR[14] pKa[14] Rf[14]
≪OP.≫ Hunnemannia fumariaefolia[14]
≪C. & So.≫ Prisms
≪Sa. & D.≫ B–HCl: 266～268° B–O,O–DiMe=tetrahydropalmatine
≪SD.≫ 14)
≪S. & B.≫

Discretamine $C_{19}H_{21}O_4N$≡327.37

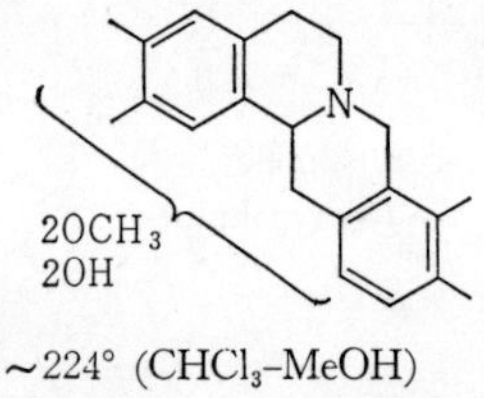

2OCH$_3$
2OH

≪mp≫ 221～224° ($CHCl_3$-MeOH)
≪[α]$_D$≫ −368° (Py)
≪Ph.≫ CR[199] UV[199]
≪OP.≫ Xylopia discreta[199]
≪C. & So.≫ Prisms
≪Sa. & D.≫ B–O,O–DiMe=tetrahydropalmatine
≪SD.≫ 199)
≪S. & B.≫

Discretinine $C_{20}H_{23}O_4N$≡341.39

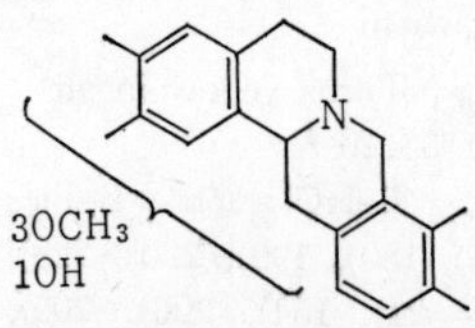

3OCH$_3$
1OH

≪mp≫ 212～214° ($CHCl_3$-Me_2CO)
≪[α]$_D$≫ −371° (Py)
≪Ph.≫ CR[199] UV[199]
≪OP.≫ Unona discreta[237], Xylopia discreta[199]
≪C. & So.≫ Fine needles
≪Sa. & D.≫ B–OMe=tetrahydropalmatine
≪SD.≫ 199)
≪S. & B.≫

Alkaloid F 51 $C_{20}H_{23}O_4N \equiv 341.39$

≪mp≫ 171°
≪[α]D≫ 0°
≪Ph.≫
≪OP.≫ Corydallis pallida[194]
≪C. & So.≫
≪Sa. & D.≫ B–OMe=*dl*–tetrahydropalmatine
≪SD.≫
≪S. & B.≫

Cyclanoline $C_{20}H_{24}O_4N^{\oplus} \equiv 342.39$

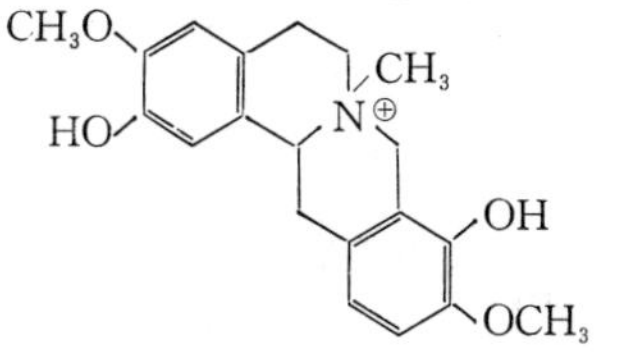

≪mp≫
≪[α]D≫ B⊕–Cl⊖ −116° (MeOH)
≪Ph.≫ UV[239] IR[239,240] Rf[240]
≪OP.≫ Cissampelos (Cyclea) insularis[238,239], Aristalochia debilis[240]
≪C. & So.≫
≪Sa. & D.≫ B⊕–Cl⊖ : 214∼215° B⊕–I⊖ : 184° B–Pic : 162° B⊕–O,O–DiMe-I⊖ : 208° (d)
≪SD.≫ 239), 241)
≪S. & B.≫

Steponine $C_{20}H_{24}O_4N^{\oplus} \equiv 342.39$

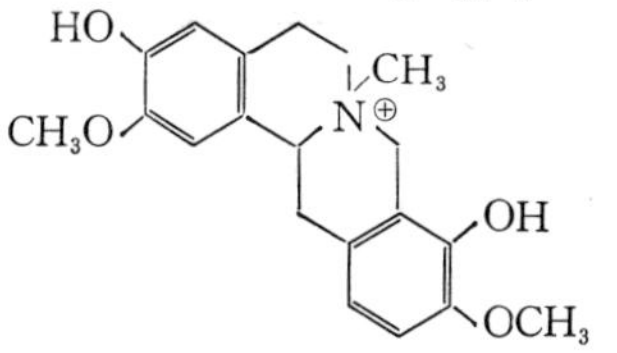

≪mp≫
≪[α]D≫ B⊕Cl⊖ −130° (MeOH)
≪Ph.≫ UV[242,243] IR[242,243]
≪OP.≫ Stephania japonica[242]
≪C. & So.≫
≪Sa. & D.≫ B⊕–Cl⊖ : 235° B⊕–I⊖ : 178°
≪SD.≫ 242), 243)
≪S. & B.≫

***α-l-N*-Methylcanadine** $C_{21}H_{24}O_4N^{\oplus} \equiv 354.40$

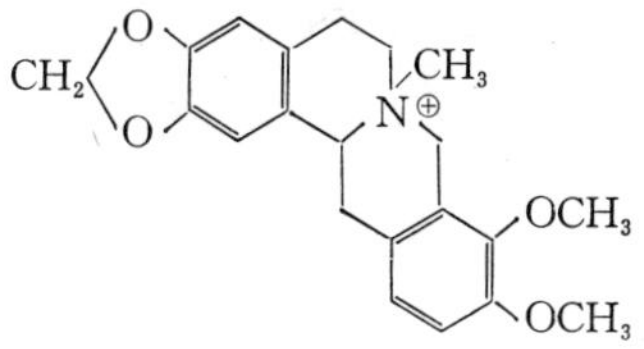

≪mp≫
≪[α]D≫ B⊕–Cl⊖ +134.2°
≪Ph.≫ IR[244]
≪OP.≫ Fagara rhoifolia[244]
≪C. & So.≫
≪Sa. & D.≫ B⊕–Cl⊖ : 235∼237°
≪SD.≫ 244)
≪S. & B.≫

Coreximine $C_{19}H_{21}O_4N \equiv 327.37$

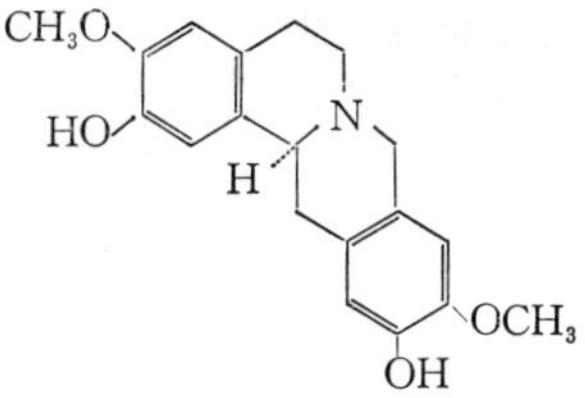

≪mp≫ 262° ($CHCl_3$)
≪[α]D≫ −411° (Py)
≪Ph.≫ UV[245,249] IR[245] NMR[245] Rf[248,249]
≪OP.≫ Asimina triloba[245], Dicentra eximia[196]
≪C. & So.≫ Colourless prisms
≪Sa. & D.≫ B–O,O–DiMe=xylopinine B–O,O–DiEt : 131° B–O,O–DiAc : 192∼195°
≪SD.≫ 199), 200), 246), 247)
≪S. & B.≫ 200), 247), 248), 249), 250), 251)

Discretine $C_{20}H_{23}O_4N \equiv 341.39$

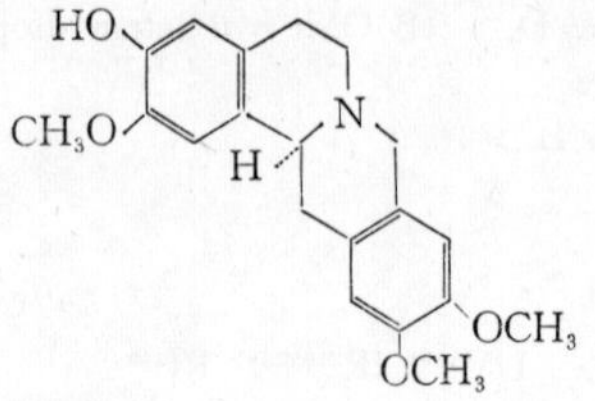

≪mp≫ 180~181°
≪$[\alpha]_D$≫ −300° ($CHCl_3$)

≪Ph.≫ CR[198] UV[199]
≪OP.≫ Xylopia (Unona) discreta[237][249]
≪C. & So.≫
≪Sa. & D.≫ B–OMe=xylopinine
≪SD.≫ 199), 252)
≪S. & B.≫

Xylopinine (*l*-Norcoralydine) $C_{21}H_{25}O_4N \equiv 355.42$

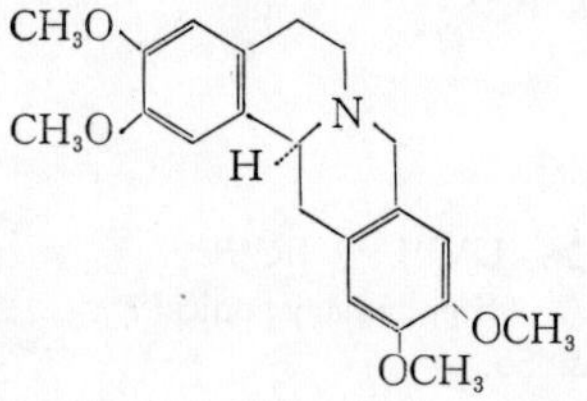

≪mp≫ 182° or 157~158° (EtOH)
≪$[\alpha]_D$≫ −297° ($CHCl_3$)

≪Ph.≫ CR[199] UV[199][256] IR[245][248] NMR[245] Mass[227]
≪OP.≫ Xylopia (Unona) discreta[199][237]
≪C. & So.≫ Colourless leaflets
≪Sa. & D.≫ B–HCl: 219° B–HBr: 225° B–Pic: 138°
≪SD.≫ 199), 229)
≪S. & B.≫ 21), 229), 231), 248), 253), 254), 255)

Phellodendrine $C_{20}H_{24}O_4N^{\oplus} \equiv 342.39$

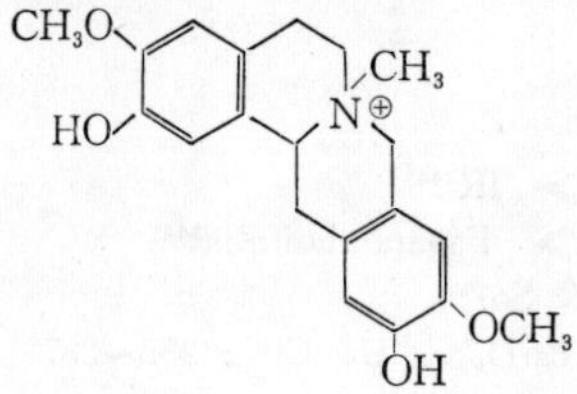

≪mp≫
≪$[\alpha]_D$≫ B⊕–I⊖ −132° (MeOH)

≪Ph.≫ CR[91] UV[91] IR[91][248] Rf[91][248]
≪OP.≫ Phellodendron amurense[91][257], P. amurense var. sachaliense[258]
≪C. & So.≫
≪Sa. & D.≫ B⊕–I⊖: 258~258.5 (d)
≪SD.≫ 91), 258), 259)
≪S. & B.≫ 248), 260)

Stepharotine $C_{21}H_{25}O_5N \equiv 371.42$

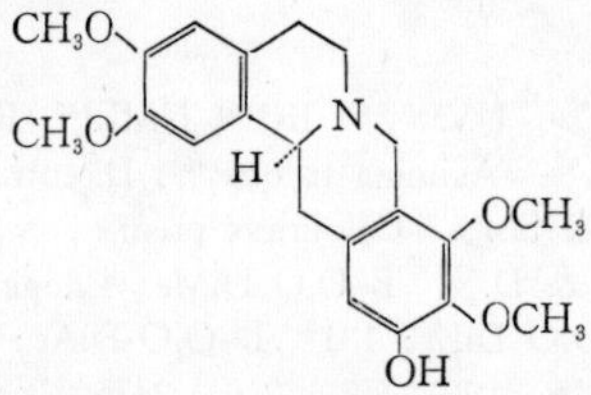

≪mp≫
≪$[\alpha]_D$≫ B–HBr −203° (MeOH)

≪Ph.≫ CR[227] UV[227] IR[227] NMR[227] Mass[227]
≪OP.≫ Stephania rotunda[227]
≪C. & So.≫ Oil
≪Sa. & D.≫ B–HBr: 227~229° B–OMe: 133~135°
≪SD.≫ 227)
≪S. & B.≫

Capaurimine $C_{20}H_{23}O_5N \equiv 357.39$

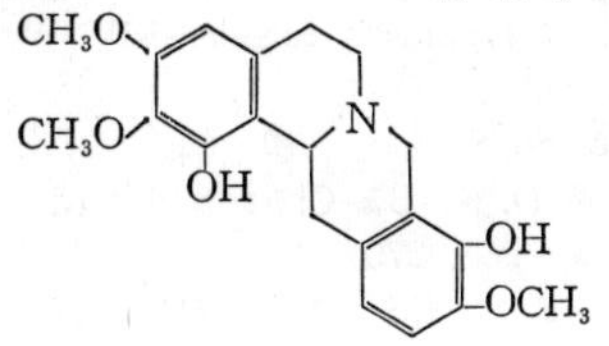

≪mp≫ 212° ($CHCl_3$–MeOH)
≪$[\alpha]_D$≫ −287° ($CHCl_3$)

≪Ph.≫
≪OP.≫ Corydalis montana[190], C. pallida[194]
≪C. & So.≫ Stout prisms
≪Sa. & D.≫ B–O,O–DiMe=O–methylcapaurine
≪SD.≫ 261)
≪S. & B.≫

Capaurine $C_{21}H_{25}O_5N \equiv 371.42$

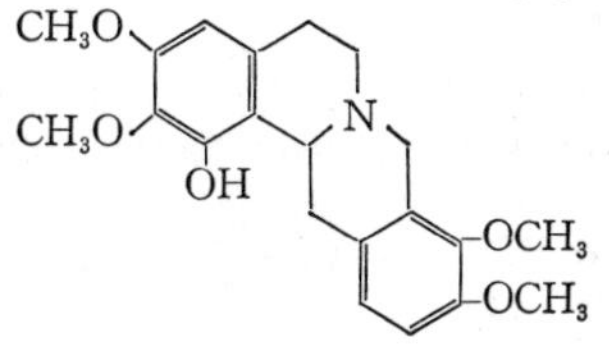

≪mp≫ 164° ($CHCl_3$–MeOH)
≪$[\alpha]_D$≫ −271° ($CHCl_3$)

≪Ph.≫
≪OP.≫ Corydalis aurea[195,262], C. crystallina[189], C. montana (aurea)[190,262], C. micrantha[60,189,190,262], C. pallida[194,262]
≪C. & So.≫ Large stout polyhedra, Sol. in $CHCl_3$, Spar. sol. in MeOH
≪Sa. & D.≫ B–OMe: 152° B–OEt: 134°
≪SD.≫ 189), 262)
≪S. & B.≫ 232)

Capauridine (*dl*-Capaurine)
≪mp≫ 208° ($CHCl_3$–MeOH)
≪$[\alpha]_D$≫ 0°
≪Ph.≫
≪OP.≫ Corydalis aurea[195], C. micrantha[189], C. montana[190], C. pallida[194]
≪C. & So.≫ Colourless prisms, Spar. sol. in $CHCl_3$
≪Sa. & D.≫ B–OMe: 142°
≪SD.≫ → capaurine
≪S. & B.≫ → capaurine

Ophiocarpine $C_{20}H_{25}O_5N \equiv 355.38$

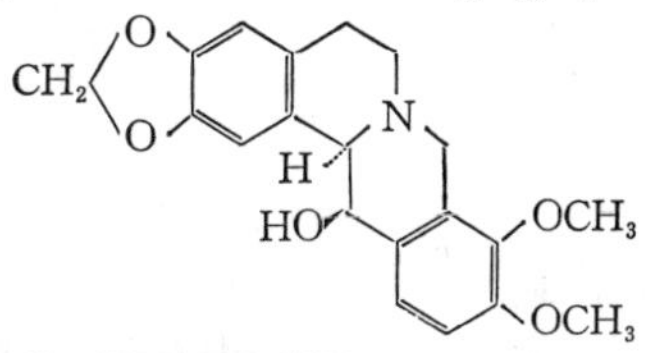

≪mp≫ 188° ($CHCl_3$–MeOH)
≪$[\alpha]_D$≫ −284° ($CHCl_3$)

≪Ph.≫ CR[74] UV[210,263,267] IR[210,263,265,267] NMR[210] pKa[210,263]
≪OP.≫ Corydalis ophiocarpa[74]
≪C. & So.≫ Stout prisms, Spar. sol. in MeOH
≪Sa. & D.≫ B–MeI: 271°
B–OAc: 141.3° (165∼167°)
≪SD.≫ 74), 232), 263), 264)
≪S. & B.≫ 232), 265), 266)

Thalidastine $C_{19}H_{16}O_5N^{\oplus} \equiv 338.32$

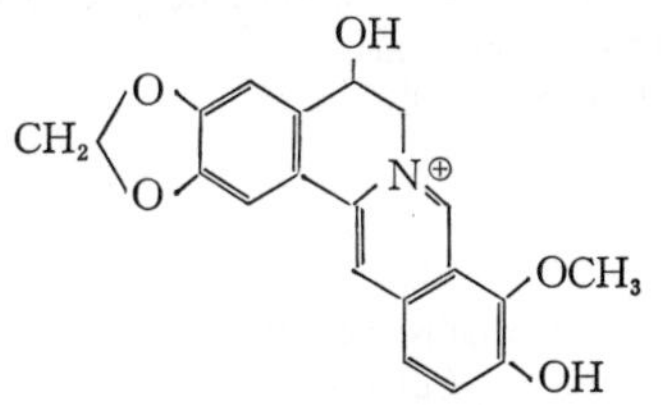

≪mp≫
≪$[\alpha]_D$≫ B⊕–Cl⊖ +138° (MeOH)

≪Ph.≫ CR[268] UV[268] NMR[268] Mass[268] ORD[268]
≪OP.≫ Thalictrum fendleri[268]
≪C. & So.≫
≪Sa. & D.≫ B⊕–Cl⊖: >200° (d)
≪SD.≫ 268), 269)
≪S. & B.≫

Berberastine $C_{20}H_{18}O_5N^{\oplus} \equiv 352.34$

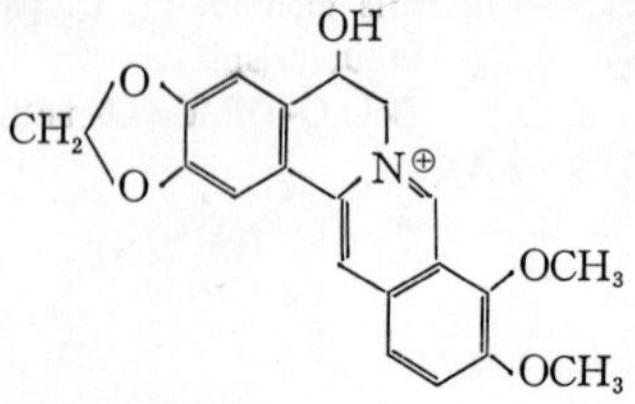

≪mp≫
≪$[\alpha]_D$≫ B⊕–I⊖ +107° (EtOH)
B⊕–Cl⊖ +138° (MeOH)

≪Ph.≫ CR[270] UV[268,270] IR[270]
≪OP.≫ Hydrastis canadensis[270,271], Thalictrum fendleri[268]
≪C. & So.≫
≪Sa. & D.≫ B⊕–Cl⊖ : >230° (d)
≪SD.≫ 268, 270)
≪S. & B.≫ BS: 118), 271)

Worenine $C_{20}H_{16}O_4N^{\oplus} \equiv 334.33$

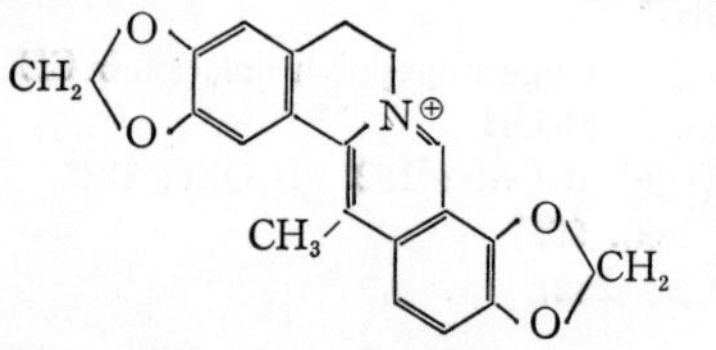

≪mp≫ 212∼213° (EtOH)
≪$[\alpha]_D$≫ 0°

≪Ph.≫ CR[145]
≪OP.≫ Coptis japonica (orientalis)[145]
≪C. & So.≫ Needles
≪Sa. & D.≫ B⊕–Cl⊖ : 265°
≪SD.≫ 145°
≪S. & B.≫

Dehydrothalictrifoline $C_{21}H_{20}O_4N^{\oplus} \equiv 338.36$

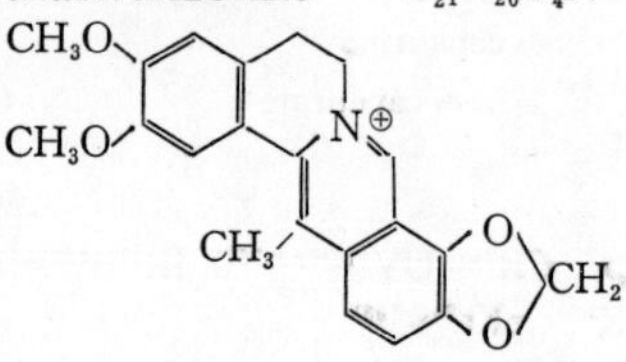

≪mp≫
≪$[\alpha]_D$≫ 0°

≪Ph.≫ UV[272] IR[272]
≪OP.≫ Corydalis thalictrifolia[177]
≪C. & So.≫
≪Sa. & D.≫ B⊕–Cl⊖ : 271°
Tetrahydro–Der. = *dl*–thalictrifoline
≪SD.≫ 272)
≪S. & B.≫

Dehydrocorydaline $C_{22}H_{24}O_4N^{\oplus} \equiv 366.41$

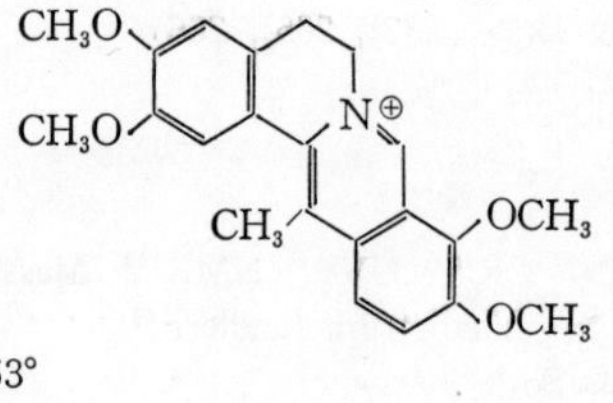

≪mp≫ 163°
≪$[\alpha]_D$≫ 0°

≪Ph.≫ Rf[3]
≪OP.≫ Berberis floribunda (aristata)[40], Corydalis ambigua[6], C. aurea[190], C. cava[273], C. montana[190], C. decumbens[222]
≪C. & So.≫
≪Sa. & D.≫ B⊕–I⊖ : 245∼6° (d)
B⊕–NO_3⊖ : 237° B⊕–$HAuCl_4$⊖ : 219°
≪SD.≫ →corydaline
≪S. & B.≫

Tetrahydrocorysamine $C_{20}H_{19}O_4N \equiv 337.36$

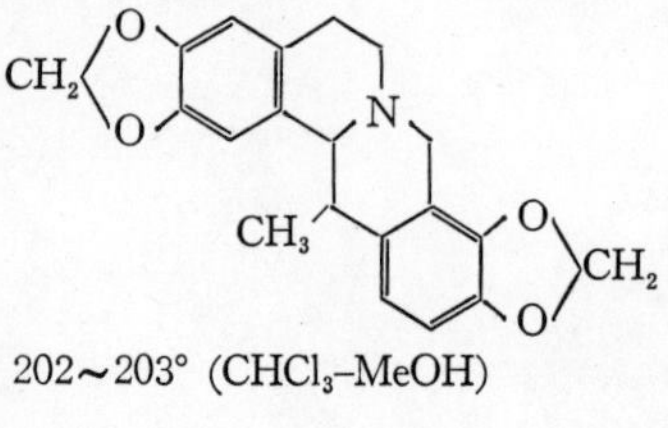

≪mp≫ 202∼203° ($CHCl_3$–MeOH)
≪$[\alpha]_D$≫

≪Ph.≫ CR[274] UV[274] IR[275]
≪OP.≫ Coydalis incisa[274]
≪C. & So.≫ Colourless scales
≪Sa. & D.≫ B–Pic: 189∼190°(d) B–MeI: 266°(d)
≪SD.≫ 275), 276)
≪S. & B.≫ 275)

Thalictricavine $C_{21}H_{23}O_4N \equiv 353.40$

≪mp≫ 149° ($CHCl_3$–MeOH)
≪$[\alpha]_D$≫ +292° ($CHCl_3$)

≪Ph.≫ UV[272,279] IR[272,277,278,279] NMR[277,279] ORD[277]
≪OP.≫ Corydalis tuberosa[224]
≪C. & So.≫ Colourless needles
≪Sa. & D.≫
≪SD.≫ 224), 277), 278)
≪S. & B.≫ 275), 279)

Thalictrifoline $C_{21}H_{23}O_4N \equiv 353.40$

≪mp≫ 155° (MeOH)
≪$[\alpha]_D$≫ +218° (MeOH)

≪Ph.≫ UV[272] IR[272] Rf[272]
≪OP.≫ Corydalis thalictrifolia[177]
≪C. & So.≫ Colourless prisms
≪Sa. & D.≫
≪SD.≫ 177), 278)
≪S. & B.≫

Base II $C_{21}H_{23}O_4N \equiv 353.40$

≪mp≫ 144～149°
≪$[\alpha]_D$≫ +306° ($CHCl_3$)

≪Ph.≫ CR[272] UV[272] IR[272] NMR[272] Rf[272]
≪OP.≫ Corydalis ambigua var. amurense[272]
≪C. & So.≫
≪Sa. & D.≫
≪SD.≫ 272), 277)
≪S. & B.≫

Isocorybulbine $C_{21}H_{25}O_4N \equiv 355.42$

≪mp≫ 187.5～188.5° (EtOH)
≪$[\alpha]_D$≫ +301° ($CHCl_3$)

≪Ph.≫ IR[277,278] NMR[277,278]
≪OP.≫ Corydalis cava (tuberosa)[226,280]
≪C. & So.≫ Colourless plates
≪Sa. & D.≫ B–MeI: 218～221°
B–OMe=corydaline
≪SD.≫ 277), 281), 282), 283), 284), 285), 286)
≪S. & B.≫ 284)

Corybulbine $C_{21}H_{25}O_4N \equiv 355.42$

≪mp≫ 237～238° (MeOH)
≪$[\alpha]_D$≫ +303.3° ($CHCl_3$)

≪Ph.≫ IR[277,278] NMR[278]
≪OP.≫ Corydalis ambigua[6], C. cava (tuberosa)[224,287,288], C. platycarpa[179]
≪C. & So.≫ Needles, Sol. in Me_2CO, $CHCl_3$, Spar. sol. in Et_2O, AcOEt, Insol. in H_2O
Unstable in light
≪Sa. & D.≫ B–HCl: 245～50° (d)
B–OEt: 129～130°
≪SD.≫ 139), 277), 278), 281), 282), 283), 284), 285), 289), 290)
≪S. & B.≫ cf. 281)

Corydaline $C_{22}H_{27}O_4N \equiv 369.44$

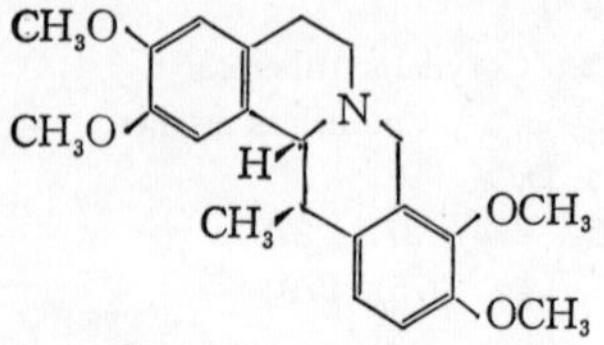

≪mp≫ 135° (EtOH)
≪$[\alpha]_D$≫ +295° (EtOH), +300° ($CHCl_3$)

≪Ph.≫ UV[3,300] IR[3,256,272,277,278,300] NMR[277,278] ORD[277] Rf[3]
≪OP.≫ Corydalis ambigua[6], C. aurea[71,196], C. cava (tuberosa)[280,291,292,293], C. decipiens (pumlia)[226], C. montana[190], C. platycarpa[179], C. nobilis[169], C. solida[176]
≪C. & So.≫ Prisms, Sol. in $CHCl_3$, Mod. sol. in Et_2O, Spar. sol. in MeOH, EtOH, Insol. in H_2O
≪Sa. & D.≫ B–HCl: 206° B–HNO_3: 198° B–$HAuCl_4$: 207° B–H_2PtCl_6: 227° B–MeI: 228°
≪SD.≫ 253), 256), 277), 278), 280), 286), 293), 294)~301)
≪S. & B.≫ 283), 293), 299), 300), 302)

Alborine $C_{22}H_{22}O_6N^{\oplus} \equiv 396.40$

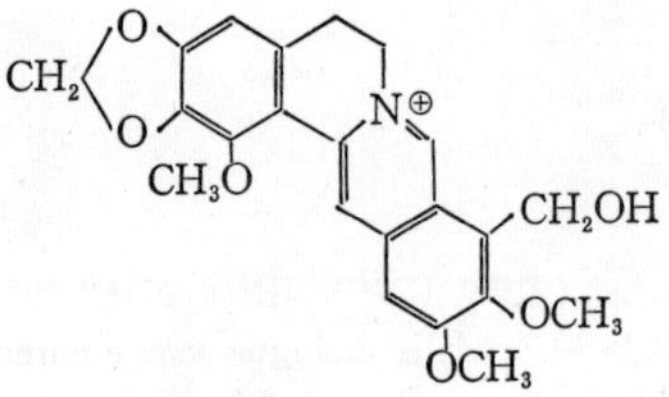

≪mp≫ 238~240° (MeOH–Et_2O)
≪$[\alpha]_D$≫ 0°

≪Ph.≫ UV[303]
≪OP.≫ Papaver alboroseum[303]
≪C. & So.≫ Yellow solid, Sol. in MeOH, Insol. in Et_2O
≪Sa. & D.≫ B⊕–Cl⊖: >360° B⊕–I⊖: >300°
≪SD.≫ 303°
≪S. & B.≫

Coralydine $C_{22}H_{27}O_4N \equiv 369.44$

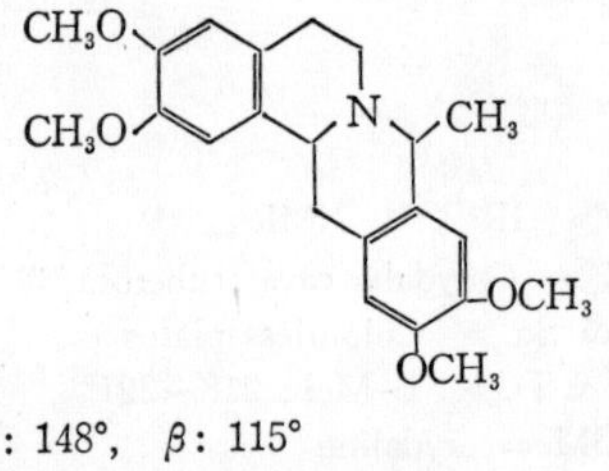

≪mp≫ α: 148°, β: 115°
≪$[\alpha]_D$≫

≪Ph.≫ UV[300] IR[300] NMR[235]
≪OP.≫
≪C. & So.≫
≪Sa. & D.≫
≪SD.≫ 300)
≪S. & B.≫ 56), 231), 297), 301)

Burasaine $C_{21}H_{23}O_4N \equiv 353.40$

≪mp≫
≪$[\alpha]_D$≫ 0°
≪Ph.≫ UV[304]
≪OP.≫ Burasaia madagascariensis[141,304]

≪C. & So.≫
≪Sa. & D.≫
≪SD.≫ 141), 304)
≪S. & B.≫

Thalictrine $C_{20}H_{27}O_4N \equiv 345.42$

≪mp≫ 208° (aq. EtOH)
≪$[\alpha]_D$≫ +308° (H_2O)
≪Ph.≫
≪OP.≫ Thalictrum foliolosum[95]

≪C. & So.≫
≪Sa. & D.≫ B–Pic: 207~208°
≪SD.≫ 95)
≪S. & B.≫

1) L. Slavikova, J. Slavik: Coll. Czech. Chem. Comm., **21**, 211 (1956).
2) W. Awe: Arzneimittel Forsch., **1**, 287 (1951).
3) I. Imaseki, H. Taguchi: J. Pharm. Soc. Japan, **82**, 1214 (1962).
4) Z. Kitasato: Proc. Imp. Acad. (Tokyo), **2**, 124 (1926).
5) R. Chatterjee, M. P. Guha, A. Chatterjee: J. Indian Chem. Soc., **29**, 97 (1952).
6) K. Makoshi: Arch. Pharm., **246**, 381 (1908).
7) Huang-Minlon: Ber., **69**., 1737 (1936).
8) C. Tani, N. Takao, S. Takao: J. Pharm. Soc. Japan, **82**, 748 (1962).
9) J. Slavik: Coll. Czech. Chem. Comm., **24**, 2506 (1959).
10) L. Slavikova, J. Slavik: Chem. Listy, **51**, 1923 (1957).
11) J. Slavik, L. Slavikova: Coll. Czech. Chem. Comm., **22**, 279 (1957).
12) J. Slavik, L. Slavikova: Coll. Czech. Chem. Comm., **28**, 2530 (1963).
13) L. Slavikova, J. Slavik: Coll. Czech. Chem. Comm., **31**, 3362 (1966).
14) L. Slavikova, J. Slavik: Coll. Czech. Chem. Comm., **31**, 1355 (1966).
15) J. Slavik, L. Slavikova: Coll. Czech. Chem. Comm., **20**, 356 (1955).
16) S. Pfeifer: Pharmazie, **21**, 378 (1966).
17) J. Slavik, L. Slavikova: Coll. Czech. Chem. Comm., **25**, 1667 (1960).
18) L. Slavikova, Tschu, J. Slavik: Coll. Czech. Chem. Comm., **25**, 756 (1960).
19) J. Slavik: Coll. Czech. Chem. Comm., **25**, 1663 (1960).
20) E. Späth, R. Posega: Ber., **62**, 1029 (1929).
21) C.K. Bradsher, N. L. Dutta: J. Org. Chem., **26**, 2231 (1961).
22) E. Späth, N. Polgar: Ber., **52**, 117 (1929).
23) G. Frerichs: Arch. Pharm., **248**, 276 (1910).
24) G. Frerichs, P. Stoepel: Arch. Pharm., **251**, 321 (1916).
25) E. Späth, G. Burger: Ber., **59**, 1486 (1926).
26) E. Späth, W. Leithe: Ber., **63**, 3007 (1930).
27) T.R. Govindachari, S. Rajadurai, C. V. Ramadas: J. Sci. Ind. Res., **18 B**, 533 (1959). [C. A., **54**, 21169 (1960)].
28) M. Shamma, M. A. Greenberg, B. S. Dudock: Tetrahedron Letters, **1965**, 3595.
29) A.C. Santos: Univ. Philippines Nat. and Appl. Sci. Bull., **1**, 153 (1931). [C., **1931**, II, 3218].
30) P.A. Foote: J. Am. Pharm. Assoc., **21**, 246 (1932).
31) A.C. Santos, P. Adkilen: J. Am. Chem. Soc., **54**, 2923 (1932).
32) A.G. Perkin: J. Chem. Soc., **71**, 1194 (1897).
33) M. Tomita, T. Kugo: J. Pharm. Soc. Japan, **75**, 753 (1955).
34) Parson: Pharm. J., [3], **13**, 46 (1882).
35) H. Pommerhne: Arch. Pharm., **233**, 127 (1895).
36) R. Chatterjee, A.Banerjee, A.K. Barua, A.K. DasGupta: J. Indian Chem. Soc., **31**, 83 (1954).
37) F. Richert: Rev. centro estud. agronemia y. veterinaria univ., Buenos Aires, **11**, 11 (1918).
38) Arata: Répert. pharm., **1892**, 45.
39) B. T. Cromwell: Biochem. J., **27**, 860 (1933).
40) R. Chatterjee: J. Indian Chem. Soc., **28**, 225 (1951).
41) M. Tomita, T. Abe: J. Pharm. Soc. Japan, **72**, 773 (1952).
42) A. Orekhov: Arch. Pharm., **271**, 323 (1933).
43) R. Chatterjee, M.P. Guha, A.K. DasGupta: J. Indian Chem. Soc., **29**, 921 (1952).
44) R. Chatterjee: J. Am. Pharm. Assoc., **30**, 247 (1941).
45) M. Tomita, T. Abe: J. Pharm. Soc. Japan, **72**, 735 (1952).
46) T.H. Yang, S.T. Lu: J. Pharm. Soc. Japan, **80**, 849 (1960).
47) R. Chatterjee, A. Banerjee: J. Indian Chem. Soc., **30**, 705 (1953).
48) L.Gurguel, O. de A. Costa, R. D. Silva: Bol. assoc. brasil. farm., **15**, 11 (1934). [C. A., **28**, 3521 (1934)].
49) R. Chatterjee: J. Indian Chem. Soc., **19**, 233 (1942).
50) R. Chatterjee: Sci. and Cult., **7**, 619 (1942).
51) R. Chatterjee: J. Am. Pharm. Assoc., **33**, 210 (1944).
52) T.R. Govindachari, B.R. Pai, S. Rajadurai, C.V. Ramadas, U.R. Rao: Proc. Indian Acad. Sci., **47 A**, 41 (1958). [C.A., **52**, 14630 (1958)].
53) Neppach: Am. J. Pharm., **1878**, 373.
54) G.A. Greathouse, E. Rigler: Plant. Physiol., **15**, 563 (1940).
55) G.A. Greathouse, G. M. Watkins: Am. J. Botany, **25**, 743 (1938).
56) H.H. S. Fong, J. L. Beal, M. P. Cava: Lloydia, **29**, 94 (1966). [C.A., **65**, 7229 (1966)].
57) H. Kondo, M. Tomita: Arch. Pharm., **268**, 549 (1930).
58) Y. Murayama, K. Shinozaki: J. Pharm. Soc. Japan, **530**, 299 (1926).
59) R. Chatterjee: J. Indian Chem. Soc., **17**,

289 (1940).
60) M. Tomita, T. Kugo: J. Pharm. Soc. Japan, **79**, 317 (1959).
61) J.A. Buchner, C. A. Buchner: Ann., **24**, 228 (1837).
62) T. Fleitmann: Ann., **59**, 160 (1846).
63) O. Hesse: Ber., **19**, 3190 (1886).
64) J. L. Beal, E. Ramstad: Naturwiss., **47**, 206 (1960).
65) J. O. Schlofferbeck: Am. J. Pharm., **79**, 584 (1902).
66) J. Gadamer: Apozeker Ztg., **39**, 1569 (1924).
67) W.H. Perkin, Jr.: J. Chem. Soc., **55**, 63 (1889).
68) C. E. Mollett, B. V. Christensen: J. Am. Pharm. Assoc., **23**, 310 (1934).
69) J. D. Perrins: J. Chem. Soc., **15**, 334 (1862).
70) D. Hooper: Pharm. J., [4], **34**, 482 (1912).
71) Schultz: J. Pharm., [3], **14**, 973 (1884).
72) Gordon: Arch. Pharm., **240**, 146 (1902).
73) R. H. F. Manske: Can. J. Res., **20B**, 57 (1942).
74) R. H. F. Manske: Can. J. Res., **17 B**, 51 (1939).
75) M. Tomita, C. Tani: J. Pharm. Soc. Japan, **61**, 251 (1941).
76) J. D. Perrins: Phil. Mag., [4], **4**, 99 (1852).
77) J. Stenhouse: J. Chem. Soc., **20**, 187 (1867).
78) R. Child, W. R. N. Nathanael: Current Sci. (India), **12**, 255 (1933).
79) N. S. Varier, P. P. Pillai: Current Sci. (India), **12**, 228 (1933).
80) A. G. Perkin, J. J. Humel: J. Chem. Soc., **67**, 413 (1895).
81) Mahla: Am. J. Sci., [2], **33**, 43 (1862).
82) R. Chatterjee, M. P. Guha: J. Am. Pharm. Assoc., **39**, 577 (1950).
83) R. Chatterjee, M. P. Guha, S. K. Sen: J. Am. Pharm. Assoc., **40**, 36 (1951).
84) R. Chatterjee, M. P. Guha: J. Am. Pharm. Assoc., **40**, 181 (1951).
85) R. Chatterjee, M. P. Guha: J. Am. Pharm. Assoc., **40**, 229 (1951).
86) E. R. Castro, A. C. Santos, P. Valenzuela: Univ. Philippines Natl. and Appl. Sci. Bull., **2**, 401 (1932). [C., **1933**, II, 1358].
87) M. Tomita, Y. Inubushi, S. Ishii, H. Yamagata: J. Pharm. Soc. Japan, **71**, 381 (1951).
88) W. Egels: Planta Med., **7**, 92 (1959). [C. A., **53**, 15475 (1959)].
89) K. Shimo: Sci. Rep. Tohoku Univ., (1), **10**, 331 (1921).
90) Y. Murayama, J. Takeda: J. Pharm. Soc. Japan, **550**, 1035 (1927).
91) M. Tomita, J. Kunitomo: J. Pharm. Soc. Japan, **80**, 880 (1960).
92) M. Tomita, J. Kunitomo: J. Pharm. Soc. Japan, **78**, 1444 (1958).
93) V.I. Frolova, A.I. Bankowskiĭ, M.B. Wolyskaya: Med. Ind. USSR., **12**, No.6, 16 (1958).
94) H.-Y. Hsu: J. Taiwan. Pharm. Assoc., **7**, 2 (1955).
95) S. K. Vashistka, S. Siddiqui: J. Indian Chem. Soc., **18**, 641 (1941).
96) J. Stenhouse: Ann., **95**, 108 (1855); **105**, 360 (1858).
97) M. Chevalier, G. Pelletan: J. Chim. Med., **2**, 314 (1826).
98) K. Ito: J. Pharm. Soc. Japan, **80**, 705 (1960).
99) E. Schmidt, C. Shilbach: Arch. Pharm., **225**, 164 (1887).
100) W. H. Perkin, Jr.: J. Chem. Soc., **57**, 992 (1890).
101) J. Gadamer: Arch. Pharm., **239**, 648 (1901).
102) J. Gadamer: Chem. Ztg., **26**, 291 (1902).
103) J. Gadamer: Arch. Pharm., **243**, 31 (1905).
104) W. H. Perkin, Jr., R. Robinson: J. Chem. Soc., **97**, 305, 318 (1910).
105) J. W. McDavid, W. H. Perkin, Jr., R. Robinson: J. Chem. Soc., **101**, 1218 (1912).
106) A. Pictet, A. Gams: Ber., **44**, 2480 (1911).
107) R. D. Haworth, W. H. Perkin, Jr.: J. Chem. Soc., **125**, 1686 (1924).
108) W. H. Perkin, Jr., J. N. Roy, R. Robinson: J. Chem. Soc., **127**, 740 (1925).
109) I. Sallay, R. H. Ayres: Tetrahedron, **19**, 1397 (1963).
110) I. Imaseki, R. Oneyama, M. Tajima: J. Pharm. Soc. Japan, **80**, 1802 (1960).
111) J. R. Gear, I. D. Spenser: Nature, **191**, 1393 (1961).
112) I. D. Spenser, J. R. Gear: Proc. Chem. Soc., **1962**, 228.
113) J. R. Gear, I. D. Spenser: J. Am. Chem. Soc., **84**, 1059 (1962).
114) D. H. R. Barton, R. H. Hesse, G. W. Kirby: Proc. Chem. Soc., **1963**, 267.
115) A. R. Battersby, R. J. Francis, M. Hirst, J. Staunton: Proc. Chem. Soc., **1963**, 268.
116) I. Monkovic, I. D. Spenser: Proc. Chem. Soc., **1964**, 223.
117) R. N. Gupta, I. D. Spenser: Can. J. Chem., **43**, 133 (1965).
118) I. Monkovic, I. D. Spenser: Can. J. Chem., **43**, 2017 (1965).
119) D. H. R. Barton, R. H. Hesse, G. W. Kirby: J. Chem. Soc., **1965**, 6379.
120) R. Robinson: "The Structural Relations of Natural Products", Clarendon Press, Oxford.

(1955).
121) R. Chatterjee: J. Indian Chem. Soc., **19**, 233 (1942).
122) B. Skinner: J. Chem. Soc., **1950**, 823.
123) B. Witkop: J. Am. Chem. Soc., **78**, 2873 (1956).
124) T. Takemoto, Y. Kondo: J. Pharm. Soc. Japan, **82**, 1673 (1962).
125) M. Tomita, J. Niimi: J. Pharm. Soc. Japan, **79**, 1023 (1959).
126) W. H. Perkin, Jr.: J. Chem. Soc., **113**, 492 (1918).
127) J. S. Buck, W. H. Perkin, Jr.: J. Chem. Soc., **125**, 1673 (1925).
128) R. D. Haworth, W. H. Perkin, Jr.: J. Chem. Soc., **1926**, 1769.
129) A. Resplandy: Compt. rend., **247**, 2428 (1958). [C. A., **53**, 13510 (1959)].
130) E. Günzel: Arch. Pharm., **244**, 257 (1905).
131) K. Feist: Arch. Pharm., **245**, 586 (1906).
132) K.Feist, G. Sandstedt: Arch. Pharm., **256**, 1 (1918).
133) E. Späth, E. Mosettig: Ber., **60**, 383 (1927).
134) T.-H. Yang, S.-T. Lu: J. Pharm. Soc. Japan, **80**, 705 (1960).
135) M. Tomita, T. Kikuchi: J. Pharm. Soc. Japan, **76**, 597 (1956).
136) C. Tani, N. Takao: J. Pharm. Soc. Japan, **77**, 805 (1957).
137) M. Tomita, C. Tani: J. Pharm. Soc. Japan, **61**, 241 (1941).
138) R. H. F. Manske: J. Am. Chem. Soc., **72**, 3207 (1950).
139) E. Späth, E. Mosettig: Ber., **58**, 2133(1925).
140) E. Späth, H. Quietensky: Ber., **58**, 2267 (1925).
141) A. Resplandy: Compt. rend., **245**, 725(1957). [C. A., **51**, 18486 (1957)].
142) T. Y. Chou,P. K. Chiang: Acta Chem. Sinica, **21**, 168 (1955).
143) W.-I. Huang, Y.-C. Chen: Acta Chem. Sinica, **23**, 230 (1957). [C. A., **52**, 15827 (1958)].
144) L. Beauquense: Bull. Sci. Pharmacol., **45**, 7 (1938).
145) Z. Kitasato: J. Pharm. Soc. Japan, **542**, 315 (1927).
146) G. Seitz: Naturwiss., **46**, 263 (1959).
147) A.Buzas, M. Osowiecki, G. Régnier: Compt. rend., **248**, 1397 (1959). [C. A., **53**, 18078 (1959).
148) J.-H. Chu, J.-C. Chen, S.-T. Sheng: Hua Hsueh Hsueh Pao, **28**, 59 (1962). [C. A., **60**, 6887 (1964)].
149) J. Gadamer: Arch. Pharm., **240**, 450(1902).
150) A. M. Kuck, S. M. Albonico, V. Deulofeu: Chem. & Ind., **1966**, 945.
151) M. Tomita, T. Asada, Y. Watanabe: J. Pharm. Soc. Japan, **72**, 203 (1952).
152) G.R. Chaudhry, S. Siddiqui: J. Sci. Ind. Res., **9 B**, 79 (1950). [C. A., **45**, 823 (1951).
153) G.R. Chaudhry, S. Siddiqui: J. Sci. Ind. Res., **11 B**, 337 (1952). [C. A., **48**, 14117 (1954)],
154) R. D.Haworth, J. B. Koepfli, W. H. Perkin, Jr.: J. Chem. Soc., **1927**, 548.
155) R. Chatterjee, P. C. Maiti: J. Indian Chem. Soc., **32**, 609 (1955).
156) H. Schmid, P. Karrer: Helv. Chim. Acta, **32**, 960 (1949).
157) M. Birsch: Arch. Pharm., **283**, 192 (1950).
158) S. Bose: J. Indian Chem. Soc., **32**, 450 (1955).
159) R. Chatterjee, P. C. Maiti: J. Indian Chem. Soc., **32**, 605 (1955).
160) R. Mirza: J. Chem. Soc., **1957**, 4400.
161) W. Awe, H. Wichmann, R. Buerhop: Ber., **90**, 1997 (1957).
162) T.-H. Yang: J. Pharm. Soc. Japan, **80**, 1302 (1960).
163) M. Tomita, T.-H. Yang: J. Pharm. Soc. Japan, **80**, 845 (1960).
164) W. H. Perkin, Jr., R. Robinson: J. Chem. Soc., **101**, 263 (1912).
165) J. Slavik: Coll. Czech. Chem. Comm., **20**, 198 (1955).
166) T. Q. Chou: Chinese J. Physiol., **2**, 203 (1928); **3**, 69, 301 (1929); **7**, 35 (1933); **8**, 155 (1934); **10**, 507 (1936).
167) R. H. F. Manske: Can. J. Res., **18 B**, 97 (1940).
168) R. H. F. Manske: Can. J. Res., **17 B**, 89 (1939).
169) R. H. F. Manske: Can. J. Res., **18 B**, 288 (1940).
170) J. Go: J. Pharm. Soc. Japan, **50**, 940 (1930).
171) R. H. F. Manske: Can. J. Res., **21 B**, 117 (1943).
172) R. H. F. Manske: Can. J. Res., **20 B**, 53 (1942).
173) R. H. F. Manske: Can. J. Res., **16 B**, 438 (1938).
174) E. Späth, P. L. Julian: Ber., **64**, 1131 (1931).
175) R. H. F. Manske: Can. J. Res., **24 B**, 66 (1946).
176) R. H. F.Manske: Can. J. Chem., **34**, 1 (1956).
177) R. H. F. Manske: Can. J. Res., **21 B**, 111 (1943).

178) F. S. Bandelin, W. Malesh: J. Am. Pharm. Assoc., **45**, 704 (1956).
179) R. H. F. Manske: Can. J. Res., **21 B**, 13 (1943).
180) W. Leithe: Ber., **64**, 2827 (1931).
181) A. R. Battersby, R. J. Francis, E. A. Ruveda, J. Staunton: Chem.Comm., **1965**, 89.
182) J. F. Eijkman: Rec. trav. chim., **3**, 197 (1884).
183) Z. Kitasato: J. Pharm. Soc. Japan, **522**, 695 (1925).
184) Z. Kitasato: Acta Phytochim. Japan, **3**, 175 (1927).
185) R. H. F. Manske: Can. J. Res., **18 B**, 100 (1940).
186) R. H. F. Manske: Can. J. Res., **14 B**, 747 (1936).
187) R. H. F. Manske: Can. J. Res., **14 B**, 354 (1936).
188) R. H. F. Manske, M. R. Miller: Can. J. Res., **16 B**, 153 (1938).
189) R. H. F. Manske: Can. J. Res., **17 B**, 57 (1939).
190) R. H. F. Manske: Can. J. Res., **20 B**, 49 (1942).
191) E. Brochman-Hanssen, B. Nielsen: Tetrahedron Letters, **1966**, 2261.
192) J. Gadamer, E. Späth, E. Mosettig: Arch. Pharm., **265**, 675 (1927).
193) R. H. F. Manske: Can. J. Res., **18 B**, 414 (1940).
194) R. H. F. Manske: Can. J. Res., **18 B**, 80 (1940).
195) R. H. F. Manske: Can. J. Res., **9**, 436 (1933).
196) R. H. F. Manske: Can. J. Res., **16 B**, 81 (1938).
197) R. H. F. Manske: Can. J. Res., **18 B**, 75 (1940).
198) R. H. F. Manske: Can. J. Res., **17 B**, 399 (1939).
199) J. Schmutz: Helv. Chim. Acta, **42**, 335 (1959).
200) A. R. Battersby, R. Southgate, J. Staunton, M. Hirst: J. Chem. Soc. (C), **1966**, 1052.
201) R. H. F. Manske: Can. J. Res., **20 B**, 57 (1942).
202) J. Go: J. Pharm. Soc. Japan, **49**, 801 (1929).
203) E. Schmidt: Arch. Pharm., **232**, 136 (1894).
204) F. L. Pyman: J. Chem. Soc., **103**, 817 (1931).
205) A. Voss, J. Gadamer: Arch. Pharm., **248**, 43 (1910).
206) T. R. Govindachari, S. Rajadurai, N. Viswanathan: J. Sci. Ind. Res., **B 18**, 176 (1959).
207) C. K. Bradsher, N. L. Dutta: J.Am. Chem. Soc., **82**, 1145 (1960).
208) C. K. Bradsher, N. L. Dutta: J. Org. Chem., **26**, 2231 (1961).
209) Roussel-UCLAE: Neth. Appl., 6,501,747. [C. A., **64**, 2135 (1966)].
210) M. Ohta, H. Tani, S. Morozumi: Chem. Pharm. Bull., **12**, 1072 (1964).
211) M. Tomita, C. Tani: J. Pharm. Soc. Japan, **61**, 251 (1941).
212) K. Goto, H. Sudzuki: Bull. Chem. Soc. Japan, **4**, 220 (1929).
213) K. Goto, Z. Kitasato: J. Chem. Soc., **1930**, 1234.
214) E. Späth, E. Mosettig: Ber., **64**, 2048 (1931).
215) C. K. Bradsher, N. L. Dutta: Nature, **184**, 1943 (1959).
216) S. Pfeifer, J. Teige: Pharmazie, **17**, 692 (1962).
217) R. H. F. Manske: Can. J. Res., **17 B**, 95 (1939).
218) T. R. Govindachari, S. Rajadurai, C. V. Ramadas: Ber., **92**, 1654 (1959).
219) R. H. F. Manske: Can. J. Res., **10**, 765 (1934).
220) E. Späth, E. Mosettig, O. Tröthandl: Ber., **56**, 875 (1923).
221) C. Schöpf, M. Schweickert: Ber., **98**, 2566 (1965).
222) Y. Asahina, S. Motigase: J. Pharm. Soc. Japan, **463**, 766 (1920).
223) M. Osada: J. Pharm. Soc. Japan, **47**, 99 (1927).
224) R. H. F. Manske: J. Am. Chem. Soc., **75**, 4928 (1953).
225) K.-C. Fang, I. I. Fadeeva, T. N. Ilinskaya: Khim. Prirodn. Soedin., Akad. Nauk Uz. SSR., **1965**, 392. [C. A., **64**, 14236 (1966)].
226) H. G. Boit, H. Ehmke: Naturwiss., **46**, 427 (1959).
227) M. Tomita, M. Kozuka, S. Uyeo: J. Pharm. Soc. Japan, **86**, 460 (1966).
228) E. Späth, E. Mosettig: Ber., **59**, 1496 (1926).
229) H. Coroddi, E. Hardegger: Helv. Chim. Acta, **39**, 889 (1956).
230) J. C. Craig, S. K. Roy: Tetrahedron, **21**, 401 (1965).
231) E. Späth, E. Kruta: Monatsh., **50**, 341 (1928).
232) T. R. Govindachari, S. Rajadurai, M. Subramanian, N. Viswanathan: J. Chem. Soc., **1957**, 2943.
233) N. L. Dutta, C. K. Bradsher: J. Org. Chem., **27**, 2213 (1962).

234) T.-C. Sun, S.-H. Lu, J.-Y. Chi: Yao Hsueh Hsueh Pao, **12**, 314 (1965). [C. A., **63**, 7062 (1965)].
235) L. Pohl, W. Wiegrebe: Z. Naturforsch., **b 20**, 1032 (1965).
236) G. G. Lyle: J. Org. Chem., **25**, 1779 (1960).
237) E. N. Porabirsing: Dissertation, 1952.
238) M. Tomita, T. Kikuchi: J. Pharm. Soc. Japan, **77**, 67 (1957).
239) M. Tomita. T. Kikuchi: J. Pharm. Soc. Japan, **77**, 73 (1957).
240) M. Tomita, K. Fukagawa: J. Pharm. Soc. Japan, **82**, 1673 (1962).
241) M. Tomita, T. Kikuchi: J. Pharm. Soc. Japan, **77**, 69 (1957).
242) M. Tomita, Y. Watanabe, M. Fuse: J. Pharm. Soc. Japan, **77**, 274 (1957).
243) Y. Watanabe: J. Pharm. Soc. Japan, **77**, 278 (1957).
244) J. M. Calderwood, F. Fish: Chem. & Ind., **1966**, 237.
245) M. Tomita, M. Kozuka: J. Pharm. Soc. Japan, **85**, 77 (1965).
246) R. H. F. Manske: J. Am. Chem. Soc., **72**, 4796 (1950).
247) R. H. F. Manske, W. R. Ashford: J. Am. Chem. Soc., **73**, 5144 (1951).
248) M. Tomita, J. Kunitomo: J. Pharm. Soc. Japan, **80**, 1238 (1960).
249) A. R. Battersby, O. J. LeCount, S. Garratt, R. I. Thrift: Tetrahedron, **14**, 46 (1961).
250) A. R. Battersby, R. Southgate, J. Staunton, M. Hirst: J. Chem. Soc. (C), **1966**, 1052.
251) T. Kametani, M. Ihara: J. Chem. Soc. (C), **1966**, 2010.
252) F. Bernoulli, H. Linde, K. Meyer: Helv. Chim. Acta, **46**, 323 (1963).
253) A. Pictet, T. Q. Chou: Ber., **49**, 370 (1916).
254) A. R. Battersby, R. Binks, P. S. Uzzell: Chem. & Ind., **1955**, 1039.
255) D. W. Brown, S. F. Dyke: Tetrahedron Letters, **1964**, 3587.
256) H. W. Bersch: Arch. Pharm., **291**, 595 (1958).
257) M. Tomita, T. Nakano: Pharm. Bull., **5**, 10 (1957).
258) M. Tomita, J. Kunitomo: J. Pharm. Soc. Japan, **78**, 1444 (1958).
259) M. Tomita, J. Kunitomo: J. Pharm. Soc. Japan, **80**, 885 (1960).
260) M. Tomita, J. Kunitomo: J. Pharm. Soc. Japan, **80**, 1245 (1960).
261) R. H. F. Manske: J. Am. Chem. Soc., **69**, 1800 (1947).
262) R. H. F. Manske, H. L. Holmes: J. Am. Chem. Soc., **67**, 95 (1945).
263) M. Ohta, H. Tani, S. Morozumi: Tetrahedron Letters, **1963**, 859.
264) W.-K. Kuei, C.C. Change, K.S. Lin: Hua Hsueh Hsueh Pao, **28**, 59 (1962). [C.A., **60**, 6887 (1964)].
265) T.R. Govindachari, S. Rajadurai: J. Chem. Soc., **1957**, 557.
266) I. Iwasa, S. Naruto: J. Pharm. Soc. Japan, **86**, 534 (1966).
267) T. Takemoto, Y. Kondo: J. Pharm. Soc. Japan, **82**, 1413 (1962).
268) M. Shamma, B.S. Dudock: Tetrahedron Letters, **1965**, 3825.
269) H.F. Andrew, C.K. Bradsher: Tetrahedron Letters, **1966**, 3069.
270) M.M. Nijland: Pharm. Weekblad., **98**, 301 (1963). [C.A., **59**, 10142 (1963)].
271) I. Monkovic, I.D. Spenser: J. Am. Chem. Soc., **87**, 1137 (1965).
272) H. Taguchi, I. Imazeki: J. Pharm. Soc. Japan, **84**, 955 (1964).
273) E. Schmidt: Arch. Pharm., **246**, 577 (1908).
274) C. Tani, N. Takao: J. Pharm. Soc. Japan, **82**, 598 (1962).
275) C. Tani, N. Takao, S. Takao: J. Pharm. Soc. Japan, **82**, 748 (1962).
276) C. Tani, N. Takao, S. Takao: J. Pharm. Soc. Japan, **82**, 751 (1962).
277) Y. Kondo: J. Pharm. Soc. Japan, **83**, 1017 (1963).
278) P.W. Jeffs: Experientia, **21**, 690 (1965).
279) T. Takemoto, Y. Kondo: J. Pharm. Soc. Japan, **82**, 1408 (1962).
280) M. Freund, W. Josephy: Ber., **25**, 2411 (1892).
281) D. Bruns: Arch. Pharm., **241**, 631 (1903).
282) F. Von Bruchhausen, K. Salwai: Arch. Pharm., **263**, 602 (1925).
283) E. Späth, A. Dobrowsky: Ber., **58**, 1274 (1925).
284) E. Späth, H. Holter: Ber., **59**, 2800 (1926).
285) J. Gadamer, K. Salwai: Arch. Pharm., **264**, 401 (1926).
286) F. von Bruchhausen, H. Stippler: Arch. Pharm., **265**, 152 (1927).
287) M. Freund, W. Josephi: Ann., **277**, 1 (1893).
288) J. Gadamer, H. Ziegenbein, H. Wagner: Helv. Chim. Acta, **240**, 19, 81 (1902).
289) J. J. Dobbie, A. Lauder, P. G. Paliatseas: J. Chem. Soc., **79**, 87 (1901).
290) J. Gadamer, D. Bruns: Arch. Pharm., **239**, 39 (1901).
291) Wackenrroder: Kostner's Arch., 8, 423 (1826).
292) Wicke: Ann., **137**, 274 (1866).

293) J. B. Koefli, W. H. Perkin, Jr.: J. Chem. Soc., **1928**, 2989.
294) J. J. Dobbie, A. Lauder: J. Chem. Soc., **81**, 145 (1902).
295) J. Gadamer, W. Klee: Arch. Pharm., **254**, 295 (1916).
296) E. Späth, N. Lang: Ber., **54**, 3074 (1921).
297) J. Gadamer, F. von Bruchhausen: Arch. Pharm., **259**, 245 (1921).
298) E. Späth, E. Mosettig: Ann., **433**, 138 (1923).
299) F. von Bruchhausen: Arch. Pharm., **261**, 28 (1923).
300) W. Awe, J. Thum, H. Wichmann: Arch. Pharm., **293**, 907 (1960).
301) A. Pictet, S. Malinowsky: Ber., **46**, 2688 (1913).
302) E. Späth, E. Kruta: Ber., **62**, 1024 (1929).
303) S. Pfeifer, D. Thomas: Pharmazie, **21**, 701 (1966).
304) A. Resplandy: Mem. Inst. Sci. Madgascar. Ser. D., **10**, 37 (1961). [C. A., **58**, 14018 (1963)].
305) F. Schwarz, H. Doehnert: Pharmazie, **21**, 443 (1966).

Chapter 11 Protopine Alkaloids

Protopine (Macleyne, Fumarine)

$C_{20}H_{19}O_5N \equiv 353.36$

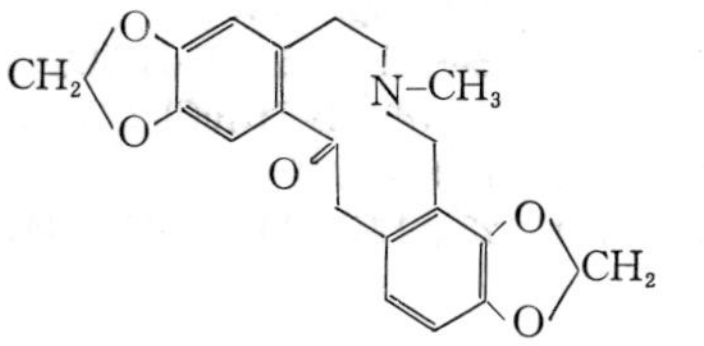

≪mp≫ 207～208° (MeOH)
≪$[\alpha]_D$≫ 0°
≪Ph.≫ CR[78] UV[126,128,129] IR[126,129,130,133] Mass[131]
Rf[101]

≪OP.≫ Adlumia cirrhosa (fungosa)[1,2,3,4], Argemone alba[5], A. mexicana[6,7,8], A. munita subsp. rotunda[9], A. ochroleuca[10], A. platyceras[11], A. squarrosa[12], Bocconia arborea[13,14,15]*
≪C. & So.≫ Prisms, Sol. in $CHCl_3$, Spar. sol. in MeOH, EtOH, Me_2CO, C_6H_6, Insol. in H_2O
≪Sa. & D.≫ B–$HClO_4$: 231～232°
B–$HAuCl_4$: 198° B–Pic : 270° (d)
B–MeI : 217° (d)
≪SD≫ 78), 122), 123), 124), 125), 126)
≪S. & B.≫ 90), 127), BG : 134)
BS : 135), 136), 137)

Hunnemannine

$C_{20}H_{21}O_5N \equiv 355.38$

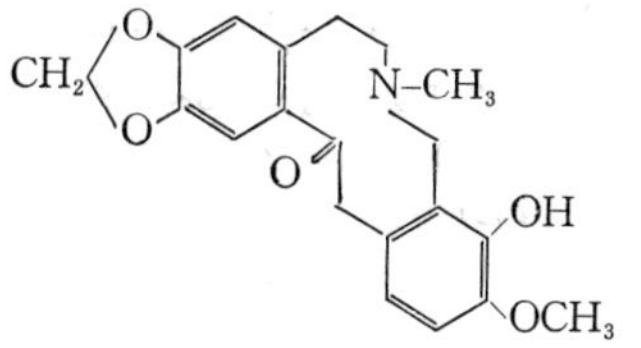

≪mp≫ 209° ($CHCl_3$–MeOH)
≪$[\alpha]_D$≫ 0°

≪Ph.≫ CR[103] IR[133] Mass[131]
≪OP.≫ Hunnemannia fumariaefolia[103,104]
≪C. & So.≫ Colourless stout polyhedra, Sol. in $CHCl_3$, Spar. sol. in MeOH, Insol. in Et_2O
≪Sa. & D.≫ B–OMe=allocryptopine
B–OEt : 168°
≪SD.≫ 103)
≪S. & B.≫ 103)

---

* B. (Macleaya) cordata[16~21], B. frutescens[22,33,34], B. latisepala[23], B. microcarpa[20], B. pearcei[24], Chelidonium majus[22,25~32], Corydalis ambigua[35,36], C. angustifolia[37], C. aurea (montana)[38,39,40], C. caseana[41], C. cava (ternata)[42], C. cheilantheifolia[43], C. claviculata[44], C. cornata[45], C. crystallina[46], C. decipiens (pumila)[37], C. deumbens[35,47], C. gortschakovii[48], C. govaniana[49], C. incisa[34], C. lutea[50], C. micrantha[46], C. kudle[51], C. nobilis[52], C. ochotensis[53], C. ochroleuca[54], C. ophiocarpa[55], C. pallida[56], C. platycarpa[57], C. pseudoadunca[48], C. scouleri[58], C. sempervirens (glauca)[59], C. sibirica[60], C. solida[61,62,63], C. ternata[51,64], C. thalictrifolia[65], Dactylicapnos (Dicentra) macrocapnos[66], Dendromecon rigidum[67], Dicentra canadensis[68], D. chrysantha[69], D. cucullaria[70,71], D. eximia[72], D. formosa[73,74], D. ochroleuca[69], D. oregana[75], D. pusilla[76], D. spectabilis[61,62,77,78], Dicranostigma (Stylophorum) diphyllum[26,79,80], D. franchetianum[80,81], D. lactucoides[24], Eschscholtzia californica[82~86], Fumaria ograria[87], F. capreolata[88], F. micrantha (densiflora)[89], F. muralis[87], F. officinalis[90,91], F. palviflora[92], F. schleicheri[88], F. vaillantii[88], Glaucium corniculatum[22,89,93], G. elegans[94], G. fimbrilligerum[95], G. luteum (flavum)[96,97,98,99], G. oxylobum[100], G. serpieri[80], G. squamigerum[88,101], G. vitellinum[102], Hunnemannia fumariaefolia[103,104], Hypecoum leptocarpum[88], H. pendulum[89], H. procumbens[105], H. trilobum[106], Meconopsis aculeata[107], M. betonicifolia[107], M. horridula[107], M. latifolia[107], M. paniculata[107], M. rudis[107], Nandina domestica[108], Papaver alpinum[31], P. anomalum[31], P. arenarium[9], P. argemone[11,109], P. atlanticum[108], P. californicum[109], P. cateritium[111], P. dubium spp. albiflorum[112], P. dubium spp. lecoquii[112], P. gracile[9], P. macrostomum[9], P. monathum[9], P. nudicaule ssp. rubroaurantiacum[31], P. nudicaule ssp. xanthopetallum[31], P. nudicaule var. amurense[110], P. nudicaule var. croceum[11], P. orientale[113,114], P. pavonium[89], P. rhoeas var. decaisnei[9], P. rhoeas var. flore-allo[9], P. pilosum[9], P. somniferum[115], Platycapnos spicatum (Fumaria spicata)[88], Pteridophyllum racemosum[116], Roemeria hybrid (vislacea)[89], R. refracta[9], Romneya courteri var. trichocalyx[117], Sanguinaria canadensis[82,118~121], Papaver nudicaule var. croceum[167]

Cryptopine (Cryptocavine) $C_{21}H_{23}O_5N \equiv 369.40$

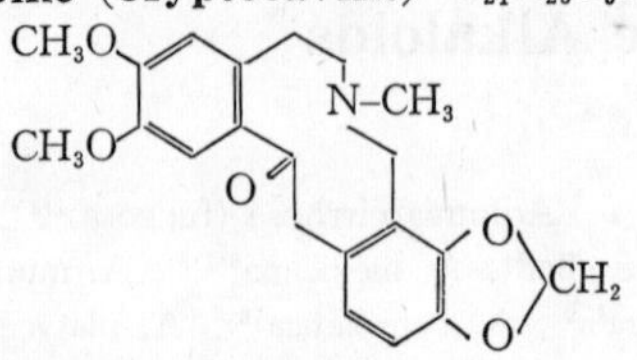

≪mp≫ 223° (MeOH)
≪$[\alpha]_D$≫ 0°

≪Ph.≫ CR[75)143)] UV[128)] IR[133)145)146)147)] Mass[131)]
≪OP.≫ Corydalis nobilis[52)], C. ochotensis[53)], C. ophiocarpa[55)], C. scouleri[58)], C. sempervirens[59)], C. sibirica[60)], Dicentra chrysantha[69)], D. cucullaria[71)], D. ochroleuca[69)]*
≪C. & So.≫ Prisms or plates, Sol. in $CHCl_3$, Spar. sol. in most org. solv.
≪Sa. & D.≫ B–HCl: 235° B–$HClO_4$: 226° B–H_2PtCl_6: 204° (d) B–$HAuCl_4$: 205° (d) B–Pic: 163° or 181～183°
≪SD.≫ 78), 123), 124), 140), 141), 142), 143), 144)
≪S. & B.≫ 127)

Fagarine-II $C_{21}H_{23}O_5N \equiv 369.40$

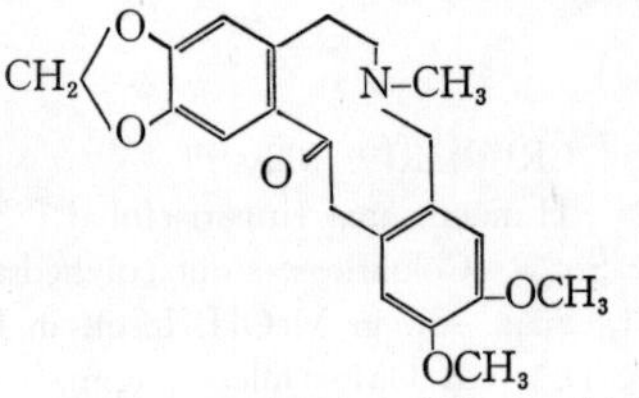

≪mp≫ 200～201° (EtOH, AcOEt or C_6H_6)
≪$[\alpha]_D$≫ 0°

≪Ph.≫ UV[149)151)] IR[150)]
≪OP.≫ Fagara (Xanthoxylum) brachyacanthum[148)], F. coco[149)]
≪C. & So.≫ Colourless needles
≪Sa. & D.≫ B–HCl: 200～202° (d) B–HBr: 208～210° (d) B–Pic: 214° B–MeI: 234° B–$HAuCl_4$: 218～219°
≪SD.≫ 149), 150), 151)
≪S. & B.≫ 151)

α-Allocryptopine (β-Homochelidonine, γ-Fagarine) $C_{21}H_{23}O_5N \equiv 369.40$

≪mp≫ 163° (AcOEt or $CHCl_3$–MeOH)
≪$[\alpha]_D$≫ 0°

≪Ph.≫ UV[149)] IR[133)153)] Mass[131)] Rf[101)112)]
≪OP.≫ Adlumia fungosa[3)], Argemone alba[5)], A. mexicana[8)], A. ochroleuca[10)], A. platyceras[11)], A. squarrosa[12)], Bocconiaarborea[15)], B. cordata[14)16)～19)], B. latispeale[152)], B. microcarpa[20)], B. frutescens[22)33)34)], B. pearcei[24)], Chelidonium majus[22)25)26)32)], Corydalis ambigua[36)]**
≪C. & So.≫ Colourless prisms, Sol. in $CHCl_3$, AcOEt, Spar. sol. in EtOH, Et_2O
≪Sa. & D.≫ B–HCl: 190° B–HI: 192～193° B–$HAuCl_4$: 190～192° B–Pic: 203～204° B–MeI α: 185° β: 211°
≪SD.≫ 156)
≪S. & B.≫ 157), 158)

---

* Eschscholtzia californica[86)], Fumaria officinalis[91)], Macleaya (Bocconia) microcarpa[20)], Papaver alpinum[138)], P. somniferum[139)]

** C. aurea (montana)[38)39)], C. caseana[41)], C. cheilantheifolia[43)], C. ophiocarpa[55)], C. scouleri[58)], C. solida[63)], C. cornata[45)], C. ternata[64)], Dactylicapnos (Dicentra) macrocapnos[66)], Dendromecon rigidum[67)], Dicentra chrysantha[69)], D. cucullaria[70)71)], D. oregana[75)], Dicranostigma franchetianum[81)], Eschscholtzia californica[82)～86)], Fagara rhoifolia[153)], Glaucium corniculatum[89)93)], G. elegans[94)], G. fimbrilligerum[95)], G. luteum (flavum)[96)～99)], G. oxylobum[100)], G. squamigerum[88)101)], G. vitellinum[102)], Hunnemannia fumariaefolia[103)104)], Papaver anomalum[31)], P. alpinum[31)], P. dubium ssp. albiflorum[112)], P. nudicaule ssp. rubroaurantiacum[31)], P. nudicaule ssp. xanthopetallum[31)], P. pavonium[89)], Pteridophyllum racemosum[116)], Sanguinaria canadensis[82)119)], Xanthoxylum brachyacanthum[108)], X. (Fagara) coco[149)], X. venificium[154)], X. planispium[155)]

β-Allocryptopine (γ-Homochelidonine)

≪mp≫ 168° (AcOEt or $CHCl_3$-MeOH)
≪$[\alpha]_D$≫ 0°
≪Ph.≫
≪OP.≫ → α-allocryptopine
≪C. & So.≫ Tablets, Sol. in $CHCl_3$, Spar. sol. in Et_2O, EtOH
≪Sa. & D.≫ B-HCl: 175° (d)
B-$HAuCl_4$: 187°
≪SD.≫ → α-allocryptopine
≪S. & B.≫ → α-allocryptopine

Muramine (Cryptopalmatine) $C_{22}H_{27}O_5N \equiv 385.44$

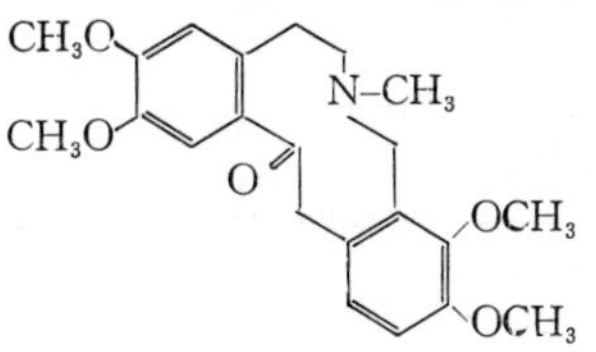

≪mp≫ 176~177°
≪$[\alpha]_D$≫ 0°
≪Ph.≫ UV[160] IR[159)160] NMR[159)160] Mass[159] Rf[160]
≪OP.≫ Argemone munita ssp. rotunda[9], Papaver alpinum[31], P. anomalum[31], P. nudicaule var. amurense[110], P. nudicaule ssp. rubroaurantiacum[31], P. nudicaule ssp. xanthopetallum[31], P. pilosum[9], P. nudicaule var. croceum[167]
≪C. & So.≫
≪Sa. & D.≫ B-MeI: 192~194°
≪SD.≫ 31), 159), 160)
≪S. & B.≫ 160), 161)

Ochrobirine $C_{20}H_{19}O_6N \equiv 369.36$

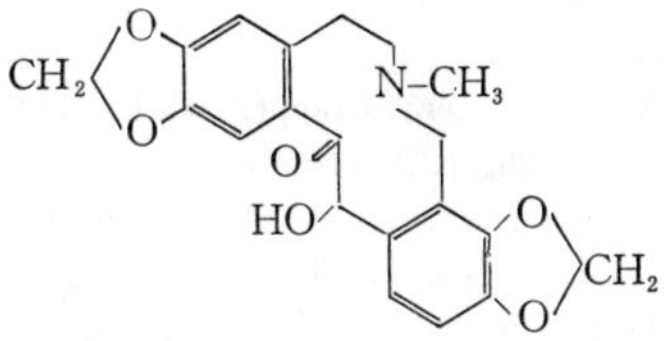

≪mp≫ 198° (MeOH), cf. 139° [B+MeOH]
≪$[\alpha]_D$≫ +36° ($CHCl_3$)
≪Ph.≫ CR[32] IR[133]
≪OP.≫ Corydalis lutea[50], C. ochroleuca[54], C. sibirica[60]
≪C. & So.≫
≪Sa. & D.≫ B-OAc: 177°
≪SD.≫ 32), 50)
≪S. & B.≫

Oxomuramine (Alpinone)[166] $C_{22}H_{25}O_6N \equiv 399.43$

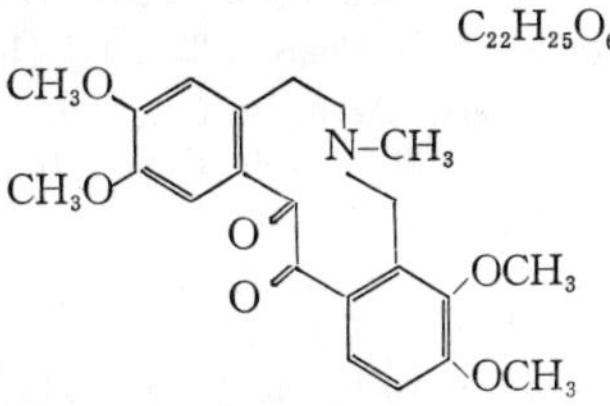

≪mp≫ 204~205° (Me_2CO)
≪$[\alpha]_D$≫ 0°
≪Ph.≫ UV[167] IR[167] Mass[167]
≪OP.≫ Papaver nudicaule var. croceum[167]
≪C. & So.≫ Yellow prisms
≪Sa. & D.≫
≪SD.≫ 166), 167)
≪S. & B.≫ 167)

Coulteropine $C_{21}H_{21}O_6N \equiv 383.39$

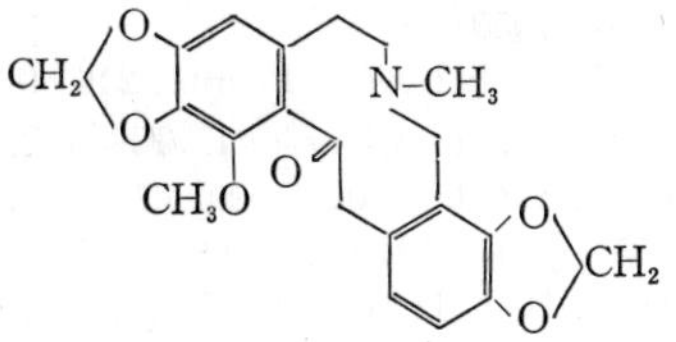

≪mp≫ 167~170°
≪$[\alpha]_D$≫ 0°
≪Ph.≫ UV[162] IR[162] NMR[162] Mass[162] X-ray[162] Rf[162]
≪OP.≫ Romneya couteri var. trichocalyx[162]
≪C. & So.≫ Colourless crystals
≪Sa. & D.≫ B-HCl: 149~150°
B-MeI: 198~200°
≪SD.≫ 162)
≪S. & B.≫

d-Corycavidine $C_{22}H_{25}O_5N \equiv 383.43$

≪mp≫ 212～213° (MeOH)
≪$[\alpha]_D$≫ +210.8° (MeOH)
≪Ph.≫
≪OP.≫ Corydalis cava (tuberosa)[42)163)164)]
≪C. & So.≫ Unstable in light, Insol. in EtOH
≪Sa. & D.≫ B–MeI: 207～210° (d)
≪SD.≫ 163), 164), 165)
≪S. & B.≫

dl-Corycavidine
≪mp≫ 193～195° (Et_2O)
≪$[\alpha]_D$≫ 0°
≪Ph.≫
≪OP.≫ Corydalis cava (tubelosa)[42)163)164)]
≪C. & So.≫ Prisms
≪Sa. & D.≫
≪SD.≫ → d–corycavidine
≪S. & B.≫

1) J.O. Schlotterbeck: Am. Chem. J., **24**, 244 (1900).
2) J.O. Schlotterbeck, H.C. Watkins: J. Am. Chem. Soc., **25**, 596 (1903).
3) J.O. Schlotterbeck, H.C. Watkins: Pharm. Arch., **6**, 17 (1903).
4) R. H. F. Manske: Can. J. Res., **8**, 210 (1933).
5) L. Slavikova, T. Shun, J. Slavik: Coll. Czech. Chem. Comm., **25**, 756 (1960).
6) J.O. Schlotterbeck: J. Am. Chem. Soc., **24**, 238 (1902).
7) A.C. Santos, P. Adkilen: J. Am. Chem. Soc., **54**, 2923 (1932).
8) J. Slavik, L. Slavikova: Coll. Czech. Chem. Comm., **21**, 211 (1956).
9) Vl. Pleininger, P. Vácha, B. Sula, F. Santavy: Planta Med., **10**, 124 (1962).
10) F. Giral, A. Sotelo: Ciencia, **19**, 67 (1959). [C.A., **54**, 3855 (1960)].
11) H.G. Boit, H. Flentje: Naturwiss., **47**, 323 (1960).
12) T.O. Soine, R.E. Willette: J. Am. Pharm. Assoc., **49**, 368 (1960).
13) E.A. Armendaritz: Datos para la materia medica Mexico, Vol. I, p. 15 (1895). Sria. de Foment, Mexico.
14) P. Murrill, J. O. Schlotterbeck: Pharm. J., **64**, 35 (1925).
15) R.H.F. Manske: Can. J. Res., **21B**, 140 (1943).
16) J. F. Eijkmann: Rec. trav. chim., **3**, 182 (1881).
17) K. Hopfgartner: Monatsh., **19**, 179 (1898).
18) P. Murrill, J. O. Schlotterbeck: Ber., **33**, 2802 (1900).
19) J.O. Schlotterbeck, W.H. Blome: Pharm. Review., **23**, 310 (1905).
20) J. Slavik, L. Slavikova: Coll. Czech. Chem. Comm., **20**, 1356 (1955).
21) C. Tani, N. Takao: J. Pharm. Soc. Japan, **82**, 755, (1962).
22) J.A. Battandier: Compt. rend., **114**, 1122 (1892); **120**, 1276 (1895).
23) L.G. Armendaritz, A. Alcala, J. Quevedo, D. Rojas: Can. J. Chem., **43**, 679 (1965).
24) R.H.F. Manske: Can. J. Chem., **32**, 83 (1954).
25) E. Schmidt, F. Selle: Arch. Pharm., **228**, 441 (1890).
26) E. Schmidt: Arch. Pharm., **231**, 136 (1893).
27) G. König: Arch. Pharm., **231**, 174 (1893).
28) F. Selle: Arch. Pharm., **228**, 441 (1890).
29) M. Wintgen: Arch. Pharm., **239**, 438 (1901).
30) J. Slavik: Coll. Czech. Chem. Comm., **20**, 198 (1955).
31) M. Maturova, B.K. Moza, J. Sitar, F. Santavy: Planta Med., **10**, 345 (1962).
32) E. Seoane: Anales Reat. Soc. Espan. Fis. Quim. (Madrid). Ser., **61** B, 747 (1965). [C.A., **63**, 18187 (1965)].
33) E.R. Miller: J. Am. Pharm. Assoc., **18**, 12 (1929).
34) R.H.F. Manske: J. Am. Chem. Soc., **72**, 3207 (1950).
35) K. Makoshi: Arch. Pharm., **246**, 401 (1908).
36) T.Q. Chou: Chinese J. Physiol., **2**, 203 (1928); **3**, 69, 301 (1929); **7**, 35 (1933); 8, 155 (1934); **10**, 507 (1936).
37) H.G. Boit, H. Ehmke: Naturwiss., **46**, 427 (1959).
38) R.H.F. Manske: Can. J. Res., **9**, 436 (1934).
39) R.H.F. Manske: Can. J. Res., **16B**, 153 (1938).

40) R. H. F. Manske: Can. J. Res., **20 B**, 49 (1942).
41) R. H. F. Manske, M. R. Miller: Can. J. Res., **16B**, 153 (1938).
42) J. Gadamer: Arch. Pharm., **249**, 224 (1911).
43) R. H. F. Manske: Can. J. Res., **20B**, 57 (1942).
44) R. H. F. Manske: Can. J. Res., **18B**, 97 (1940).
45) R. H. F. Manske: Can. J. Res., **24B**, 66 (1946).
46) R. H. F. Manske: Can. J. Res., **17B**, 57 (1939).
47) Y. Asahina, S. Motigase: J. Pharm. Soc. Japan, **40**, 766 (1920).
48) M. S. Yunusov, S. T. Akramov, S. Yu. Yunusov: Dokl. Akad. Nauk SSSR., **162**, 607 (1965). [C. A., **63**, 5695 (1966)].
49) O. E. Edwards, K. L. Handa: Can. J. Chem., **39**, 1801 (1961).
50) R. H. F. Manske: Can. J. Res., **17B**, 89 (1939).
51) J. Go: J. Pharm. Soc. Japan, **49**, 801 (1929).
52) R. H. F. Manske: Can. J. Res., **18B**, 288 (1940).
53) R. H. F. Manske: Can. J. Res., **18 B**, 75 (1940).
54) R. H. F. Manske: Can. J. Res., **17 B** 95 (1939).
55) R. H. F. Manske: Can. J. Res., **17 B**, 51 (1939).
56) R. H. F. Manske: Can. J. Res., **18 B**, 80 (1940).
57) R. H. F. Manske: Can. J. Res., **21 B**, 13 (1943).
58) R. H. F. Manske: Can. J. Res., **14B**, 347 (1936).
59) R. H. F. Manske: Can. J. Res., **8**, 407 (1933).
60) R. H. F. Manske: Can. J. Res., **14B**, 354 (1936).
61) O. Haars: Arch. Pharm., **243**, 154 (1905).
62) G. Heyl: Apoth. Ztg., **25**, 36 (1910).
63) R. H. F. Manske: Can. J. Chem., **34**,1(1956).
64) J. Go: J. Pharm. Soc. Japan, **49**, 125, 814 (1929); **50**, 937, 940 (1930).
65) R. H. F. Manske: Can. J. Res., **21B**, 111 (1943).
66) R. H. F. Manske: Can. J. Res., **21B**, 117 (1943).
67) R. H. F. Manske: Can. J. Res., **27B**, 653 (1949).
68) R. H. F. Manske: Can. J. Res., **7**, 758(1932).
69) R. H. F. Manske: Can. J. Res., **15B**, 274 (1937).
70) R. Fischer, O. A. Sölle: Pharm. Arch., **5**, 121 (1902).
71) R. H. F. Manske: Can. J. Res., **7**, 264(1932).
72) R. H. F. Manske: Can. J. Res., **8**, 529(1933).
73) G. Heyl: Arch. Pharm., **241**, 313 (1903).
74) R. H. F. Manske: Can. J. Res., **10**, 521(1934).
75) R. H. F. Manske: Can. J. Res., **10**, 765 (1934).
76) Y. Asahina: Arch. Pharm., **247**, 201 (1909).
37) J. Gadamer: Apoth. Ztg., **16**, 621 (1901).
78) P. W. Danckwortt: Arch. Pharm., **260**, 94 (1922); **250**, 590 (1912).
79) J. O. Schlotterbeck, H. C. Watkins: Ber., **35**, 7 (1902).
80) R. H. F. Manske: Can. J. Res., **20B**, 53 (1942).
81) L. Slavikova, J. Slavik: Coll. Czech. Chem. Comm., **24**, 559 (1959).
82) E. Schmidt, G. König, W. Tietz: Arch. Pharm., **231**, 136 (1893).
83) R. Fischer: Arch. Pharm., **239**, 421 (1901).
84) R. Fischer, M. E. Tweeden: Pharm. Arch., **5**, 117 (1902).
85) J. Slavik, L. Slavikova: Coll. Czech. Chem. Comm., **20**, 27 (1955).
86) R. H. F. Manske, K. H. Shin: Can. J. Chem., **43**, 2180 (1965).
87) J. Susplugas, M. Lalaurie, G. Privat, P. Chicaya: Trav. soc. pharm. Montpellier, **17**, 134 (1957). [C. A., **50**, 11439 (1956)].
88) R. H. F. Manske: The Alkaloids, Vol. IV. p. 158 (1954).
89) T. F. Platonova, P. S. Massagetov, A. D. Kozovkov, L. M. Utkin: Zhur. Obshchei Khim., **26**, 173 (1956). [C., **1956**, 11439].
90) R. D. Haworth, W. H. Perkin, Jr., J. S. Stevens: J. Chem. Soc., **1926**, 1764.
91) R. H. F. Manske: Can. J. Res., **16B**, 438 (1938).
92) T. R. Govindachari, K. Nagarajan, B. R. Pai, S. Rajappa: J. Sci. Ind. Res., **17B**, 73(1958). [C. A., **52**, 18674 (1958)].
93) J. Slavik, L. Slavikova: Coll. Czech. Chem. Comm., **22**, 279 (1957).
94) J. Slavik: Coll. Czech. Chem. Comm., **25**, 1698 (1960).
95) R. A. Konovalova, S. Yu. Yunusov, A. P. Orechov: Zhur. Obshchei Khim., **9**, 1939 (1939). [C., **1941**, I. 2530].
96) R. Fischer: Arch. Pharm., **239**, 426 (1901).
97) R. H. F. Manske: Can. J. Res., **17B**, 399 (1939).
98) J. Slavik: Coll. Czech. Chem. Comm., **20**, 32 (1954).
99) J. Slavik, L. Slavikova: Coll. Czech. Chem. Comm., **24**, 3141 (1959).
100) J. Slavik, L. Slavikova: Coll. Czech. Chem. Comm., **28**, 2530 (1963).

101) L. Slavikova: Coll. Czech. Chem. Comm., **31**, 4181 (1966).
102) J. Slavik: Coll. Czech. Chem. Comm., **24**, 3999 (1959).
103) R. H. F. Manske, L. Marion, E. Ledingham: J. Am. Chem. Soc., **64**, 1659 (1942).
104) L. Slavikova, J. Slavik: Coll. Czech. Chem. Comm., **31**, 1355 (1966).
105) E. Schmidt: Arch. Pharm., **239**, 395 (1901).
106) S. Yu. Yunusov, S. T. Akramov, G. P. Sidyakin: Dokl. Akad. Nauk Uz. SSR., **1957**, No. 7, 23. [C. A., **53**, 3606 (1959)].
107) J. Slavik: Coll. Czech. Chem. Comm., **25**, 1663 (1960).
108) J. Ohta: J. Pharm. Soc. Japan, **69**, 502 (1949).
109) F. Santavy, M. Maturova, A. Nemeckova, H. B. Schröter: Planta Med., **8**, 160 (1960).
110) H. G. Boit, H. Flentje: Naturwiss., **47**, 180 (1960).
111) H. B. Schröter, M. Maturova, F. Santavy: Planta Med., **7**, 298 (1959). [C. A., **54**, 5012 (1960)].
112) J. Slavik: Coll. Czech. Chem. Comm., **29**, 1314 (1964).
113) W. Klee: Arch. Pharm., **253**, 211 (1914).
114) J. Gadamer: Arch. Pharm., **252**, 274 (1914).
115) O. Hesse: Ber., **4**, 693 (1871); Ann., **8**, 261 (1872).
116) K. Kojima, Y. Ando: J. Pharm. Soc. Japan, **71**, 625 (1951).
117) F. R. Stermitz, L. Chen, J. I. White: Tetrahedron, **22**, 1095 (1966).
118) G. König: Zeitschr. Naturw., **63**, 369 (1893).
119) G. König, W. Tietz: Arch. Pharm., **231**, 145 (1893).
120) R. Fischer: Arch. Pharm., **239**, 409 (1901).
121) T. Kozniewski: Anzeiger Akad. Wiss. Krakau, **1910**, A, 235. [C., **1910**, II, (1932)].
122) J. Gadamer: Habilitations Schrift, Breslau, p. 25, 1912.
123) W. H. Perkin, Jr.: J. Chem. Soc., **109**, 815 (1916).
124) W. H. Perkin, Jr.: J. Chem. Soc., **115**, 713 (1919).
125) J. Gadamer, H. Kollmar: Arch. Pharm., **261**, 153 (1923).
126) N. J. Leonard, R. R. Sauers: J. Org. Chem., **22**, 63 (1957).
127) R. D. Haworth, W. H. Perkin, Jr.: J. Chem. Soc., **1926**, 1769.
128) P. M. Oertreicher, C. G. Farmilo, L. Levi: Bull. Narcotics, U. N. Dept. Social Affairs, **7**, 42 (1954).
129) F. A. L. Anet, L. Marion: Can. J. Chem., **32**, 452 (1954).
130) E. H. Mottus, H. Schwarz, L. Marion: Can. J. Chem., **31**, 1144 (1953).
131) L. Dolejs, V. Hanus, J. Slavik: Coll. Czech. Chem. Comm., **29**, 2479 (1964).
132) L. Slavikova, J. Slavik: Coll. Czech. Chem. Comm., **29**, 1314 (1964).
133) L. Marion, D. A. Ramsey, R. N. Jones: J. Am. Chem. Soc., **73**, 305 (1951).
134) R. Robinson: "The Stractural Relations of Natural Products," Clarendon Press, Oxford, 1955.
135) M. Sribney, S. Kirkwood: Nature, **171**, 931 (1953).
136) D. H. R. Barton, R. H. Kirby, G. W. Kirby: Proc. Chem. Soc., **1963**, 267.
137) D. H. R. Barton, R. H. Kirby, G. W. Kirby: J. Chem. Soc., **1965**, 6379.
138) M. Maturova, B. K. Moza, J. Sitar, F. Santavy: Planta Med., **7**, 298 (1959).
139) J. Smiles: Pharm. J., [2], **8**, 595 (1867).
140) D. R. Brown, W. H. Perkin, Jr.: Proc. Chem. Soc., **7**, 166 (1891).
141) W. H. Perkin, Jr.: J. Chem. Soc., **113**, 492 (1918).
142) A. Pictet, G. H. Kramers: Ber., **43**, 1329 (1910).
143) R. H. F. Manske, L. Marion: J. Am. Chem. Soc., **62**, 2042 (1940).
144) R. Mirza: Experientia, **8**, 258 (1952).
145) F. A. L. Anet, A. S. Bailey, R. Robinson: Chem. & Ind., **1953**, 944.
146) N. J. Leonard: J. Am. Chem. Soc., **76**, 630 (1954).
147) A. F. Thomas, L. Marion, R. H. F. Manske: Can. J. Chem., **33**, 570 (1955).
148) H. A. D. Jowett, F. L. Pyman: J. Chem. Soc., **103**, 290 (1913).
149) C. E. Redemann, B. B. Wisegarver, G. A. Alles: J. Am. Chem. Soc., **71**, 1030 (1949).
150) J. Comin, V. Deulofeu: Tetrahedron, **6**, 63 (1959).
151) D. Giacopello, V. Deulofeu, J. Comin: Tetrahedron, **20**, 2971 (1964).
152) L. G. Armendaritz, A. Alcale, J. Quevedo, D. Rojas: Can. J. Chem., **43**, 679 (1965).
153) J. M. Calderwood, F. Fish: Chem. & Ind., **1966**, 237.
154) J. R. Cannon, G. K. Hughes, E. Ritchie, W. C. Taylor: Austral. J. Chem., **6**, 86 (1953).
155) H. Ishii: J. Pharm. Soc. Japan, **81**, 238 (1961).
156) J. Gadarmer: Arch. Pharm., **257**, 298 (1919).
157) F. L. Pyman: J. Chem. Soc., **103**, 817 (1913).
158) R. D. Haworth, W. H. Perkin, Jr.: J. Chem. Soc., **1926**, 445.

159) A. D. Cross, L. Dolejs, V. Hanus, M. Maturova, F. Santavy: Coll. Czech. Chem. Comm., **1965**, 1335.
160) D. Giacopello, V. Deulofeu: Tetrahedron Letters, **1966**, 2859.
161) R. D. Haworth, J. B. Koepfli, W. H. Perkin, Jr.: J. Chem. Soc., **1927**, 2261.
162) F. R. Stermitz, L. Chen, J. I. White: Tetrahedron, **22**, 1095 (1966).
163) J. Gadamer: Arch. Pharm., **249**, 30 (1911).
164) F. von Bruchhausen: Arch. Pharm., **263**, 570 (1925).
165) T. R. Govindachari, S. Rajadurai: J. Sci. Ind. Res., **16B**, 506 (1957).
166) M. Maturova, D. Paulaskova, F. Santavy: Planta Med., **14**, 22 (1966).
167) H. Flentje, W. Döpke: Pharmazie, **21**, 321 (1966).

Chapter 12 Phthalide Alkaloids

Adlumidine $C_{20}H_{17}O_6N \equiv 367.34$

<mp> 237° ($CHCl_3$-MeOH)
<$[\alpha]_D$> +116° ($CHCl_3$)

<Ph.> IR[9,13] NMR[14] ORD[9] Rf[9] pKa[9]
<OP.> Adlumina fungosa (cirrhosa)[1,5,6], Corydalis incisa[2], C. thalictrifolia[3], C. sempervirens[4]
<C. & So.> Rhombic plates, Spar. sol. in EtOH, Et_2O
<Sa. & D.> B–HCl: 259°
<SD.> 1), 2), 7), 8), 9)
<S. & B.>

Capnoidine (*l*-Adlumidine)
<mp> 238° ($CHCl_3$-MeOH)
<$[\alpha]_D$> −113.2° ($CHCl_3$)
<Ph.> → adlumidine
<OP.> Corydalis crystallina[10], C. scouleri[11], C. sempervirens (glauca)[4]
<C. & So.> Stout polyhedra or prisms
<Sa. & D.> B–HCl: 244°
<SD.> 2), 4), 8), 11), 12)
<S. & B.>

Bicuculline $C_{20}H_{17}O_6N \equiv 367.34$

<mp> 215° or 177° (Solidifies and remelts at 193 ~195°) ($CHCl_3$–MeOH)
<$[\alpha]_D$> +130° ($CHCl_3$)

<Ph.> IR[9,13,25] NMR[14,25] ORD[9] Rf[9] pKa[9]
<OP.> Adlumina fungosa (cirrhosa)[1,7], Corydalis aurea[15], C. caseana[16,17], C. crystallina[10], C. nobilis[17], C. ochroleuca[18], C. platycarpa[19], C. scoureli[11], C. sempervirens[4,8], C. sibirica[12], Dicentra chrysantha[20], D. cucullaria[8,21], D. ochroleuca[20]
<C. & So.> Plates, Sol. in $CHCl_3$, Spar. sol. in MeOH, Et_2O
<Sa. & D.> B–HCl: 259° B–Pic: 112~113°
<SD.> 2), 7), 8), 9), 11), 14)
<S. & B.> 22), 23), 24)

Corlumidine $C_{20}H_{19}O_6N \equiv 369.36$

<mp> 236° ($CHCl_3$-MeOH)
<$[\alpha]_D$> +80° ($CHCl_3$)

<Ph.>
<OP.> Corydalis scouleri[11]
<C. & So.> Prisms
<Sa. & D.> B–OMe = corlumine
<SD.> 8), 11), 12), 26)
<S. & B.>

l-Adlumine $C_{21}H_{21}O_6N \equiv 383.39$

≪mp≫ 180° or 188° ($CHCl_3$-MeOH)
≪$[\alpha]_D$≫ −42° ($CHCl_3$)

≪Ph.≫ IR[9,13] NMR[14] ORD[9] Rf[9] pKa[9]
≪OP.≫ Adlumina fungosa (cirrhosa)[1,27], Corydalis ophiocarpa[28], C. sempervirens[8]
≪C. & So.≫ Plates
≪Sa. & D.≫
≪SD.≫ 1), 8), 9), 12), 14), 26), 27), 29), 30), 31), 32), 33), 34)
≪S. & B.≫

***d*-Adlumine**

≪mp≫ 180° or 188°
≪$[\alpha]_D$≫ +42° ($CHCl_3$)
≪Ph.≫
≪OP.≫ Corydalis ophiocarpa[28], C. scouleri[11], C. sempervirens (glauca)[8]
≪C. & So.≫
≪Sa. & D.≫
≪SD.≫ → *l*-adlumine
≪S. & B.≫

Corlumine $C_{21}H_{21}O_6N \equiv 383.39$

≪mp≫ 158∼160° (d) (MeOH)
≪$[\alpha]_D$≫ +77.3° (MeOH)

≪Ph.≫ IR[9,25] NMR[14,25] ORD[9] Rf[9] pKa[9]
≪OP.≫ Corydalis nobilis[17], C. scouleri[11], C. sibirica[11,12,32], Dicentra cucullaria[8,11,17,32]
≪C. & So.≫ Prisms
≪Sa. & D.≫
≪SD.≫ 8), 9), 11), 12), 14), 26), 29), 30), 31), 32)
≪S. & B.≫

Hydrastine $C_{21}H_{21}O_6N \equiv 383.39$

≪mp≫ 132° (EtOH)
≪$[\alpha]_D$≫ −49.8° (EtOH), −68° ($CHCl_3$), cf. B-HCl +127.3° (H_2O)

≪Ph.≫ CR[83] UV[66,84–90] IR[9] NMR[14,69,91] ORD[9,67,68] Rf[9] pKa[9]
≪OP.≫ Berberis laurina[34,35,36], Hydrastis canadensis[37]
≪C. & So.≫ Prisms, Sol. in Et_2O, $CHCl_3$, C_6H_6, Me_2CO, Spar. sol. in EtOH, Insol. in Petr. ether
≪Sa. & D.≫ B-HCl: 116° B-H_2PtCl_6: 207° B-$HAuCl_4$: 110° B-Pic: 184° B-Picrol: 220° B-MeI: 208° B-EtI: 205∼206°
≪SD.≫ 5), 9), 14), 39), 40), 41), 42)∼69)
≪S. & B.≫ 22), 24), 70)∼76), cf. 77), BS: 78), 79), 80), BG: 81), 82)

Narcotoline $C_{21}H_{21}O_7N \equiv 399.39$

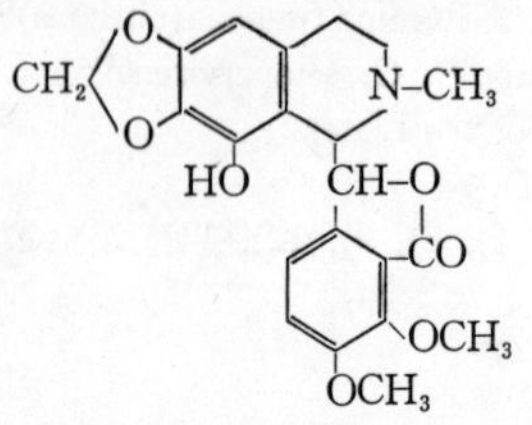

≪mp≫ 202° (189°) (aq. MeOH)
≪$[\alpha]_D$≫ −189° ($CHCl_3$), +5.8° (0.1N-HCl)

≪Ph.≫ UV[90,95] IR[9] NMR[97] ORD[9,68,97] Rf[9] pKa[9]
≪OP.≫ Papaver somniferum[92,93,94]
≪C. & So.≫ Rectangular rods
≪Sa. & D.≫ B–HCl: 161° B–Pic: 140° B–Picrol: 202∼203° B–OMe=narcotine B–OAc: 208∼209°
≪SD.≫ 9), 90), 93), 95), 96), 97), 98)
≪S. & B.≫ 97), 99)

Narcotine (Opianine) $C_{22}H_{23}O_7N \equiv 413.41$

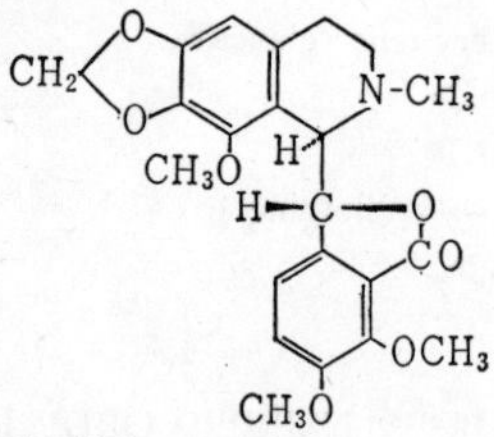

≪mp≫ 176° (EtOH)
≪$[\alpha]_D$≫ −198° ($CHCl_3$), +50° (1 % HCl)
cf. 105), 108), 109), 110)

≪Ph.≫ CR[83,140,141] UV[66,84,85,86,88,89,90,124,125,134,142,143] IR[9,144] NMR[14,97,135,145] ORD[9,68,97,135,137] X-ray[146] Rf[9,147,148] pKa[9]
≪OP.≫ Papaver somniferum (setigerum)[93,99], P. somniferum[100~104], Rauwolfia heterophylla[105]
≪C. & So.≫ Needles, Sol. in $CHCl_3$, Mod. sol. in Et_2O, EtOH, Spar. sol. in C_6H_6, Me_2CO, AcOEt, Insol. in H_2O, Polymorph.
≪Sa. & D.≫ B–Pic: 174° B–Oxal: 174° B–MeI: 154∼156°
≪SD.≫ 9), 14), 22), 48), 56), 62), 63), 65), 66), 67), 68), 70), 71), 72), 77), 97), 98), 111)∼137)
≪S. & B.≫ 91), cf. 126), 129)
BS: 97), 99), 135), 138), 139)

***dl*-Narcotine (α-Gnoscopine)**

≪mp≫ 232∼233° (EtOH)
≪$[\alpha]_D$≫ 0°
≪Ph.≫ UV[84]
≪OP.≫ Papaver somniferum[93,149]
≪C. & So.≫ Needles, Sol. in $CHCl_3$, Spar. sol. in C_6H_6, Insol. in EtOH
≪Sa. & D.≫ B–HCl: 220° or 238° B–Pic: 188∼189° B–MeI: 210∼212°
≪SD.≫ 71), 98)
≪S. & B.≫

β-Gnoscopine $C_{22}H_{23}O_7N \equiv 413.41$

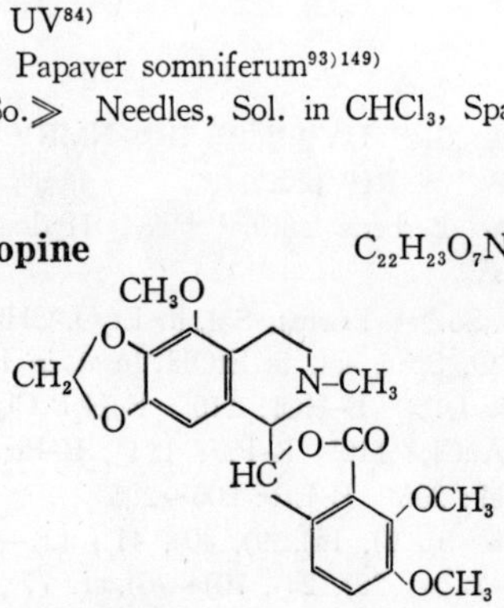

≪mp≫ 180° (MeOH)
≪$[\alpha]_D$≫

≪Ph.≫ UV[84]
≪OP.≫ Hydrastis canadensis[150], Papaver somniferum[150], Natural Product?[150]
≪C. & So.≫ Prisms
≪Sa. & D.≫ B–HCl: 86∼88° (on standing → 224∼226°) B–Pic: 199∼201°
≪SD.≫ 98), 150)
≪S. & B.≫ 74)

Cordrastine $C_{22}H_{25}O_6N \equiv 399.43$

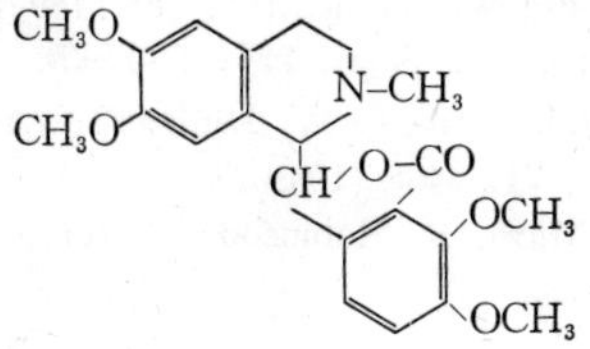

≪mp≫ 196°
≪[α]D≫
≪Ph.≫
≪OP.≫ Corydalis aurea[8)15)]
≪C. & So.≫
≪Sa. & D.≫
≪SD.≫ 8), 32)
≪S. & B.≫ 24), 76)

Oxynarcotine (Nornarceine) $C_{22}H_{25}O_8N \equiv 431.43$

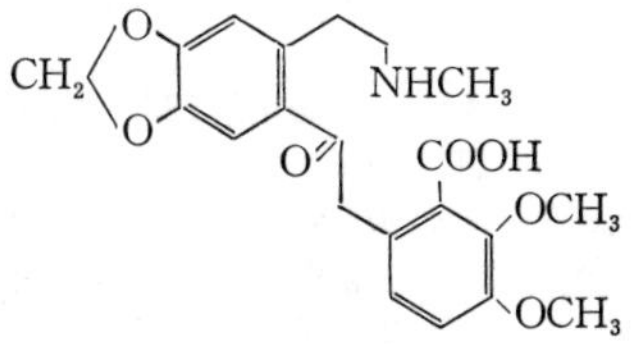

≪mp≫ 147° or 229°
≪[α]D≫ 0°
≪Ph.≫
≪OP.≫ Papaver somniferum[93)151)]
≪C. & So.≫
≪Sa. & D.≫ B–HCl : 144°
≪SD.≫ 150), 152)
≪S. & B.≫

Narceine $C_{23}H_{27}O_8N \equiv 445.45$

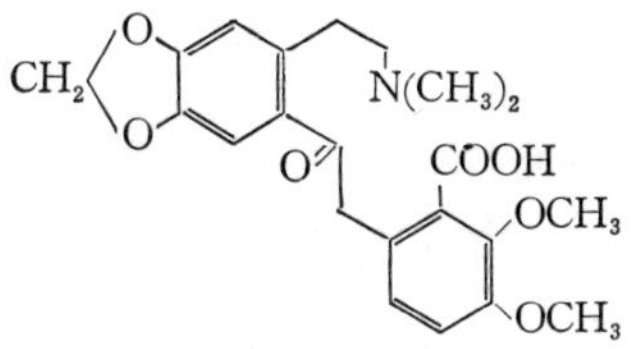

≪mp≫ 140～145°
≪[α]D≫ 0°
≪Ph.≫ UV[84)86)134)143)154)158)] Chromato[159)160)]
≪OP.≫ Papaver somniferum[93)151)153)154)]
≪C. & So.≫ Needles
≪Sa. & D.≫ B–HCl : 192～193°
B–H_2PtCl_6 : 196° B–Pic : 195°
≪SD.≫ 118), 134), 154), 155), 156), 157)
≪S. & B.≫ cf. 154)

1) R. H. F. Manske : Can. J. Res., 8, 201 (1933).
2) R. H. F. Manske : J. Am. Chem. Soc., **72**, 3207 (1950).
3) R. H. F. Manske : Can. J. Res., **21B**, 111 (1943).
4) R. H. F. Manske : Can. J. Res., 8, 407 (1933).
5) J. O. Schlotterbeck, H. C. Watkins : J. Am. Chem. Soc., **25**, 596 (1903).
6) J. O. Schlotterbeck : Am. Chem. J., **24**, 249 (1900).
7) R. H. F. Manske : Can. J. Res., 8, 142 (1933).
8) R. H. F. Manske : Can. J. Res., **16B**, 81 (1938).
9) K. Blaha, T. Hrbek, Jr., J. Kovar, L. Pijewska, F. Santavy : Coll. Czech. Chem. Comm., **29**, 2328 (1964).
10) R. H. F. Manske : Can. J. Res., **17B**, 57 (1939).
11) R. H. F. Manske : Can. J. Res., **14B**, 347 (1936).
12) R. H. F. Manske : Can. J. Res., **14B**, 354 (1936).
13) L. Marion, D. A. Ramsay, R. N. Jones : J. Am. Chem. Soc., **73**, 305 (1951).
14) S. Safe, R. Y. Moir : Can. J. Chem., **42**, 160 (1964).
15) R. H. F. Manske : Can. J. Res., **9**, 436 (1934).
16) R. H. F. Manske : Can. J. Res., **16B**, 153 (1938).
17) R. H. F. Manske : Can. J. Res., **18B**, 288 (1940).
18) R. H. F. Manske : Can. J. Res., **17B**, 95 (1939).
19) R. H. F. Manske : Can. J. Res., **21B**, 13 (1943).
20) R. H. F. Manske : Can. J. Res., **15B**, 274 (1937).
21) R. H. F. Manske : Can. J. Res., **7**, 265 (1932).
22) M. A. Marshall, F. L. Pyman, R. Robinson : J. Chem. Soc., **1934**, 1315.
23) P. W. G. Groenwoud, R. Robinson : J. Chem. Soc., **1936**, 199.
24) R. D. Haworth, A. R. Pinder, R. Robinson : Nature, **165**, 529 (1950).
25) O. E. Edwards, K. L. Handa : Can. J. Chem., **39**, 1801 (1961).
26) R. H. F. Manske : Can. J. Res., **15B**, 159 (1937).
27) J. O. Schletterbeck, H. C. Watkins : Pharm.

Arch., **6**, 17 (1903).
28) R. H. F. Manske: Can. J. Res., **17B**, 51 (1939).
29) F. L. Pyman: J. Chem. Soc., **95**, 1266 (1909).
30) W.H. Perkin, Jr., V.M. Trikojus: J. Chem. Soc., **1926**, 2925.
31) R. H. F. Manske: Can. J. Res., **8**, 404 (1933).
32) R.H.F. Manske: Can. J. Res., **14B**, 47 (1936).
33) R.H.F. Manske: Can. J. Res., **14B**, 325 (1936).
34) O. deA. Costa, R. Diaz deSilva: Rev. soc. brasil. chim., **4**, 199 (1933). [C.A., **28**, 2909 (1934)].
35) L. Gurguel, O. deA. Costa, R. D. deSilva: Bol. assoc. brasil. farm., **15**, 11 (1934). [C.A., **28**, 3521 (1934)].
36) O. deA. Costa, V. Lucas: Rev. quim. farm. (Santiago, Chili), **4**, 63 (1939). [C.A., **33**, 7957 (1939)].
37) Durand: Am. J. Pharm., **23**, 112 (1851).
38) J.D. Perrins: Pharm. J., [2], **3**, 546 (1862).
39) F. Mahla: J. prakt. Chem., [1], **91**, 248 (1864).
40) F.B. Power: Pharm. J., [3], **15**, 297 (1884).
41) M. Freund, W. Will: Ber., **19**, 2797 (1886).
42) H. Decker: Ger. Pat. 270,859, 281,546, 281,547.
43) J. Eijkmann: Rec. trav. chim., **5**, 290 (1886).
44) M. Freund, W. Will: Ber., **20**, 88 (1887).
45) M. Freund, W. Will: Ber., **20**, 2400 (1887).
46) E. Schmidt, F. Wilhelm: Arch. Pharm., **226**, 346 (1888).
47) W. Roser: Ann., **249**, 159 (1888).
48) E. Schmidt, W. Kerstein: Arch. Pharm., **228**, 49 (1890).
49) J.J. Dobbie, C.K. Tinkler: J. Chem. Soc., **85**, 1005 (1904).
50) C. Liebermann, F. Kropf: Ber., **37**, 211 (1904).
51) W.H. Perkin, Jr., R. Robinson: J. Chem. Soc., **91**, 1073 (1907).
52) E. Hope, R. Robinson: J. Chem. Soc., **99**, 2114 (1901).
53) H. Thomas, W. Siebeling: Ber., **44**, 2134 (1911).
54) M. Freund: Ber., **22**, 456, 1156, 2329 (1889).
55) H. Decker, P. Becker: Ann., **395**, 358 (1913).
56) F.L. Pyman, F.G.P. Ramfry: J. Chem. Soc., **101**, 1885 (1915).
57) T.S. Stevens, M.C. Roberts: J. Chem. Soc., **1927**, 2790.
58) S.N. McGeogh, T. S. Stevens: J. Chem. Soc., **1934**, 1465.
59) M. Freund: Ann., **281**, 383 (1892).
60) P. Fritsch: Ann., **286**, 1 (1895).
61) N.J. Leonard, G.W. Leubner: J. Am. Chem. Soc., **71**, 3408 (1949).
62) R. Mirza, R. Robinson: Nature, **166**, 271 (1950).
63) W.M. Whaley, M. Meadow: J. Org. Chem., **19**, 666 (1954).
64) M.M. Janot, H. Dourrat: Bull. Soc. Chim. France, **1955**, 823.
65) K. Molnárová, L. Molnár: Acta Chim. Acad. Sci. Hang., **18**, 93 (1959). [C.A., **54**. 6783 (1960)].
66) J. Knabe: Arch. Pharm., **293**, 121 (1960).
67) M. Ohta, H. Tani, S. Morozumi: Tetrahedron Letters, **1963**, 859.
68) M. Ohta, H. Tani, S. Morozumi, S. Kodaira, S. Kuriyama: Tetrahedron Letters, **1963**, 1857.
69) M. Ohta, H. Tani, S. Morozumi: Chem. Pharm. Bull., **12**, 1072 (1964).
70) C. Liebermann: Ber., **29**, 183, 2040 (1896).
71) W.H. Perkin, Jr., R. Robinson: J. Chem. Soc., **99**, 775 (1911).
72) E.G. Jones, W.H. Perkin, Jr., R. Robinson: J. Chem. Soc., **101**, 257 (1912).
73) E. Hope, R. Robinson: Proc. Chem. Soc., **28**, 16 (1912).
74) E. Hope, R. Robinson: J. Chem. Soc., **105**, 2085 (1914).
75) E. Hope, F.L. Pyman, F.G.P. Ramfry, R. Robinson: J. Chem. Soc., **1931**, 236.
76) R.D. Haworth, A.R. Pinder: J. Chem. Soc., **1950**, 1776.
77) E.V. Seshacharyula, S. Dutt: Proc. Acad. Sci. United Provinces Agra and Oudh, India, **4**, 159 (1934). [C. A., **29**, 7989 (1935)].
78) J.R. Gear, I.D. Spener: Nature, **191**, 1393 (1961).
79) I.D. Spenser, J.R. Gear: J. Am. Chem. Soc., **84**, 1059 (1962).
80) R.N. Gupta, I.D. Spenser: Can. J. Chem., **43**, 133 (1965).
81) R. Robinson: "The Structural Relations of Natural Products", Clarendon Press, Oxford (1955).
82) E. Wenkert: Experientia, **15**, 165 (1959).
83) J.A. Labat: Bull. soc. chim. biol., **15**, 1344 (1933). [C. A., **28**, 2300 (1934)].
84) J.J. Dobbie, A. Lauder. J. Chem. Soc., **83**, 605 (1903).
85) P. Steiner: Compt. rend., **176**, 244 (1923).
86) Z. Kitasato: Acta Phytochim. Japan, **3**, 229 (1927).

87) A. Andant: Bull. Sci. Pharmacol., **37**, 89 (1930). [C. A., **24**, 3441 (1930)].
88) P. Csókan: Z. anal. Chem., **124**, 344 (1942).
89) R.H. Pinyazhko: Farmatsevi Zh. (Kiev.), **19**, 12 (1964). [C. A., **64**, 6701 (1960)].
90) B. Goeber, S. Pfeifer: Arch. Pharm., **299**, 196 (1966).
91) N.S. Bhacca, L.F. Johnson, J.N. Shoolery: NMR Spectra Catalog, Spectrum No. 347. Varian Associates.
92) F. Wrede: Ar. Pth., **184**, 331 (1937). [C., **1937**, II. 2611].
93) F. Wrede: Forschungen u. Fortsh., **14**, 173 (1938).
94) S. Pfeifer: Pharmazie, **21**, 378 (1966).
95) S. Pfeifer, F. Weiss: Arch. Pharm., **289**, 24 (1956).
96) S. Pfeifer: Arch. Pharm., **290**, 261 (1957).
97) A.R. Battersby, H. Spencer: J. Chem. Soc., **1965**, 1087.
98) W.-K. Kuei, C.-C. Chang, K.-S. Lin: Hua Hsueh Hsueh Pao, **31**, 470 (1965). [C. A., **64**, 15936 (1966)].
99) H. Asahina, T. Kawakani, M. Ono, S. Fujita: Bull. Narcotics, **9**, No. 2, 20 (1957). [C. A., **51**, 17094 (1957)].
100) C. Derosne: Ann. Chim., [i] **45**, 274 (1804).
101) M. Robiquet: Ann. Chim., [ii] **5**, 275 (1817).
102) J.N. Rakshit: Analyst, **46**, 481 (1921). [C. A., **16**, 611 (1922)].
103) Z. Arima, M. Iwakiri: Rept. Inst. Sci. Res., Manchoukus, **2**, 221 (1938). [C. A., **33**, 1875 (1939)].
104) E. Both: Apotheker im Osten, **1**, 137 (1942). [C. A., **38**, 3086 (1944)].
105) C. Djerassi, M. Gorman, A.L. Nussbaum, J. Reyncso: J. Am. Chem. Soc., **75**, 5446 (1953).
106) P. Gaubert: Compt. rend., **156**, 1161 (1913).
107) R. Kremann, N. Schniderschitz: Monatsh., **35**, 1423 (1914).
108) H.E. Annett: J. Chem. Soc., **123**, 376 (1923).
109) J.N. Rakshit: J. Phys. Chem., **34**, 2539 (1930).
110) F.W. Bollinger: J. Am. Pharm. Assoc., **44**, 580 (1955).
111) F. Wöhler: Ann., **50**, 1 (1844).
112) T. Anderson: Ann., **86**, 179 (1853).
113) A. Matthiessen, F.C. Foster: Jahresbericht, **1861**, 539.
114) E. von Gerichten: Ber., **13**, 1635 (1880).
115) E. von Gerichten: Ann., **210**, 79 (1881).
116) W. Roser: Ann., **234**, 116 (1886).
117) W. Roser: Ann., **249**, 156 (1888).
118) W. Roser: Ann., **254**, 334, 359 (1889).
119) H. Decker: J. prakt. Chem., (2), **47**, 222 (1893).
120) R. Fritsch: Ann., **301**, 352 (1893).
121) A. Hantzsch, H. Kalb: Ber., **32**, 3109 (1899).
122) A. Hantzsch, H. Kalb: Ber., **33**, 2201 (1900).
123) M. Freund, F. Becker: Ber., **36**, 1521 (1903).
124) J.J. Dobbie, A. Lauder, C.K. Tinkler: J. Chem. Soc., **83**, 598 (1903).
125) J.J. Dobbie, A. Lauder, C.K. Tinkler: J. Chem, Soc., **85**, 121 (1904).
126) A. H. Salway: J. Chem. Soc., **95**, 1204 (1909).
127) W.H. Perkin, Jr., R. Robinson, F. Thomas: J. Chem. Soc., **95**, 1977 (1909).
128) A. H. Salway: J. Chem. Soc., **97**, 1208 (1910).
129) G. A. Edwards, W. H. Perkin, Jr., F. W. Stoyle: J. Chem. Soc., **127**, 195 (1925).
130) J.N. Rakshit: J. Chem. Soc., **1913**, 466.
131) G.S. Ahluwalia, B.D. Kochlar, J.N. Ray: J. Indian Chem. Soc., **9**, 215 (1932).
132) K. Topchiev: Zhur. Priklado. Khim., **6**, 529 (1933). [C. A., **28**, 2718 (1934)].
133) B. Skinner: J. Chem. Soc., **1950**, 823.
134) M.S. Dyer, A.J. McBay: J. Am. Pharm. Assoc., **44**, 156 (1955).
135) A. R. Battersby, H. Spencer: Tetrahedron Letters, **1964**, 11.
136) M. Ohta, H. Tani, S. Morozumi, S. Kodaira, K. Kuriyama: Tetrahedron Letters, **1964**, 693.
137) M. Ohta, H. Tani, S. Morozumi, S. Kodaira: Chem. Pharm. Bull., **12**. 1080 (1964).
138) A.R. Battersby, D.J. McCaldin: Proc.Chem. Soc., **1962**, 365.
139) A.R. Battersby, M. Hirst: Tetrahedron Letters, **1965**, 669.
140) L. van Itallie, A.J. Stenhauer: Arch. Pharm., **265**, 696 (1927).
141) L. Ekkert: Magyar. Gyogyszeresztud. Torsasag Ertesitöze, **5**, 21 (1929). [C., **1929**, I, 1990].
142) P. Steiner: Ann. med. légale criminol et palice sci., **2**, 338 (1922). [C. A., **20**, 1178 (1926)].
143) P.M. Oertreicher, C.G. Farmilo, L. Levi: Bull. Narcotics, **7**, 42 (1954).
144) L.H. Briggs, L.D. Colebrook, H.M. Fales, W. C. Wildman: Anal. Chem., **29**, 904 (1957).
145) Y. Sasaki: J. Pharm. Soc. Japan, **80**, 241 (1960).
146) W. H. Barnes: Can. J. Chem., **33**, 444 (1955).
147) W. Doepke: Arch. Pharm., **295**, 605 (1962).

148) W. Debska: Poznan. Towarz. Przyjaciol Nauk Wydzial Lekar., Prace Komisji, Farm., **2**, 23 (1962). [C. A., **59**, 6451 (1963)].
149) T. Smith, H. Smith: Ber., **26**, 593 R (1893); Pharm. J., (3), **9**, 82 (1878).
150) P. Rabe, A. McMillan: Ann., **377**, 223 (1910).
151) G. H. Beckett, C. R. Wright: J. Chem. Soc., **29**, 461 (1876).
152) P. Rabe: Ber., **40**, 3280 (1907).
153) J. Pelletier: Ann. Chim. Phys., **50**, 240, 262 (1832).
154) C. R. Addinall, R. T. Major: J. Am. Chem. Soc., **55**, 1202, 2153 (1933).
155) W. Roser: Ann., **247**, 167 (1888).
156) M. Freund, G. B. Frankforter: Ann., **277**, 20 (1893).
157) M. Freund, P. Oppenheim: Ber., **42**, 1084 (1909).
158) L. Levi: Bull. Narcotics, **7**, 43 (1955).
159) N. A. Izmairov, A. K. Franke, I. S. Simon: Med. Prom. SSSR., **13**, No. 11, 36 (1959). [C. A., **55**, 15838 (1961)].
160) M. Danilovic, N. Ristic: Arkiv. Farm., **12**, No. 2, 112 (1962). [C. A., **58**, 10038 (1963)].

Chapter 13 Ochotensine and Related Alkaloids

Ochotensine (**F 17**)[2] $C_{21}H_{21}O_4N \equiv 351.39$

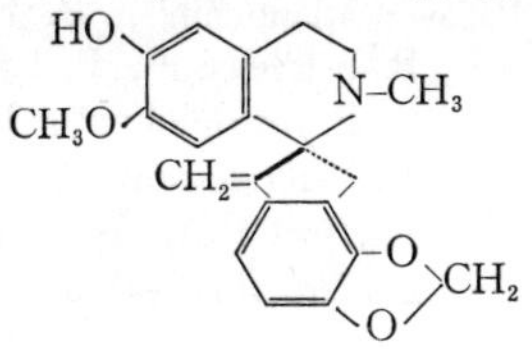

HO, CH_3O, N–CH_3, CH_2=, O, O, CH_2

≪mp≫ 252° ($CHCl_3$)

≪[α]D≫ +51.7° ($CHCl_3$), +63.9° (0.1N-HCl)

≪Ph.≫ X-ray[4]

≪OP.≫ Corydalis ochotensis[1], C. sibirica[2], Dicentra cucullaria[3]

≪C. & So.≫ Colourless prisms

≪Sa. & D.≫ B–MeI: 215°

B–OMe = ochotensimine

≪SD.≫ 1), 4)

≪S. & B.≫

Ochotensimine $C_{22}H_{23}O_4N \equiv 365.41$

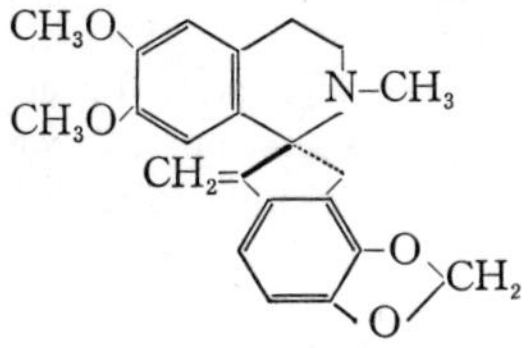

CH_3O, CH_3O, N–CH_3, CH_2=, O, O, CH_2

≪mp≫

≪[α]D≫ +49.2° (MeOH)

≪Ph.≫ UV[5] NMR[5]

≪OP.≫ Corydalis ochotensis[1]

≪C. & So.≫ Yellowish-brown amorph

≪Sa. & D.≫ B–MeI: 225° (d)

B–Dihydro-Der.: [α]D +112°

≪SD.≫ 1), 5)

≪S. & B.≫

---

1) R. H. F. Manske: Can. J. Res., **18B**, 75 (1940).
2) R. H. F. Manske: Can. J. Res., **14B**, 354 (1936).
3) R. H. F. Manske: Can. J. Res., **16B**, 81 (1938).
4) S. McLean, M.-S. Lin, A. C. MacDonald, J. Trotter: Tetrahedron Letters, **1966**, 185.
5) S. McLean, M.-S. Lin: Tetrahedron Letters, **1964**, 3819.

Chapter 14 Rhoeadine and Related Alkaloids

Papaverrubine A

$C_{20}H_{19}O_6N \equiv 369.36$

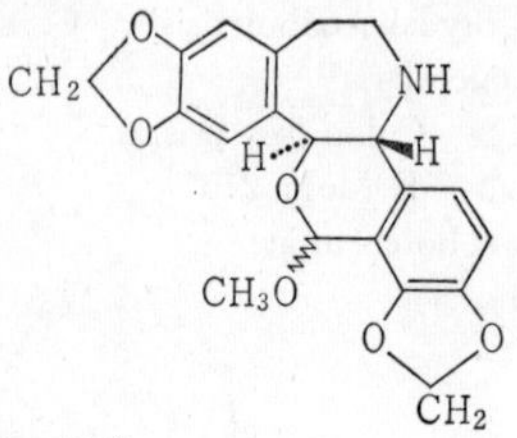

≪mp≫ 223~224°

≪[α_D]≫ +410±10° (CHCl$_3$)

≪Ph.≫ UV[3] Mass[4]

≪OP.≫ Papaver atlanticum[1], P. caucasicum[1], P. dubium[1], P. heldreichii[1], P. lateritum[1], P. macrostomum[1], P. oreophilum[1], P. pilosum[1], P. rhoeas[1][2][3], P. sterigosum[1]

≪C. & So.≫ Sol. in CHCl$_3$, Spar. sol. in MeOH

≪Sa. & D.≫ B-Picrol: 170~173°

≪SD.≫ 4)

≪S. & B.≫

Papaverrubine E

$C_{20}H_{19}O_6N \equiv 369.36$

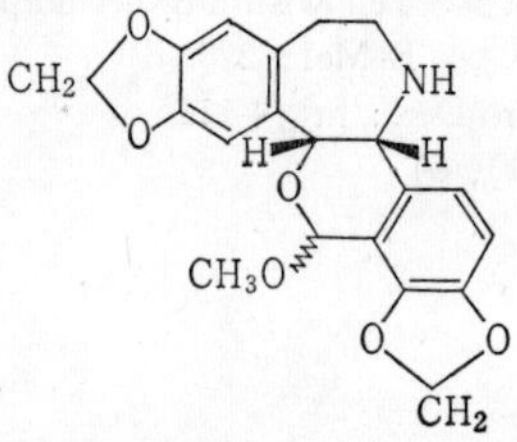

≪mp≫ 231°

≪[α]$_D$≫ +331±5° (CHCl$_3$)

≪Ph.≫ UV[3] Mass[4]

≪OP.≫ Papaver alpinum[1], P. argemone[1], P. atlanticum[1], P. bracteatum[1], P. caucasicum[1], P. dubium[1], P. heldreichii[1], P. feddei[1], P. fugax[1], P. hybridum[1], P. lateretium[1]*

≪C. & So.≫ Sol. in CHCl$_3$, Spar. sol. in MeOH

≪Sa. & D.≫ B-Pic: 155~157°
B-Picrol: 176~178°

≪SD.≫ 4)

≪S. & B.≫

Porphyroxine (Papaverrubine D)

$C_{20}H_{19}O_6N \equiv 383.89$

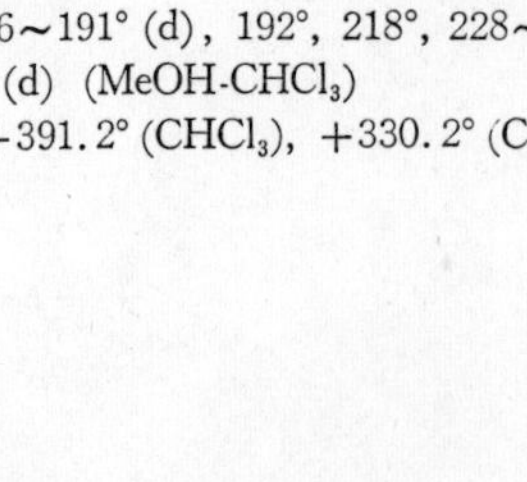

≪mp≫ 186~191° (d), 192°, 218°, 228~231° (d), 234~236° (d) (MeOH-CHCl$_3$)

≪[α]$_D$≫ +391.2° (CHCl$_3$), +330.2° (CHCl$_3$)

≪Ph.≫ CR[5][6][7][8] UV[6][7][8][10] IR[6][7][8] NMR[7] Mass[4] Rf[5][6][7][8] pKa[8]

≪OP.≫ Papaver alpinum[1], P. argemone[1], P. atlanticum[1], P. bracteatum[1], P. caucasicum[1], P. dubium[1], P. feddi[1], P. fugax[1], P. glaucium[1][5], P. heldreichii[1], P. hybridum[1], P. lateritinum[1]**

≪C. & So.≫ Plates, Sol. in CHCl$_3$, CH$_2$Cl$_2$, Spar. sol. in MeOH, EtOH, Et$_2$O

≪Sa. & D.≫ B-Pic: 241~244°
B-OMe: 202~203° or 241~243°
B-MeI: 150~152° B-Me$_2$SO$_4$: 205°

≪SD.≫ 4), 6), 7), 9)

≪S. & B.≫

---

* P. macrostomum[1], P. oreophilium[1], P. orientale[1], P. pavonium[1], P. pilosum[1], P. pseudocanescens[1], P. radicatum[1], P. rhoeas[1][3], P. setigerum[1], P. strigosum[1]

** P. macrostonum[1], P. nudicaule[1], P. oreophilum[1], P. orientale[1], P. pavonium[1], P. perisicum[1], P. pilosum[1], P. pseudscanecens[1], P. radicatum[1], P. rhoeas[1], P. rupifragum[1], P. setigerum[1], P. somniferum[1][2][6][8][34][35], P. strigosum[1]

Rhoeadine $C_{21}H_{21}O_6N$≡383.39

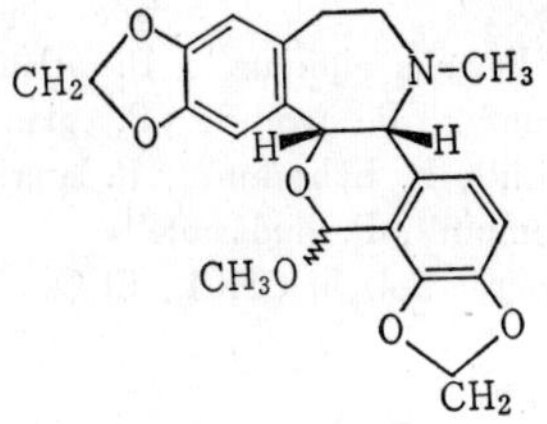

≪mp≫ 245~247°, 249°, 251~253°, 252~254°, 259°, 272~274° (Et_2O)

≪$[\alpha]_D$≫ +235±2° ($CHCl_3$), +174±2° (Py), +237±2° (AcOH)

≪Ph.≫ CR[14,18] UR[12,22,24,25] IR[12,15,22,23,24,26] NMR[24,25,27] Mass[4,11,24,25,28,29] Rf[16,24,30] Polaro[22]

≪OP.≫ Papaver hybridum[11], P. lateritium[12], P. rhoeas[11,13,14,15,16,17,18], isolation[42]

≪C. & So.≫ Needles, Sol. in C_6H_6, Insol. in AmOH

≪Sa. & D.≫ B–HCl: 224~226° B–HBr: 228~230° B–$HClO_4$: 181~182.5° B–MeI: 196~206°

≪SD.≫ 4), 9), 12), 13), 14), 18), 19), 20), 21), 22), 23), 24), 43)

≪S. & B.≫ cf. 25), BG: 24)

Rhoeagenine $C_{20}H_{19}O_6N$≡369.36

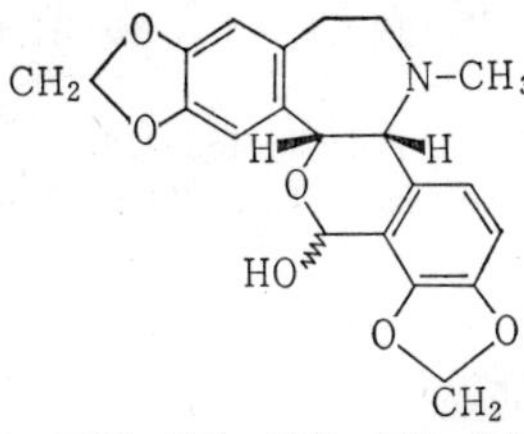

≪mp≫ 236~238°, 237~239°, 243~245°

≪$[\alpha]_D$≫ +134° (Py), +154° (AcOH)

≪Ph.≫ CR[31] UV[24] IR[23,24,26] NMR[24] Mass[24,29] Rf[16,24,30,31]

≪OP.≫ hydrolysis product of rhoeadine

≪C. & So.≫ Prisms

≪Sa. & D.≫ B–MeI: 225~228° (d) B–$HClO_4$: 161~163° (d)

≪SD.≫ 22), 23), 29)

≪S. & B.≫ BG: 24), 44)

Dubirheine $C_{22}H_{23}O_6N$≡394.71

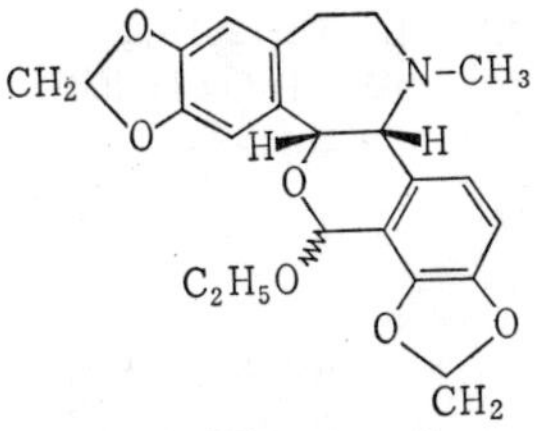

≪mp≫ 236~237° (AcOEt)

≪$[\alpha]_D$≫ +236°

≪Ph.≫ Mass[28] Rf[28,32]

≪OP.≫ Papaver dubium[28,32]

≪C. & So.≫ Colourless needles, Sol. in $CHCl_3$, Insol. in EtOH

≪Sa. & D.≫

≪SD.≫ 28), 33)

≪S. & B.≫

Isorhoeadine $C_{21}H_{21}O_6N$≡383.89

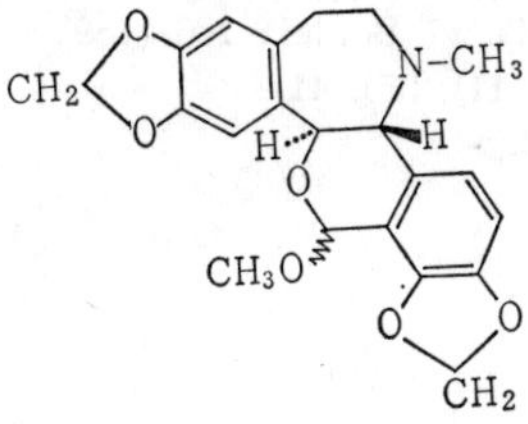

≪mp≫ 159~161° (MeOH), 237~239° (AcOEt-Et_2O)

≪$[\alpha]_D$≫ +314±3° ($CHCl_3$), 155±4° (AcOH)

≪Ph.≫ CR[31] UV[24,25] IR[15,24] NMR[24,25,27] Mass[4,24,25] Rf[16,24,31,36]

≪OP.≫ Papaver dubium[31], P. rhoeas[9,15,16,36]

≪C. & So.≫ Needles or prisms, Sol. in $CHCl_3$, Insol. in MeOH

≪Sa. & D.≫

≪SD.≫ 4), 24)

≪S. & B.≫ BG: 24)

Papaverrubine B $C_{21}H_{23}O_6N \equiv 385.40$

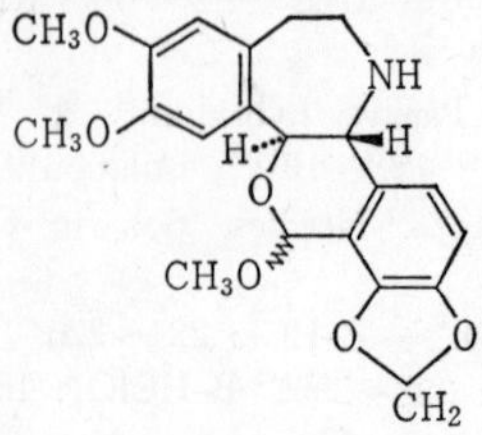

≪mp≫ 202~204° (MeOH)
≪[α]$_D$≫

≪Ph.≫ UV[5,7,37,38] IR[5,7,38] NMR[38] Mass[4,38] Rf[5,7,37]
≪OP.≫ Papaver alpinum[1], P. atlanticum[1], P. caucasicum[1], P. feddi[1], P. glaucum[1,7], P. heldreichii[1], P. hybridum[1], P. lateritinum[1], P. macrostomum[1], P. nudicaule[1]*
≪C. & So.≫ Sol. in $CHCl_3$, Et_2O, Spar. sol. in MeOH
≪Sa. & D.≫
≪SD.≫ 4), 7), 37), 38)
≪S. & B.≫

Glaudine $C_{22}H_{25}O_6N \equiv 399.43$

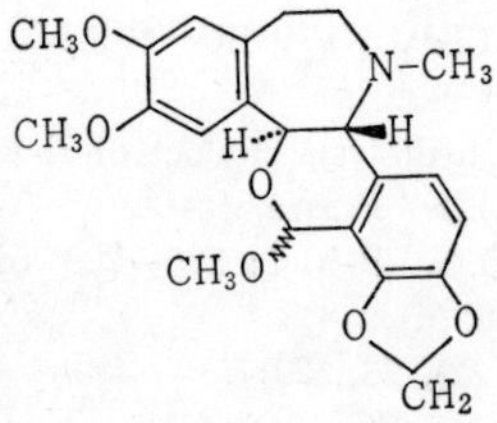

≪mp≫ 103~105° (n-heptane)
≪[α]$_D$≫ +455±5° ($CHCl_3$)

≪Ph.≫ UV[5,7,39] IR[5,7] NMR[27] Mass[4,5,27] Rf[5,7]
≪OP.≫ Papaver glaucum[1,5,29,39]
≪C. & So.≫ Plates, Sol. in MeOH, C_6H_6, $CHCl_3$, Et_2O
≪Sa. & D.≫ B–HCl: 185~187°
B–HI: 180~182° B–$HClO_4$: 173~181°
B–Oxal: 152° B–MeI: 174~176°
≪SD.≫ 4), 7), 27), 39)
≪S. & B.≫

Glaucamine $C_{21}H_{23}O_6N \equiv 385.40$

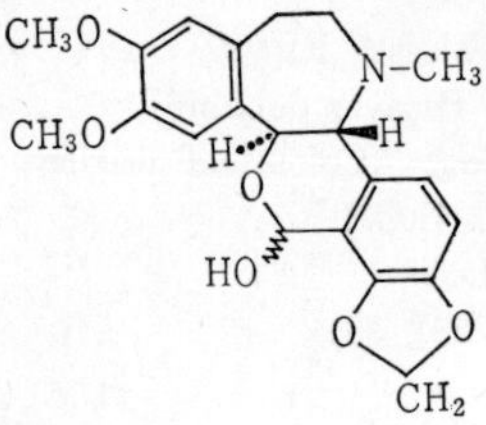

≪mp≫ 222~223°
≪[α]$_D$≫ +298±3° ($CHCl_3$)

≪Ph.≫ UV[5,39] NMR[40] Mass[27,29] Rf[5]
≪OP.≫ Papaver glaucum[5,29,39]
≪C. & So.≫
≪Sa. & D.≫ B–HCl: 201~203°
B–HI: 203~205° B–Oxal: 160°
≪SD.≫ 27), 29), 40)
≪S. & B.≫

Oreodine $C_{22}H_{25}O_6N \equiv 394.43$

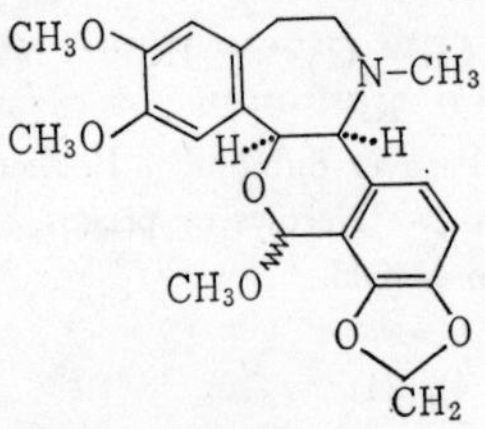

≪mp≫ 184~186° (Et_2O)
≪[α]$_D$≫ +224±5° ($CHCl_3$)

≪Ph.≫ UV[35] NMR[27] Mass[27,41] Rf[35]
≪OP.≫ Papaver oreophilum[27,35]
≪C. & So.≫ Sol. in MeOH, $CHCl_3$
≪Sa. & D.≫ B–MeI: 186~189°
≪SD.≫ 11), 35), 41)
≪S. & B.≫

---

* P. oreophilum[1], P. orientale[1], P. pilosum[1], P. pseudocanascens[1], P. radicatum[1], P. rhoeas[1,37], P. rupifragum[1], P. setigerum[1], P. somniferum[1,2,38], P. strigosum[1]

Oreogenine $C_{21}H_{23}O_6N \equiv 385.40$

≪mp≫ 177～178° (MeOH)

≪[α]D≫ $-254 \pm 5°$ ($CHCl_3$)

≪Ph.≫ UV[35] NMR[27] Mass[27)41] Rf[35]

≪OP.≫ Papaver oreophilum[27)35]

≪C. & So.≫

≪Sa. & D.≫ B-MeI: 207～209°

≪SD.≫ 11), 35), 41)

≪S. & B.≫

---

1) S. Pfeifer, S. K. Banerjee: Pharmazie, **19**, 286 (1964)
2) S. Pfeifer: Pharmazie, **20**, 240 (1965).
3) S. Pfeifer, S. K. Banerjee: Arch. Pharm., **298**, 385 (1965).
4) S. Pfeifer, S. K. Banerjee: Pharmazie, **20**, 45 (1965).
5) S. Pfeifer, I. Mann: Pharmazie, **20**, 643 (1965).
6) K. Genest, C. G. Farmilo: J. Pharm. Pharmacol., **15**, 197 (1963).
7) S. Pfeifer: J. Pharm. Pharmacol., **18**, 133 (1966).
8) S. Pfeifer, J. Teige: Pharmazie, **17**, 692 (1962).
9) W. Winkler: Arch. Pharm., **290**, 367 (1957)
10) D. L. Klaymann: Dissertation, New Jearsey, 1965.
11) V. Davesi: Affi. seale ist. Dotan. Univ. Pavia, **9**, (1906). [C., **1906**, I, 690]
12) F. Santavy, M. Maturova, A. Nemeckova, M. Hrak: Coll. Czech. Chem. Comm., **24**, 3493 (1959).
13) E. Späth, L. Schmid, H. Sternberg: Monatsh., **68**, 33 (1936).
14) W. Awe: Arch. Pharm., **274**, 439 (1936).
15) W. Awe, W. Winkler: Arch. Pharm., **290**, 367 (1957).
16) J. Slavik: Chem. Listy, **52**, 1957 (1958).
17) J. Slavik: Coll. Czech. Chem. Comm., **24**, 2506 (1959).
18) W. Awe, W. Winkler: Nature, **47**, 107 (1960); cf. Arzneimittel-Forsch., **9**, 773 (1959).
19) W. Awe, W. Winkler: Arch. Pharm., **291**, 642 (1958).
20) W. Awe, W. Winkler: Bull. Narcotics U.S. Dept. Social Affairs, **10**, 20 (1958).
21) W. Awe, W. Winkler: Arch. Pharm., **292**, 293 (1959).
22) F. Santavy, M. Maturova, A. Nemeckova, M. Horak: Coll. Czech. Chem. Comm., **25**, 1901 (1960).
23) J. Suszko, M. D. Rozwadowska: Bull. Akad. Polon. Sci. Ser. Sci. Chem., **11**, 513 (1963).
24) F. Santavy, J. L. Kaul, L. Hruban, L. Doleys, V. Hanus, K. Blaha: Coll. Czech. Chem. Comm., **30**, 3479 (1965).
25) F. Santavy, J. L. Kaul, L. Hruban, L. Doleys, V. Hanus, K. Blaha, A. D. Cross: Coll. Czech. Chem. Comm., **30**, 335 (1965).
26) J. Suszko, M. D. Rozwadowska: Bull. Akad. Polon. Sci., Ser. Sci. Chem., **12**, 767 (1964). [C. A., **62**, 10471 (1965)].
27) A. D. Cross, I. Mann, S. Pfeifer: Pharmazie, **21**, 181 (1966).
28) J. Slavik, V. Hanus, K. Vokac, L. Doleys: Coll. Czech. Chem. Comm., **30**, 2464 (1965).
29) J. Slavik, L. Doleys, K. Vokac, V. Hanus: Coll. Czech. Chem. Comm., **30**, 2864 (1965).
30) A. Namekova, F. Santavy: Coll. Czech. Chem. Comm., **30**, 912 (1965).
31) A. Namekova, F. Santavy: Coll. Czech. Chem. Comm., **27**, 1210 (1962).
32) J. Slavik: Coll. Czech. Chem. Comm., **28**, 1738 (1963).
33) W. Awe: Arch. Pharm., **214**, 439 (1936).
34) J. N. Rakshit: Ber., **59**, 2473 (1921).
35) S. Pfeifer, I. Mann: Pharmazie, **19**, 786 (1964).
36) F. Santavy, A. Namekova, H. Potesilova: Acta Chim. Acad. Sci. Hung., **18**, 457 (1959). [C. A., **53**, 19301 (1959)].
37) S. Pfeifer: Pharmazie, **19**, 786 (1964).
38) W. D. Hughes, C. G. Familo: J. Pharm. Pharmacol., **117**, 757 (1965).
39) S. Pfeifer: Phramazie, **19**, 724 (1964).
40) A. D. Cross, J. Slavik: Coll. Czech. Chem. Comm., **31**, 1425 (1966).
41) S. Pfeifer, I. Mann, L. Doleys, V. Hanus: Pharmazie, **20**, 585 (1965).
42) O. Hesse: Ann., **140**, 145 (1866); **149**, 35 (1869); **185**, 329 (1877); Arch. Pharm., **228** 7 (1890).
43) W. Awe: Dtsh. Apoth. Ztg., **97**, 559 (1957); Pharmazie, **12**, 633 (1957).
44) L. Kühn, S. Pfeifer: Pharmazie, **20**, 659(1965).

Chapter 15 Morphine and Related Alkaloids

Norsinoacutine $C_{18}H_{19}O_4N \equiv 313.34$

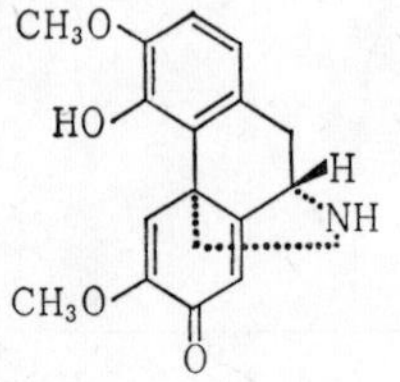

≪Ph.≫ UV[1] IR[1] NMR[1]
≪OP.≫ Croton balsamifera[1]
≪C. & So.≫
≪Sa. & D.≫
≪SD.≫ 1)
≪S. & B.≫

≪mp≫
≪$[\alpha]_D$≫

Amurine $C_{19}H_{19}O_4N \equiv 325.34$

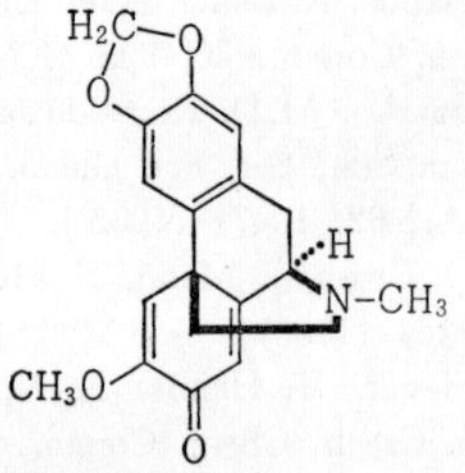

≪Ph.≫ CR[7,8] UV[5,7,8,9] IR[1,6,7,8,9] NMR[5] Mass[7,9] CD[10] Polaro[7]
≪OP.≫ Papaver feddei[2], P. nudicaule var. amurense[3,4,5], P. nudicaule var. aurantiacum[4], P. nudicaule var. croceum[4], P. nudicaule var. xanthopetalum[6]
≪C. & So.≫
≪Sa. & D.≫ B–HI: 216～218°
B–$HClO_4$: 215～217° B–Pic: 226～227° (d)
B–MeI: 203～205°
≪SD.≫ 3), 5)
≪S. & B.≫

≪mp≫ 213～215° (Me_2CO)
≪$[\alpha]_D$≫ +10° ($CHCl_3$)

Nudaurine $C_{19}H_{21}O_4N \equiv 327.36$

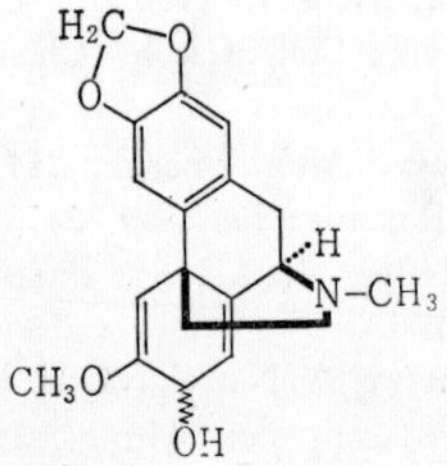

≪Ph.≫ UV[5] IR[5] NMR[5] CD[10]
≪OP.≫ Papaver nudicaule var. aurantiacum[4,5]
≪C. & So.≫
≪Sa. & D.≫ B–HI: 190° (d) B–$HClO_4$: 188° (d)
B–Pic: 150°
≪SD.≫ 4), 5)
≪S. & B.≫

≪mp≫ 201～202°
≪$[\alpha]_D$≫ −52° ($CHCl_3$)

Sinoacutine $C_{19}H_{21}O_4N \equiv 327.36$

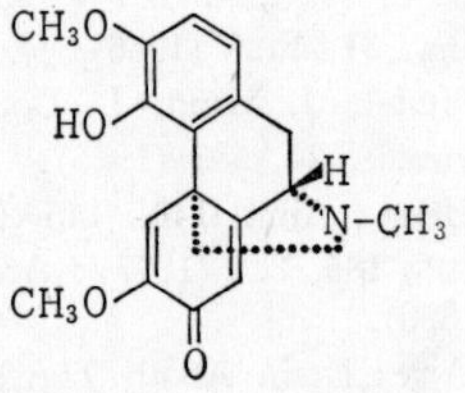

≪Ph.≫ CR[12,13] UV[12,13] IR[12,13] CD[10]
≪OP.≫ Croton balsamifera[1], Sinomenium acutum[11,12,13]
≪C. & So.≫
≪Sa. & D.≫ B–$HClO_4$: 177° B–Pic: 222° (d)
B–Tart: 203～204° B–MeI: 216° B–OAc: 175°
B–Oxime: 223° cf. Sinoactinol: 226°
≪SD.≫ 13)
≪S. & B.≫ BS: 17)

≪mp≫ 198°
≪$[\alpha]_D$≫ −81.6° ($CHCl_3$), −112° (EtOH)

Salutaridine $C_{19}H_{21}O_4N \equiv 327.36$

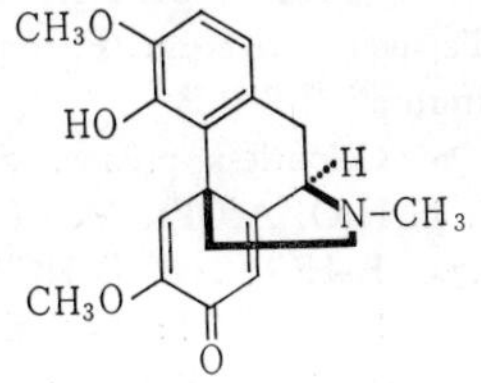

≪mp≫ 197～198° (AcOEt)
≪$[\alpha]_D$≫ +111° (EtOH), +104° (AcOEt)

≪Ph.≫ CR[14] UV[14] IR[1,14] NMR[18] Rf[1]
≪OP.≫ Croton balsamifera,[1] C. salutaris[14], Papaver orientale[14,15]
≪C. & So.≫ Rods
≪Sa. & D.≫ B-Pic: 212～216°(d) B-OAc: 171° cf. salutaridinol-I: 227～229° (d) II: 132～134°
≪SD.≫ 14), 16)
≪S. & B.≫ 14), 16), BS: 14), 15), BG:→morphine

Sinomenine $C_{19}H_{23}O_4N \equiv 329.38$

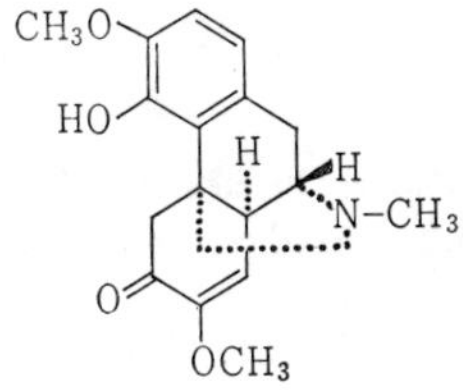

≪mp≫ 182° (C_6H_6), cf. 162° [$B+H_2O$]
≪$[\alpha]_D$≫ −70.6° (EtOH)

≪Ph.≫ CR[25,39,86] UV[11,12,24,87] IR[1] NMR[14]
≪OP.≫ Menispermum dauricum[19,20], Sinomenium acutum[11,12,21,22,23,24], S. diversifolius (Cocculus diversifolius)[21,25]
≪C. & So.≫ Needles, Sol. in EtOH, $CHCl_3$, Me_2CO, Mod. sol. in Et_2O, C_6H_6, Spar. sol. in Petr. ether, H_2O
≪Sa. & D.≫ B-HCl: 229～231° B-HBr: 231° B-HI: 236° B-HNO_3: 214～218° B_2-H_2SO_4: 216～217° B-Pic: 159～160° B-Picrol: 94～95° B-MeI: 255°(d) B-OBz: 225° B-Oxime: 254°
≪SD.≫ 19), 22), 23), 25)～85)
≪S. & B.≫ cf. 61), 66), BS: 14), 17), 88), BG: 88)

Isosinomenine $C_{19}H_{23}O_4N \equiv 329.38$

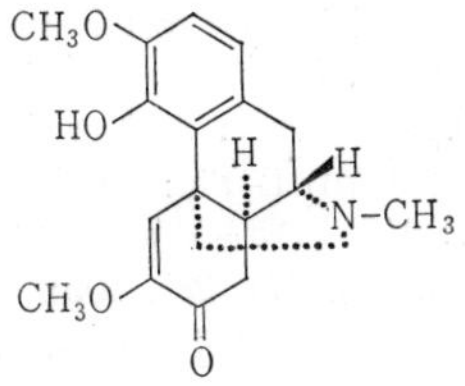

≪mp≫ 210°～212° (EtOH)
≪$[\alpha]_D$≫ +54° (EtOH)

≪Ph.≫ UV[11,24,89] IR[14,89] NMR[89]
≪OP.≫ Sinomenium acutum[24]
≪C. & So.≫ Colourless needles, sublimation.
≪Sa. & D.≫ Dihydro-Der.: 176～178°
≪SD.≫ 11), 24), 89)
≪S. & B.≫

Disinomenine (Dehydrosinomenine) $C_{38}H_{44}O_8N_2 \equiv 656.75$

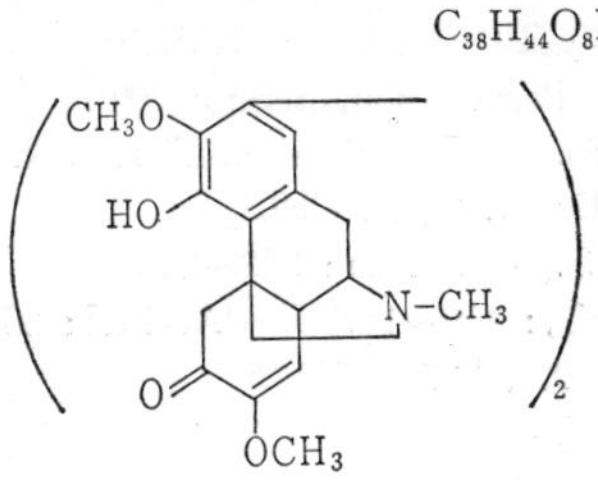

≪mp≫ 222°
≪$[\alpha]_D$≫ +150°

≪Ph.≫
≪OP.≫ Sinomenium acutum[90]
≪C. & So.≫
≪Sa. & D.≫ B-2MeI: 263° B-Dioxime: 265° Tetrahydro-Der.: 252°
≪SD.≫ 40), 91), 92)
≪S. & B.≫

Morphine $C_{17}H_{19}O_3N$≡285.33

≪mp≫ 253～254° [B+H_2O] (aq. EtOH)
≪$[\alpha]_D$≫ −130.9° (MeOH)

≪Ph.≫ CR[86)99)273)] UV[87)274)275)276)] NMR[277)278)] Mass[279)280)281)] ORD[282)] CD[283)284)] X-ray[213)] Rf[278)]
≪OP.≫ Papaver setigerum (P. somniferum)[93)94)], P. somniferum[95)96)97)98)99)]
≪C. & So.≫ Colourless prisms, Sol. in EtOH, Mod. sol. in H_2O, AcOEt, Me_2CO, $CHCl_3$, C_6H_6
≪Sa. & D.≫ B–HCl: 200° B–$HClO_4$: 150°
B–Pic: 163～165° B–MeI: 286° (d)
B–O, O–DiMe: 140～141°
≪SD.≫ 9), 16), 26), 45), 56), 57), 58), 59), 61), 83), 100)～220)
≪S. & B.≫ 16), 221), 222), 223), 224), cf. 64), 225)～243), BS: 14), 16), 224～264), cf. 88), 265)～269), BG: 88), 270), 271), 272)

Oripavine $C_{18}H_{19}O_3N$≡297.34

≪mp≫ 201～202° (EtOH)
≪$[\alpha]_D$≫ −212° ($CHCl_3$)

≪Ph.≫
≪OP.≫ Papaver bracteatum[285)286)], P. orientalis[286)]
≪C. & So.≫ Needles, Sol. in $CHCl_3$, Spar. sol. in EtOH, Me_2CO
≪Sa. & D.≫ B–HCl: 258～259° (d)
B–MeI: 207～208° B–OMe=thebaine
≪SD.≫ 286)
≪S. & B.≫

Codeine $C_{18}H_{21}O_3N$≡299.36

≪mp≫ 156～157°
≪$[\alpha]_D$≫ −111° ($CHCl_3$), −137.8 (EtOH)

≪Ph.≫ CR[86)99)273)] UV[87)276)] NMR[277)324)] Mass[279)] ORD[282)] X-ray[218)325)] Rf[278)]
≪OP.≫ Papaver setigerum (P. somniferum)[94)], P. somniferum[97)98)99)287)288)]
≪C. & So.≫ Rhombic octahedra, Sol. in C_6H_6, EtOH, $CHCl_3$, Mod. sol. in Et_2O
≪Sa. & D.≫ B–HCl: 264° or 287°
B–MeI: 270° (d) B–Pic: 197°
B–Styph: 115° (195～197°) B–OAc: 133°
≪SD.≫ 9), 26), 27), 28), 103), 106), 108), 111), 112), 115), 116), 117), 118), 122), 125), 131), 132), 142), 147), 149), 158), 164), 169), 171), 173), 174), 177), 179), 180), 183), 184), 187), 189), 193), 195), 207), 211), 219), 289)～323)
≪S. & B.≫ →morphine

Neopine (β-Codeine) $C_{18}H_{21}O_3N$≡299.36

≪mp≫ 127～127.5° (Petr. ether)
≪$[\alpha]_D$≫ −28.1° ($CHCl_3$)

≪Ph.≫ CR[86)97)] UV[87)97)] NMR[324)]
≪OP.≫ Papaver somniferum[97)326)]
≪C. & So.≫ Colourless needles, Sol. in H_2O, Et_2O, EtOH, C_6H_6, $CHCl_3$
≪Sa. & D.≫ B–HBr: 282～283°
B_2–H_2SO_4: 166～167° B–OAc·MeI: 256～257°
≪SD.≫ 97), 327), 328)
≪S. & B.≫

10-Hydroxycodeine $C_{18}H_{21}O_4N \equiv 315.36$

≪mp≫ 207°
≪$[\alpha]_D$≫ −132° (EtOH)
≪Ph.≫
≪OP.≫ Opium[329]
≪C. & So.≫
≪Sa. & D.≫ B–HCl: 254° B–Pic: 166°
B–O,O–DiAc: 160°
≪SD.≫ 97), 143), 297), 301), 323)
≪S. & B.≫

Thebaine $C_{19}H_{21}O_3N \equiv 311.37$

≪mp≫ 193° (EtOH) cf. Sublime at 91°/0.01
≪$[\alpha]_D$≫ −218.6° (EtOH), −229° ($CHCl_3$)
≪Ph.≫ CR[86)99)168] UV[87)276] IR[334] NMR[277]
Mass[281]
≪OP.≫ Papaver intermedium (P. rhoeas)[330], P. orientale[168], P. setigerum[94],
P. somniferum[97)98)99)167)285)331], P. strigosum[330]
≪C. & So.≫ Prisms or plates, Very sol. in EtOH, $CHCl_3$, Sol. in H_2O, C_6H_6
≪Sa. & D.≫ B–Pic: 217° B–Tart: 223°
B–MeI: 224°
≪SD.≫ 28), 45), 68), 83), 130), 138), 139), 140), 141), 142), 168), 174), 183), 186), 187), 188), 193), 199), 200), 202), 211), 291), 295), 298), 313), 318), 319), 332)∼336)
≪S. & B.≫ →morphine

Pseudomorphine $C_{34}H_{36}O_6N_2 \equiv 568.64$

≪mp≫ 327°
≪$[\alpha]_D$≫ −115° (HCl)
≪Ph.≫ UV[87)275]
≪OP.≫ Opium[331]
≪C. & So.≫
≪Sa. & D.≫ B–O,O–DiMe: 155∼156°
≪SD.≫ 275), 337), 338), 339), 340)
≪S. & B.≫

1) C. Chambers, L. J. Haynes, K. L. Stuart: Chem. Comm., **1966**, 449.
2) S. Pfeifer, I. Mann: Pharmazie, **20**, 643 (1965).
3) H.-G. Boit, H. Flentje: Naturwiss., **46**, 514 (1959).
4) H.-G. Boit, H. Flentje: Naturwiss., **47**, 180 (1960).
5) H. Flentje, W. Doepke, P. W. Jeffs: Naturwiss., **52**, 259 (1965).
6) M. Maturova, B. K. Moza, J. Sitar, F. Santavy: Planta Med., **10**, 345 (1962). [C. A., **58**, 5988 (1962)].
7) M. Maturova, L. Hruban, F. Santavy, W. Wiegrebe: Arch. Pharm., **298**, 209 (1965).
8) L. Kühn, S. Pfeifer: Pharmazie, **20**, 659 (1965).
9) R. E. Lutz, L. F. Small: J. Am. Chem. Soc., **56**, 1741 (1934).
10) G. Snatzke, G. Wollenberg: J. Chem. Soc. (C), **1966**, 1681.
11) Y. Sasaki, U. Ueda: J. Pharm. Soc. Japan, **78**, 44 (1958).
12) J.-H. Chu, S.-Y. Lo, Y.-L. Chou: Hua Hsueh Hsueh Pao, **30**, 265 (1964).
13) J.-H. Hsu, S.-Y. Lo., J.-H. Chu: Sci. Sinica, **13**, 2016 (1964). [C. A., **62**, 9183 (1965)].
14) D. H. R. Barton, G. W. Kirby, W. Steglich, G. M. Thomas, A. R. Battersby, T. A. Dobson, H. Ramuz: J. Chem. Soc., **1965**, 2423.
15) A. R. Battersby, T. H. Brown: Chem. Comm., **1966**, 170.
16) D. H. R. Barton, G. W. Kirby, W. Steglich, G. M. Thomas: Proc. Chem. Soc., **1963**, 203.
17) D. H. R. Barton, A. J. Kirby, G. W. Kirby: Chem. Comm., **1965**, 52.
18) L. J. Haynes, K. L. Stuart, D. H. R. Barton,

G. W. Kirby: J. Chem. Soc. (C), **1966**, 1676.
19) M. Ohta, Z. Kitasato: Arch. Exptle. Med., **6**, 283 (1925).
20) T. N. Iljnskaja: Dokl. Akad. Nauk USSR., **108**, 1081 (1956). [C. A., **51**, 4408 (1957)].
21) N. Ishiwari: Chugai Iji Shimpo, No. 959, 1, (1920). [C. A., **15**, 1598 (1921)].
22) H. Kondo, E. Ochiai, T. Nakajima: J. Pharm. Soc. Japan, **497**, 511 (1923).
23) H. Kondo, E. Ochiai: Ann., **470**, 224 (1929).
24) Y. Sasaki, S. Ueda: J. Pharm. Soc. Japan, **78**, 44 (1958).
25) K. Goto: J. Chem. Soc. Japan, **44**, 795 (1923).
26) L. Knorr: Ber., **36**, 3074 (1903).
27) C. Mannich, H. Löwenheim: Arch. Pharm., **258**, 295 (1920).
28) J. Gadamer, F. Knoch: Arch. Pharm., **259**, 135 (1921).
29) E. Speyer, S. Siebert: Ber., **54**, 1519 (1921).
30) M. Freund, W. W. Melber, E. Schlesinger: J. prakt. Chem., (2), **101**, 1 (1921).
31) H. Kondo, E. Ochiai: J. Pharm. Soc. Japan, **44**, 8 (1924).
32) E. Ochiai: J. Pharm. Soc. Japan, **503**, 8 (1924).
33) K. Goto: Proc. Imp. Acad. Tokyo, **2**, 7 (1926).
34) K. Goto: Proc. Imp. Acad. Tokyo, **2**, 167 (1926).
35) H. Kondo, E. Ochiai: J. Pharm. Soc. Japan, **538**, 1009, 1015 (1927).
36) H. Kondo, E. Ochiai: J. Pharm. Soc. Japan, **539**, 17 (1927).
37) H. Kondo, E. Ochiai: J. Pharm. Soc. Japan, **539**, 913 (1927).
38) K. Goto: J. Agr. Chem. Soc. Japan, **5**, 53 (1929).
39) K. Goto: Bull. Chem. Soc. Japan, **4**, 103 (1929).
40) K. Goto, H. Suzuki: Bull. Chem. Soc. Japan, **4**, 107 (1929).
41) K. Goto, H. Sudzuki: Bull. Chem. Soc. Japan, **4**, 163 (1929).
42) K. Goto, H. Sudzuki: Bull. Chem. Soc. Japan, **4**, 244 (1929).
43) K. Goto, H. Sudzuki: Bull. Chem. Soc. Japan, **4**, 271 (1929).
44) E. Ochiai, K. Hakozaki: J. Pharm. Soc. Japan, **50**, 360 (1930).
45) C. Schöpf, T. Pfeifer: Ann., **483**, 157 (1930).
46) H. Kondo, E. Ochiai: Ber., **63**, 646 (1930).
47) H. Kondo, Z. Narita: Ber., **63**, 2420 (1930).
48) K. Goto, R. Inaba: Bull. Chem. Soc. Japan, **5**, 93 (1930).
49) K. Goto, T. Namba: Bull. Chem. Soc. Japan, **5**, 165 (1930).
50) K. Goto, T. Namba, R. Inaba: Bull. Chem. Soc. Japan, **5**, 223 (1930).
51) K. Goto, S. Mitsui: Bull. Chem. Soc. Japan, **5**, 282 (1930).
52) K. Goto, H. Shishido: Bull. Chem. Soc. Japan, **5**, 311 (1930).
53) K. Goto, H. Shishido, R. Inaba: Bull. Chem. Soc. Japan, **5**, 315 (1930).
54) K. Goto, R. Inaba, H. Shishido: Ann., **485**, 247 (1931).
55) K. Goto, K. Takubo, S. Mitsui: Ann., **489**, 86 (1931).
56) K. Goto, H. Sudzuki: Bull. Chem. Soc. Japan, **6**, 33 (1931).
57) K. Goto, H. Shishido: Bull. Chem. Soc. Japan, **6**, 79 (1931).
58) K. Goto, K. Takubo: Bull. Chem. Soc. Japan, **6**, 126 (1931).
59) K. Goto, S. Mitsui: Bull. Chem. Soc. Japan, **6**, 197 (1931).
60) K. Goto, S. Mitsui: Bull. Chem. Soc. Japan, **7**, 228 (1932).
61) C. Schöpf, T. Pfeifer, H. Hirsch: Ann., **492**, 213 (1932).
62) K. Goto, K. Takubo, S. Mitsui: Ann., **494**, 1 (1932).
63) K. Goto, K. Takubo, S. Mitsui: Ann., **495**, 122 (1932).
64) K. Goto, H. Shishido, K. Takubo: Ann., **497**, 289 (1932).
65) K. Goto, K. Takubo: Ann., **499**, 169 (1932).
66) K. Goto, H. Shishido: Ann., **501**, 304 (1933).
67) K. Goto, H. Shishido: Ann., **507** 297 (1933).
68) K. Goto, H. Michinaka, H. Shishido: Ann., **515**, 297 (1935).
69) K. Goto, H. Shishido: Bull. Chem. Soc. Japan, **10**, 597 (1935).
70) K. Goto: Proc. Imp. Acad. Tokyo, **16** 403 (1940).
71) K. Goto, H. Shishido: Bull. Chem. Soc. Japan, **16**, 170 (1941).
72) K. Goto, T. Arai: Ann., **547**, 194 (1941).
73) K. Goto, T. Arai: Bull. Chem. Soc. Japan, **17**, 304 (1941).
74) K. Goto, M. Nagai: Bull. Chem. Soc. Japan, **18**, 143 (1943).
75) K. Goto, M. Nagai: Bull. Chem. Soc. Japan, **18**, 218 (1943).
76) K. Goto, K. Michi: Acta Phytochimica, **15**,

187 (1949).
77) K. Goto, K. Michi: Bull. Chem. Soc. Japan, **22**, 262 (1949).
78) K. Goto, T. Arai, M. Kono: Bull. Chem. Soc. Japan, **23**, 17 (1950).
79) K. Goto, I. Yamamoto: Proc. Japan. Acad., **29**, 210 (1953).
80) K. Goto, I. Yamamoto: Proc. Japan. Acad., **29**, 457 (1953).
81) K. Goto, I Yamamoto: Proc. Japan. Acad., **30**, 769 (1954).
82) K. Goto, I Yamamoto: Bull. Chem. Soc. Japan, **19**, 110 (1955).
83) J. Kalvoda, P. Buchschacher, O. Jeger: Helv. Chim. Acta, **38**, 1847 (1955).
84) K. Goto, I. Yamamoto: Proc. Japan. Acad., **32**, 45 (1956).
85) K. Goto, I. Yamamoto: Proc. Japan. Acad., **29**, 513 (1953).
86) K.W. Bentley: Chemistry of Carbon Compounds, Vol. IVc. p. 2060. Edited by E.H. Rodd, Elsevier Publishing Company (1960).
87) P.M. Dertreicher, C.G. Farmilo, L. Levi: Bull. Narcotics U. N. Dept. Social Affairs, **7**, 42 (1954).
88) D.H.R. Barton: Chemistry of Natural Products, Vol. III, p. 35.
89) Y. Sasaki, K. Okabe: J. Pharm. Soc. Japan, **83**, 418 (1963).
90) K. Goto: J. Chem. Soc. Japan, **44**, 795 (1923), cf. Proc. Imp. Acad. Tokyo, **2**, 7 (1926).
91) K. Goto: J. Agr. Chem. Soc. Japan, **1**, 3, 50, 89 (1925).
92) K. Goto, I. Yamamoto, S. Matsumoto: Proc. Japan. Acad., **30**, 883 (1954).
93) C. G. Farmilo, H. L. J. Rhodes, H. L. R. Hart, H. Taylor: Bull. Narcotics U.N. Dept. Social Affairs, **5**, No. 1, 26 (1953).
94) H. Asahina, T. Kanatani, M. Ono. S. Fujita: Bull. Narcotics U.N. Dept. Social Affairs, **9**, No. 2, 20 (1957).
95) M.A. Séguin: Ann. Chim., **92**, 225 (1814).
96) F.W. Sertürner: Gilbert's Ann. Phys., **55**, 56 (1817); Ann. Chim. Phys., (2), **5**, 21 (1817); (Trommsdorff's) J. Pharmazie, **13** (1), 234 (1805); **14** (1), 47 (1806); **20** (1), 99 (1811).
97) J. J. Dobbie, A. Lauder: J. Chem. Soc., **99**, 34 (1911).
98) W. R. Heymann: Bull. Narcotics U. N. Dept. Social Affairs, **9**, No. 2, 34 (1957). [C. A., **51**, 17094 (1957)].
99) G. Kleinschmidt: Planta Med., **7**, 471 (1957).
100) M. J. Liebig: Ann. Chim. Phys., (2), **47**, 147 (1831).
101) M. A. Laurent: Ann. Chim. Phys., (3), **19**, 359 (1847).
102) E. L. Meyer: Ber., **4**, 121 (1871).
103) C. R. A. Wright: J. Chem. Soc., **27**, 1031 (1874).
104) K. Polstorff: Ber., **13**, 86, 91, 92, 93, 98 (1880).
105) C. R. A. Wright, E. A. Rennie: J. Chem. Soc., **77**, 609 (1880).
106) E. Grimaux: Compt. rend., **92**, 1140 (1881).
107) E. von Gerichten, H. Schrötter: Ann., **210**, 396 (1881).
108) E. von Gerichten, H. Schrötter: Ber., **15**, 1484, 2179 (1882).
109) O. Hesse: Ann., **222**, 203 (1883).
110) L. Barth, H. Weidel: Monatsh., **4**, 700 (1883).
111) E. von Gerichten: Ber., **19**, 65 (1886).
112) O. Fischer, E. von Gerichten: Ber., **19**, 792 (1886).
113) J. F. Eykmann: Z. physik. Chem., **2**, 964 (1888).
114) N. v. Klubukow: Z. physik. Chem., **3**, 476 (1889).
115) G. H. Beckett, C. R. A. Wright: J. Chem. Soc., **28**, 312 (1875).
116) L. Knorr, R. Pschorr: Ber., **22**, 181 (1889).
117) L. Knorr: Ber., **22**, 1113 (1889).
118) E. von Gerichten: Ber., **31**, 2924 (1898).
119) G. N. Vis: J. prakt. Chem., (2), **47**, 584 (1893).
120) L. Knorr: Ber., **27**, 1144 (1894).
121) E. von Gerichten: Ber., **30**, 354 (1897).
122) E. von Gerichten: Ber., **30**, 2439 (1897).
123) J. Herzig, H. Meyer: Monatsh., **18**, 379 (1897).
124) E. von Gerichten: Ber., **31**, 51 (1898).
125) E. von Gerichten: Ber., **31**, 3198 (1898).
126) H. Causse: Compt. rend., **126**, 1799 (1898); **128**, 181 (1898).
127) L. Knorr, H. Matthes: Ber., **32**, 1047 (1899).
128) E. von Gerichten: Ber., **32**, 1521 (1898).
129) E. von Gerichten: Ber., **32**, 2379 (1898).
130) R. Pschorr: Ber., **33**, 176 (1900).
131) E. von Gerichten: Ber., **33**, 352 (1900).
132) R. Pschorr, C. Summuleanu: Ber., **33**, 1810 (1900).
133) S. B. Schryver, F. H. Lees: J. Chem. Soc., **77**, 1024 (1900).
134) E. von Gerichten: Ber., **34**, 1162 (1901).
135) S. B. Schryver, F. H. Less: J. Chem. Soc., **79**, 563 (1901).
136) R. Pschorr, B. Jaeckel, H. Fecht: Ber., **35**, 4377 (1902).

137) R. Pschorr, H. Vogtherr: Ber., **35**, 4412 (1902).
138) L. Knorr: Ber., **38**, 3143 (1905).
139) R. Pschorr, A. Plaff, F. Herschmann: Ber., **38**, 3160 (1905).
140) L. Knorr, R. Pschorr: Ber., **38**, 3172 (1905).
141) M. Freund: Ber., **38**, 3234 (1905).
142) L. Knorr, H. Hörlein: Ber., **39**, 1409 (1906).
143) L. Knorr, W. Schneider: Ber., **39**, 1414 (1906).
144) R. Pschorr: Ber., **39**, 3124 (1906).
145) R. Pschorr: Ber., **39**, 3130 (1906).
146) L. Knorr, H. Hörlein: Ber., **39**, 4409 (1906).
147) E. Mosettig, E. Meitzner: J. Am. Chem. Soc., **56**, 2738 (1934).
148) H.T. Bucherer, F. Seyde, J. Schwenkel: Angew. Chem., **20**, 1672 (1907).
149) L. Knorr, H. Hörlein: Ber., **40**, 376 (1907).
150) L. Knorr, H. Hörlein: Ber., **40**, 2042 (1907).
151) R. Pschorr, H. Einbeck: Ber., **40**, 1980 (1907).
152) R. Pschorr: Ber., **40**, 1984 (1907).
153) R. Pschorr, O. Spangenberg: Ber., **40**, 1995 (1907).
154) R. Pschorr, H. Einbeck, O. Spangenberg: Ber., **40**, 1998 (1907).
155) L. Knorr, H. Hörlein, C. Grimme: Ber., **40**, 3844 (1907).
156) L. Knorr, R. Waentig: Ber., **40**, 3860 (1907).
157) L. Knorr, H. Hörlein: Ber., **40**, 4883 (1907).
158) L. Knorr, H. Hörlein: Ber., **40**, 4889 (1907).
159) F.H. Lees: Proc. Chem. Soc., **1907**, I, 352.
160) F.H. Lees: J. Chem. Soc., **91**, 1408 (1907).
161) A. Oppé: Ber., **41**, 975 (1908).
162) L. Knorr, H. Hörlein, F. Staubach: Ber., **42**, 3511 (1909).
163) R. Pschorr, F. Dickhäuser: Ann., **373**, 80 (1910).
164) A. Skita, H.H. Franck: Ber., **44**, 2862 (1911).
165) H. Wieland, P. Kappelmeier: Ann., **382**, 306 (1911).
166) M. Fround, E. Speyer: Angew. Chem., **24**, 1122 (1911).
167) J. Gadamer: Angew. Chem., **26**, 625 (1913).
168) V. Klee: Arch. Pharm., **252**, 211 (1914).
169) J.v. Braun: Ber., **47**, 2312 (1914).
170) J. Gadamer: Arch. Pharm., **253**, 266 (1915).
171) J.v. Braun: Ber., **49**, 977 (1916).
172) M. Freund, E. Speyer: Ber., **49**, 1287 (1916).
173) C. Mannich: Arch. Pharm., **254**, 349 (1916).
174) M. Freund, E. Speyer: J. prakt. Chem., **94**, 135 (1916).
175) J.v. Braun, E. Aust: Ber., **50**, 43 (1917).
176) G. Barger: J. Chem. Soc., **113**, 218 (1918).
177) J.v. Braun, K. Kindler: Ber., **49**, 2655 (1916).
178) C. Mannich, H. Löwenhein: Arch. Pharm., **258**, 295 (1920).
179) M. Freund, W.W. Melber, E. Schlesinger: J. prakt. Chem., (2), **101**, 1 (1921).
180) R.S. Cahn, R. Robinson: J. Chem. Soc., **1926**, 908.
181) J.M. Gulland, R. Robinson: J. Chem. Soc., **123**, 980 (1923).
182) H. Wieland, E. Koralek: Ann., **433**, 267 (1923).
183) E. Speyer, K. Sarre: Ber., **57**, 1404 (1924).
184) E. Speyer, K. Sarre: Ber., **57**, 1427 (1924).
185) H. Wieland, M. Kotake: Ann., **444**, 69 (1925).
186) E. Speyer, H. Rosenfield: Ber., **58**, 1125 (1925).
187) H. Wieland, P. Garbsch: Ber., **59**, 2490 (1926).
188) J.v. Braun, R.S. Cahn: Ann., **451** 55 (1927).
189) C. Schöpf: Ann., **452**, 211 (1927).
190) H. Avenarius, R. Pschorr: Ber., **62**, 321 (1929).
191) E. Späth, O. Hromatka: Ber., **62**, 325 (1929).
192) J.M. Gulland, R. Robinson: Nature, **115**, 625 (1925).
193) H. Emde: Helv. Chim. Acta, **13**, 1035 (1930).
194) L.F. Small, F.L. Cohen: J. Am. Chem. Soc., **53**, 2214 (1931).
195) R.E. Lutz, L.F. Small: J. Am. Chem. Soc., **56**, 1741 (1934).
196) K.W. Bentley: "The Chemistry of Morphine Alkaloids" (1954).
197) E. Mossetig, E. Meitzner: J. Am. Chem. Soc., **56**, 2738 (1934).
198) K. Folkers: J. Am. Chem. Soc., 58, 1814 (1936).
199) L.F. Small, E.M. Fry: J. Org. Chem., **3**, 509 (1939).
200) L.F. Small, G.L. Browning: J. Org. Chem., **3**, 618 (1939).
201) L.F. Small, B.F. Faris, J.E. Mallonee: J. Org. Chem., **5**, 334 (1940).
202) H. L. Holmes, L. W. Trevoy: Can. J. Res., **22 B**, 56 (1944).
203) H.L. Holmes, C.C. Lee: J. Am. Chem. Soc., **69**, 1996 (1947).
204) H.L. Holmes, C.C. Lee, A. Mooradian: J. Am. Chem. Soc., **69**, 1998 (1947).

205) H. Rapoport, G. B. Payne: J. Org. Chem., **15**, 1093 (1950).
206) H. Rapoport, G. B. Payne: J. Org. Chem., **15**, 1350 (1950).
207) H. Papoport: J. Org. Chem., **13**, 714 (1948).
208) H. Rapoport, G. B. Payne: J. Am. Chem. Soc., **74**, 2630 (1952).
209) I. R. C. Bick: Nature, **169**, 756 (1952).
210) H. Papoport, J. B. Lavigne: J. Am. Chem. Soc., **75**, 5329 (1953).
211) A. K. Bose: Chem. & Ind., **1954**, 130.
212) K. W. Bentley, H. M. E. Cardwell: J. Chem. Soc., **1955**, 3252.
213) M. McKay, D. C. Hodgkin: J. Chem. Soc., **1955**, 3261.
214) D. Ginsburg: "The Opium Alkaloids", Interscience Publishers (1962).
215) F. Gautschi, O. Jeger, v. Prelog, R. B. Woodward: Helv. Chim. Acta, **38**, 296 (1955).
216) G. Stork, R. K. Hill: J. Am. Chem. Soc., **79**, 495 (1957).
217) H. Corrodi, E. Hardegger: Helv. Chim. Acta, **38**, 2038 (1955).
218) J. M. Lindsey, W. H. Barnes: Acta Cryst., **8**, 227 (1955). [C. A., **49**, 9350 (1955)].
219) R. E. Lutz, L. F. Small: J. Am. Chem. Soc., **56**, 2466 (1934).
220) J. Kalvoda, P. Buchschacher, O. Jeger: Helv. Chim. Acta, **38**, 1847 (1955).
221) M. Gates, G. Tschudi: J. Am. Chem. Soc., **74**, 1109 (1952).
222) D. Elad, D. Ginsburg: J. Am. Chem. Soc., **76**, 312 (1954).
223) D. Elad, G. Ginsburg: J. Chem. Soc., **1954**, 3052.
224) M. Gates, G. Tschudi: J. Am. Chem. Soc., **78**, 1380 (1956).
225) R. Grewe: Naturwiss., **33**, 333 (1946).
226) R. Grewe, A. Mondon: Ber., **81**, 279 (1948).
227) M. Gates: J. Am. Chem. Soc., **70**, 2261 (1948).
228) R. Grewe, A. Mondon, E. Nolte: Ann., **564**, 161 (1949).
229) M. Gates: Experientia, **5**, 235 (1949).
230) M. Gates: J. Am. Chem. Soc., **72**, 228 (1950).
231) M. Gates: J. Am. Chem. Soc., **72**, 4839 (1950).
232) R. Grewe: Ber., 84, 527 (1951).
233) D. Ginsburg, R. Pappo: J. Chem. Soc., **1951**, 938.
234) O. Schnider, A. Grüssner: Helv. Chim. Acta, **34**, 2211 (1951).
235) O. Schnider, J. Hellerbach: Helv. Chim. Acta, **34**, 2218 (1951).
236) J. A. Barltrop, J. E. Saxton: J. Chem. Soc., **1952**, 1038.
237) D. Ginsburg, R. Pappo: J. Chem. Soc., **1953**, 1524.
238) C. F. Koelsch, N. F. Albertson: J. Am. Chem. Soc., **75**, 2095 (1953).
239) R. Grewe, R. Hamann, G. Jacobsen, E. Nolte, K. Riecke: Ann., **581**, 85 (1953).
240) H. Henecke: Ann., **583**, 110 (1953).
241) O. Schnider, A. Brossi, K. Vogler: Helv. Chim. Acta, **37**, 710 (1954).
242) J. Hellerbach, A. Grüssner, O. Schnider: Helv. Chim. Acta, **39**, 429 (1956).
243) H. Corrodi, J. Hellerbach, A. Züst, E. Hardegger, O. Schnider: Helv. Chim. Acta, **42**, 212 (1959).
244) G. Rosenfeld, L. C. Leeper, S. Udenfriend: Arch. Biochem. Biophys., **74**, 252 (1958).
245) A. R. Battersby, B. J. T. Harper: Tetrahedron Letters, **1960**, No. 27, 21.
246) A. R. Battersby, B. J. T. Harper: Chem. & Ind., **1958**, 365.
247) E. Leete: Chem. & Ind., **1958**, **977**.
248) G. Kleinschmidt, K. Mothes: Z. Naturforsch., **14b**, 52 (1959).
249) E. Leete: J. Am. Chem. Soc., **81**, 3948 (1959).
250) A. R. Battersby, R. Binks, D. J. LeCount: Proc. Chem. Soc., **1960**, 287.
251) A. R. Battersby, R. Binks: Proc. Chem. Soc., **1960**, 360.
252) A. R. Battersby, B. J. T. Harper: Tetrahedron Letters, **1960**, No. 27, 21.
253) H. Rapoport, F. R. Stermitz, D. R. Baker: J. Am. Chem. Soc., **82**, 2765 (1960).
254) F. R. Stermitz, H. Rapoport: Nature, **189**, 310 (1961).
255) S. Pfeifer, K. Heydenrich: Naturwiss., **48**, 222 (1961).
256) P. R. Stermitz, H. Rapoport: J. Am. Chem. Soc., **83**, 4045 (1961).
257) H. Rapoport, N. Levy, F. R. Stermitz: J. Am. Chem. Soc., **83**, 4298 (1961).
258) A. R. Battersby, R. Binks, B. J. T. Harper: J. Chem. Soc., **1962**, 3534.
259) A. R. Battersby, R. Binks, D. M. Foulkes, R. J. Francis, D. J. McCaldin, H. Ramuz: Proc. Chem. Soc., **1963**, 203.
260) E. Leete, J. B. Murrill: Tetrahedron Letters, **1964**, 147.
261) A. R. Battersby, R. Binks, R. J. Francis, D. J. McCaldin, H. Ramuz: J. Chem. Soc., **1964**, 3600.
262) E. Brochmann-Hanssen, D. Nielsen: Tetrahedron Letters, **1965**, 271.

263) A. R. Battersby, G. W. Zvans, R. O. Martin, M. E. Warren, H. Rapoport: Tetrahedron Letters, **1965**, 1275.
264) A. R. Battersby, D. M. Foulkers, R. Binks: J. Chem. Soc., **1965**, 3323.
265) A. R. Battersby: Quart. Revs., **11**, 259 (1961).
266) K. Mothes, H. R. Schütte: Angew. Chem., **75**, 357 (1963).
267) A. R. Battersby: Proc. Chem. Soc., **1963**, 189.
268) D. H. R. Barton: Proc. Chem. Soc., **1963**, 293.
269) L. Kühn, S. Pfeifer: Pharmazie, **20**, 659 (1965).
270) R. Robinson: "The Structural Relations of Natural Products," Clarendon Press, Oxford (1955).
271) K. W. Bentley: Experientia, **12**, 251 (1956).
272) D. H. R. Barton, T. Cohen: "Festschrift Arthur Stoll," Birkhauser, Basel, p. 117 (1957).
273) H. Wachsmuth, L. van Koeckhoven: J. pharm. Belg., **14**, 215 (1959). [C. A., **54**, 3474 (1960)].
274) N. D. Coggenshall, A. S. Glessner: J. Am. Chem. Soc., **71**, 3150 (1949).
275) K. W. Bentley, S. F. Dyke: J. Chem. Soc., **1959**, 2574.
276) R. M. Pinyazhko: Farmatsevi. Zh. (Kiev.), **19**, 21 (1964). [C. A., **64**, 6701 (1966)].
277) T. Ruell: Bull. Soc. Chim. France, **1963**, 586.
278) W. Debska: Poznan. Towarz. Przyaciol. Nauk, Wydzial Lekar, Prace Komisji Farm., **2**, 23 (1962). [C. A., **59**, 6451 (1963)].
279) H. Audier, M. Fetizon, D. Ginsburg, A. Mandelbaum, Th. Rüll: Tetrahedron Letters, **1963**, 13.
280) H. Nakata, Y. Hirata, A. Tatematsu, H. Tada, Y. K. Sawa: Tetrahedron Letters, **1965**, 829.
281) A. Mandelbaum, D. Ginsburg: Tetrahedron Letters, **1965**, 2479.
282) J. M. Bobbitt, U. Weiss, D. D. Hanessian: J. Org. Chem., **24**, 1582 (1959).
283) U. Weiss, T. Ruell: Bull. Soc. Chim. France, **1965**, 3707.
284) T. Ruell: Bull. Soc. Chim. France, **1965**, 3715.
285) R. Konowalova, S. Yu. Yunusov, A. Orechoff: Ber., **68**, 2158 (1935).
286) V. V. Kisseler, R. A. Konowalowa: Zhur. Obshchei Khim., **18**, 142 (1948). [C. A., **42**, 5037 (1948)].
287) R. Robiquet: Ann., **5**, 82 (1833).
288) H. K. Heydenreich, R. Miram, S. Pfeifer: Sci. Pharm., **29**, 221 (1961). [C. A., **56**, 13257 (1962)].
289) T. Anderson: Trans. roy. soc. Edinburg, **20**, 57 (1850).
290) E. Grimaux: Compt. rend., **93**, 591 (1881).
291) M. Freund, E. Göbel: Ber., **28**, 941 (1895).
292) M. Freund, H. Michaelis, E. Göbel: Ber., **30**, 1357 (1897).
293) E. von Gerichten: Ber., **33**, 1824 (1900).
294) L. Knorr, S. Smiles: Ber., **35**, 3009 (1902).
295) R. Pschorr, C. Seydel, W. Stöhler: Ber., **35**, 4400 (1902).
296) R. Pschorr, H. Vogtherr: Ber., **35**, 4412 (1902).
297) F. Ach, L. Knorr: Ber., **36**, 3067 (1903).
298) L. Knorr: Ber., **36**, 3074 (1903).
299) L. Knorr, W. Schnider: Ber., **39**, 1414 (1906).
300) E. von Gerichten, O. Dittmer: Ber., **39**, 1718 (1906).
301) L. Knorr, H. Hörlein: Ber., **39**, 3252 (1906).
302) F. H. Lees: J. Chem. Soc., **91**, 1408 (1907).
303) L. Knorr, H. Hörlein: Ber., **40**, 2032 (1907).
304) L. Knorr, H. Hörlein: Ber., **40**, 2042 (1907).
305) L. Knorr, H. Hörlein: Ber., **40**, 3341 (1907).
306) L. Knorr, H. Hörlein: Ber., **40**, 4883 (1907).
307) L. Knorr, H. Hörlein: Ber., **40**, 4889 (1907).
308) L. Knorr, H. Hörlein: Ber., **41**, 969 (1908).
309) R. Pschorr, G. Knöffler: Ann., **382**, 50 (1911).
310) R. Pschorr: Ann., **391**, 40 (1912).
311) W. W. Melker, E. Schlesinger: J. prakt. Chem., **101**, 1 (1921).
312) H. Wieland, E. Koralek: Ann., **433**, 267 (1923).
313) J. M. Gulland, R. Robinson: Mem. Proc. Manchester Lit. Phil. Soc., **69**, 79 (1925). [C. A., **20**, 765 (1926)].
314) R. S. Cahn, R. Robinson: J. Chem. Soc., **1926**, 908.
315) C. Schöpf: Ann., **452**, 211 (1927).
316) T. Ruell: Bull. Soc. Chim. France, **1962**, 1337.
317) R. E. Lutz, L. Small: J. Am. Chem. Soc., **57**, 2651 (1935).
318) W. Sandermann: Ber., **71**, 648 (1938).
319) C. Schöpf, K. v. Gottberg, W. Petri: Ann., **536**, 216 (1938).
320) H. L. Holmes, C. C. Lee: J. Am. Chem. Soc., **69**, 1996 (1947).
321) S. P. Findley, L. F. Small: J. Am. Chem. Soc., **72**, 3247 (1950).
322) S. P. Findley, L. F. Small: J. Am. Chem.

Soc., **73**, 4001 (1951).

323) H. Rapoport, G. W. Stevenson: J. Am. Chem. Soc., **76**, 1796 (1954).

324) T. J. Batterham, K. H. Bell, U. Weiss: Austral. J. Chem., **18**, 1799 (1965).

325) G. Kartha, F. R. Ahmed, W. H. Barnes: Acta Cryst., **15**, 326 (1962). [C. A., **57**, 1652 (1962)].

326) A. H. Homeyer, W. L. Shilling: J. Org. Chem., **12**, 356 (1947).

327) C. F. van Duin, R. Robinson, J. C. Smith: J. Chem. Soc., **1926**, 903.

328) H. Conroy: J. Am. Chem. Soc., **77**, 5960 (1955).

329) L. F. Small, B. Witkop, R. Goodwin: J. Am. Chem. Soc., **75**, 3371 (1953).

330) F. Santavy, M. Maturova, A. Nemeckova, H. B. Schröter, H. Potesilova, J. Preininger: Planta Med., **8**, 167 (1960). [C. A., **54**, 18846 (1960)].

331) P. Pelletier: J. Pharm., (2), **21**, 555 (1835); Ann., **16**, 27 (1835).

332) T. Anderson: J. prakt. Chem., **57**, 358 (1852).

333) M. Freund: Ber., **27**, 2961 (1884).

334) G. Stork: J. Am. Chem. Soc., **74**, 768 (1952).

335) G. Stork, F. H. Clarke: J. Am. Chem. Soc., **78**, 4619 (1956).

336) H. Rapoport, H. N. Reist, C. H. Lovell: J. Am. Chem. Soc., **78**, 5128 (1956).

337) K. Goto, Z. Kitasato: Ann., **481**, 81 (1930).

338) L. F. Small, B. F. Faris: J. Am. Chem. Soc., **56**, 1930 (1934).

339) K. W. Bentley, S. F. Dyke: Chem. & Ind., **1957**, 398.

340) K. W. Bentley, S. F. Dyke: J. Chem. Soc., **1959**, 2574.

Chapter 16 Hasubanonine and Related Alkaloids (Hasubanan Alkaloids)

Metaphanine $C_{19}H_{23}O_5N \equiv 345.38$

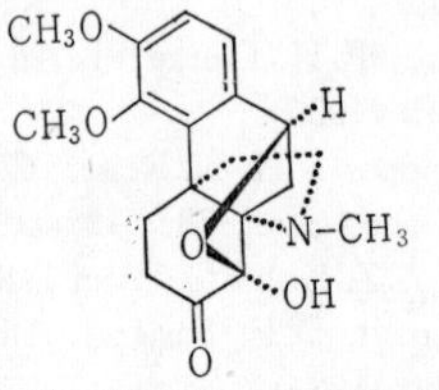

≪mp≫ 232° (d) (Me_2CO or $CHCl_3$–Et_2O), cf. 222° (d) [$B+H_2O$]
≪$[\alpha]_D$≫ −17° ($CHCl_3$)[4], −41.1° ($CHCl_3$)[5]

≪Ph.≫ CR[1] UV[4,25] IR[4,5,8,25] NMR[8,25] Mass[8,10,25] pKa[4,8] Rf[5]
≪OP.≫ Stephania japonica[1,2,3,4,5,6], S. abyssinica[24]
≪C. & So.≫ Colourless needles or prisms, cf. bp 170∼5°/2×10^{-4}, Sublimation
≪Sa. & D.≫ B–HCl: 225° (d) B–HBr: 235° (d) B–HI: 230° (d) B_2–H_2SO_4: 174° B–MeI: 199∼200° (d) B–OAc: 214° (d) B–Oxime: 215° 2,4–D: 157°
≪SD.≫ 1), 4), 7), 8), 9), 25)
≪S. & B.≫ cf. 27)

Prometaphanine $C_{20}H_{25}O_5N \equiv 359.41$

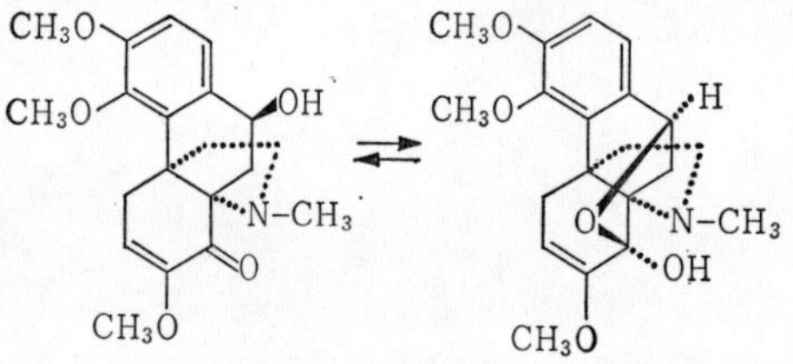

≪mp≫
≪$[\alpha]_D$≫ B–MeI −32° (MeOH)

≪Ph.≫ UV[11] IR[11] NMR[11]
≪OP.≫ Stephania japonica[11]
≪C. & So.≫ Amorph.
≪Sa. & D.≫ B–MeI: 207°
≪SD.≫ 11)
≪S. & B.≫

Homostephanoline $C_{20}H_{25}O_5N \equiv 359.41$

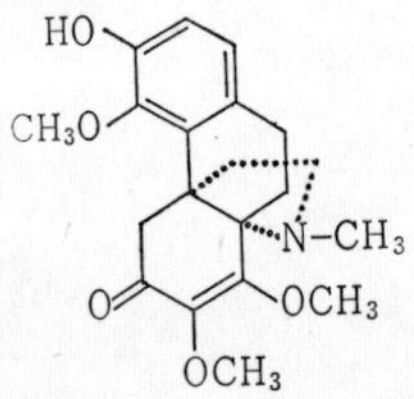

≪mp≫ 233° (MeOH)
≪$[\alpha]_D$≫ −247.8° ($CHCl_3$)

≪Ph.≫ CR[12,13] IR[14,15] NMR[15]
≪OP.≫ Stephania japonica[6,12]
≪C. & So.≫ Prisms, cf. bp 225∼40°/0.09
≪Sa. & D.≫ B–HCl: 238° (d) B–OMe=hasbanonine B–OEt·HBr: 194∼196°
≪SD.≫ 13), 14), 15)
≪S. & B.≫

Hasubanonine $C_{21}H_{27}O_5N \equiv 373.43$

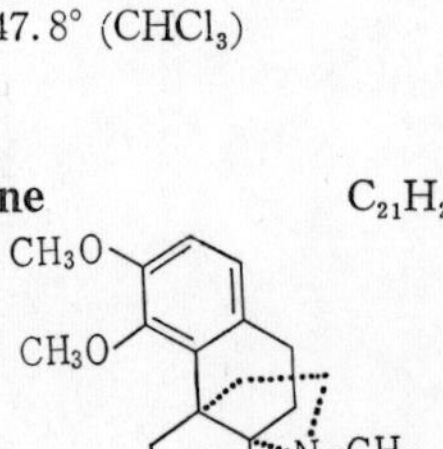

≪mp≫ 116∼117° (MeOH)
≪$[\alpha]_D$≫ −219.5° (EtOH)

≪Ph.≫ CR[16] UV[17,19,22,26] IR[13,19,22] NMR[13,22,26] Mass[10,26] ORD[13,22] pKa[4]
≪OP.≫ Stephania japonica[3,6,16,26]
≪C. & So.≫ Colourless prisms, Sol. in most org. solv, Spar. sol. in Petr. ether, cf. bp 205∼220°/1
≪Sa. & D.≫ B–HCl: 207∼208° B–HBr: 207∼208° (d) B–HNO_3: 222° B–MeI: 177∼178° B–Pic: 210° B–Oxime: 142° B–Semicarb: 228° (d)
≪SD.≫ 16), 17), 18), 19), 20), 21), 22)
≪S. & B.≫ BG: 23), BS: 26)

1) H. Kondo, T. Sanada: J. Pharm. Soc. Japan, **44**, 1034 (1924).
2) H. Kondo, T. Sanada: J. Pharm. Soc. Japan, **47**, 177 (1927).
3) H. Kondo, T. Watanabe: J. Pharm. Soc. Japan, **58**, 268 (1938).
4) K. Takeda: Ann. Rept. ITSUU Lab., **11**, 61 (1960).
5) M. Tomita, T. Ibuka: J. Pharm. Soc. Japan, **83**, 996 (1963).
6) M. Tomita, Y. Watanabe, K. Okui: J. Pharm. Soc. Japan, **76**, 856 (1956).
7) M. Tomita, T. Ibuka, Y. Inubushi, K. Takeda: Tetrahedron Letters, **1964**, 3605.
8) M. Tomita, T. Ibuka, Y. Inubushi, K. Takeda: Chem. Pharm. Bull., **13**, 695 (1965).
9) M. Tomita, T. Ibuka, Y. Inubushi, K. Takeda: Chem. Pharm. Bull., **13**, 704 (1965).
10) M. Tomita, A. Kato, T. Ibuka: Tetrahedron Letters, **1965**, 1019.
11) M. Tomita, T. Ibuka, Y. Inubushi: Tetrahedron Letters, **1964**, 3617.
12) H. Kondo, T. Sanada: J. Pharm. Soc. Japan, **48**, 1141 (1928).
13) M. Tomita, T. Ibuka, Y. Inubushi, Y. Watanabe, M. Matsui: Chem. Pharm. Bull., **13**, 538 (1965).
14) M. Tomita, Y. Watanabe: J. Pharm. Soc. Japan, **76**, 856 (1956).
15) Y. Watanabe, M. Matsui, K. Ido: J. Pharm. Soc. Japan, **85**, 584 (1965).
16) H. Kondo, M. Satomi, T. Odera: Ann. Rept. ITSUU Lab., **2**, 1 (1951).
17) M. Satomi: Ann. Rept. ITSUU Lab., **3**, 1 (1952).
18) H. Kondo, M. Satomi, T. Odera: Ann. Rept. ITSUU Lab., **4**, 1 (1953).
19) H. Kondo, M. Satomi: Ann. Rept. ITSUU Lab., **8**, 6 (1957).
20) K.H. Bentley: Experientia, **12**, 251 (1961).
21) Y. Watanabe, H. Matsumura: J. Pharm. Soc. Japan, **83**, 991 (1963).
22) M. Tomita, T. Ibuka, Y. Inubushi, M. Matsui: Tetrahedron Letters, **1964**, 2937.
23) D.H.R. Barton: "Chemistry of Natural Products," Vol. III, p. 46. (1964).
24) H.L. deWaal, E. Weideman: Tydskr. Naturwet, **2**, 12 (1962).
25) H.L. deWaal, B.J. Prislow, R.R. Arndt: Tetrahedron Letters, **1966**, 6109.
26) D.H.R. Barton, G.W. Kirby, A. Wiechers: J. Chem. Soc. (C), **1966**, 2313.
27) M. Tomita, T. Ibuka, M. Kitano: Tetrahedron Letters, **1966**, 6233.

Chapter 17 Emetine and Related Alkaloids

Protoemetine $C_{19}H_{27}O_3N \equiv 317.41$

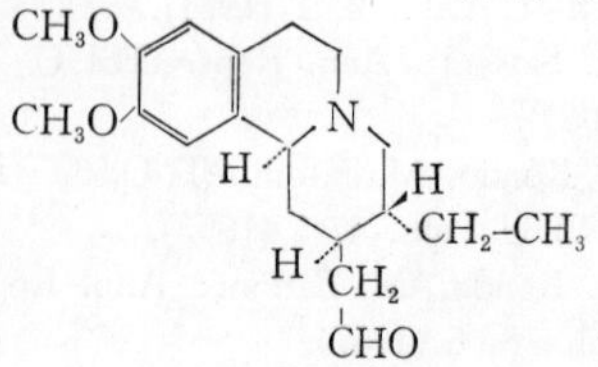

CH_3O, CH_3O, N, H, H, H, CH_2-CH_3, CH_2, CHO

≪mp≫

≪$[\alpha]_D$≫ B-$HClO_4$ −11° (EtOH)

≪Ph.≫ UV[1)6)] IR[1)6)] pKa[6)]

≪OP.≫ Psychotria granadensis (Uragoga granatensis)[1)2)], P. ipecacuanha (Cephaelis ipecacuanha, Uragoga ipecacuanha)[1)2)]

≪C. & So.≫

≪Sa. & D.≫ B-$HClO_4$: 193~195°

cf. B-$HClO_4 \cdot H_2O$: 140~142°

B-Semicarb: 168~169°

≪SD.≫ 1), 2), 3), 4), 5), 6), 7), 8), 9) → emetine

≪S. & B.≫ 4), 8), 9), 10), 11), 13), 14), 15)

BG: 12)

Protoemetinol (Dihydroprotoemetine) $C_{19}H_{29}O_3N \equiv 319.43$

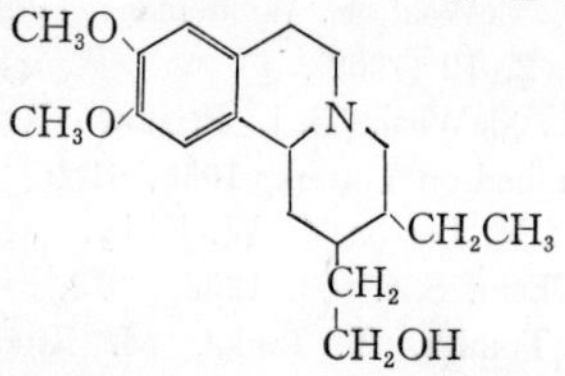

CH_3O, CH_3O, N, CH_2CH_3, CH_2, CH_2OH

≪mp≫

≪$[\alpha]_D$≫

≪Ph.≫ IR[6)] Mass[16)17)]

≪OP.≫ Alangium lamarckii[16)17)]

≪C. & So.≫

≪Sa. & D.≫ B-$HClO_4$: 199~200°

≪SD.≫ 16) 17)

≪S. & B.≫ cf. 6)

Ankorine $C_{19}H_{29}O_4N \equiv 335.43$

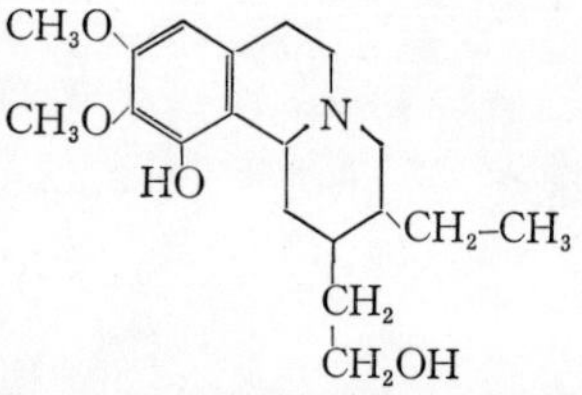

CH_3O, CH_3O, HO, N, CH_2-CH_3, CH_2, CH_2OH

≪mp≫ 174~176° (Me_2CO)

≪$[\alpha]_D$≫ −62° or −53.11° ($CHCl_3$)

≪Ph.≫ UV[17)18)] IR[17)18)] NMR[17)18)] Mass[17)]

≪OP.≫ Alangium lamarckii[17)18)]

≪C. & So.≫

≪Sa. & D.≫ B-HCl: 233~234°

B-HBr: 220~222°(d) B_2-H_2SO_4: 286°(d)

B-Oxal: 190~191°(d)

≪SD.≫ 17), 18)

≪S. & B.≫

Ipecoside $C_{27}H_{35}O_{12}N \equiv 565.56$

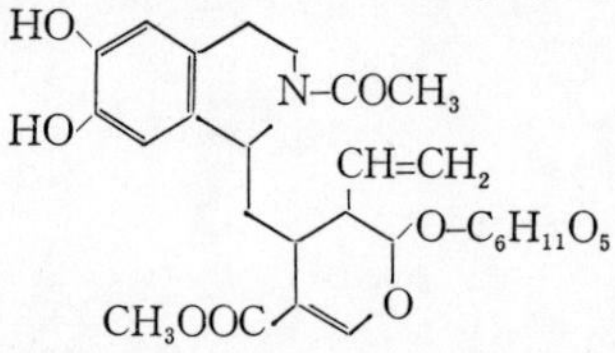

HO, HO, N-$COCH_3$, CH=CH_2, O-$C_6H_{11}O_5$, CH_3OOC, O

≪mp≫ 175°

≪$[\alpha]_D$≫ −185±3° (MeOH)

≪Ph.≫

≪OP.≫ Psychotria (Cephaelis) ipecacuanha[19)]

≪C. & So.≫

≪Sa. & D.≫ B-O,O-DiMe: 136~138° B-hexa Ac: 125~128°

≪SD.≫ 20)

≪S. & B.≫

So-called Rotundine $C_{17}H_{21}O_3N \equiv 287.35$

≪mp≫ 140～141°
≪$[\alpha]_D$≫ −262° (EtOH)
≪Ph.≫ UV[21)23)] IR[21)23)] Rf[23)]
≪OP.≫ Stephania japonica[21)]
≪C. & So.≫ Colourless plates
≪Sa. & D.≫ B–HCl: 261～262°(d)[21)] or 222～223°(d)[23)] B–MeI: 258～259°(d)[21)] or 248～250°[23)]
≪SD.≫ 21), 22), 23) This structure was corrected to tetrahydropalmatine by Sugasawa[23)]
≪S. & B.≫

Psychotrine $C_{28}H_{36}O_4N \equiv 464.58$

≪mp≫ 128° (sint. 120°) (aq. MeOH or aq. Me_2CO)
≪$[\alpha]_D$≫ +69.3° (EtOH)
≪Ph.≫ CR[29)38)39)] UV[37)] IR[37)] NMR[37)] Mass[24)37)] ORD[37)] Rf[37)]
≪OP.≫ Alangium lamarckii[24)], Bothriospora corymbosa[25)], Caperona decorticans[25)], Ferdinandusa elliptica[25)], Psychotria (Uragoga) grandensis[26)], P. (Cephaelis or Uragoga) ipecacuanha[26)], Tocoyena longiflora[25)]
≪C. & So.≫ Yellow prisms, Sol. in EtOH, Me_2CO, $CHCl_3$, Spar. sol. in C_6H_6 Et_2O, Petr. ether
≪Sa. & D.≫ B–2HI: 200～222°(d), B–$2HNO_3$: 184～187° B–H_2SO_4: 214～217° B–Oxal: 130～145° B–OMe: 123～124°
≪SD.≫ 10), 27), 28), 29), 30), 31), 32), 33), 34), 35), 36), 37) → emetine
≪S. & D.≫ 36), 37)

Cephaeline $C_{28}H_{38}O_4N \equiv 466.60$

≪mp≫ 115～116° (Et_2O), cf. 120～130°
≪$[\alpha]_D$≫ −43.4° ($CHCl_3$), −21.2° (EtOH)
≪Ph.≫ CR[29)38)39)42)] UV[44)] IR[16)] NMR[16)]
≪OP.≫ Alangium lamarckii[16)24)], Bothriospora corymbosa[25)], Caperona decorticans[25)], Ferdinandusa elliptica[25)], Psychotria (Uragoga) grandensis[40)], P. (Cephaelis or Uragoga) ipecacuanha[40)41)], Ramijia amazonica[25)], Tocoyena eongiflora[25)]
≪C. & So.≫ Needles, Sol. in MeOH, EtOH, Me_2CO, $CHCl_3$, Spar. sol. in Et_2O, Petr. ether, Insol. in H_2O
≪Sa. & D.≫ B–2HCl: 249～250°[27)] or 270°[28)] B–2HBr: 266～293° B–H_2SO_4: 230～235° B–2MeI: 225～228° B–O,N–DiAc: 116° B–O,N–DiBz: 169–170° B–NMe: 194° B–OMe=emetine
≪SD.≫ 10), 27), 28), 29), 30), 35), 36), 42), 43), 44), 45), 46), 47) →emetine
≪S. & B.≫ 14), 36), BS: 41)

Emetamine $C_{29}H_{36}O_6N_2 \equiv 476.6°$

≪mp≫ 155～156° (EtOAc), 135～137° [B+Et_2O] (Et_2O–Petr. ether)
≪$[\alpha]_D$≫ +14° (EtOH), +10° ($CHCl_3$)
≪Ph.≫ CR[31)38)] UV[49)]
≪OP.≫ Psychotria (Uragoga) grandensis[31)], P. (Cephaelis or Uragoga) ipecacuanha[31)]
≪C. & So.≫ Needles, Sol. in MeOH, EtOH, AcOEt, Me_2CO, C_6H_6, Spar. sol. in Et_2O, Insol. in Petr. ether.
≪Sa. & D.≫ B–2HCl: 218～223° B–2HBr: 210～225° B–$2HNO_3$: 165～166°(d) B–Oxal: 170～171°(d) B–2pic: 173° B–2MeI: 238°
≪SD.≫ 10), 19), 31), 33), 34), 36), 48), 49)
≪S. & B.≫

O-Methylpsychotrine $C_{29}H_{38}O_4N_2 \equiv 478.61$

≪mp≫ 123~124° (Et_2O)
≪$[\alpha]_D$≫ +43.9° ($EtOH-H_2SO_4$)
≪Ph.≫ UV[50,53] IR[2,51]
≪OP.≫ Psychotria (Uragoga) grandensis[31], P. (Cephaelis or Uragoga) ipecacuanha[31]
≪C. & So.≫ Prisms
≪Sa. & D.≫ B-2HBr: 190~200° B-$2HNO_3$: 210° B-H_2SO_4: 247° B-Pic: 142~175° B-Oxal: 162~163°(d) B-NBz: 135°
≪SD.≫ 3), 7), 10), 27), 28), 30), 31), 32), 33), 34), 35), 36), 50), 51)
≪S. & B.≫ 6), 36), 52), 53), 54), 55), 56)

Almarckine $C_{29}H_{40}O_4N_2 \equiv 480.63$

≪mp≫
≪$[\alpha]_D$≫
≪Ph.≫
≪OP.≫ Alangium lamarckii[24,57]
≪C. & So.≫
≪Sa. & D.≫
≪SD.≫ 24)
≪S. & B.≫

Emetine $C_{20}H_{40}O_4N_2 \equiv 480.63$

≪mp≫ 74°
≪$[\alpha]_D$≫ −50° ($CHCl_3$), −22° (EtOH)
≪Ph.≫ CR[29,38,42] UV[44,53,89,90,104,113,114,115] IR[2,7,52,53,81,90] Mass[24,116] ORD[52] X-ray[52] Rf[90]
≪OP.≫ Alangium lamarckii[24], Borrena verticillata[58], Bothriospora corymbosa[25], Caperona decorticans[25], Ferdinandusa elliptica[25], Hillia illustris[25], Psychotria (Uragoga) grandensis[26,59], P. (Cephaelis or Uragoga) ipecacuanha[26,59,60], Remijia amazonica[25], Tocoyena longiflora[25]
≪C. & So.≫ Colourless amorph., Sensitive to light, Sol. in MeOH, EtOH, Me_2CO, AcOEt, Et_2O, $CHCl_3$, Spar. sol. in C_6H_6, Petr. ether, Insol. in H_2O
≪Sa. & D.≫ B-2HCl: 235~255° B-2HBr: 253~255° B-2HI: 235~238° B-$2HNO_3$: 245° (sint: 188°) B-H_2SO_4: 205~245° B-Oxal: 166° B-H_2PtCl_6: 248~249° B-NAc: 97~99° B-NBz: 185~186° B-NMe: 213~216°
≪SD.≫ 3)~11), 27)~31), 33)~36), 42)~46), 48), 52), 59)~93)
≪S. & B.≫ 13), 14), 35), 36), 52), 53), 54), 55), 56), 84), 85), 88), 92), 94)~112)

No Name $C_{28}H_{35}O_3N_3 \equiv 461.58$

≪mp≫ 198~200°
≪$[\alpha]_D$≫ −51.9° (Py)
≪Ph.≫ UV[117]
≪OP.≫ Alangium lamarckii[117]
≪C. & So.≫
≪Sa. & D.≫ B-OMe=tubulosine
≪SD.≫ 117)
≪S. & B.≫

Deoxytubulosine $C_{29}H_{37}O_2N_3 \equiv 459.61$

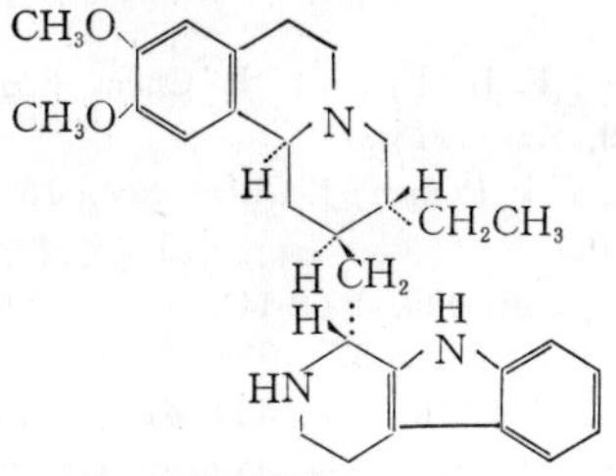

≪mp≫ 230～232°
≪$[\alpha]_D$≫ −24±1° ($CHCl_3$)

≪Ph.≫ UV[118,119] IR[118,119] NMR[118] Mass[118,119]
≪OP.≫ Alangium lamarckii[117,118], A. salviifolium[118], Cassinopsis ilicifolia[119], Pogonopus tubulosus[119]
≪C. & So.≫
≪Sa. & D.≫ B–NTos: 145～150°
≪SD.≫ 118), 119)
≪S. & B.≫ 118)

Tubulosine $C_{29}H_{37}O_3N_3 \equiv 475.61$

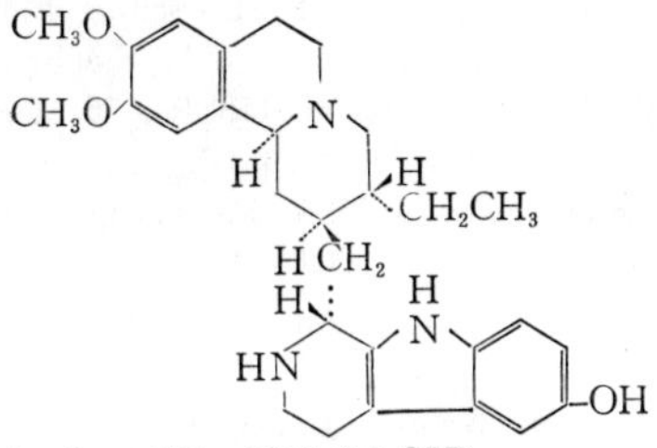

≪mp≫ 259～261° (MeOH)
≪$[\alpha]_D$≫ −65.9°, −72° (Py)

≪Ph.≫ UV[16,119,121,122] IR[16,121] NMR[16,121] Mass[119,121] Rf[123] pKa[121]
≪OP.≫ Alangium lamarckii[16,117,120], Pogonopus tubulosus[119,121]
≪C. & So.≫
≪Sa. & D.≫ B–Ac: 184～186°
B–O, N–DiAc: 150～155°
≪SD.≫ 16), 119), 121)
≪S. & B.≫ 122)

Isotubulosine $C_{29}H_{37}O_3N_3 \equiv 475.61$

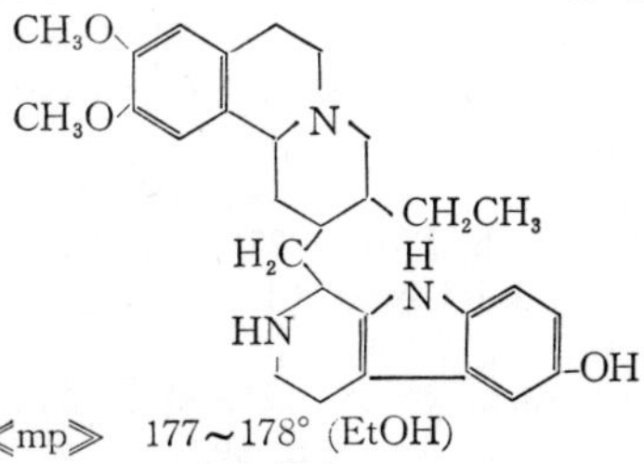

≪mp≫ 177～178° (EtOH)
≪$[\alpha]_D$≫ −84° (Py)

≪Ph.≫ UV[123] IR[123] NMR[123] Mass[123] Rf[123]
≪OP.≫ Alangium lamarckii[123]
≪C. & So.≫
≪Sa. & D.≫
≪SD.≫ 123)
≪S. & B.≫

Alangimarckine $C_{29}H_{37}O_3N_3 \equiv 475.61$

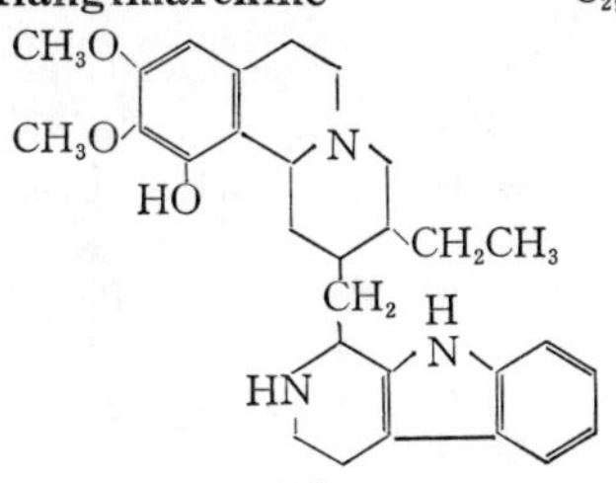

≪mp≫ 184～186°
≪$[\alpha]_D$≫ −67.7° (Py)

≪Ph.≫ UV[17] IR[17] NMR[17] Mass[17]
≪OP.≫ Alangium lamarkii[17]
≪C. & So.≫
≪Sa. & D.≫
≪SD.≫ 17)
≪S. & B.≫

1) A. R. Battersby, G. C. Davidson, B. J. T. Harper: Chem. & Ind., **1957**, 983.
2) A. R. Battersby, G. C. Davidson, B. J. T. Harper: J. Chem. Soc., **1957**, 1744.
3) A. R. Battersby, R. Binks, D. Davidson, G. C. Davidson, T. P. Edwards: Chem. & Ind., **1957**, 982
4) A. R. Battersby, S. Cox: Chem. & Ind., **1957**, 983.
5) A. R. Battersby, S. Garratt: Proc. Chem. Soc., **1959**, 86.
6) A. R. Battersby, B. J. T. Harper: J. Chem. Soc., **1959**, 1748.
7) A. R. Battersby, R. Binks, G. C. Davidson: J. Chem. Soc., **1959**, 2704.
8) A. R. Battersby, S. Garratt: J. Chem. Soc., **1959**, 3512.
9) E. E. Van Tamelen, P. E. Aldrich, J. B. Hester, Jr.: J. Am. Chem. Soc., **81**, 6214 (1959)
10) E. E. Van Tamelen, P. E. Aldrich, J. B. Hester, Jr.: J. Am. Chem. Soc., **79**, 4817 (1957).
11) E. E. Van Tamelen, J. B. Hester, Jr.: J. Am. Chem. Soc., **81**, 507 (1959).
12) This book: Chapter. 1, p. 19.
13) Cs. Szantay, L. Töke: Tetrahedron Letters, **1963**, 1323.
14) Cs. Szantay, L. Töke, P. Kolonits: J. Org. Chem., **31**, 1447 (1966).
15) L. Szabo, L. Töke, K. Honty, Cs. Szantay: Tetrahedron Letters, **1966**, 2975.
16) J. D. Albright, J. C. V. Meter, L. Goldman: Lloydia, **28**, 212 (1965). [C. A., **63**, 18189 (1965)].
17) A. R. Battersby, P. S. Kapil, B. S. Bhakuni, S. P. Popli, J. R. Merchant, S. S. Salgar: Tetrahedron Letters, **1966**, 4965.
18) B. Das Gupta: J. Pharm. Sci., **54**, 481(1965).
19) A. R. Battersby: Spec. Publ. Chem. Soc. [London], **3**, 36 (1955). [C. A., **50**, 8684 (1956)].
20) A. R. Battersby: IUPAC, "International Symposium of Natural Products," Stockholm Sweden, 1966.
21) H. Kondo, T. Matsuno; J. Pharm. Soc. Japan, **64**, No. 2, 113 (1944).
22) T. Matsuno: J. Pharm. Soc. Japan, **64**, No. 2, 274 (1944).
23) M. Kawanishi, S. Sugasawa: Chem. Pharm. Bull., **13**, 522 (1965).
24) H. Budzikiewicz, S.C. Pakrashi, H. Vorbrueggen: Tetrahedron, **20**, 399 (1964).
25) F. W. Freise: Pharm. Zentralhalle, **76**, 223 (1935).
26) B. H. Paul, A. J. Cownley: Pharm. J., [3] **25**, 690 (1894/1895).
27) F. H. Carr, F. L. Pyman: J. Chem. Soc. (Proc.), **29**, 226 (1913).
28) F. H. Carr, F. L. Pyman: J. Chem. Soc., **105**, 1591 (1914).
29) O. Hesse: Ann., **405**, 1 (1914).
30) P. Karrer: Ber., **49**, 2057 (1916).
31) F. L. Pyman: J. Chem. Soc., **111**, 419 (1917).
32) F. L. Pyman: J. Chem. Soc., **113**, 222 (1918).
33) W. H. Brindley, F. L. Pyman: J. Chem. Soc., **130**, 1067 (1927).
34) P. Karrer, C. H. Eugster, O. Rüttner: Helv. Chim. Acta, **31**, 1219 (1948).
35) R. P. Jewstignejewa, R. S. Liwschitz, L. I. Sacharkin, M. S. Bainowa, N. A. Preobrazhenskii: Ber. Akad. Wiss. USSR., **75**, 539 (1950). [C. A., **45**, 7577 (1951)].
36) R. P. Jewstignejewa, N. A. Preobrazhenskii: Tetrahedron, **4**, 223 (1958).
37) S. Teitel, A. Brossi: J. Am. Chem. Soc., **88**, 4068 (1966).
38) S. Palkin, H. Wales: J. Am. Chem. Soc., **47**, 2005 (1925).
39) F. L. Pyman: J. Am. Chem. Soc., **68**, 836 (1946).
40) B. H. Paul, A. J. Cownley: Pharm. J., [3] **25**, 111 (1894).
41) A. R. Battersby, R. Binks, W. Lawrie, G. V. Parry, B. R. Webster: J. Chem. Soc., **1965**, 7459.
42) O. Keller: Arch. Pharm., **249**, 512 (1911).
43) O. Keller: Arch. Pharm., **251**, 701 (1913).
44) J. J. Dobbie, J. J. Fox: J. Chem. Soc., **105**, 1639 (1914).
45) O. Keller: Arch. Pharm., **263**, 401 (1925).
46) E. Späth, W. Leithe: Ber., **60**, 688 (1927).
47) M. Pailer, K. Porschinski: Monatsh., **80**, 101 (1949).
48) A. Ahl, T. Reichstein: Helv. Chim. Acta, **27**, 366 (1944).
49) A. R. Battersby, G. C. Davidson, J. C. Turner: J. Chem. Soc., **1961**, 3899.
50) H. T. Openshaw, H. C. S. Wood: J. Chem. Soc., **1952**, 391.
51) A. R. Auterhoff, K. Merz: Arch. Pharm., **291**, 326 (1958).
52) A. R. Battersby, J. C. Turner: Chem. & Ind., **1958**, 1324.
53) A. Brossi, H. Baumann, O. Schnider: Helv. Chim. Acta, **42**, 1515 (1959).
54) A. R. Battersby, J. C. Turner: J. Chem. Soc., **1960**, 717.
55) H. T Openshaw, N. Whittaker: J. Chem. Soc., **1963**, 1461.

56) A. R. Battersby: Brit. Pat., 895, 910 [C. A., **58**, 6878 (1963)].
57) A. V. Subbaratnam, S. Siddiqui: J. Sci. Ind. Res., **15B**, 432 (1956). [C. A. **51**, 3090 (1957)].
58) O. O. Orazi: Rev. facultad cienc. quim., **19**, 17 (1946). [C. A., **41**, 2210 (1947)].
59) P. Pelletier, Magendie: Ann. chim. phys., [2], **4**, 172 (1817).
60) H. Kunz: Arch. Pharm., **225**, 461 (1887).
61) J. B. Dumas, J. Pelletier: Ann. chim. phys., [2], **24**, 180 (1823).
62) Podwyssotzki: Arch. exptle. Pharmkol., **11**, 236 (1879).
63) H. Kunz-Krause: Arch. Pharm., **232**, 466 (1894).
64) O. Hesse: Pharm. J., [4], **7**, 98 (1898).
65) A. Windaus, L. Hermanns: Ber., **47**, 1470 (1914).
66) L. Hermanns: Dissertrtion, Freiburg. (1915).
67) H. Staub: "A Century of Chemical Research on Ipecacuanha Alkaloids and Thesis," Zurich, 1927.
68) H. Staub: Helv. Chim. Acta, **10**, 826 (1927).
69) R. Robinson: Nature, **162**, 524 (1948).
70) E. Späth, M. Pailer: Monatsh., **78**, 348 (1948).
71) M. Pailer: Monatsh., **79**, 127 (1949).
72) M. Pailer, L. Bilek: Monatsh., **79**, 135 (1949).
73) M. Pailer: Monatsh., **79**, 331 (1949).
74) A. R. Battersby, H. T. Openshaw, H. C. S. Wood: Experientia, **5**, 114 (1949).
75) A. R. Battersby, H. T. Openshaw: Experientia, **5**, 398 (1949).
76) M. Pailer, K. Porschinski: Monatsh., **80**, 94 (1949).
77) G. Norcross, H. T. Openshaw: J. Chem. Soc., **1949**, 1174.
78) A. R. Battersby, H. T. Openshaw: J. Chem. Soc., **1949**, 3207.
79) A. R. Battersby, H. T. Openshaw: J. Chem. Soc., **1949**, 59, 67.
80) P. Karrer, O. Rütter: Helv. Chim. Acta, **33**, 291 (1950).
81) R. N. Hazlett, W. E. McEwen: J. Am. Chem. Soc., **73**, 2578 (1951).
82) R. F. Tietz, W. E. McEwen: J. Am. Chem. Soc., **75**, 4945 (1953).
83) H. T. Openshaw: Spec. Publ. Chem. Soc. [London], **3**, 28 (1955).
84) M. Pailer, G. Beier: Monatsh., **88**, 830 (1957).
85) J. M. Osbond: Chem. & Ind., **1959**, 257.
86) Y. Ban, T. Terashima, O. Yonemitsu: Chem. & Ind., **1959**, 568.
87) Y. Ban, T. Terashima, O. Yonemitsu: Chem. & Ind., **1959**, 569.
88) A. W. Burgstahler, B. J. Bithos: J. Am. Chem. Soc., **81**, 503 (1959).
89) Y. Ban, O. Yonemitsu, T. Terashima: Chem. Pharm. Bull., **8**, 183 (1960).
90) M. Terashima: Chem. Pharm. Bull., **8**, 517 (1960).
91) A. R. Battersby, R. Binks, T. P. Edwards: J. Chem. Soc., **1960**, 3474.
92) A. Brossi, F. Burkhardt: Experientia, **18**, 211 (1962).
93) A. Brossi, M. Bawmann, F. Burkhardt, R. Richle, J. R. Frey: Helv. Chim. Acta, **45**, 2219 (1962).
94) R. P. Jewstignejewa, R. S. Liwschitz, M. S. Sacharkin, M. S. Bainowa, N. A. Preobrazhenski: Zhur. Obshchei Khim., **22**, 1467 (1952). [C. A., **47**, 5949 (1953)].
95) Y. Ban: Pharm. Bull., **3**, 53 (1955).
96) M. Barash, J. M. Osbond: Chem. & Ind., **1958**, 490.
97) M. Barash, J. M. Osbond, J. C. Wickens: J. Chem. Soc., **1959**, 3530.
98) A. Brossi, M. Baumann, L. H. Chopard-Dit-Jean, J. Würsch, F. Schneider, O. Schnider: Helv. Chim. Acta, **42**, 772 (1959).
99) A. Grüssner, E. Jäger, J. Hellerbach, O. Schnider: Helv. Chim. Acta, **42**, 2431 (1959).
100) A. W. Burgstahler, Z. J. Bithos: J. Am. Chem. Soc., **82**, 5466 (1960).
101) Roche Products Ltd: Brit. Pat., 821,332 [C. A., **58**, 6878 (1960)].
102) D. E. Clark, R. F. K. Meredith, R. C. Ritchie, T. Walker: J. Chem. Soc., **1962**, 2490.
103) A. C. Ritchie, D. R. Preston, T. Walker, K. D. E. Whiting: J. Chem. Soc., **1962**, 3385.
104) A. Brossi, O. Schnider: Helv. Chim. Acta, **45**, 1899 (1962).
105) H. T. Openshaw, N. Whittaker: J. Chem. Soc., **1963**, 144.
106) E. E. Van Tamelen, G. P. Schiemenz, H. L. Arons: Tetrahedron Letters, **1963**, 1005.
107) Glaxo Group Ltd: Fr. Pat., 1,351,409. [C.A., **60**, 15929 (1964)].
108) F. Hoffman-LaRoche & Co. A.-G: Fr. Pat., 1,351,814. [C. A, **60**, 15930 (1964)].
109) Chinoin Gyogyszer es Vegyeszeti Termekek Gyara Rt.: Hung. Pat., 151,712 [C.A., **62**, 7822 (1965)].
110) Glaxo Group Ltd.: Brit. Pat., 1,387,814. [C. A., **62**, 16318 (1965)].
111) Wellcome Fundation Ltd.: Brit. Pat., 999,096. [C. A., **63**, 16402 (1965)].
112) Glaxo Group Ltd: Neth. Appl., 299,887.

[C. A. **64**, 5157 (1966)].
113) W. F. Elvidge: Quart. J. Pharm. Pharmacol., **13**, 219 (1940).
114) P. Bayard: Bull. soc. roy. sci. Liege, **47**, 2005 (1925).
115) R. M. Pinyazhko: Farmatsevi. Zh(kiev)., **19**, 12 (1964). [C. A., **64**, 6701 (1966)].
116) G. Spiteller, M. Spiteller-Friedmann: Tetrahedron Letters, **1963**, 153.
117) A. Popelak, E. Haack, H. Spingler: Tetrahedron Letters, **1966**, 1081.
118) A. R. Battersby, J. R. Merchant, E. A. Ruveda, S. S. Salgar: Chem. Comm., **1965**, 315.
119) H. Monteiro, H. Budzikiewicz, C. Djerassi: Chem. Comm., **1965**, 317.
120) S. C. Pakraski: Indian J. Chem., **2**, 468 (1964).
121) P. Brauchli, V. Deulofeu, H. Budzikiewicz, C. Djerassi: J. Am. Chem. Soc., **86**, 1895 (1964).
122) H. T. Openshaw, N. Whittaker: Chem. Comm., **1966**, 131.
123) A. Popelak, E. Haack, H. Spingler: Tetrahedron Letters, **1966**, 5077.

Chapter 18 Erythrina Alkaloids

Cocculolidine $C_{15}H_{19}O_3N \equiv 261.31$

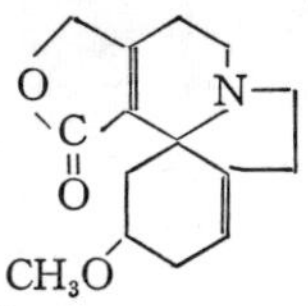

≪mp≫ 144～146°
≪$[\alpha]_D$≫ +273°($CHCl_3$)

≪Ph.≫ UV[1] IR[1] NMR[1] Mass[1]
≪OP.≫ Cocculus trilobus[1]
≪C. & So.≫
≪Sa. & D.≫ Dihydro-Der.: 121～121.5°
Tetrahydro-Der.: 187～189°
≪SD.≫ 1)
≪S. & B.≫

***α*-Erythroidine** $C_{16}H_{19}O_3N \equiv 273.32$

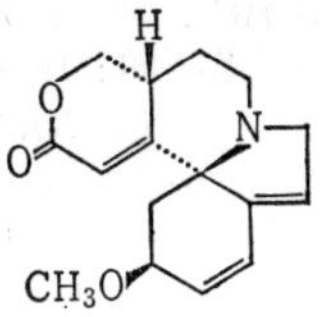

≪mp≫ 58～60°
≪$[\alpha]_D$≫ +136°(H_2O)
≪Ph.≫ UV[16)28)] ORD[37)38)] X-ray[35)36)]

≪OP.≫ Erythrina americana (carnea)[2)3)17)] E. berteroana[3)], E. costaricensis[3)], E. tholboniana[5)]
≪C. & So.≫ Sol. in C_6H_6, MeOH, EtOH, $CHCl_3$, H_2O, Mod. sol. in Et_2O
≪Sa. & D.≫ B-HCl: 228～229°(d)
B-HBr: 220～222° B-HI: 212° B-$HClO_4$: 208～208.5° B-MeI: 220°
≪SD.≫ 3), 5), 6), 7), 9), 14), 15), 17), 19), 20), 28), 29), 31), 32), 34), 37), 38)
≪S. & B.≫ BS: 44)

***β*-Erythroidine** $C_{16}H_{19}O_3N \equiv 273.32$

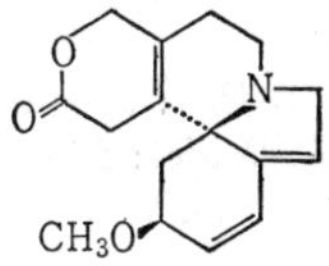

≪mp≫ 99～100°
≪$[\alpha]_D$≫ +88.8°(H_2O)

≪Ph.≫ →α-erythroidine
≪OP.≫ →α-erythroidine
≪C. & So.≫
≪Sa. & D.≫ Salts[4)], B-HCl: 232°(d)
B-HBr: 230° B-HI: 206°
B-$HClO_4$: 203～203.5°
≪SD.≫ 3), 5)～16), 18)～27), 30)～33), 35), 36), 38)～42)
≪S. & B.≫ cf. 32), 43), BS: 44)

Erysopine $C_{17}H_{19}O_3N \equiv 285.333$

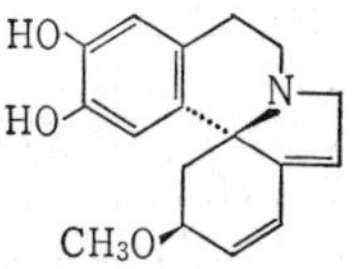

≪mp≫ 241～242°(EtOH)
≪$[\alpha]_D$≫ +265.2°(EtOH-glycerin), + 225° (Morpholine), +276° (HCl-H_2O)

≪Ph.≫ CR[45)47)49)51)68)] UV[69)] pKa[69)]
≪OP.≫ Erythrina abyssinica[45)46)], E. acanthocarpa[47)], E. arborescens[48)], E. costaricensis[48)], E. dominguezii[47)49)], E. excelsa[48)], E. falcata[49)]*
≪C. & So.≫ Unstable in alkali, Spar. sol. in H_2O, $CHCl_3$
≪Sa. & D.≫ B-HCl: 260～265°(270°)
B-2OMe: 98°
≪SD.≫ 15), 19), 20), 21), 22), 45), 50), 51), 52), 53), 54), 55), 56), 57), 58), 59), 60), 61), 62), 63), 64)
≪S. & B.≫ cf. 63), 65), 66), 67),

---

* E. flabelliformis[45)], E. fuscata[47)], E. glauca[45)], E. herbacea[45)], E. macrophylla[47)48)], E. orophila[5)], E. pallida,[48)] E. rubrinervia[47)], E. sandwicensis[45)], E. senegalensis[47)], E. suburbrans[47)], E. thilloniana[5)]

Erysonine $C_{17}H_{19}O_3N \equiv 285.33$

≪mp≫ 236~237°(d) (EtOH)
≪$[\alpha]_D$≫ +285~288° (aq. HCl), +272° (Morpholine)
≪Ph.≫ pKa[69]
≪OP.≫ Erythrina costaricensis[47]
≪C. & So.≫
≪Sa. & D.≫
≪SD.≫ 15), 47), 70), 71)
≪S. & B.≫

Erythraline $C_{18}H_{19}O_3N \equiv 297.34$

≪mp≫ 106~108° (EtOH), cf. 120°[50]
≪$[\alpha]_D$≫ +211.8° (EtOH)
≪Ph.≫ UV[14,50,74] X-ray[77] pKa[69]
≪OP.≫ Erythrina abyssinica[46,50], E. crystagalli[68], E. folkersii[72], E. fusca[72], E. glauca[72], E. grisebanchii[72], E. macrophylla[72] E. orophila[5], E. varieyata var. orientalis[72], E. velutina[72]
≪C. & So.≫
≪Sa. & D.≫ B–HCl: 249° B–HBr: 243° B–HI: 252~253° B–MeI: 185~187°
≪SD.≫ 9), 15), 16), 17), 24), 50), 54), 55), 56), 72), 73), 74), 75), 76)
≪S. & B.≫ cf. 77)

Erythramine $C_{18}H_{21}O_3N \equiv 299.36$

≪mp≫ 103~104° (Et_2O–Petr. ether), bp 125°/3.9×10^{-4}
≪$[\alpha]_D$≫ +227° (EtOH)
≪Ph.≫ UV[14,50,74] pKa[33]
≪OP.≫ Erythrina crista-galli[68], E. glauca[72], E. sandwicensis[73], E. subumbrans[73]
≪C. & So.≫ Free base: unstable, Sol. in MeOH, EtOH, AcOEt, C_6H_6, Mod. sol. in Et_2O, Insol. in Petr. ether
≪Sa. & D.≫ B–HCl: 250°(d) B–HBr: 228° B–HI: 249°(d) B–MeI: 96~98°
≪SD.≫ 9), 14), 50), 74)
≪S. & B.≫

Erysodine $C_{18}H_{21}O_3N \equiv 299.36$

≪mp≫ 204~205° (EtOH)
≪$[\alpha]_D$≫ +248° (EtOH), +231±4° ($CHCl_3$)
≪Ph.≫ UV[8,50] NMR[79] ORD[37] pKa[69]
≪OP.≫ Erythrina abyssinica[45,46,50], E. americana (carnea)[45], E. arborescens[48], E. berteroana[48], E. costaricensis[47], E. crista-galli[47,49,68], E. cubensis[48], E. dominguezii[47], E. excelsa[48], E. falcata[49], E. flabelliformis[45], E. folkersii[48], E. fusca[47]*
≪C. & So.≫ Needles, Sol. in $CHCl_3$, Mod. Sol. in EtOH, Et_2O
≪Sa. & D.≫ B–MeI: 229° B–Pic: 200° B–OAc: 136° B–OBz: 199° B–OMe: 98°
≪SD.≫ 15), 16), 17), 19), 20), 21), 22), 45), 50), 51), 52), 53), 54), 55), 56), 57), 58), 59), 61), 62), 63), 71), 76), 78,) 79)
≪S. & B.≫ 71), cf. 32), 38), 42), 43), 53) 57), 58), 59), 63), 65), 66), 67), 77), 78), 80), 81), 82), 83), 84), 85), 86), BG: 69), 87), 88), 89), 90), 91), 92)

* E. glauca[45], E. herbacea[45], E. macrophylla[47], E. orophila[5], E. pallida[48], E. poeppigiana[45], E. sandwicensis[45], E. senegalensis[47], E. subumbrans[47], E. tholloniana[5], E. velutina[48]

Erysovine $C_{18}H_{21}O_3N \equiv 299.36$

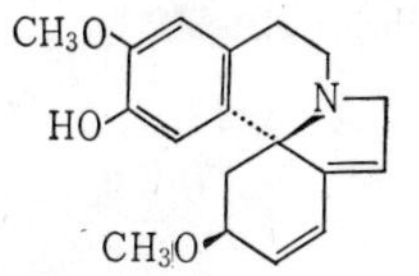

≪mp≫ 178~179°(Et_2O)
≪$[\alpha]_D$≫ +234°(EtOH)
≪Ph.≫ UV[69] NMR[79] pKa[69]
≪OP.≫ Erythrina americana(carnea)[45], E. arborescens[48], E. berteroana[45)48], E. costaricensis[47], E. crista-galli[49)68], E. cubensis[48], E. dominguezii[49], E. excelsa[48], E. falcata[49]*
≪C. & So.≫
≪Sa. & D.≫ B–HCl: 237~238° B–HBr: 151° B–HI: 162° B–OMe: 98°
≪SD.≫ 15), 19), 20), 21), 22), 45), 50), 51), 52), 53), 54), 55), 56), 57), 58), 59), 61), 62), 63), 71), 78), 79)
≪S. & B.≫ → erysodine

Dihydroerysodine $C_{18}H_{23}O_3N \equiv 301.37$

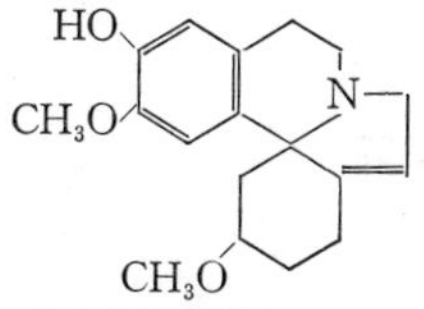

≪mp≫ 208~209°(Me_2CO)
≪$[\alpha]_D$≫ +224°(MeOH)
≪Ph.≫ CR[93] UV[50)93] IR[93] pKa[69]
≪OP.≫ Cocculus laurifolius[93)94]
≪C. & So.≫ Colourless needles
≪Sa. & D.≫ B–MeI: 282°(d) B–OMe·Pic: 163~164°
≪SD.≫ 93)
≪S. & B.≫

No name $C_{18}H_{19}O_4N \equiv 313.34$

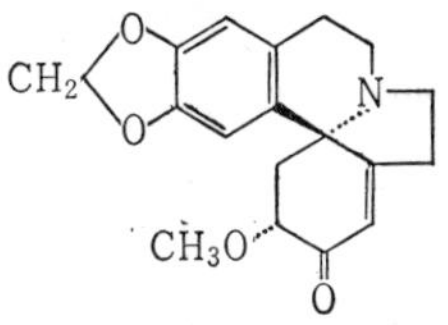

≪mp≫
≪$[\alpha]_D$≫
≪Ph.≫ NMR[79]
≪OP.≫ Erythrina crysta-galli[79]
≪C. & So.≫
≪Sa. & D.≫
≪SD.≫ 79)
≪S. & B.≫

Erythratine $C_{18}H_{21}O_4N \equiv 315.36$

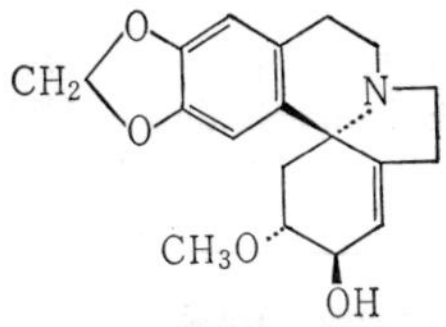

≪mp≫ 170~171°(EtOH or Et_2O–Petr. ether)
≪$[\alpha]_D$≫ +145.5°(EtOH), +110°(H_2O)
≪Ph.≫ UV[14]
≪OP.≫ Erythrina crista-galli[68], E. glauca[72]
≪C. & So.≫
≪Sa. & D.≫ B–HCl: 250° B–HBr: 241° B–HI: 242° B–OAc: 129°
≪SD.≫ 14), 55), 72), 79), 98)
≪S. & B.≫

Erythratidine $C_{19}H_{25}O_4N \equiv 331.40$

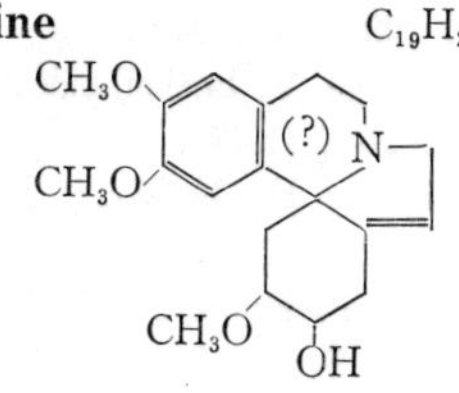

≪mp≫ 120~121°
≪$[\alpha]_D$≫ +279.5 ±3° (EtOH)
≪Ph.≫
≪OP.≫ Erythrina falcata[95]
≪C. & So.≫ Needles
≪Sa. & D.≫ B–HI: 201~202° B–MeI: 183° B–Pic: 222~224°
≪SD.≫ 95)
≪S. & B.≫

* E. flabelliformis[45], E. folkersii[48)49], E. glauca[48], E. pallida[48], E. poeppigiana[45], E. sandwicensis[45], E. velutina[48]

Erysothiopine $C_{19}H_{21}O_7NS \equiv 407.96$

CH_3O-
$HOOCCH_2SO_2O-$
CH_3O
N

≪mp≫ 168～169°
≪$[\alpha]_D$≫ +193° (EtOH)
≪Ph.≫
≪OP.≫ Erythrina glauca[96)]
≪C. & So.≫
≪Sa. & D.≫
≪SD.≫ 96)
≪S. & B.≫

Erysothiovine $C_{20}H_{23}O_7SN \equiv 421.46$

CH_3O-
$HOOCCH_2SO_2O-$
CH_3O
N

≪mp≫ 187°
≪$[\alpha]_D$≫ +208° (EtOH)
≪Ph.≫
≪OP.≫ Erythrina giana[96)], E. glauca[96)], E. pallida[96)], E. poeppi[96)]
≪C. & So.≫
≪Sa. & D.≫
≪SD.≫ 96), 97)
≪S. & B.≫

Glaucoprysodine $C_{24}H_{31}O_8N \equiv 461.50$

$C_6H_{11}O_5O$
CH_3O
CH_3O
N

≪mp≫ 135～137°
≪$[\alpha]_D$≫ +127° (H_2O)
≪Ph.≫
≪OP.≫ Erythrina abyssinica[5)46)]
≪C. & So.≫
≪Sa. & D.≫
≪SD.≫ 5), 46)
≪S. & B.≫

Erysocine $C_{18}H_{21}O_3N \equiv 299.36$

≪mp≫ 160～161° (EtOHor Et_2O)
≪$[\alpha]_D$≫ +238.1° (EtOH)
≪Ph.≫ CR[45)]
≪OP.≫ Erythrina americana[45)], E. cubensis[48)], E. falcata[49)], E. flabelliformis[45)], E. poeppigiana[45)], E. sandwicensis[45)]
≪C. & So.≫ Needles, Sol. in $CHCl_3$, Mod. sol. in EtOH, Et_2O
≪Sa. & D.≫
≪SD.≫ mixt. of erysodine and erysovine
≪S. & B.≫

1) K. Wada, S. Marumo, K. Munakata: Tetrahedron Letters, **1966**, 5179.
2) K. Folkers, R.T. Major: J. Am. Chem. Soc., **59**, 1580 (1937).
3) K. Folkers, R.T. Major: U.S. Pat., 2,385,266. [C.A., **39**, 5408 (1945)].
4) K. Folkers, R.T. Major: U.S. Pat., 2,373,952. [C.A., **39**, 3544 (1945)].
5) C. Lapiére: Dissertation, Liegé, 1952.
6) Merco & Co., and R.T. Major: Brit. Pat., 543,187. [C.A., **37**, 232 (1943)].
7) K. Folkers, F. Koniuszy; U.S. Pat., 2,730,651. [C.A., **39**, 4511 (1945)].
8) K. Folkers, F. Koniuszy: Brit. Pat., 596,976. [C.A., **42**, 3914 (1948)].
9) K. Folkers, F. Koniuszy: J. Am. Chem. Soc., **61**, 3053 (1939).
10) F. Koniuszy, K. Folkers: J. Am. Chem. Soc., **72**, 5579 (1950).
11) G.L. Sauvage, F.M. Berger, V. Boekelheide: Science, **109**, 627 (1949).
12) G.L. Sauvage, V. Boekelheide: J. Am. Chem. Soc., **72**, 2062 (1950).
13) F. Koniuszy, K. Folkers: J. Am. Chem. Soc., **73**, 333 (1951).
14) K. Folker, F. Koniuszy, J. Shaval, Jr.: J. Am. Chem. Soc., **64**, 2146 (1942).
15) M. Carmack, B.C. McKusick, V. Prelog: Helv. Chim. Acta, **34**, 1601 (1951).
16) V. Boekelheide, J. Weinstock, M.F. Grundon, G.L. Sauvage, E.J. Agnello: J. Am. Chem. Soc., **75**, 2550 (1953).

17) V. Boekelheide, M.F. Grundon: J. Am. Chem. Soc., **75**, 2563 (1953).
18) C. Lapière, R. Robinson: Chem. & Ind., **30**, 650 (1951).
19) M.F. Grundon, V. Boekelheide: J. Am. Chem. Soc., **74**, 2637 (1952).
20) J.R. Merchant: Ph. D. Thesis, Eidgenössische Technische Hochschule, Zürich, 1953.
21) M.F. Grundon, V. Boekelheide: J. Am. Chem. Soc., **75**, 2637 (1953).
22) M.F. Grundon, G.L.Sauvage, V. Boekelheide: J. Am. Chem. Soc., **75**, 2541 (1953).
23) B.D. Astill, V. Boekelheide: J. Am. Chem. Soc., **77**, 4079 (1955).
24) W.J. Gall, B.D. Astill, V. Boekelheide: J. Org. Chem., **20**, 1538 (1955).
25) W.G. Gall, V. Boekelheide: J. Org. Chem., **19**, 504 (1954).
26) V. Boekelheide, E.G. Agnello: J. Am. Chem. Soc., **73**, 2286 (1951).
27) J. Weinstock, V. Boekelheide: J. Am. Chem. Soc., **75**, 2546 (1953).
28) J.C. Godfrey, D.S. Tarbell, V. Boekelheide: J. Am. Chem. Soc., **77**, 3342 (1955).
29) V. Boekelheide, G.C. Morrison: J. Am. Chem. Soc., **80**, 3905 (1958).
30) V. Boekelheide, A.E. Anderson, Jr., G.L, Sauvage: J. Am. Chem. Soc., **75**, 2558 (1953).
31) G. Stork: J. Am. Chem. Soc., **74**, 768 (1952).
32) M. Müller, T.T. Grossnickle, V. Boekelheide: J. Am. Chem. Soc., **81**, 3959 (1959).
33) V. Boekelheide, M.F. Grundon, J. Weinstock: J. Am. Chem. Soc., **74**, 1866 (1952).
34) R.K. Hill, W.R. Schearer: J. Org. Chem., **27**, 921 (1962).
35) A.W. Hanson: Proc. Chem. Soc., **1963**, 52.
36) G.R. Wenzinger, V. Boekelheide: Proc. Chem. Soc., **1963**, 53.
37) U. Weiss, H. Ziffer: Experientia, **19**, 108, 660 (1963).
38) V. Boekelheide, M.Y. Chang: J. Org. Chem., **29**, 1303 (1964).
39) V. Boekelheide, G.R. Wenzinger: J. Org. Chem., **29**, 1307 (1964).
40) J. Blake, J.R. Tretter, H. Rapoport: J. Am. Chem. Soc., **87**, 1397 (1965).
41) J. Blake, J.R. Tretter, G.J. Juhasz, W. Bonthone, H. Rapoport: J. Am. Chem. Soc., **88**, 4061 (1966).
42) B. Belleau: Can. J. Chem., **35**, 663 (1957).
43) V. Boekelheide, M. Müller, J. Jack, T.T. Grossnickle, M. Chang: J. Am. Chem. Soc., **81**, 3955 (1959).
44) E. Leete, A. Ahmad: J. Am. Chem. Soc., **88**, 4722 (1966).
45) K. Folkers, F. Koniuszy: J. Am. Chem. Soc., **62**, 1677 (1940).
46) C. Lapiere: J. Pharm. Belg., **6**, 71 (1951). [C.A., **45**, 9806 (1951)].
47) K. Folkers, J. Shavel, Jr., F. Koniuszy: J. Am. Chem. Soc., **63**, 1544 (1941).
48) K. Folkers, J. Shavel, Jr.: J. Am. Chem. Soc., **64**, 1892 (1942).
49) R.A. Gentile, R. Labriola: J. Org. Chem., **7**, 136 (1942).
50) V. Prelog, K. Wiesner, H.G. Khorana, G.W. Kenner: Helv. Chim. Acta, **32**, 453 (1949).
51) F. Koniuszy, P.F. Wiley, K. Folkers: J. Am. Chem. Soc., **71**, 875 (1949).
52) Z. Valenta, K. Wiesner: Chem. & Ind., **1954**, 402.
53) K. Wiesner, Z. Valenta, A.J. Manson, F.W. Stonner: J. Am. Chem. Soc., **77**, 675 (1955).
54) G.W. Kenner, H.G. Khorana, V. Prelog: Helv. Chim. Acta, **34**, 1969 (1951).
55) K. Folkers, F. Koniuszy, J. Shavel, Jr.: J. Am. Chem. Soc., **73**, 589 (1951).
56) V. Prelog, B.C. McKusick, J.R. Merchant, S. Julia, M. Wilhelm: Helv. Chim. Acta, **39**, 498 (1956).
57) B. Belleau: Chem. & Ind., **1956**, 410.
58) B. Belleau: Angew. Chem., **68**., 578 (1956).
59) B. Belleau: Can. J. Chem., **35**, 651 (1957).
60) A. Mondon, H.U. Menz: Tetrahedron, **20**, 1279 (1964).
61) R.A. Labiola, V. Deulofeu, B. Berinzaghi: J. Org. Chem., **16**, 90 (1951).
62) A. Mondon: Ber., **92**, 1461 (1959).
63) A. Mondon: Ber., **92**, 1472 (1959).
64) A. Mondon, H.J. Nestler, H.G. Vilhuber, M. Ehrhardt: Ber., **98**, 46 (1965).
65) A. Mondon, H.J. Nestler: Angew. Chem., **76**, 651 (1964).
66) A. Mondon: Ann., **628**, 123 (1959).
67) A. Mondon, J. Zander, H.U. Menz: Ann., **667**, 126 (1963).
68) V. Deulofeu, R. Labriola, E. Hug, M. Fondovila, A. Kauffmann: J. Org. Chem., **12**, 486 (1947).
69) V. Prelog: Angew. Chem., **69**, 33 (1957).
70) K. Folkers, J. Shavel, Jr.: U.S. Pat., 2, 391,015. [C.A., **40**, 1284 (1946)].
71) A. Mondon, M. Ehrhardt: Tetrahedron Letters, **1966**, 2557.
72) K. Folkers, F. Koniuszy: J. Am. Chem. Soc., **62**, 436 (1940).
73) K. Folkers, F. Koniuszy: J. Am. Chem. Soc., **61**, 1231 (1939).
74) K. Folkers, F. Koniuszy: J. Am. Chem.

Soc., **62**, 1673 (1940).
75) K. Wiesner, F. H. Clarke, S. Kairys: Can. J. Res., **28B**, 234 (1950).
76) G. L. Sauvage, V. Boekelheide: J. Am. Chem. Soc., **72**, 2062 (1950).
77) A. J. Manson, K. Wiesner: Chem. & Ind., **1953**, 641.
78) V. Prelog, A. Langemann, O. Rodig, M. Ternbah: Helv. Chim. Acta, **42**, 1301 (1959).
80) J. E. Gervay, F. McCapra, T. Money, G. H. Sharma, A. I. Scott: Chem. Comm., **1966**, 142.
81) S. Chiavarelli, E. F. Rogers, G. B. Marini-Bottolo: Gazz. chim. ital., **83**, 347 (1953).
82) V. Boekelheide: "The Alkaloids," Vol. VII, p. 201. Edited by R. H. F. Manske. Academic Press, New Yosk (1960).
83) B. Belleau: J. Am. Chem. Soc., **75**, 5765 (1953).
84) A. Mondon: Abstract, Nordwestdeutsche Chemiedozenten-Tagung, Münster, April, 1–3 1957.
85) A. Mondon: Angew. Chem., **70**, 406 (1958).
86) D. Megirian, D. E. Leary, I. H. Slater: J. Pharm. Exptle. Therap., **113**, 212 (1955).
87) V. Boekelheide, V. Prelog: Progr. in Org. Chem., **3**, 242 (1955).
88) V. Boekelheide: Record of Chem. Progr. Kresge-Hooker Sci. Lib., **16**, 227 (1955).
89) E. Wenkert: Chem. & Ind., **1953**, 1088.
90) R. Robinson: Chem. & Ind., **1953**, 1317.
91) B. Witkop, S. Goodwin: Experientia, **8**, 377 (1952).
92) D. H. R. Barton, T. Cohen: "Festshrift Arthur Stoll," p. 127. Birkhäuser, Basel, (1957).
93) M. Tomita, H. Yamaguchi: Pharm. Bull., **4**, 225 (1956).
94) J. Kunitomo: J. Pharm. Soc. Japan, **81**, 1261 (1961).
95) V. Deulofeu: Ber., **85**, 620 (1952).
96) K. Folkers, F. Koniuszy, J. Shavel, Jr.: J. Am. Chem. Soc., **66**, 1082 (1944).
97) K. Unna, J. G. Greslin: J. Pharmacol., **80**, 53 (1944).
98) K. Folkers, F. Koniuszy, J. Shavel, Jr.: J. Am. Chem. Soc., **64**, 2146 (1942).

Chapter 19 Protostephanine Alkaloids

Protostephanine $C_{21}H_{27}O_4N \equiv 357.43$

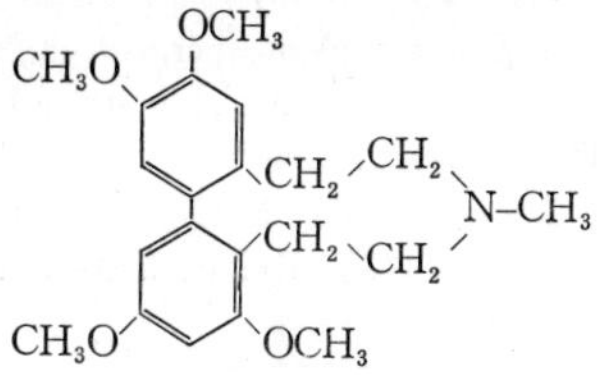

OCH_3 CH_3O CH_2 CH_2 $N-CH_3$ CH_2 CH_2 CH_3O OCH_3

≪mp≫ 90∼95° (MeOH), cf. 70∼75° [B+MeOH]

≪$[\alpha]_D$≫ 0° or +3.44°[1] (EtOH)

≪Ph.≫ CR[1] UV[3,9,10,13] IR[9,12] NMR[14,17] Rf[14]

≪OP.≫ *Stephania japonica*[1,2,17]

≪C. & So.≫ Colourless prisms

≪Sa. & D.≫ B–HCl: 150°(d) B–HBr: 185° B–Pic: 209° B–H_2PtCl_6: 223° B–MeI: 220∼221°

≪SD.≫ 1), 2), 3), 4), 5), 6), 7), 8), 9), 10), 11), 12), 13), 16)

≪S. & B.≫ 14), BS: 17), BG: 15)

---

1) H. Kondo, T. Sanada: J. Pharm. Soc. Japan, **47**, 177 (1927).
2) H. Kondo, T. Watanabe: J. Pharm. Soc. Japan, **58**, 268 (1938).
3) H. Kondo, T. Watanabe: Ann. Rep. ITSUU Lab., **1**, 1 (1950).
4) H. Kondo, T. Nakamura, M. Fujii, T. Kato: Ann. Rep. ITSUU Lab., **1**, 5 (1950).
5) H. Kondo, M. Satomi, T. Kodera: Ann. Rep. ITSUU Lab., **1**, 9 (1950).
6) H. Kondo, T. Watanabe: Ann. Rep. ITSUU Lab., **1**, 12 (1950).
7) H. Kondo, T. Watanabe, K. Takeda: Ann. Rep. ITSUU Lab., **3**, 7 (1952).
8) H. Kondo, K. Takeda: Ann. Rep. ITSUU Lab., **4**, 6 (1953).
9) H. Kondo, K. Takeda: Ann. Rep. ITSUU Lab., **5**, 1 (1954).
10) H. Kondo, K. Takeda: Ann. Rep. ITSUU Lab., **7**, 30 (1956).
11) K. Takeda: Bull. Agr. Chem. Soc. Japan, **20**, 165 (1956).
12) H. Kondo, K. Takeda: Ann. Rep. ITSUU Lab., **9**, 33 (1958).
13) K. Takeda: Ann. Rep. ITSUU Lab., **9**, 24 (1958).
14) K. Takeda: Ann. Rep. ITSUU Lab., **13**, 45 (1963).
15) D.H.R. Barton: "Chemistry of Natural Products", Vol. Ⅲ. p. 35 (1964).
16) B. Pecherer, A. Brossi: Helv. Chim. Acta, **49**, 2261 (1966).
17) D.H.R. Barton, G.W. Kirby, A. Wiechers: J. Chem. Soc.(C), **1966**, 2313.

Chapter 20 Tyrophorine and Cryptopleurine Alkaloids

Cryptopleurine $C_{24}H_{27}O_3N \equiv 377.46$

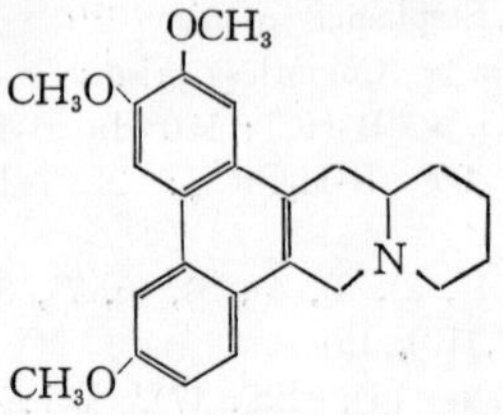

≪mp≫ 197～198° (Me_2CO, EtOH, MeOH)
≪$[\alpha]_D$≫ −106° ($CHCl_3$)

≪Ph.≫ UV[4] IR[4)6] X-ray[4)5]
≪OP.≫ Cryptocarya pleurosperma[1]
≪C. & So.≫ Long prisms, skin irritant
≪Sa. & D.≫ B-HCl: 262° (d)
B-HBr: 258～260° (d) B-HI: 256～258° (d)
B-$HClO_4$: 177～178° B-Pic: 220～221° (d)
B-MeI: 215～217°
≪SD.≫ 2), 3), 4), 5)
≪S. & B.≫ 6), 7), 8), BG: 23)

Tylophorinine $C_{23}H_{25}O_4N \equiv 379.44$

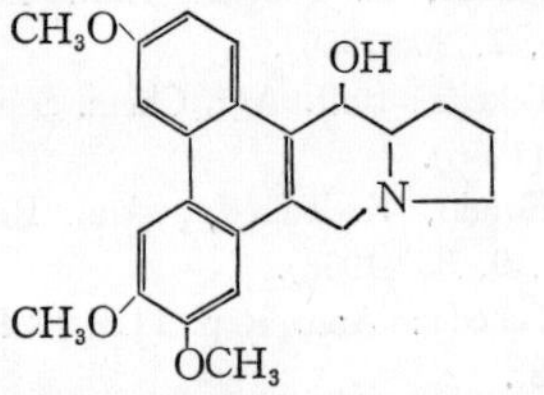

≪mp≫ 248～249° ($CHCl_3$-EtOH)
≪$[\alpha]_D$≫ −14.2° ($CHCl_3$)

≪Ph.≫ UV[11)12)13)14] IR[12)14] Rf[13]
≪OP.≫ Tyrophora asthmatica[9)10)11],
Ficus septica[22]
≪C. & So.≫
≪Sa. & D.≫ B-HCl: 257° (d)
B-HNO_3: 240～242° B-MeI: 243～245° (d)
B-OAc: 222～223°
≪SD.≫ 10), 12), 13)
≪S. & B.≫ 14), BG: 23)

Tylophorine $C_{24}H_{27}O_4N \equiv 393.46$

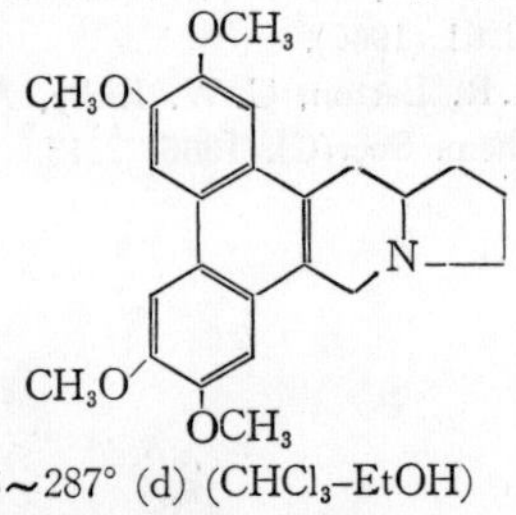

≪mp≫ 286～287° (d) ($CHCl_3$-EtOH)
≪$[\alpha]_D$≫ −11.6° ($CHCl_3$)

≪Ph.≫ UV[11)21] LR[19)20)21] X-ray[21] Rf[21]
≪OP.≫ Tylophora asthmatica[9)11)15],
T. crebriflora[21]
≪C. & So.≫
≪Sa. & D.≫ B-HCl: 276～278°
B-HI: 243～245° B-HNO_3: 265～267°
B-MeI: 280°
≪SD.≫ 9), 11), 16), 17), 18)
≪S. & B.≫ 19), 20), BG: 23)

***l*-Tylocrebrine** $C_{24}H_{27}O_4N \equiv 393.46$

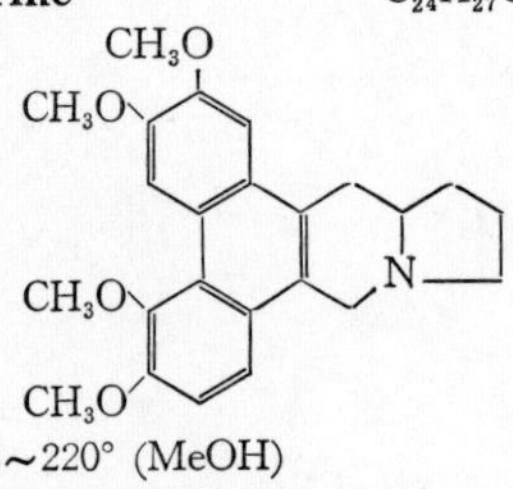

≪mp≫ 218～220° (MeOH)
≪$[\alpha]_D$≫ −45±2° ($CHCl_3$)

≪Ph.≫ UV[21] IR[21] Rf[21] pKa[21]
≪OP.≫ Tylophora crebriflora[21]
≪C. & So.≫
≪Sa. & D.≫ B-HI: 214～217° (d)
B-$HClO_4$: 262～264° (d) B-Pic: 134～136°
B-MeI: 255～258°
≪SD.≫ 21)
≪S. & B.≫ 21)

d-Tylocrebrine

≪mp≫ 220～222°
≪$[\alpha]_D$≫ +20.5 ±2°
≪Ph.≫
≪OP.≫ Ficus septica[22]
≪C. & So.≫
≪Sa. & D.≫ B–Pic: 144～146°
B–MeI: 271～274°
≪SD.≫ → *l*–tylocrebrine
≪S. & B.≫ → *l*–tylocrebrine

---

1) I.S. De la Lande: Austral. J. Exptl. Bio. Med. Sci., **26**, 181 (1948). [C. A., **42**, 7490 (1948)].
2) E. Gellert, N.V. Riggs: Austral. J. Chem., **7**, 113 (1954).
3) E. Gellert: Austral. J. Chem., **9**, 489 (1956).
4) J. Fridrichsons, A. M. Mithieson: Nature, **173**, 732 (1954). [C.A., **50**, 11925 (1956)].
5) J. Fridrichsons, A.M. Mithieson: Acta Cryst., **8**, 761 (1955). [C.A., **50**, 3837 (1956)].
6) C.K. Bradsher, H. Berger: J. Am. Chem. Soc., **79**, 3287 (1957).
7) C.K. Bradsher, H. Berger: J. Am. Chem. Soc., **80**, 930 (1958).
8) P. Marchini, B. Belleau: Can. J. Chem., **36**, 581 (1958).
9) A.N. Ratnagiriswarin, K. Venkatachalam: Indian J. med. Res., **22**, 433 (1935).
10) T. R. Govindachari, B. R. Pai, I. S. Ragade, S. Rajappa, N. Viswanathan: Chem. & Ind., **1960**, 966.
11) T.R. Govindachari, B.R. Pai, K. Nagarajan: J. Chem. Soc., **1954**, 2801.
12) T.R. Govindachari, B.R. Pai, S. Rajappa, N. Viswanathan: Chem. & Ind., **1959**, 950.
13) T.R. Govindachari, B.R. Pai, I. S. Ragade, S. Rajappa, N. Viswanathan: Tetrahedron, **14**, 288 (1961).
14) T.R. Govindachari, B.R. Pai, S. Prabhakra, T.S. Savitri: Tetrahedron, **21**, 2573 (1965).
15) Hooper: Pharm. J., [3], **21**, 617 (1891).
16) T.R. Govindachari, M.V. Lakshmikantham, K. Nagarajan, B.R. Pai: Chem. & Ind., **1957**, 1484.
17) T.R. Govindachari, M.V. Lakshmikantham, K. Nagarajan, B.R. Pai: Tetrahedron, **4**, 311 (1958).
18) T.R. Govindachari, M.V. Lakshmikantham, B.R. Pai, S. Rajappa: Tetrahedron, **9**, 53 (1960).
19) T.R. Govindachari, M.V. Lakshmikantham, S. Rajadurai: Chem. & Ind., **1960**, 664.
20) T.R. Govindachari, M.V. Lakshmikantham, S. Rajadurai: Tetrahedron, **14**, 284 (1964).
21) E. Gellert, T.R. Govindachari, M.V. Lakshmikantham, I. S. Ragade, R. Rudzats, N. Viswanathan: J. Chem. Soc., **1962**, 1008.
22) J.H. Russel: Naturwiss., **50**, 443 (1963).
23) E. Wenkert: Experientia, **15**, 165 (1959).

Chapter 21 Amaryllidaceae Alkaloids

Ungeremine $C_{16}H_{11}O_3N$≡265.26

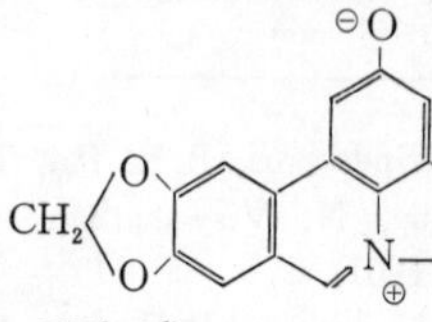

≪mp≫ 270~272° (d)
≪$[\alpha]_D$≫ 0°

≪Ph.≫ UV[1)] IR[1)]
≪OP.≫ Ungernia minor[1)]
≪C. & So.≫
≪Sa. & D.≫ B–HCl : 314~315° (d)
B–HBr : 327~329° (d) B–HNO_3 : 277~278°
B–Pic : 280~282° (d)
≪SD.≫ 1)
≪S. & B.≫ 1)

Caranine $C_{16}H_{17}O_3N$≡271.30

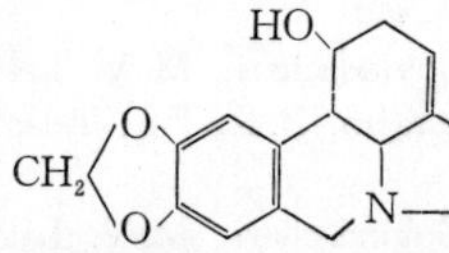

≪mp≫ 178~180° (AcOEt or Me_2CO)
≪$[\alpha]_D$≫ −197° ($CHCl_3$)
≪Ph.≫ UV[4)5)29)] IR[16)30)] NMR[26)] Mass[31)32)] CD[26)] Rf[6)33)]

≪OP.≫ Amarylis belladonna (Brunsvigia rosea)[2)3)4)5)], A. belladonna var. purpurea[3)], A. parkeri (A. belladonna X Brunsvigia josephinae)[3)]*
≪C. & So.≫ Prisms
≪Sa. & D.≫ B–$HClO_4$: ca. 270°
B–MeI : α : 255°, β : 314°
B–OAc = bellamarine
≪SD.≫ 5), 16), 18), 19)~25), 26), 272)
≪S. & B.≫ cf. 22), 25), 27), BG : 28)

Lycorine $C_{16}H_{17}O_4N$≡287.30
(Amarylline[34)], Belamarine[34)], Galanthidine, Narcissine[35)])

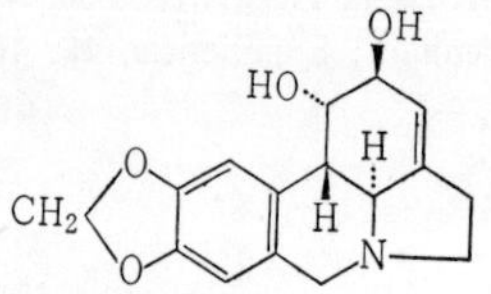

≪mp≫ 280° (MeOH)
≪$[\alpha]_D$≫ −120° (EtOH), cf. −42° (DMF)
≪Ph.≫ CR[42)91)] UV[29)53)112)119)] IR[30)114)] NMR[26)] Mass[31)32)] CD[26)272)] Rf[6)11)33)] pKa[119)] ORD[272] X-ray[271)]

≪OP.≫ Amaryllis belladonna[2)3)4)5)34)], A. belladonna var. purpurea[2)3)], A. parkeri (A. belladonna X Brunsvigia josephinae)[3)7)], Ammocharis coranica[5)6)], A. falcata[5)], Brunsdonna tubergenii[7)], Brunsvigia cooperi[37)]**
≪C. & So.≫ Prisms, Spar. sol. in Et_2O, EtOH, $CHCl_3$, Insol. in H_2O
≪Sa. & D.≫ B–HCl : 206°, 217°, 225°
B–$HClO_4$: 230(d) B–Pic : 198~200°, 230° (d)
B–MeI : α : 247°, β : 198° B–O¹–Ac : 215~216°
B–O,O–DiAc : 216~217°
≪SD.≫ 18), 19), 20), 22), 23), 24), 25), 26), 40), 42), 72), 86), 87), 88), 105~128), 271), 272)
≪S. & B.≫ cf. 126), 127), 129), BS : 130)~138)

---

* Ammocharis coranica (A. falcata)[3)5)6)], Brunsdonna tubergenii (Brunsvigia josephinae X Amaryllis belladonna)[7)], Buphane disticha[8)], Crinum defixum[9)], C. laurentii[10)], C. macrantherum[11)], C. powellii[12)], C. powellii var. album[13)], Hymenocallis festalis[14)], H. harrisiana[14)], Narcissus incomparabilis (hybr.)[15)], Nerine falcata[16)], N. laticoma (lucida)[16)], N. masonorum[17)], N. bowdeni[10)]

** B. radulosa[38)], Buphane disticha[34)37)39)40)41)], B. fischeri[42)], Calostemma purpureum[43)], Chlidanthus fragrans[16)44)], Clivia elisabethae (miniata X nobilis)[9)45)], O. miniata[34)46)], Cooperanthes (Cooperia X Zephyranthes) hortensis[9)], Cooperia drummondii[34)], C. pedunculata[47)], Crinum asiaticum[9)48)], C. asiaticum var. japonicum[49)], C. defixum[9)50)], C. erubescens[51)52)], C. fribriatulum[51)52)], C. firmifolium[53)], C. giganteum[47)], C. latifolium[54)55)], C. laurentii[9)], C. macrantherum[11)], C. moorei[2)46)56)57)], C. powellii (moorei X longifolium)[2)12)46)57)58)], C. powellii var. harlemense[57)], C. powellii var. album[13)], C. powellii var. krelagei[13)57)], C. pratense[48)], C. scabrum[59)], C. yemense (latifolium)[9)], C. zelanium[51)52)], C–Hybr. "Ellen Bosanquet"[10)], Crythanthus pallidus[34)], Elisena longipetale,[43)] Euchris amazonica[10)], E. glandiflora[48)], Eurycles ambbinensis (sylvestris)[48)60)], Eustephia yuyuensis[43)], Galanthus elwesii[58)], G. nivalis[45)61)62)], G. nivalis var. glacilis[63)64)], G. woronowii[65)66)], Haemanthus coccineus[67)], H. katherinae[10)], H. multiflorus[2)68)69)70)], H–

Norpluvine $C_{16}H_{19}O_3N \equiv 273.32$

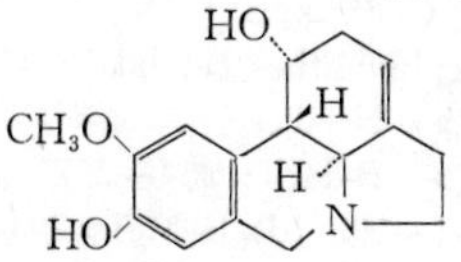

≪mp≫ 274～275° (d) (EtOH)
≪$[\alpha]_D$≫ −160° (MeOH)

≪Ph.≫
≪OP.≫ Hippeastrum aulicum var. robustum[72], Lycoris radiate[139], L. squamigera[83], Narcissus incomparabilis[10]
≪C. & So.≫ Needles or prisms
≪Sa. & D.≫ B–OMe=pluvine B–OEt: 149° B–OAc: 171～173°
≪SD.≫ 139°
≪S. & B.≫

Pseudolycorine $C_{16}H_{19}O_4N \equiv 289.32$

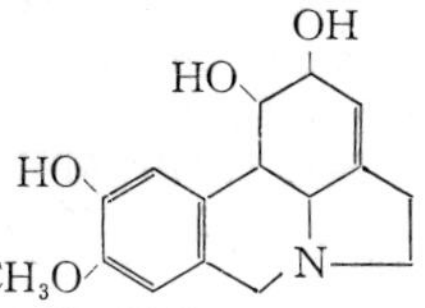

≪mp≫ 247～248° (H_2O)
≪$[\alpha]_D$≫ −62° (EtOH)

≪Ph.≫
≪OP.≫ Brunsdonna tubergenii(Brunsvigia josephinae X Amaryllis belladonna)[7], Hippeastrum aculum var. robustum[72][123], Lycoris aurea[83]*
≪C. & So.≫ Needles, Sol. in hot H_2O, Spar. sol. in Me_2CO, $CHCl_3$, MeOH
≪Sa. & D.≫ B–HCl: 266°(d) B–MeI: 250～252° (d) B–OMe: 241°(d) B–O,O,O–TriAc: 205°
≪SD.≫ 139), 140), 141), 142)
≪S. & B.≫

Golceptine $C_{16}H_{19}O_4N \equiv 289.32$

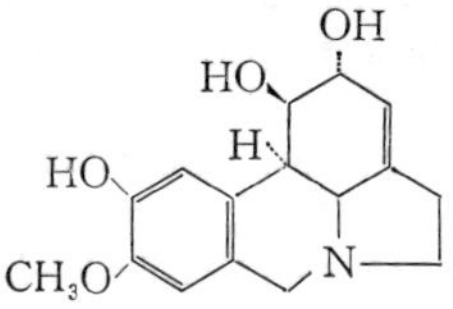

≪mp≫ 146～148° (Me_2CO)
≪$[\alpha]_D$≫ −156° ($CHCl_3$)

≪Ph.≫ CR[8] UV[8] IR[8]
≪OP.≫ Narcissus jonquilla[8][143]
≪C. & So.≫ Needles
≪Sa. & D.≫ B–Pic: 170° B–OAc: 157° B–O,O,O–TriAc: 188°
≪SD.≫ 8)
≪S. & B.≫

Hybr. "King Albert" (Puniceus X katherinae)[44][71], Hippeastrum aulicum var. robustum[72], H. bifidum[43], H. brachyandrum[73], H. candium[74], H. rutilum[73], H. reginae[48], H. vittatum (Amaryllis vittata)[44][58], H. "Orange fire," Hybr.[74], H.–Hybr. "Anna Pawlowna"[13], H. "King of Striped"[74], H.–Hybr. "Queen of the Whites"[13], H.–Hybr. "Salmon Joy"[13], Hymenocallis amancaes[23][75], H. calathiana[75], H. caymanensis[76], H. festalis[14], H. harrisiana,[14] H. littoralis[48], H. occidentalis (crassifolia)[64][76], H. rotata (lacera)[75], H. speciosa[43], Ismense(Hymenocallis)-Hybr. "Surfur Queen"[34], Leucojum aestivum[9][77][78], L. amboinensis[60], L. autumnale[9], L. vernum[79], Lycoris albiflora[80], L. aurea[15][81], L. incarnata[76][81], L. radiata[82][83][84], L. squamigera[83], Narcissus cyclamineus (hybr.)[85], N. gracilis (jonquilla X pseudonarcissus)[85][86], N. incomparabilis (hybr.)[87], N. jonquilla[8][85], N. odorus var. rugulosus[85], N. poeticus (hybr.)[79][87][88], N. pseudonarcissus (hybr.)[15][35][87][89], N. tazetta[87], N. triandrus[85], Nerine bowdenii[2][56], N. corusca (sarniensis)[81], N. falcata[16], N. flexuosa[81], N. krigeii[45][90], N. laticoma (lucida)[16], N. masanorum[17], N. sarniensis[46], N. triandrus[85], N. undulata (crispa)[44], Pancratium illyricum[81], P. longiflorum[91], P. maritimum[2][92][93], P. sickenbergeri[94], P. zeylonicum[48], Polianthes tuberosa[48], Sprekelia (Amaryllis) formosissiana[34][58], Sternbergia fischeriana[80][96], S. lutea[95], Ungernia ferganica[96], U. minor[1], U. (Lycoris) severtzovii[96][97][98][99], U. tadshicorum[96][100], U. trisphaera[101], U. victoris[96], Urceolina miniata[43], Vallota purpurea[44], Zephyranthes candida[58][102][104], Z. carinata[9], Z. citrina[9], Z. rosea[9][48], Z. texana[47], Z. tubispatha[104], Z–Hybr. "Ajax", (Candida X citrina)[10]

* L. radiata[83][139][140], L. squamigera[83], Hymenocallis harrisiana[14]

Zepyranthine $C_{16}H_{19}O_4N \equiv 289.32$

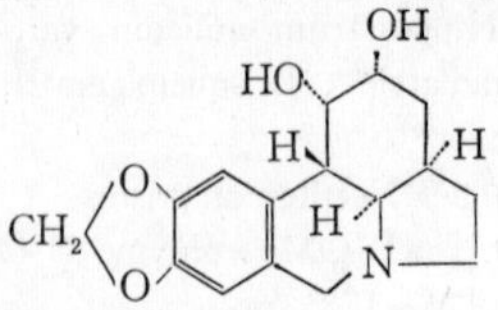

≪mp≫ 201～202° (Me_2CO)
≪[α]D≫ −43.17° ($CHCl_3$), cf +77.23° ($CHCl_3$)[103]
≪Ph.≫ CR[103] UV[102)103] IR[102)103] NMR[26)144] Mass[102)103] CD[26] pKa[102)103]
≪OP.≫ Zephyranthes candida[102)103)144]
≪C. & So.≫ Needles
≪Sa. & D.≫ B–HCl: 254～257° (cf. 247～248°) B–Pic: 194～195 (d) B–MeI: 294～295° (d) B–OAc: 68～76° B–O,O–DiAc: 206～207° (d)
≪SD.≫ 26), 102), 103), 144)
≪S. & B.≫

Pluviine $C_{17}H_{21}O_3N \equiv 287.35$

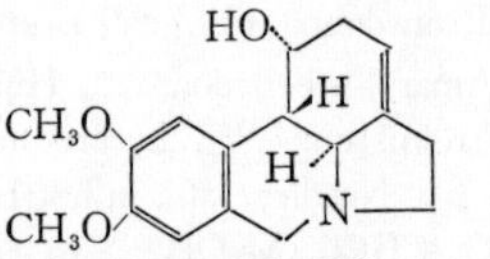

≪mp≫ 225° (MeOH or Me_2CO)
≪[α]D≫ −140° ($CHCl_3$), −170.5° (EtOH)
≪Ph.≫ UV[89)142] IR[89)142] Rf[33]
≪OP.≫ Lycoris radiata[142], L. squamigera[83], Narcissus cyclamines[85], N. incomparabilis[15)87], N. triandrus[85], N. tazetta[87], Nerine bowdeni[7]*
≪C. & So.≫ Prisms, Sol. in $CHCl_3$, EtOH, Mod. sol. in Me_2CO, AcOEt
≪Sa. & D.≫ B–HCl: 230° (d) B–HI: 237° (d) B–$HClO_4$: 260° (d) B–MeI: 259～261° (d) B–OAc: 184°
≪SD.≫ 24), 89), 142), 145), 272)
≪S. & B.≫

Methylpseudolycorine $C_{17}H_{21}O_4N \equiv 303.35$

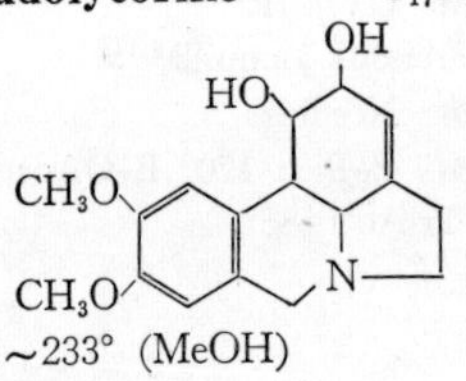

≪mp≫ 228～233° (MeOH)
≪[α]D≫ −40° (DMF)
≪Ph.≫ UV[29)141)142] IR[141] Mass[31]
≪OP.≫ Amaryllis parkeri (A. belladonna X Brunsvigia josephinae)[3], Crinum powellii (moorei X longifolium)[57)146]**
≪C. & So.≫ Prisms, Sublimation at 234～242°
≪Sa. & D.≫ B–$HClO_4$: 230° B–O,O–DiAc: 174～175° Dihydro–Der.: 192～196°
≪SD.≫ 139), 141), 142)
≪S. & B.≫

Goleptine $C_{17}H_{21}O_4N \equiv 303.35$

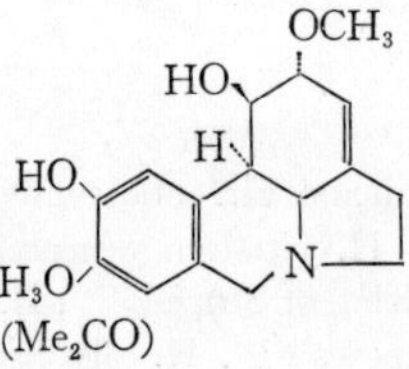

≪mp≫ 141° (Me_2CO)
≪[α]D≫ −99° ($CHCl_3$)
≪Ph.≫ CR[8] UV[8] IR[8]
≪OP.≫ Narcissus jonquilla[8)143]
≪C. & So.≫ Needles
≪Sa. & D.≫ B–$MeClO_4$: 211° (d) B–Pic: 174° B–OMe=galanthine Dihydro–Der.: 178～180°
≪SD.≫ 8)
≪S. & B.≫

* Hippeastrum–Hybr. "Queen of the Whites"[14], Nacissus pseudonarcissus[15)85)89]
** Narcissus pseudonarcissus[3)141], Nerine bowdeni[72], Sternbergia fischeriana[10]

Galanthine $C_{17}H_{23}O_4N \equiv 305.36$

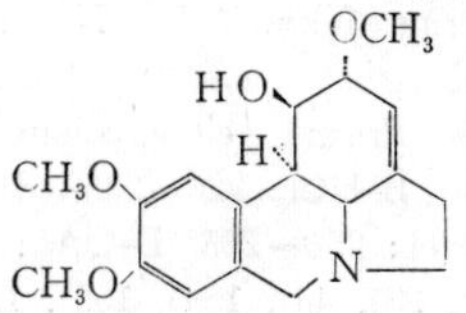

≪mp≫ 162～164° (Me_2CO or EtOH)
cf. 132～134° [$B+H_2O$]
≪[α_D]≫ −85° ($CHCl_3$), −82° (EtOH)
≪Ph.≫ UV[29)141)] IR[148)] Rf[33)]

≪OP.≫ Amaryllis belladonna[3)], A. belladonna var. purpurea[3)], Brunsdonna tubergenii (Brunsvigia josephinae X Amaryllis belladonna)[7)]*
≪C. & So.≫ Pale green prisms, Sol. in $CHCl_3$, EtOH, MeOH, AcOEt, Spar. sol. in Et_2O
≪Sa.&D.≫ B–HCl: 198～199° B–HBr: 187～208° B–HI: 165～167° B–$HClO_4$: 218° B–Pic: 199～200° (d) Dihydro–Der.: 146～148°
≪SD.≫ 5), 23), 66), 89), 147), 148)
≪S.&B.≫ BS: 149)

Ungminorine $C_{17}H_{19}O_5N \equiv 317.33$

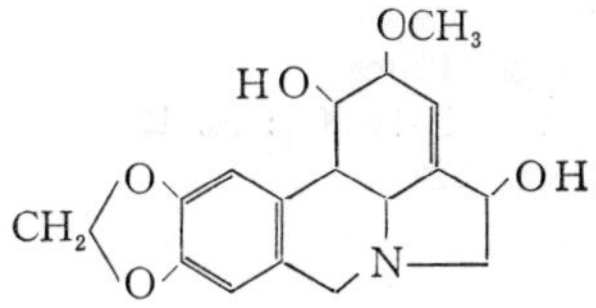

≪mp≫ 206～208° (d) (Me_2CO)
≪[α]$_D$≫ −28.8° ($CHCl_3$), −48.6° (EtOH)

≪Ph.≫ UV[1)27)] IR[1)]
≪OP.≫ Ungernia minor[1)], U. severtzovii[98)150)]
≪C. & So.≫
≪Sa. & D.≫ B–Pic: 174～175° B–MeI: 246～248°(d) B–O,O–DiAc: 173～174° Dihydro–Der.: 170～171°
≪SD.≫ 1), 27)
≪S. & B.≫

Parkacine $C_{17}H_{21}O_5N \equiv 319.35$

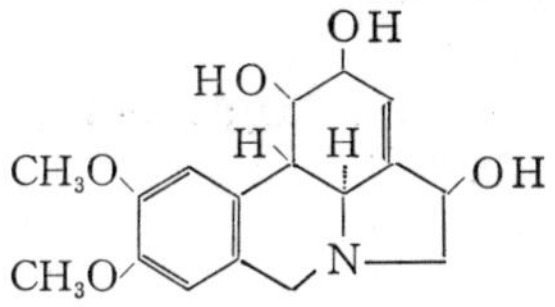

≪mp≫ 223～224° (d) (Me_2CO)
≪[α]$_D$≫ −58° ($CHCl_3$)

≪Ph.≫ UV[8)] IR[151)]
≪OP≫ Amaryllis belladonna var. blanda,[151)] Brunsvigia josephinae[151)]
≪C. & So.≫ Needles
≪Sa. & D.≫ B–$HClO_4$: 240° (d) B–Pic: 197° B–MeI: 272° (d) B–O,O,O–TriAc: 198° (d) Dihydro–Der.: 89～90°
≪SD.≫ 151)
≪S. & B.≫

Cocculine $C_{17}H_{21}O_2N \equiv 271.35$

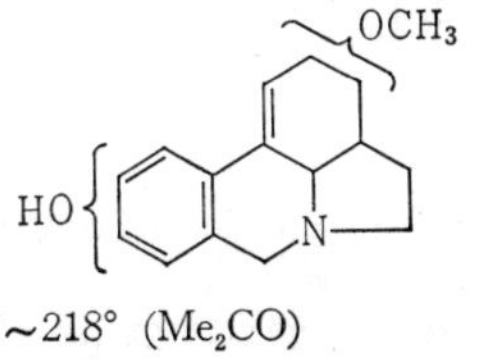

≪mp≫ 217～218° (Me_2CO)

≪[α]$_D$≫ +271,1° (MeOH)
≪Ph.≫
≪OP.≫ Cocculus laurifolius[152)]
≪C. & So.≫
≪Sa. & D.≫ B–HCl: 106～110° or 222～223° B–HNO_3: 196～197° B–OMe=cocculidine
≪SD.≫ 152), 153)
≪S. & B.≫

* Crinum defixum[9)], C. laurentii[9)], C. powelli var. harlemense[57)], C.–Hybr. "Ellen Bosanquet"[16)], Eustephia yuyuensis[43)], Galanthus woronowii[65)147)], Hippeastrum aulicum var. robustum[72)123)], H.–Hybr. "Queen of the Whites"[13)], Hymenocallis amancaes[75)], H. festalis[14)], H. harisiana[14)], Ismene (Hymenocallis)–Hybr. "Sulfur Queen"[14)], Narcissus cyclamineus[85)], N. incomparabilis[15)87)], N. poeticus[87)], N. pseudonarcissus[15)87)89)141)148)], N. tazetta[87)], Nerine bowdeni[10)], N. flexuosa[72)], Zephyranthes carinata[9)]

Falcatine $C_{17}H_{19}O_4N \equiv 301.33$

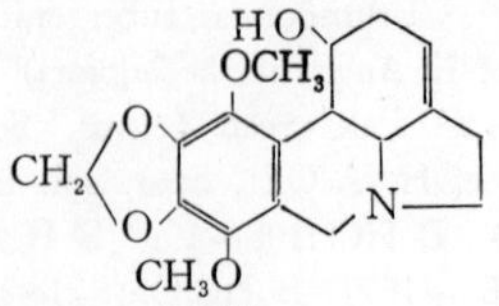

≪mp≫ 127°~128° (Et_2O)
≪$[\alpha]_D$≫ −198° ($CHCl_3$)
≪Ph.≫ UV[16,155] IR[16,30] NMR[155]
≪OP.≫ Nerine falcata[16,23], N. laticonia (lucida)[16,23]
≪C. & So.≫ Prisms, yellow colour in air or light
≪Sa. & D.≫ B–HCl: 220~235° (d) B–Pic: 182~185° B–MeI: 250~255° B–OAc: 201~202°
≪SD.≫ 16), 23), 44), 154), 155), 156)
≪S. & B.≫

Nerispine $C_{17}H_{19}O_4N \equiv 301.33$

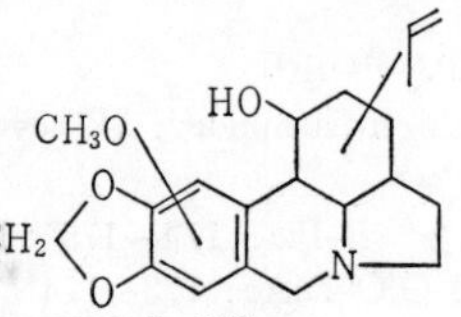

≪mp≫ 194~195° (Me_2CO)
≪$[\alpha]_D$≫ −210° ($CHCl_3$)
≪Ph.≫ Rf[33]
≪OP.≫ Nerine undulata[44], Zephyranthes tubispitha[104]
≪C. & So.≫ Plates
≪Sa. & D.≫ B–$HClO_4$: ca. 120°
≪SD.≫ 44), 72)
≪S. & B.≫

Amaryllidine $C_{17}H_{19}O_5N \equiv 317.33$

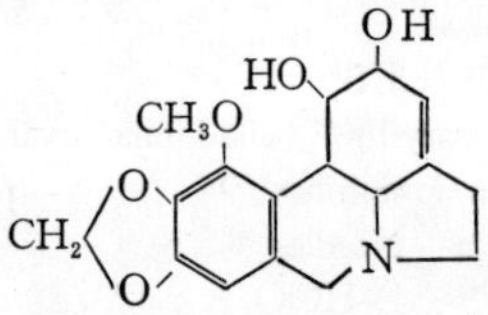

≪mp≫ 204°
≪$[\alpha]_D$≫ +64° ($CHCl_3$)
≪Ph.≫ CR[3] IR[3]
≪OP.≫ Amaryllis belladonna,[2,3] A, belladonna var. purpurea[3,7]
≪C. & So.≫
≪Sa. & D.≫ B–$HClO_4$: 134~135°
≪SD.≫ 2), 3)
≪S. & B.≫

Magnarcine (Base M)[15,48] $C_{17}H_{21}O_4N \equiv 303.35$
Methoxy isomer or diastereomer of methylpseudolycorine

≪mp≫ 221° (MeOH), 253~4° (Me_2CO)
≪$[\alpha]_D$≫ −40° (DMF)
≪Ph.≫
≪OP.≫ Galanthus nivalis[72], Narcissus incomparabilis[15], N. pseudonarcissus[15], Nerine undulate[157]
≪C. & So.≫
≪Sa. & D.≫ B–$HClO_4$: 285°
≪SD.≫ 3), 15)
≪S. & B.≫

Anhydromethyl-pseudolycorine $C_{17}H_{17}O_2N \equiv 267.31$

≪mp≫
≪$[\alpha]_D$≫ 0°
≪Ph.≫ UV[141]
≪OP.≫ Narcissus pseudonarcissus[141]
≪C. & So.≫
≪Sa.&D.≫ B–Pic: 234~252° (d)
≪SD.≫ 141)
≪S. & B.≫

Cocculidine $C_{18}H_{23}O_2N \equiv 285.37$

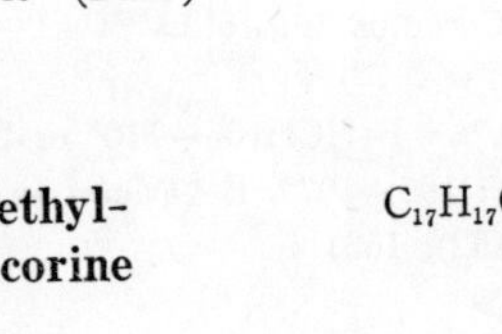

≪mp≫ 86~87° (Petr. ether)
≪$[\alpha]_D$≫ +251° ($CHCl_3$)
≪Ph.≫
≪OP.≫ Cocculus laurifolius[152]
≪C. & So.≫
≪Sa. & D.≫ B–HI: 174~175° B–HNO_3: 137~138° B–MeI: 238~239°
≪SD.≫ 152), 153)
≪S. & B.≫

Parkamine $C_{18}H_{21}O_5N \equiv 331.36$

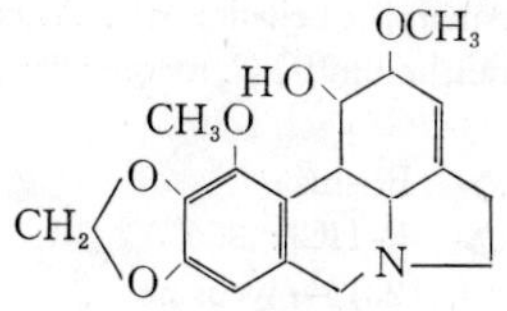

≪mp≫ 251°~253° (d) (MeOH)
≪$[\alpha]_D$≫ +69° ($CHCl_3$)
≪Ph.≫ IR[3] Rf[33]
≪OP.≫ Amaryllis parkeri (A. belladonna X Brunsvigia josephinae)[3)157], Vallota purpurea[10]
≪C. & So.≫ Leaflets
≪Sa. & D.≫ B–$HClO_4$: 245°(d) B–Pic: 178° (d) B–Picrol: 167°(d) B–MeI: 298°(d) B–$MeClO_4$: 216~217° (d)
≪SD.≫ 3), 157)
≪S. & B.≫

Narcissidine $C_{18}H_{23}O_5N \equiv 333.37$

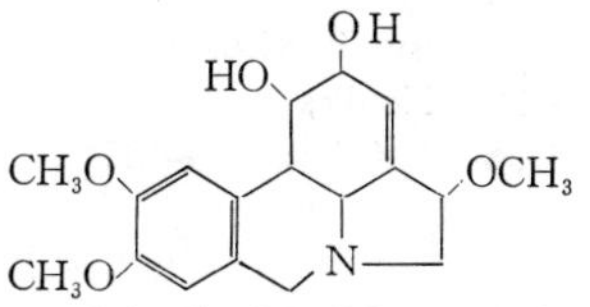

≪mp≫ 201~203° (d) (Me_2CO)
≪$[\alpha]_D$≫ −32° ($CHCl_3$)
≪Ph.≫ UV[23)31] Mass[31] Rf[33]
≪OP.≫ Crinum powelli (moorei X longifolium)[146], Hippeastrum aulicum var. robustum[72)123], H. brachyandrum[73], H–Hybr. "Queen of the whites"[113]*
≪C. & So.≫ Prisms, Spar. sol. in H_2O, Insol. in Et_2O
≪Sa. & D.≫ B–HI: 253~254° (d) B–Pic: 192° (d) B–Picrol: 204~205° (d) B–MeI: 266~267° (d) B–O,O–DiAc: 172° (d)
≪SD.≫ 23), 31), 88)
≪S. & B.≫

Bellamarine $C_{18}H_{19}O_4N \equiv 313.34$

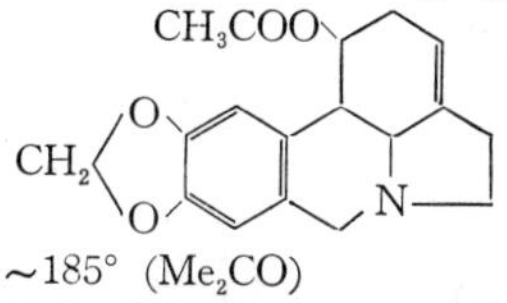

≪mp≫ 184~185° (Me_2CO)
≪$[\alpha]_D$≫ −177° ($CHCl_3$)
≪Ph.≫ CR[6] Mass[31] Rf[6)11]
≪OP.≫ Amaryllis belladonna (Brunsvigia rosea)[2)3)5], A. belladonna var. purpurea[3], Ammocharis coranica (falcata)[5)6]**
≪C. & So.≫ Needles
≪Sa. & D.≫ B–$HClO_4$: 250~261°
α–Dihydro–Der.: 193~195°
≪SD.≫ 150), cf. 5)→caranine
≪S.&B.≫

Aulamine $C_{18}H_{19}O_5N \equiv 329.34$

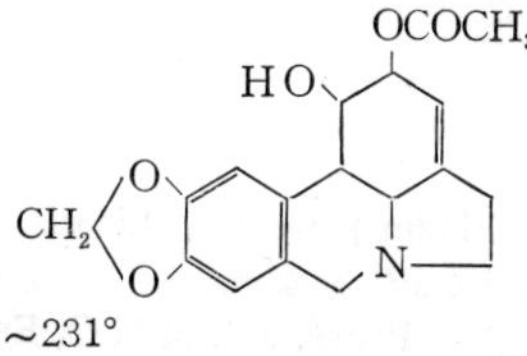

≪mp≫ 230~231°
≪$[\alpha]_D$≫ −40° ($CHCl_3$), +20° (EtOH)
≪Ph.≫
≪OP.≫ Hippeastrum aulicum var. robustum[69)72]
≪C. & So.≫
≪Sa. & D.≫ B–Picrol: 156~158°
B–OAc=diacetyllycorine
≪SD.≫ 22), 72)
≪S. & B.≫

* Hymenocallis amancaes[23], H. festalis[14], H. harrisiana[14], Ismene (Hymenocalis)–Hybr. "Sulfur Queen"[14], Narcissus cyclamineus[85], N. incomparabilis[15)87)89], N. poeticus[87)88], N. pseudonarcissus[15], N. tazetta[87], Nerine masonorum[17], N. undulata[10]

** Crinum macrantherum[11], C. moorei[57], Nerine bowdeni[72]

O^1-Acetyllycorine $C_{18}H_{19}O_5N \equiv 329.34$

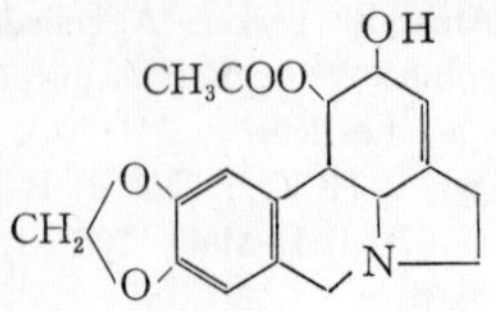

≪mp≫ 220～221° (Me_2CO)
≪[α]D≫ −96° ($CHCl_3$), −66° (EtOH)

≪Ph.≫ UV[29)56)] IR[56)] Mass[31)32)]
≪OP.≫ Amaryllis belladonna[4)], A. coranica[6)], Crinum macrantherum[11)], C. moorei[56)57)], Nerine bowdeni[56)]
≪C. & So.≫ Prisms
≪Sa. & D.≫ B–HCl: 266°
≪SD.≫ 56), 123) → lycorine
≪S. & B.≫

Poetamine $C_{18}H_{19}O_5N \equiv 329,34$

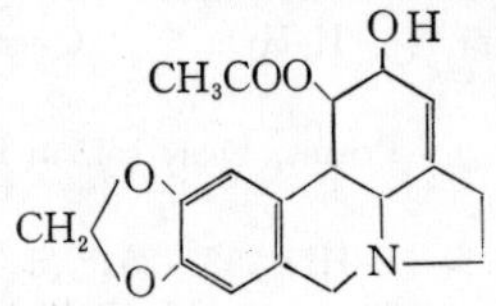

≪mp≫ 221～223°

≪[α]D≫ +100° ($CHCl_3$)
≪Ph.≫ UV[98)128)] IR[98)128)] cf. Mass[91)]
≪OP.≫ Narcissus poeticus[128)]
≪C. & So.≫ Needles
≪Sa. & D.≫ B–OAc: 215～217°
≪SD.≫ 128)
≪S. & B.≫

Jonquilline $C_{18}H_{17}O_5N \equiv 327.32$

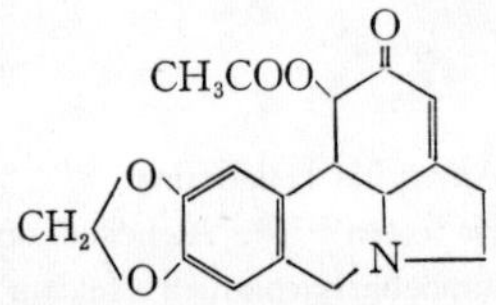

≪mp≫ 188～190° (d) (MeOH)

≪[α]D≫ −325° ($CHCl_3$)
≪Ph.≫ UV[143)] IR[143)]
≪OP.≫ Narcissus jonquilla[143)]
≪C. & So.≫ Needles
≪Sa. & D.≫
≪SD.≫ 143)
≪S. & B.≫

Nartazine $C_{20}H_{23}O_6N \equiv 373.39$

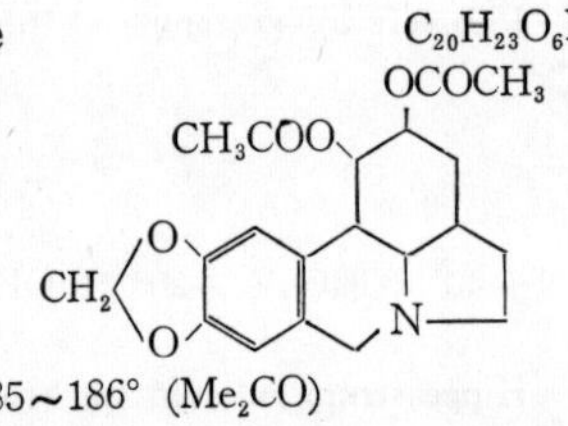

≪mp≫ 185～186° (Me_2CO)

≪]α]D≫ −120° ($CHCl_3$)
≪Ph.≫ IR[87)] Rf[33)]
≪OP.≫ Galanthus nivalis[72)], Narcissus tazetta[87)]
≪C. & So.≫ Needles
≪Sa. & D.≫ B–MeI: 209～210°
≪SD.≫ 72), 87)
≪S. & B.≫

Pseudohomolycorine $C_{17}H_{25}O_3N \equiv 291.38$

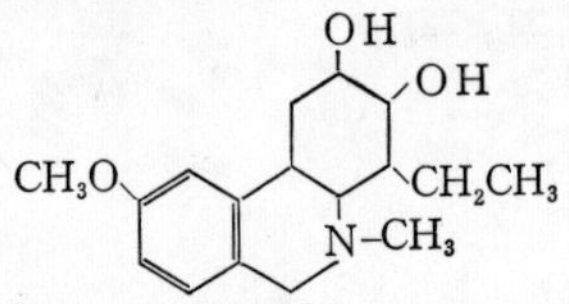

≪mp≫ 120～121° (Me_2CO)
≪[α]D≫ −98.2° ($CHCl_3$)
≪Ph.≫

≪OP.≫ Crinum powellii[7)], Licoris radiata[140)160)], L. squamigera[83)]
≪C. & So.≫ Plates, Sol. in H_2O, EtOH, Me_2CO, $CHCl_3$, Insol. in Petr.ether
≪Sa. & D.≫ B–$HClO_4$: 245° (d) B–Pic: 108～109° B–MeI: 308°(d) B–OAc: 88° B–O,O–DiAc: 95°
≪SD.≫ 140), 160)
≪S. & B.≫

Crinidine (Crinine) $C_{16}H_{17}O_3N \equiv 271.30$

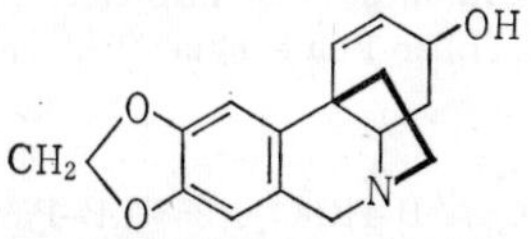

≪mp≫ 208~210° (Me_2CO), cf. 214~215° (d)
≪$[\alpha]_D$≫ −21° (EtOH), −23° ($CHCl_3$)

≪Ph.≫ UV[4,5,29,38,41,42] IR[30,41,81] NMR[170,171,173] Rf[33,41] pKa[164,165]
≪OP.≫ Brunsvigia rosea[5], Buphane disticha[41,161], B. fischeri[42], Calostemma purpureum[43], Crinum asiaticum[9,39], C. defixum[9], C. moorei[2,30,43,46,56,57,58,161,162], C. powellii var. album[13]*
≪C. & So.≫ Prisms or needles, Sol. in MeOH, Spar. sol. in $CHCl_3$
≪Sa. & D.≫ B–$HClO_4$: 135~137° B–Pic: 237~239° B–MeI: 198~199° B–OAc: 145~146° Dihydro–Der.: 212~214°
≪SD.≫ 5), 15), 44), 46), 52), 56), 69), 70), 163)~168)
≪S. & B.≫ 160), cf. 170), 171), 172)

Vittatine (*d*-Crinidine)

≪mp≫ 207~208° (Me_2CO)
≪$[\alpha]_D$≫ −38~−10° ($CHCl_3$), +38° ($CHCl_3$)
≪Ph.≫ → crinidine
≪OP.≫ Hypeastrum bifidum[43], H. vittattum[44] Hymenocallis calathina[75], Lycoris radiata[174]**

≪C. & So.≫ Prisms or needles
≪Sa. & D.≫ B–Pic: 234~235° B–MeI: 198~199°
≪SD.≫ 44), 81)
≪S. & B.≫

***d*-Epicrinine (Epivittatine)** $C_{16}H_{17}O_3N \equiv 271.30$

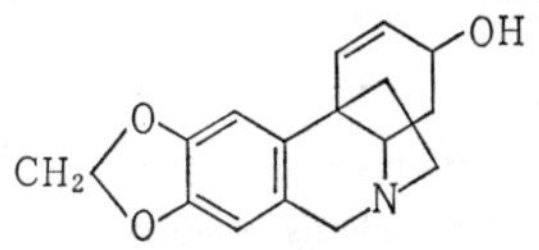

≪mp≫ 209~210° ($CHCl_3$–Me_2CO)
≪$[\alpha]_D$≫ +139° ($CHCl_3$), −142° ($CHCl_3$)

≪Ph.≫ CR[6] UV[165] Rf[6]
≪OP.≫ Ammocharis coranica[6], Nerine bowdeni[56]
≪C. & So.≫
≪Sa. & D.≫ B–Pic: 227~229° B–$HClO_4$: 245~246°
≪SD.≫ 52), 56), 164), 165), 166), 171), 175)
≪S. & B.≫

Haemaltine $C_{16}H_{17}O_3N \equiv 271.30$

≪mp≫ 174~175° (Me_2CO)
≪$[\alpha]_D$≫ +147° ($CHCl_3$)

≪Ph.≫ IR[68] Rf[70]
≪OP.≫ Amaryllis parkeri (A. belladonna X Brunsvigia josephinae)[3], Crinum powellii (moorei X longifolium)[7]***
≪C. & So.≫ Prisms
≪Sa. & D.≫ B–HI: 102° B–$HClO_4$: 246° B–MeI: 263~264° (d) B–Pic: 208~210° (d) B–OAc-$HClO_4$: 192° (d) Dihydro–Der.: 218~220°
≪SD.≫ 52), 68), 80), 166), 176)
≪S. & B.≫

---

* C. powellii var. harlemense[57], C. powellii (moorei X longifolium)[2,12,57,58], C. powellii var. krelageii[13,14], Nerine bowdenii[2], N. flexuosa[13], N. masonorum[17], N. undulata[157], Zephyranthes-Hybr.[10], Brunsvigia radulosa[38]

** L. squamigera[83], Nerine corusca (sarniensis)[81], N. flexuosa[81], Pancreatinum illyricum[81]

*** Haemanthus katherinae[10], H. multiflorus[68,80,176], Licoris albiflora[10]

Powellamine $C_{16}H_{17}O_3N \equiv 271.30$

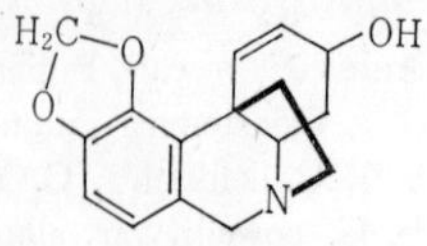

≪mp≫ 198°~200° (Me_2CO)
≪$[\alpha]_D$≫ −49° ($CHCl_3$)

≪Ph.≫ CR[146] IR[146] Rf[33]
≪OP.≫ Crinum powellii (moorei X longifolium)[146], C–Hybr. "Ellen-Bon-Sanguet"[10], Nerine flexuosa[17]
≪C. & So.≫
≪Sa. & D.≫ B–$HClO_4$: 258° B–Pic: 185~186° B–$MeClO_4$: 211° B–MeI: 224~226°
≪SD.≫ 7), 146)
≪S. & B.≫

Cripaline (Optical isomer of powellamine)
≪mp≫ 198~199°
≪$[\alpha]_D$≫ +50° ($CHCl_3$)
≪Ph.≫ IR[57]
≪OP.≫ Crinum moorei[2,57], C. powellii[2,146], C. powellii var. harlemense[57], C. powellii var. krelagei[57]
≪C. & So.≫ Needles
≪Sa. & D.≫ B–$HClO_4$: 257~258° B–Pic: 185~186°
≪SD.≫ 57)
≪S. & B.≫

Buphanisine $C_{17}H_{19}O_3N \equiv 285.33$

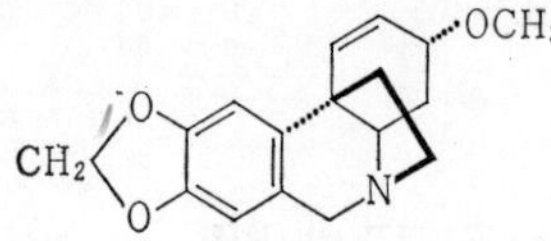

≪mp≫ 122~124° (EtOH or Et_2O)
≪$[\alpha]_D$≫ −26° (EtOH), −35° ($CHCl_3$)

≪Ph.≫ UV[4,29,41,42] IR[41,42] Mass[177] Rf[6,41]
≪OP.≫ Amaryllis coranica[6], Buphane disticha[41], B. fischeri[23,42], Crinum powellii[165], Nerine bowdeni[56], Ammocharis coranica[6]
≪C. & So.≫ Prisms
≪Sa. & D.≫ Dihydro-Der.: 218~220°, cf. 95~96°
≪SD.≫ 23), 42), 52), 165) 166), 167)
≪S. & B.≫

Epibuphanasine $C_{17}H_{19}O_3N \equiv 285.33$

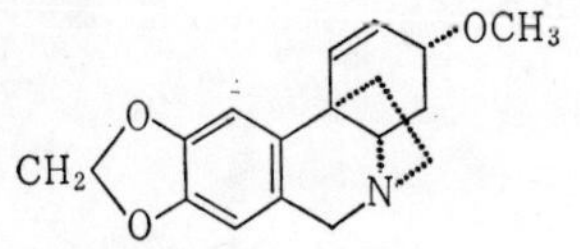

≪mp≫ 123~125° (Et_2O)
≪$[\alpha]_D$≫ +133° (EtOH), +141° ($CHCl_3$)

≪Ph.≫ CR[6] UV[6] IR[6] Rf[6]
≪OP.≫ Ammocharis coranica[6]
≪C. & So.≫ Prisms
≪Sa. & D.≫ B–$HClO_4$: 244~246° (d)
≪SD.≫ 6)
≪S. & B.≫

Buphanamine $C_{17}H_{19}O_4N \equiv 301.33$

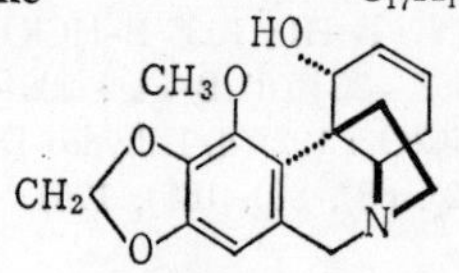

≪mp≫ 184~186° (Me_2CO), 192~194° (Me_2CO)
≪$[\alpha]_D$≫ −205° (EtOH)

≪Ph.≫ UV[4,29,41,42,69] IR[41,42,173] Rf[33,41] pKa[69]
≪OP.≫ Amaryllis belladonna[162], Buphane disticha[10,40,41], B. fischeri[42], Haemanthus toxicarius[178,179], N. flexuosa[13], N. undulata[42], Crinum-Hybr. "Ellen Bosanquet"[10], C. laurentii[10]
≪C. & So.≫ Prisms
≪Sa. & D.≫ B–HCl: 180° B–$HClO_4$: 232~234° B–HNO_3: 136~140°
≪SD.≫ 40), 42), 69), 163)
≪S. & B.≫

Powelline $C_{17}H_{19}O_4N \equiv 301.33$

≪mp≫ 197～198° (Me_2CO, C_6H_6 or AcOEt)
≪$[\alpha]_D$≫ 0° ($CHCl_3$)

≪Ph.≫ UV[4)29)165)180)] IR[156)] NMR[156)] Rf[33)] pKa[165)]
≪OP.≫ Amaryllis belladonna[162)], Calostemma purpureum[43)], Buphane disticha[161)], Crinum moorei[2)56)], C. powellii[2)12)57)58)165)], C. powellii var. album[13)]*
≪C. & So.≫ Needles
≪Sa. & D.≫ B–Pic: 223～224° (d)
B–MeI: 273～274° (d)
≪SD.≫ 23), 52), 58), 69), 156), 163), 165), 166), 167), 173)
≪S. & B.≫

Haemanthamine (Natalensine) $C_{17}H_{19}O_4N \equiv 301.33$

≪mp≫ 203～203.5° (Me_2CO)
≪$[\alpha]_D$≫ +19.7° ($CHCl_3$)

≪Ph.≫ UV[4)29)69)70)] IR[69)70)] NMR[190)] Mass[177)] Rf[33)] pKa[70)]
≪OP.≫ Brunsvigia species[69)], Buphane Species[69)], Calostemma purpureum[43)], Crinum asiaticum[9)], C. defixum[9)], C. laurentii[9)], C. powellii[12)]**
≪C. & So.≫ Prisms, Sol. in MeOH, EtOH, $CHCl_3$, Mod. sol. in Me_2CO, Spar. sol. in Et_2O
≪Sa. & D.≫ B–$HClO_4$: 231～241° (d)
B–Pic: 220～222° B–Styph: 148～151°
B–MeI: 192° B–OAc: 209° Dihydro–Der.: 225°
≪SD.≫ 52), 67), 69), 70), 71), 141), 166), 181)～186)
≪S. & B.≫ BS: 132), 135), 137), 149), 187), 188), 189)

Fiancine $C_{17}H_{19}O_4N \equiv 301.33$

≪mp≫ 239～241° (d) (Me_2CO)
≪$[\alpha]_D$≫ +75° ($CHCl_3$)

≪Ph.≫
≪OP.≫ Hippeastrum aulicum var. robustum[72)123)], Narcissus pseudonarcissus[15)], N. tazetta[87)]
≪C. & So.≫ Prisms
≪Sa. & D.≫ B–Pic: 223～225° (d)
B–OAc–$HClO_4$: 224～226°
≪SD.≫ 72), 87)
≪S. & B.≫

* C. powellii var. harlemsense[57)], C. powellii var. krelagei[13)57)], C.-Hybr. "Ellen Bosanquet"[10)], Hippeastrum–Hybr. "Orange fire"[74)], H.–Hybr. "Salmon Joy"[13)], Nerine flexuosa[13)], Zephyranthes tubispatha[104)]

** C. powellii var. album[13)], Elisena longipetale[43)], Encharis amazonica[10)], Galanthus elwesii[58)], G. nivalis[72)], Haemanthus-Hybr. "King Albert"[17)44)71)], H. katherinae[10)], H. nathalensis[67)], H. puniceus,[67)] Hippeastrum rubilum[73)], H. vittatum[44)], H.-Hybr. "Anna Pawlowna"[17)], H.-Hybr. "King of the Stripped"[74)], H.-Hybr. "Orange fire"[74)], H.-Hybr. "Salmon Joy"[13)], Hymenocallis amancaes[75)], H. calanthina[75)], H. festalis[14)], H. harrisiana[14)], H. rotata (lacera)[75)], H. speciosa[43)], Ismene (Hymenocallis)-Hybr. "Sulfur Queen"[14)], Leucojum aestivum[10)], Lycoris rodiata[174)], Narcissus canaliculatus (tazetta)[87)], N. incomparabilis[15)87)], N. jonquilla[85)143)], N. lobularis (pseudonarcissus)[15)], N. namus (pseudonarcissus)[10)], N. poeticus[87)], N. pseudonarcissus[15)87)89)], N. tazetta[87)], N. triandrus[85)], Nerine bowdenii[10)], N. corusea (sarniensis)[10)], N. flexuosa[72)], N. masonorum[17)], N. undulata[10)], Sprekalia formosissima[58)], Sternbergia lutea[10)], Urceolina miniata[43)], Vallota purpurea[44)], Zephyranthes andersonii (andersoniana)[80)], Z. candida[58)], Z. carinata[9)], Z.citrina[9)], Z.-Hybr. "Ajax" (candida X citrina)[10)]

Crinamine $C_{17}H_{19}O_4N \equiv 301.33$

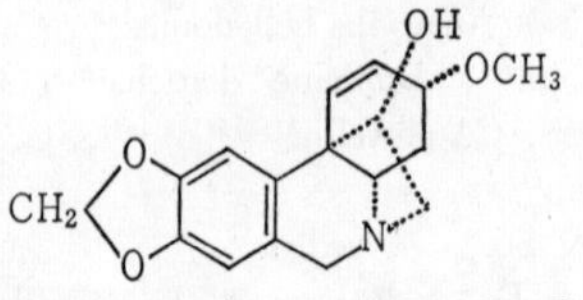

≪mp≫ 198～199° (Me_2CO or $CHCl_3$–Et_2O)
≪[α]D≫ +156.6° ($CHCl_3$), +100° (EtOH)

≪Ph.≫ UV[4,5,29,69,70] IR[30,69,70] NMR[70,190,191] Mass[177] Rf[6,11,33,41] pKa[11]
≪OP.≫ Ammocharis coranica (falcata)[5,6], Brunsvigia cooperi[37], B. radulosa[38], Crinum asiaticum[9,49], C. asiaticum var. japonicum[49]*
≪C. & So.≫ Needles, cf. bp 155～165°/3×10^{-3}, Sol. in $CHCl_3$, MeOH, EtOH, Spar. sol. in C_6H_6, Me_2CO, Et_2O
≪Sa. & D.≫ B–$HClO_4$: 201～201.5° (d)
B–Pic: 273～274° (d) B–MeI: 265°
B–OAc: 161.5～163° Dyhidro–Der.: 233°
≪SD.≫ 5), 49), 51), 52), 68), 69), 70), 164), 166), 181), 182), 190)
≪S. & B.≫ cf. 56), 164), 165), BS: 135)

Flexinine $C_{17}H_{19}O_4N \equiv 301.33$

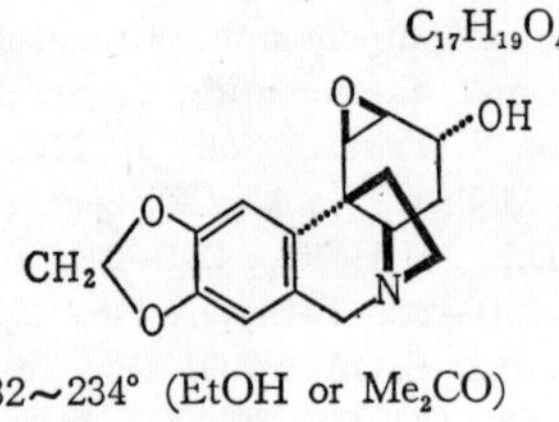

≪mp≫ 232～234° (EtOH or Me_2CO)
cf. 221～222°
≪[α]D≫ −14° ($CHCl_3$)

≪Ph.≫ UV[192] IR[81] Rf[33]
≪OP.≫ Nerine flexuosa[81,192], N. undulata[10]
≪C. & So.≫ Prisms or needles, Sol. in $CHCl_3$, Mod. sol. in MeOH, Me_2CO
≪Sa. & D.≫ B–$HClO_4$: 250° (d)
B–MeI: 223～224° B–OAc: 206～207°
≪SD.≫ 81), 192)
≪S. & B.≫

Tubispane $C_{17}H_{17}O_5N \equiv 315.31$

≪mp≫ 197～199° (Me_2CO)
≪[α]D≫ −145° ($CHCl_3$)

≪Ph.≫ IR[104]
≪OP.≫ Zephyranthes tubispatha[104]
≪C. & So.≫ Prisms
≪Sa. & D.≫
≪SD.≫ 104)
≪S. & B.≫

* C. defixum[9], C. fribriatulum[51,52], C. erubescens[51,52], C. laurentii[9], C. powelii (moorei X longifolium)[57,58,152], C. powellii var. krelagei[57], C. macrantherum[11,41], C. zelanicum[51,52], Nerine bowdeni[56,70]

Haemanthidine $C_{17}H_{19}O_5N \equiv 317.33$
(Pancratine)

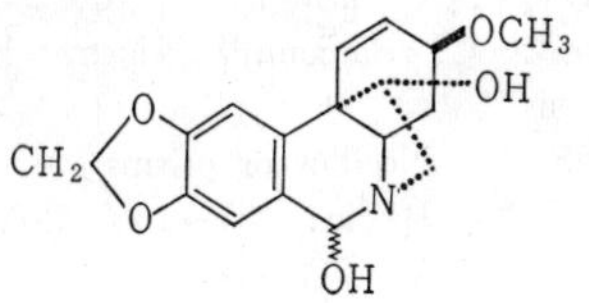

≪mp≫ 189∼190° ($CHCl_3$ or Me_2CO)
≪$[\alpha]_D$≫ −41° ($CHCl_3$)

≪Ph.≫ UV[29,193,198] NMR[190,198,199,200] Mass[177] Rf[33]
≪OP.≫ Haemanthus-Hybr. "King Albert" (puniceus X katherina),[44,71], H. multiflorus[68], H. puniceus[67], Hymenocallis karrisiona[14]*
≪C. & So.≫ Prisms, Sol. in $CHCl_3$, Me_2CO, Mod. sol. in MeOH, EtOH
≪Sa. & D.≫ B–Pic: 208°(d) B–DiAc: 200∼205°(d) Dihydro–Der.=hippawine
≪SD.≫ 13), 51), 52), 71), 184), 190), 194), 195)∼199)
≪S. & B.≫ BS: 187), 188)

Buphanidrine $C_{18}H_{21}O_4N \equiv 315.36$
(Distichine)[179,201]

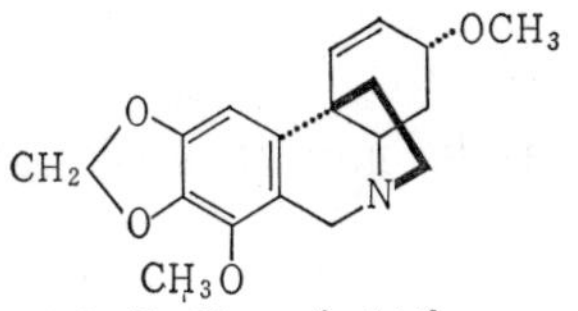

≪mp≫ 90∼92° (Et_2O), cf. 144°
≪$[\alpha]_D$≫ +0.8° (EtOH), −6.9° ($CHCl_3$)

≪Ph.≫ UV[4,29,30,41,42] IR[30,42,165] NMR[190,200] Mass[162] Rf[6,41] pKa[165]
≪OP.≫ Amaryllis coranica[4,6,23], A. belladonna X Brunsvigia gigantea[165], Brunsvigia species[69], Buphane fischeri[23,42], B. disticha[8,36,179,201], Nerine bowdeni[26]
≪C. & So.≫ Prisms or needles
≪Sa. & D.≫ B–HBr: 195∼197° or 204∼205° B–$HClO_4$: 250∼252°(d) B–Pic: 235° B–Styph: 239∼241° B–MeI: 271°
≪SD.≫ 23), 42), 52), 107), 163), 165), 166), 190), 202)
≪S. & B.≫

Hydroxycrinamine $C_{17}H_{19}O_5N \equiv 317.33$
(Epihaemanthidine)

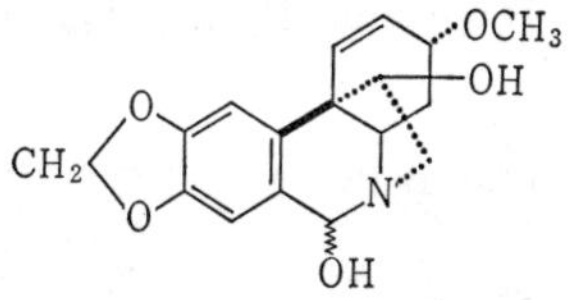

≪mp≫ 211° (Me_2CO–Petr. ether), cf. 146° [B+1/2H_2O]
≪$[\alpha]_D$≫ +46° ($CHCl_3$)

≪Ph.≫ CR[6] UV[6,12,51,52,73,197,198] IR[197] NMR[190,198,199,200] Mass[8] Rf[6]
≪OP.≫ Ammocharis coranica[6], Crinum erubescens[51,52], C. fimbriatulum[51,52], C. powellii (moorei X longifolium)[7], C. zelanicum[51,52], Haemanthus natalensis[197]
≪C. & So.≫ Prisms
≪Sa. & D.≫ B–MeI: 174° B–Methopic: 146° B–DiAc: 182∼184°
≪SD.≫ 51), 52), 166), 197), 199)
≪S. & B.≫ BS: 188)

---

* Lycoris incranata[81], L. radiata[193], L. sincarnata[81], Narcissus cyclamineus[85], Nerine corusca[81], N. flexuosa[81], Pancratium maritimum[194], P. sickenbergeri[94], Sprekelia formosissiama[58], Ungernia severtzovii[44,98], Vallota purpurea[44], Zephyranthes candida[103]

Crinamidine $C_{17}H_{19}O_5N \equiv 317.33$

≪mp≫ 235~236° (Me_2CO)
≪$[\alpha]_D$≫ +24° ($CHCl_3$)

≪Ph.≫ UV[29,46,51] IR[41] Rf[33,41]
≪OP.≫ Buphane disticha[41], Crinum moorei[46,56,192], C powellii[2,7], C. powellii var. krelagei[57], C. zelanicum[51], Haemanthus natalensis[197]*
≪C. & So.≫ Needles or prisms
≪Sa. & D.≫ B–Pic: 131~132° B–MeI: 263~265°(d) B–OAc: 205~206°
≪SD.≫ 2), 12), 46), 52), 73), 81), 173), 192)
≪S. & B.≫

Crinalbine (antipode of crinamidine)
≪mp≫ 235~236°
≪$[\alpha]_D$≫ −23° ($CHCl_3$)
≪Ph.≫
≪OP.≫ Crinum powellii var. album[12,13]

≪C. & So.≫
≪Sa.&D.≫
≪SD.≫ 12), 49)
≪S. & B.≫

Amaryllisine $C_{18}H_{23}O_4N \equiv 317.36$

≪mp≫ 225~228° (d) (MeOH)
≪$[\alpha]_D$≫ +2.4° ($CHCl_3$)

≪Ph.≫ UV[162] IR[162] NMR[162] Mass[162] ORD[162]
≪OP.≫ Brunsvigia rosea (Amaryllis belladonna)[162]
≪C. & So.≫ Prisms
≪Sa. & D.≫ B–OMe: 99~100° B–OAc: 181~182°
≪SD.≫ 162)
≪S. & B.≫

Hippawine $C_{17}H_{21}O_5N \equiv 319.35$

≪mp≫ 203°
≪$[\alpha]_D$≫ 0°

≪Ph.≫ Rf[33]
≪OP.≫ Hippeastrum-Hybr. "Queen of the Whites"
≪C. & So.≫
≪Sa. & D.≫ B–Pic: 152° B–O,O–DiAc: 168°
≪SD.≫ 13)
≪S. & B.≫

Buphanitine (Nerbowdine) $C_{17}H_{21}O_5N \equiv 319.35$

≪mp≫ 230~232° ($CHCl_3$–Me_2CO), 244~245° (d) (EtOH)
≪$[\alpha]_D$≫ −109° ($CHCl_3$)

≪Ph.≫ UV[29,41,54,56] IR[41,56,179] NMR[203] Rf[41]
≪OP.≫ Amaryllis belladonna[161,162,179], Brunsviginia-species[56,69], Buphane disticha[39,179,201], Nerine bowdeni[56]
≪C. & So.≫ Prisms, Sol. in $CHCl_3$, AcOEt, Mod. sol. in hot EtOH
≪Sa. & D.≫ B–HCl: 265~268° B–MeI: 278° (d) B–O^3–Ac: 207~209° B–O,O–DiAc: 151~152°
≪SD.≫ 69), 161), 179), 192), 203)
≪S. & B.≫

---

* Nerine bowdenii[2,56,192], N. corusca[81], N. flexuosa[81,192]

Undulatine $C_{18}H_{21}O_5N \equiv 331.36$

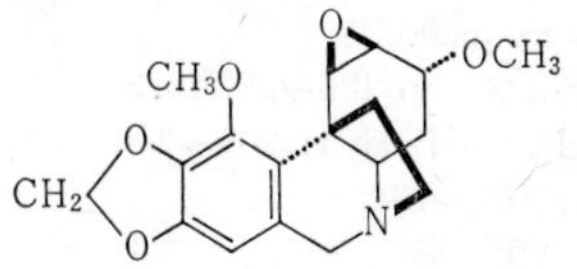

≪mp≫ 148～149° (Me_2CO)
≪[α]$_D$≫ −33° ($CHCl_3$)

≪Ph.≫ CR[44] UV[4,29,41,202] IR[4] Rf[33,41,192]
≪OP.≫ Amaryllis belladonna[4,202], Buhane disticha[41], Crinum powellii[7,57], C. powellii var. krelagei[57], C. yemense[9]*
≪C. & So.≫ Prisms or needles, Sol. in $CHCl_3$, MeOH, Me_2CO, Insol. in Et_2O
≪Sa. & D.≫ B–$HClO_4$: 229.5～230.5° (d)
B–Pic: 227～228° (d) B–MeI: 267～268° (d)
≪SD.≫ 2), 4), 44), 81), 167), 173), 192,) 202)
≪S. & B.≫

Ambelline $C_{18}H_{21}O_5N \equiv 331.36$

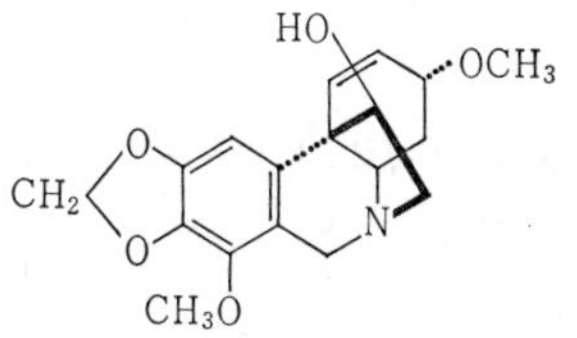

≪mp≫ 254～6° (EtOH or AcOEt)
≪[α]$_D$≫ +35±2° (EtOH)

≪Ph.≫ CR[42] UV[5,42] IR[30] Mass[56] Rf[6]
≪OP.≫ Amaryllis belladonna[2,4,5], A. belladonna var. purpurea[3], Ammocharis coranica[6], Brunsvigia species[69], Buphane disticha[161], B. fischeri[42]**
≪C. & So.≫ Plates
≪Sa. & D.≫ B–HCl: 227～230°
B–$HClO_4$: 200° (d) B–MeI: 297～298° (d)
B–OAc–Oxal: 163～164° Dyhidro–Der.: 198～199°
≪SD.≫ 5), 42), 204)
≪S. & B.≫

Acetylbowdine $C_{19}H_{23}O_6N \equiv 361.38$

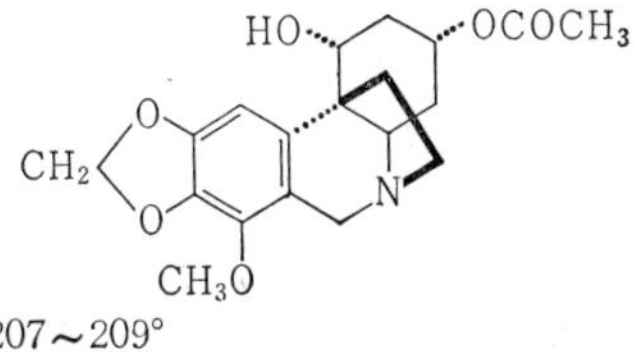

≪mp≫ 207～209°
≪[α]$_D$≫ −116° ($CHCl_3$)

≪Ph.≫ NMR[203]
≪OP.≫ Buphane disticha[41,203]
≪C. & So.≫
≪Sa. & D.≫
≪SD.≫ 41), 203)
≪S. & B.≫

***dl*-Mesembrenine** $C_{17}H_{21}O_3N \equiv 287.35$

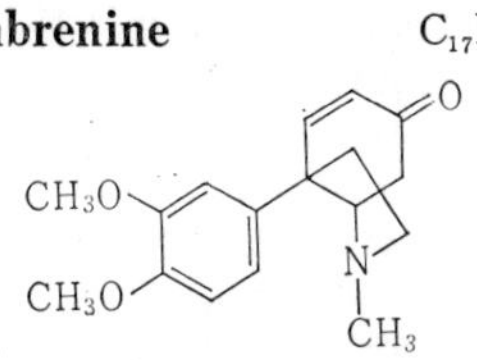

≪mp≫ 88～89°
≪[α]$_D$≫ 0°

≪Ph.≫
≪OP.≫ Mesembryanthemum tortuosum[205], Sceletium namaquense[99]
≪C. & So.≫
≪Sa. & D.≫ B–HCl: 192～193°
≪SD.≫ 99)
≪S. & B.≫

* Hippeastrum brachyandrum[73,197], Nerine bowdenii[2,56], N. corusca[81], N. flexuosa[81], N, undulata[47]

** Clivia elisabethae[9], Crinum laurentii[9], C. powellii var. karlemense[57], C. yemense[9], Hippeastrum aulicum var. robustum[72,123], H. brachyandrum[73], H.-Hybr. "Queen of the Whites"[13], Nerine bowdeni[2,56], N. corusca[81], N. flexuosa[81], N. undulata[44]

Mesembrine $C_{17}H_{23}O_3N \equiv 289.36$

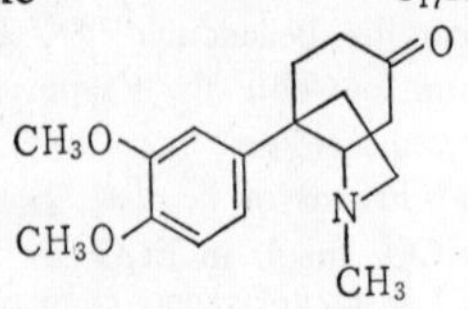

≪mp≫ cf. *dl*-compd. 183~185°
≪$[\alpha]_D$≫ −55° ($CHCl_3$)

≪Ph.≫
≪OP.≫ Mesembryanthemum tortuosum[205], Sceletium namaquense[99]
≪C. & So.≫ bp 186~190°/0.3
≪Sa. & D.≫ B–HCl: 205~206°
B–MeI: 162~164°
≪SD.≫ 99), 206), 207)
≪S. & B.≫ 208)

Masonine $C_{17}H_{17}O_4N \equiv 299.31$

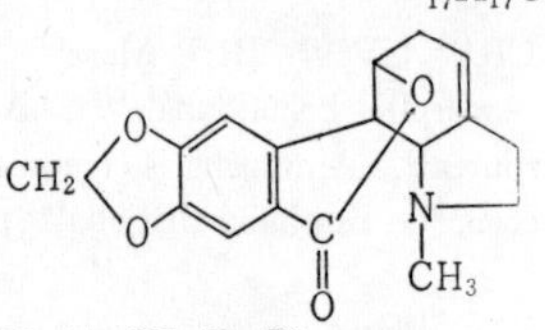

≪mp≫ 180° (AcOEt-Et_2O)
≪$[\alpha]_D$≫ +140° ($CHCl_3$)

≪Ph.≫ IR[17] Rf[33]
≪OP.≫ Galanthus nivalis[72], Narcissus jonquilla[143], N. odorus var. rugulosus[10], Nerine masonorum[17]
≪C. & So.≫
≪Sa. & D.≫ B–HI: 239° B–$HClO_4$: 281°
B–Pic: 253°
≪SD.≫ 17)
≪S. & B.≫

Demethylhomolycorine $C_{17}H_{19}O_4N \equiv 301.33$

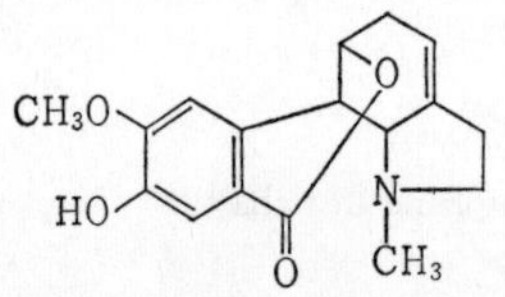

≪mp≫ 213~214° (AcOEt)
≪$[\alpha]_D$≫ +96° ($CHCl_3$)

≪Ph.≫ UV[29][139] IR[139]
≪OP.≫ Lycoris radiata[139]
≪C. & So.≫ Needles
≪Sa. & D.≫ B–OMe=homolycorine
≪SD.≫ 139) → lycorenine
≪S. & B.≫

Hippeastrine
(Trispherine)[209] $C_{17}H_{17}O_5N \equiv 315.31$

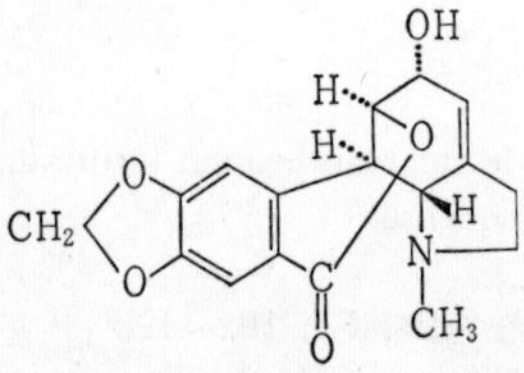

≪mp≫ 214~215° (Me_2CO)
≪$[\alpha]_D$≫ +147° (EtOH), +160° ($CHCl_3$)

≪Ph.≫ UV[29][193] IR[2] NMR[200] Mass[32] Rf[33]
≪OP.≫ Amaryllis parkeri (A. belladonna X Brunsvigia josephinae)[2][3], Crinum defixum[9], C. powellii (moorei X longifolium)[7], Clivia miniata[210]*
≪C. & So.≫ Prisms, Sol. in $CHCl_3$, MeOH, Spar. sol. in Me_2CO
≪Sa. & D.≫ B–$HClO_4$: 256° B–Pic: 256° B–MeI: 238° Dihydro–Der.: 187~188°
≪SD.≫ 2), 16), 44), 145), 200), 209), 211)
≪S. & B.≫

* Galanthus nivalis[72], Haemanthus katherinae[10], H. multiflorus[46][68], Hippeastrum bifidum[43], H. rutilum[73], H. vittatum[44], H.-Hybr. "King of the Stripped"[74], H.–Hybr. "Orange–fire"[74], Hymenocallis amancaes[23][75], H. rotata (lacera)[75], H. speciosa[43], Leucojum aestivum[10], Lycoris radiata[193], L. squamigera[83], Narcissus incomparabilis[15], N. jonquilla[85][108][143], N. odorus var. rugulosus[85], N. tazetta[87], Pancreatum maritinum[2], Sternbergia fischeriana[80], Hippeastrum-Hybr. "Queen of the Whites"[13], Nerine bowdeni[10], Ungernia frisphaera[209], U. severtzovii[92][150], U. trisphaera[209], Vallota purpurea[10]

Clivonine $C_{17}H_{19}O_5N \equiv 317.33$

≪mp≫ 199～200° (AcOEt)
≪$[\alpha]_D$≫ +41.2° ($CHCl_3$)

≪Ph.≫ UV[29,45] IR[45] Rf[33]
≪OP.≫ Clivia miniata[139]
≪C. & So.≫
≪Sa. & D.≫ B-HCl: 282～287°
B-Pic: 250～254° (d) B-OAc: 194～196°
≪SD.≫ 45), 181), 210)
≪S. & B.≫

Homolycorine (Narcipoetine) $C_{18}H_{21}O_4N \equiv 315.36$

≪mp≫ 175° (H_2O)
≪$[\alpha]_D$≫ +85° (EtOH), +93° ($CHCl_3$)

≪Ph.≫ UV[45,215] IR[29,176,214,215] NMR[200] Mass[32] Rf[33]
≪OP.≫ Clivia elisabethae (miniata X nobilis)[9], Hippeastrum bifidum[43], H. rutilum[73], H. vittatum[44], Hymenocallis calatiana[75]*
≪C. & So.≫ Prisms, Sol. in EtOH, $CHCl_3$, Me_2CO, Spar. sol. in C_6H_6, Et_2O
≪Sa. & D.≫ B-HCl: 285°, B-HI: 266°
B-$HClO_4$: 278° B-Pic 268°: B-MeI: 256°
≪SD.≫ 141), 145), 200), 212), 213), 214), 215), 216)
≪S. & B.≫ cf. 214)

Penarcine $C_{18}H_{21}O_4N \equiv 315.36$

≪mp≫ 171～172°
≪$[\alpha]_D$≫ +110° ($CHCl_3$)

≪Ph.≫ IR[176] Rf[33]
≪OP.≫ Narcissus cyclamineus[176]
≪C. & So.≫
≪Sa. & D.≫ B-HI: 256° B-$HClO_4$: 258°
B-Pic: 255° B-MeI: 256～257°
≪SD.≫ 176)
≪S. & B.≫

Ungerine $C_{18}H_{19}O_5N \equiv 329.34$

≪mp≫ 135～136° (70% EtOH)
≪$[\alpha]_D$≫ +117° ($CHCl_3$)

≪Ph.≫ UV[211] IR[211]
≪OP.≫ Ungernia (Lycoris) severtzovii[96,97,217]
≪C. & So.≫
≪Sa. & D.≫ B-HCl: 270～271°(d)
B-HBr: 287～288°(d) B-HNO_3: 260°
B-MeI: 265～266° Dihydro-Der.: 138～139°
≪SD.≫ 96), 97), 98), 211)
≪S. & B.≫

---

* H. rotata (lacern)[75], Leucojum aestivum[10], L. vernum[79], Lycoris albiflora[80], L. aurea[83], L. radiata[83,84,212], L. squamigera[83], Narcissus cyclamineus[85], N. incomparabilis[15], N. odorus var. rugulosus[85], N. poeticus[79,87,213], N. pseudonarcissus[15,87], N. tazetta[87], N. triandrus[85], N. jonquilla Hybr. Golden Sceptre[143], Pancratium martimum[94], P. sickenbergeri[94]

Nivaline $C_{18}H_{19}O_5N \equiv 329,34$

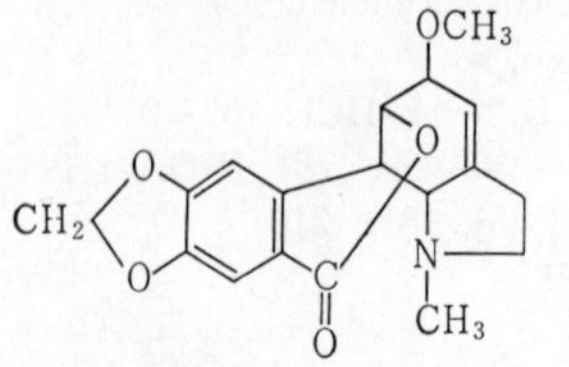

≪mp≫ 131~132° (EtOH)
≪[α]$_D$≫ +268° (EtOH)

≪Ph.≫ UV[29,45] IR[45]
≪OP.≫ Galanthus nivalis[45], Hymenocallis occidentalis (crassifolia)[45]
≪C. & So.≫
≪Sa. & D.≫
≪SD.≫ 17), 45), 81)
≪S. & B.≫

Norneronine $C_{17}H_{17}O_6N \equiv 331.32$

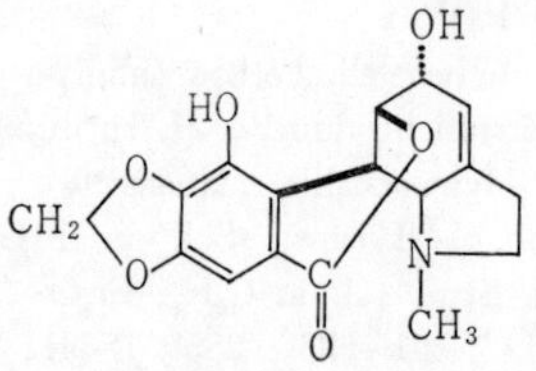

≪mp≫ 238~240° ($CHCl_3$–MeOH)
≪]α]$_D$≫ +90° (MeOH)

≪Ph.≫ CR[91] IR[91] NMR[200]
≪OP.≫ Pancratinum longiflorum[91]
≪C. & So.≫ Needles
≪Sa. & D.≫ B–OMe=neronine
B–O,O–DiAc: 248~50°(d) Dihydro–Der.: 157~158°
≪SD.≫ 91)
≪S. & B.≫

Neronine $C_{18}H_{19}O_6N \equiv 345.34$

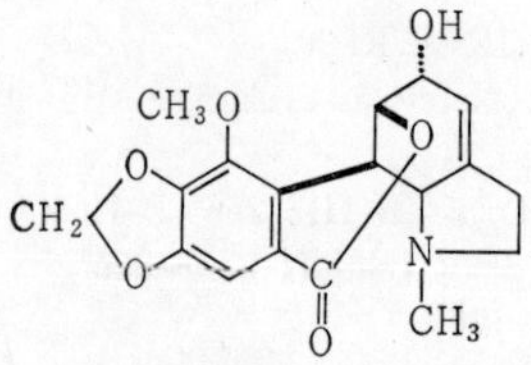

≪mp≫ 196~197° (AcOEt or C_6H_6–C_6H_{12})
≪[α]$_D$≫ +161.6° ($CHCl_3$)

≪Ph.≫ CR[91] UV[45,220] IR[45,91,220] NMR[200]
≪OP.≫ Nerine krigeii[45,90,220]
≪C. & So.≫ Prisms, hygroscopic
≪Sa. & D.≫ B–Pic: 205~209° (d)
B–MeI: 260~265° (d) B–OAc: 201~202°
≪SD.≫ 45), 91), 200), 215)
≪S. & B.≫

Candimine $C_{18}H_{19}O_6N \equiv 345.34$

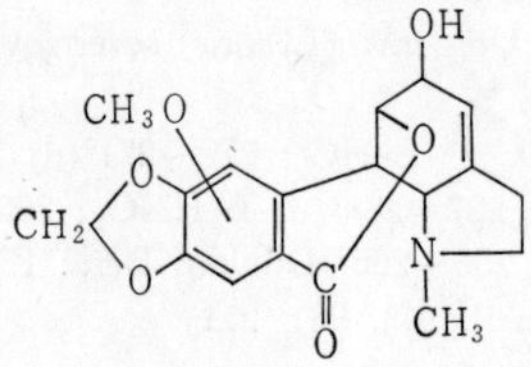

≪mp≫ 218~220° (MeOH)
≪[α]$_D$≫ +220° ($CHCl_3$)

≪Ph.≫ IR[74]
≪OP.≫ Hippeastrum candidum[74]
≪C. & So.≫ Prisms
≪Sa. & D.≫ B–$HClO_4$: 177~179° or 262° (d)
B–Pic: 220°(d) B–OAc: 239~241° (d)
≪SD.≫ 74)
≪S. & B.≫

Albomaculine $C_{19}H_{23}O_5N \equiv 345.38$

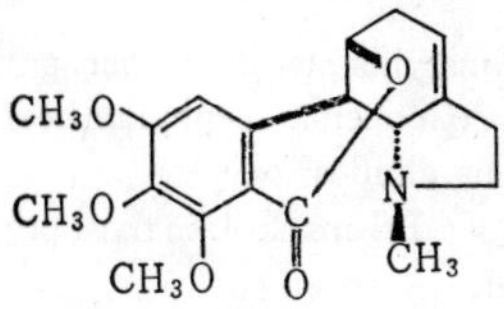

≪mp≫ 180~181° (AcOEt)
≪$[\alpha]_D$≫ +71.1° ($CHCl_3$)

≪Ph.≫ UV[45] IR[45] NMR[91]
≪OP.≫ Haemantus albmaculatus[45]
≪C. & So.≫
≪Sa.&D.≫ B–$HClO_4$: 285~289° (d)
B–Pic: 189~198° B–Methopic: 244~246° (d)
≪SD.≫ 45), 91), 181), 200)
≪S. & B.≫

Urminine $C_{19}H_{23}O_5N \equiv 345.38$

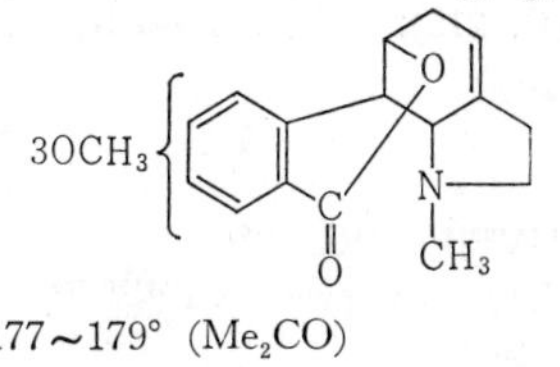

≪mp≫ 177~179° (Me_2CO)
≪$[\alpha]_D$≫ −40° ($CHCl_3$)

≪Ph.≫ CR[43] IR[43]
≪OP.≫ Amaryllis parkeri (A. belladonna X Brunsvigia josephinae)[3], Urceoline miniata[43]
≪C. & So.≫ Prisms, Sol. in MeOH, $CHCl_3$, Mod. sol. in Me_2CO
≪Sa. & D.≫
≪SD.≫ 43)
≪S. & B.≫

Clivimine $C_{43}H_{43}O_{12}N_3 \equiv 765.78$

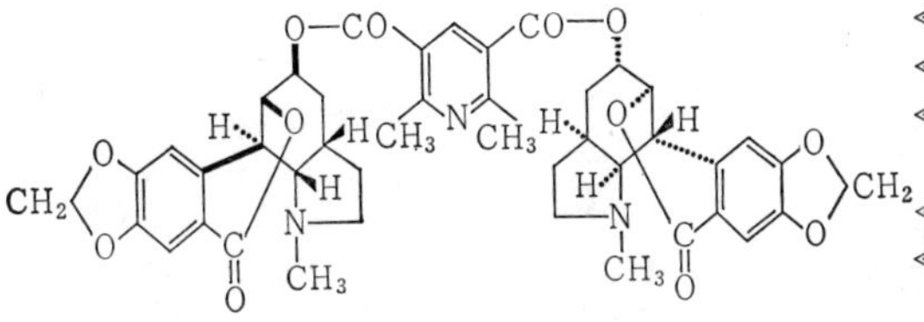

≪mp≫ 279~281°
≪$[\alpha]_D$≫ +26° ($CHCl_3$)

≪Ph.≫
≪OP.≫ Clivia miniata[10][210]
≪C. & So.≫
≪Sa. & D.≫ B–$3HClO_4$: 305~307° B–3Pic: 211~213° B–3MeI: 292°~294°
≪SD.≫ 210)
≪S. & B.≫

Oduline $C_{17}H_{19}O_4N \equiv 301.33$

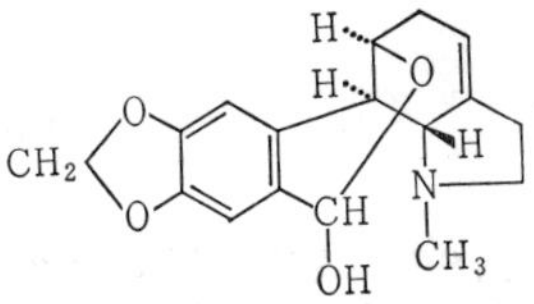

≪mp≫ 168° (Me_2CO)
≪$[\alpha]_D$≫ +239° ($CHCl_3$)

≪Ph.≫ NMR[33][231]
≪OP.≫ Narcissus incomparabilis[15], N. jonquilla[85][108][143], N. odorus var. rugulosus[85], Hippeastrum vittatum[231]
≪C. & So.≫ Prisms, Sublimation at 120°/8×10^{-3}
≪Sa. & D.≫ B–Pic: 221° (d) B–MeI: 247°
≪SD.≫ 17), 33), 85), 231)
≪S. & B.≫

Radiatine $C_{17}H_{19}O_5N \equiv 317.33$

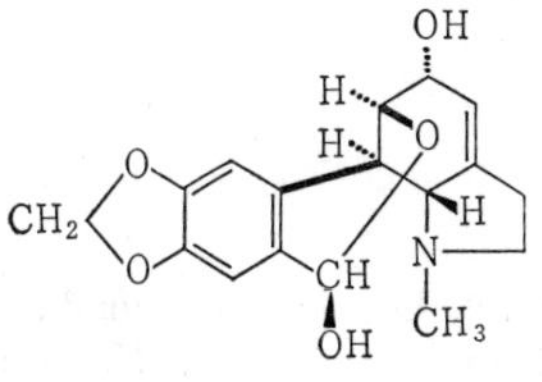

≪mp≫
≪$[\alpha]_D$≫ B–OEt: +237.4° (EtOH)

≪Ph.≫ UV[84] IR[84]
≪OP.≫ Lycoris radiata[84]
≪C. & So.≫ Amorph
≪Sa. & D.≫ B–OEt: 171~172.5 (d)
B–OEt–HCl: 259° (d)
B–OEt–Oxime: 194~195.5°
≪SD.≫ 84)
≪S. & B.≫

Lycorenine $C_{18}H_{23}O_4N \equiv 317.33$

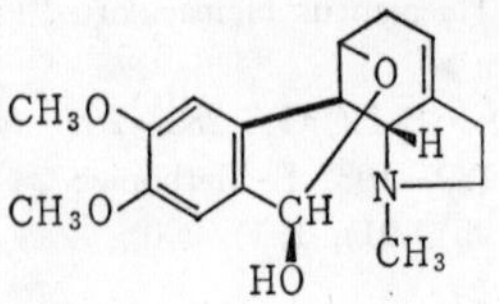

≪mp≫ 202° (Me_2CO)
≪$[\alpha]_D$≫ +179.6° (EtOH)

≪Ph.≫ UV[29)45)191)215)] NMR[191)200)228)] Mass[32)] Rf[33)]
≪OP.≫ Amaryllis belladonna var. purpurea[3)], Cooperanthus (Cooperia, Zephyranthes) hortensis[9)], Haemanthus albiflos[62)]*
≪C. & So.≫ Prisms, Sol. in most org. solv. except Petr. ether.
≪Sa. & D.≫ B–HCl: 146~147°(d) B_2–H_2PtCl_6: 210° B–MeI: 260°(d) B–OAc: 180° B–O,O–DiAc: 183~184° B–Pic: 162° (d) B–Oxime·HCl: 258° (d)
≪SD.≫ 17), 24), 81), 139), 140), 145), 200), 212), 214), 215), 216), 221)~227)
≪S. & B.≫

Unsevine $C_{18}H_{21}O_5N \equiv 331.36$

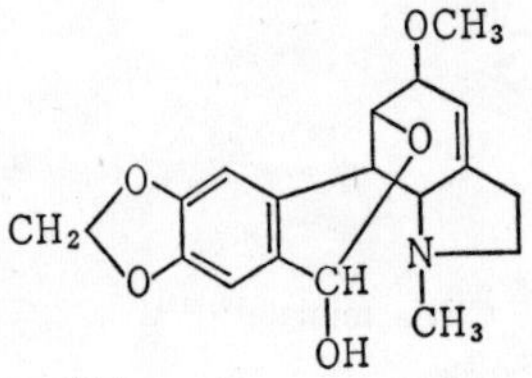

≪mp≫ 173~174°
≪$[\alpha]_D$≫ +163° ($CHCl_3$)

≪Ph.≫ UV[98)211)] IR[211)229)]
≪OP.≫ Ungernia severtzovii[98)150)229)]
≪C. & So.≫
≪Sa. & D.≫ B–HBr: 180~183° B–Oxal: 195~196° B–MeI: 249~250° Dihydro–Der.: 154~155°
≪SD.≫ 98), 150), 211)
≪S. & B.≫

Krigenamine $C_{18}H_{21}O_6N \equiv 347.36$

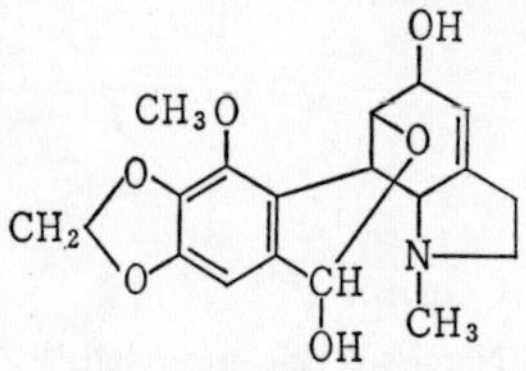

≪mp≫ 210~211° (Me_2CO, EtOH or AcOEt)
≪$[\alpha]_D$≫ +210° ($CHCl_3$)

≪Ph.≫ UV[29)90)220)] IR[90)220)] NMR[228)]
≪OP.≫ Nerine krigeii[90)220)]
≪C. & So.≫ Needles
≪Sa. & D.≫ B–$MeClO_4$: 248~249° (d) B–MeI: 235~237° (d)
≪SD.≫ 90), 220)
≪S. & B.≫

Krigeine $C_{18}H_{21}O_6N \equiv 347.36$

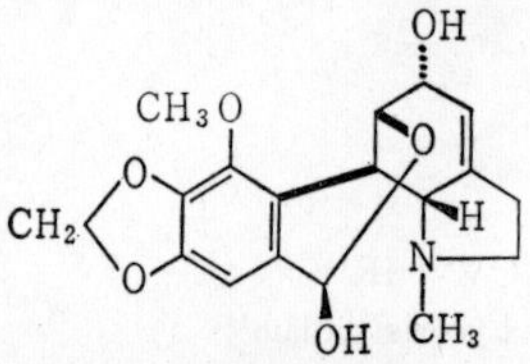

≪mp≫ 209~210° (d) (Me_2CO)
≪$[\alpha]_D$≫ +234° ($CHCl_3$)

≪Ph.≫ UV[45)220)] IR[45)] NMR[200)228)]
≪OP.≫ Nerine krigeii[45)90)]
≪C. & So.≫
≪Sa. & D.≫
≪SD.≫ 36), 44), 45), 200)
≪S. & B.≫

---

* H. albomaculatus[45)], H. katerinae[10)], Hippeastrum brachyandrum[73)], Leucojum aestivum[9)], Lycoris albiflora[80)], L. aurea[83)], L. radiata[83)84)140)], L. squamigera[83)], Nacissus cyclamineus[85)], N. incomparabilis[15)], N. jonquilla[85)143)], N. poeticus[79)87)], N. pseudonarcissus[15)87)89)], N. triandulus[85)], Nerine bowdeni[72)], Zephyranthus citrina[9)]

Nerinine $C_{19}H_{25}O_5N \equiv 347.40$

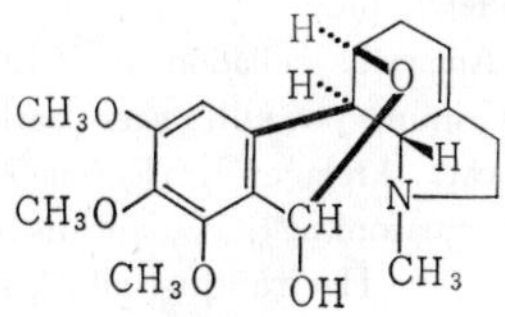

≪mp≫ 209～210° (Me_2CO)
≪$[\alpha]_D$≫ +155° ($CHCl_3$)

≪Ph.≫ UV[45),191)] NMR[191),200),228)] Rf[33)]
≪OP.≫ Hymenocallis amancaos[75)], H. calathina[75)], H. speciosa[43)], Nerine sarniensis[46)], Zephylanthus candida[58),102),103)]
≪C. & So.≫ Plates
≪Sa. & D.≫ Dihydro-Der.: 52～53°
≪SD.≫ 46), 181), 191), 200), 230)
≪S. & B.≫

Urceoline $C_{19}H_{25}O_5N \equiv 347.40$

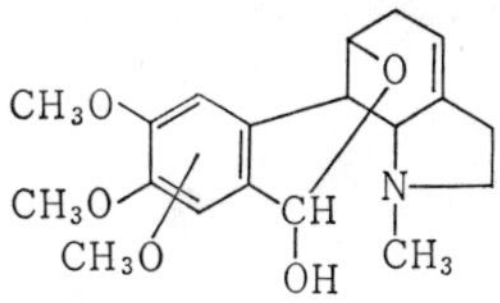

≪mp≫ 189～190° (Me_2CO)
≪$[\alpha]_D$≫ +180° ($CHCl_3$)

≪Ph.≫ CR[43)] IR[43)]
≪OP.≫ Hippeastrum brachyandrum[73)], Urceolina miniata[43)]
≪C. & So.≫ Prisms
≪Sa. & D.≫ B-Pic: 188° (d)
≪SD.≫ 43), 58)
≪S. & B.≫

Nerscine $C_{18}H_{23}O_3N \equiv 301.37$
(**Deoxylycorenine**)[17)]

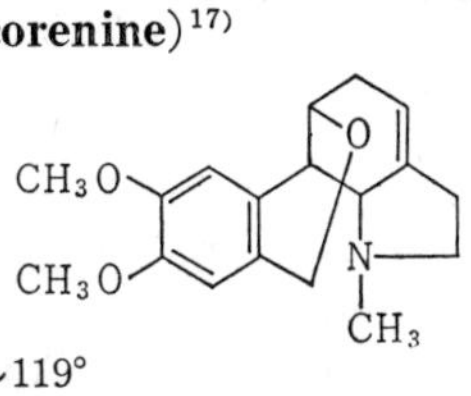

≪mp≫ 118～119°
≪$[\alpha]_D$≫ +92° ($CHCl_3$)

≪Ph.≫
≪OP.≫ Crinum powellii (moorei X longifolium)[7)], Hippeastrum aulicum var. robstrum[72)], Nerine corusia (sarniensis)[81)], N. flexuosa[72)]
≪C. & So.≫
≪Sa. & D.≫ B-HI: 247° (d) B-$HClO_4$: 212～213° (d)
≪SD.≫ 81), 215) → lycorenine
≪S. & B.≫

Macronine $C_{18}H_{19}O_5N \equiv 329.34$

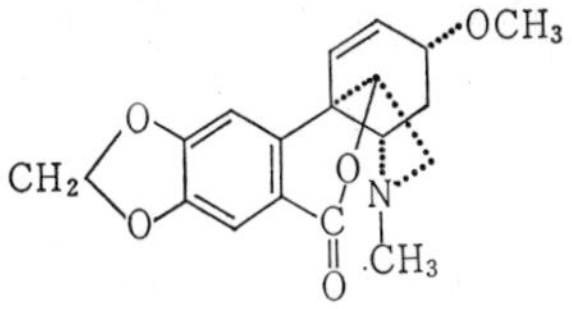

≪mp≫ 203～205° (Me_2CO)
≪$[\alpha]_D$≫ +413° $(CHCl)_3$, +380° (EtOH)

≪Ph.≫ UV[11),232)] IR[11),232)] NMR[11),232)] Rf[11)] pKa[11)]
≪OP.≫ Crinum macrantherum[11),232)]
≪C. & So.≫ Prisms
≪Sa. & D.≫
≪SD.≫ 11), 232)
≪S. & B.≫

Tazettine $C_{18}H_{21}O_5N$≡331.36
(Base VIII, Sekisamine,
Ungernine, Sekisanoline)[82,212,227]

≪mp≫ 210~211° (Me_2CO)
≪$[\alpha]_D$≫ +150.4° ($CHCl_3$), +122° (EtOH)

≪Ph.≫ UV[29,76,198,237] IR[30,242] NMR[190,198,200] Mass[177] Rf[33] pKa[227]
≪OP.≫ Amaryllis belladonna[2], Chlidanthus fragrans[44], Crinum powellii var. harlemense[57], C. powellii var. krelagei[57], Elisena longipetala[43], Eucharis amazonica[10], Galanthus elmesii[58], G. nivalis[45,61,233], Haemanthus albiflos[62,67]*
≪C. & So.≫ Prisms, Sol. in $CHCl_3$, MeOH, EtOH, Spar. sol. in Et_2O, Petr. ether
≪Sa. & D.≫ B–HCl: 217° B–$HClO_4$: 105~108° (d) B–Pic: 205~208° (d) B–MeI: 220° (d) B–OAc: 125~126°
≪SD.≫ 77), 116), 140), 145), 190), 193), 194), 195), 196), 212), 223), 227), 232), 234), 237)~253)
≪S. & B.≫

Isotazettine** $C_{18}H_{21}O_5N$≡331.36

≪mp≫
≪$[\alpha]_D$≫ +66.4° ($CHCl_3$)

≪Ph.≫
≪OP.≫ Leucojum aestivum[77]
≪C. & So.≫ Amorph or yellow glass
≪Sa. & D.≫ B–HCl: 224~225° B–HBr: 230~233°
≪SD.≫ 77)
≪S. & B.≫

Criwelline $C_{18}H_{21}O_5N$≡331.36

≪mp≫ 205~206° (211~212°)
≪$[\alpha]_D$≫ +220° ($CHCl_3$)

≪Ph.≫ UV[29,51,52,198] IR[52] NMR[190,198,200] Mass[177] Rf[11,33]
≪OP.≫ Crinum erubescens[51], C. fimbriatum[51], C. macrantherum[11], C. powellii (moorei X longiflorum)[2,12], C. powellii var. album[13], C. zeylanicum[51], Galanthus nivalis[72]
≪C. & So.≫
≪Sa. & D.≫ B–HI: 228~229° (d) B–$HClO_4$: 217~218° (d) B–Pic: 233° B–OMe: 127~128°
≪SD.≫ 2), 51), 52), 166), 190), 200)
≪S. & B.≫

---

* H. albomaculatus[45], Hippeastrum bifidum[43], H. vittatum[44,46,75], H.-Hybr. "King of the Stripped"[74], Hymenocallis amancaes[75], H. calathina,[75] H. caymensensis[76], H. festalis[14], H. harrisiana[14], H. littoralis[76], H. occidentalis (crassifolia)[45,76], H. rotata (lacera)[75], H. speciosa[43], Ismene (Hymenocallis)-Hybr. "Sulfur Queen"[14], Leucojum aestivum[76], Lycori saurea[83], L. radiata[83,140], L. squamigera[83,150], Narcissus canaliculatus (tazetta)[87], N. cycleamineus[85], N jonquilla[85,108], N. odorus var. rugulosus[85], N. pseudonarcissus[15], N. tazetta[87,234], N, triandrus[85], Nerine corusca (sarniensis)[80], N. flexuosa[81], N. masonorum[17], N. sarniensis[46], Pancreatium mantimum[2,93], P. sickenbergii[94], Sprekelia formossima[58], Sternbergia lutea[10], Ungernia ferganica[96], U. (Lycoris) severtzovii[96,98,235,236], Urceolina miniata[43], Zephyranthes candida,[58,102,103], Z. carinata[9], Z.-Hybr. "Ajax" (candida X citrina)[10]

** Isotazettine has the same structure as squamigerine described later but both are stereoisomers each other.

Squamigerine $C_{18}H_{21}O_5N \equiv 331.36$

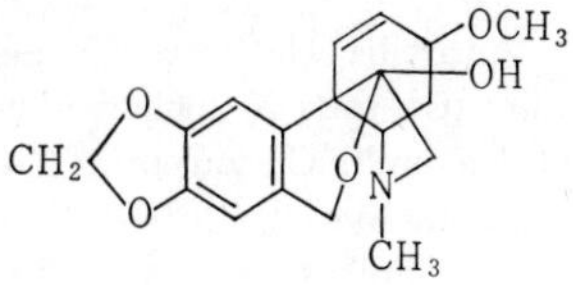

≪mp≫ 260°
≪$[\alpha]_D$≫ +165° ($CHCl_3$)

≪Ph.≫ UV[83] IR[83] Rf[83]
≪OP.≫ Lycoris squamigera[83]
≪C. & So.≫
≪Sa. & D.≫
≪SD.≫ 83)
≪S. & B.≫

Nivaridine $C_{17}H_{19}O_2N \equiv 269.33$

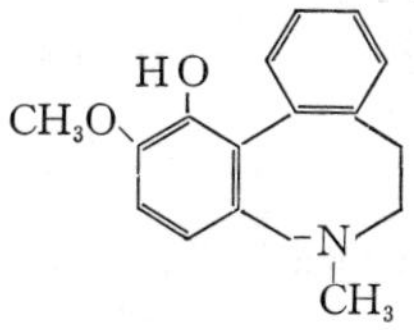

≪mp≫ 206~208° (C_6H_6)
≪$[\alpha]_D$≫

≪Ph.≫ UV[64] IR[64] Rf[64]
≪OP.≫ Galanthus nivalis var. gracilis[63][64]
≪C. & So.≫ Needles
≪Sa. & D.≫ B-HBr: 238~239° B-HI: 243~244° B-MeI: 252~253° B-OAc: 107~109.5°
≪SD.≫ 64)
≪S. & B.≫ cf. 64)

Narwedine $C_{17}H_{19}O_3N \equiv 285.33$

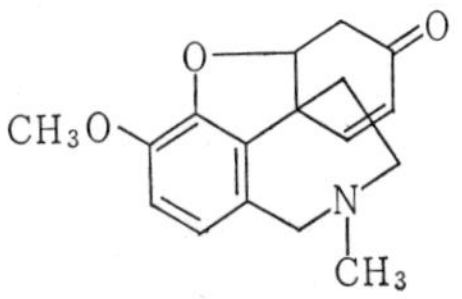

≪mp≫ 188~190° (Me_2CO)
≪$[\alpha]_D$≫ +100° ($CHCl_3$)

≪Ph.≫ CR[44] UV[254] IR[15][254] NMR[255] Rf[33]
≪OP.≫ Galanthus nivalis[72], Narcissus cyclamineus[15], N. incomparabilis[15], N. odorus var. rugulosus[10], Ungernia severtzovii[98][150]
≪C. & So.≫ Prisms, Sol. in MeOH, $CHCl_3$, Me_2CO
≪Sa. & D.≫ B-Pic: 123° B-MeI: 195~196°(d) B-Semicarb: 240°~241° (d)
≪SD.≫ 12), 15), 130), 254), 255)
≪S. & B.≫ 130), 254), BS: 130), 254), 256)

Chlidanthine $C_{17}H_{21}O_3N \equiv 287.35$
(Chlidanthamine ?)[143]

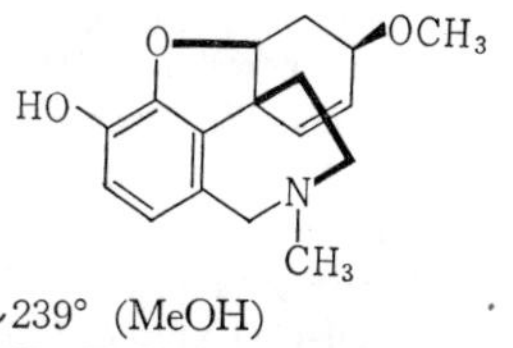

≪mp≫ 238~239° (MeOH)
≪$[\alpha]_D$≫ −140° (EtOH)

≪Ph.≫ NMR[143]
≪OP.≫ Chlidanthus fragrans[44][198], Haemanthus multiflorus[68], Hippeastrum caulium var. robustum[72]
≪C. & So.≫ Plates, Sol. in MeOH, EtOH, Spar. sol. in $CHCl_3$, Me_2CO
≪Sa. & D.≫ B-HBr: 213° B-$MeClO_4$: 255~256°(d) B-MeI: 263~264° (d)
≪SD.≫ 44), 143), 181)
≪S. & B.≫

Galanthamine $C_{17}H_{21}O_3N \equiv 287.35$
(Lycoremine)

≪mp≫ 127~129° (Me_2CO)
≪$[\alpha]_D$≫ −122° (EtOH)

≪Ph.≫ UV[29,64,141] IR[64,254] NMR[143] X-ray[263] Rf[33]
≪OP.≫ Amaryllis belladonna var. purpurea[3], Cooperanthes (Cooperia X Zephyranthes) hortensis[9], Crinum defixum[9], C. laulentii[10], C. moorei[57], C. powellii var. krelagei[13,57]*
≪C. & So.≫ Prisms
≪Sa. & D.≫ B–HCl: 254~255° (d) B–HI: 260~261° B–$HClO_4$: 225~227° (d) B–MeI: 288° (d) B–OAc: 129~130° Dihydro–Der. = lycoramine
≪SD.≫ 15), 64), 130), 140), 141), 160), 227), 237), 254), 255), 257), 258)~265)
≪S. & B.≫ 130), 149), 254), 256)

Epigalanthamine $C_{17}H_{21}O_3N \equiv 287.35$

≪mp≫ 190°
≪$[\alpha]_D$≫ −222° (MeOH)

≪Ph.≫ IR[149]
≪OP.≫ Lycoris radiata[266], L. squamigera[63,64]
≪C. & So.≫
≪Sa. & D.≫ B–HCl: 234° B–$HClO_4$: 237° B–Pic: 146° B–MeI: 275°
≪SD.≫ 12), 141), 254)
≪S. & B.≫

Irenine $C_{17}H_{23}O_3N \equiv 289.36$

≪mp≫ 128° (AcOEt)
≪$[\alpha]_D$≫ +120° ($CHCl_3$)

≪Ph.≫ IR[15]
≪OP.≫ Narcissus incomparabilis[15]
≪C. & So.≫ Prisms
≪Sa. & D.≫
≪SD.≫ 15), 130), 254)
≪S. & B.≫

Lycoramine $C_{17}H_{23}O_3N \equiv 289.36$

≪mp≫ 122~124° (Me_2CO)
≪$[\alpha]_D$≫ −98° (EtOH)

≪Ph.≫
≪OP.≫ Crinum powellii (moolei X longifolium)[7,141], Hymenocallis amanceas[23], Lycoris aurea[83], L. radiata[83], Narcissus cyclamineus[85], N. pseudonarcissus[141]
≪C. & So.≫ Prisms or plates
≪Sa. & D.≫ B–$HClO_4$: 139° B–Pic: 108~109° B–H_2PtCl_6: 245° B–MeI: 308°
≪SD.≫ 141), 237), 257), 259), 261)
≪S. & B.≫ cf. 265)

---

* C. yemense (latifolium)[43], Galanthus elwesii[58], G. nivalis[64,72], G. nivalis var. gracilis[63], G. woronowii[88,227], Haemanthus katerinae[10], Hippeastrum rutium[73], H.-Hybr. "Salmon Joy"[13], H.-Hybr. "Orange-fire"[74], Hymenocallis amancaes[75], H. calathina[75], H. rotata (lacera)[75], Leucojum aestivum[9,77,78], L. vernum[79], Lycoris albiflora[80], L. aurea[81], L. illyricium[81], L. incarnata[81], L. radiata[84,167,257], L. squamigera[83], Narcissus gracilis (jonquilla X pseudonarcissus)[85,143], N. incomparabilis[15,87], N. jonquilla[85,143], N. lobularis (pseudonarcissus)[15], N. odorus var. rugulosus[85], N. poeticus[87], N. pseudonarcissus[15,87,89], N. tazetta[87], N. triandrus[85], Nerine corusca[81], N. flexuosa[72,81], N. undulata[10], Pancratium illyricum[81], Sternbergia fischeriana[80], Ungernia victoris[96], U. severtzovii[44,98], Vallota purpurea[44], Zephyranthes andersonii[80], Z. rosea[9]

Coccinine $C_{17}H_{19}O_4N \equiv 301.33$

≪mp≫ 162~163°
≪$[\alpha]_D$≫ −188.8° (EtOH)

≪Ph.≫ UV[4,16,29,67] IR[30,38] Mass[177]
≪OP.≫ Haemanthus amarylloides[67], H. allomaculatus[67], H. coccineus[67], H. trigrinus[38,45]
≪C. & So.≫
≪Sa. & D.≫ B–$HClO_4$: 254~255° (d) B–Pic: 155~160° (d) B–MeI: 223~224° B–OMe: 122~126°
≪SD.≫ 38)
≪S. & B.≫ cf. 38)

Montanine $C_{17}H_{19}O_4N \equiv 301.33$

≪mp≫ 88~89°
≪$[\alpha]_D$≫ −98° ($CHCl_3$)

≪Ph.≫ UV[4,16,29,67] IR[30] Mass[177]
≪OP.≫ Haemanthus amarylloides[67], H. coccineus[67], H. montanus[67], H. multiflorus[69,70], H. trigrinus[38], Hippeastrum aulicum var. robustum[72]
≪C. & So.≫ Needles or prisms
≪Sa. & D.≫ B–$HClO_4$: 249~250° (d) B–Pic: 225~226° (d) B–Oxal: 227~229° (d) B_2–H_2PtCl_6: 220~221° (d) B–MeI: 252~254° or 270~272°
≪SD.≫ 38), 67)
≪S. & B.≫ cf. 38)

Manthine $C_{17}H_{19}O_4N \equiv 301.33$

≪mp≫ 114~116° (Et_2O)
≪$[\alpha]_D$≫ −71° ($CHCl_3$)

≪Ph.≫ UV[4,16,29,67]
≪OP.≫ Haemanthus amarylloides[67], H. trigrinus[38]
≪C. & So.≫
≪Sa. & D.≫
≪SD.≫ 38)
≪S. & B.≫ cf. 38)

Galanthidine $C_{14}H_{17}O_3N \equiv 247.28$

≪mp≫ 235~238° (EtOH)
≪$[\alpha]_D$≫ B–HBr: +32.3°
≪Ph.≫
≪OP.≫ Galanthus woronovi[147]
≪C. & So.≫ Spar. sol. in Et_2O, $CHCl_3$
≪Sa. & D.≫ B–HCl: 197~199° B–HBr: 213~213.5° B–MeI: 232~233°
≪SD.≫ 147)
≪S. & B.≫

Trispheridine $C_{16}H_{11}O_3N \equiv 265.26$

≪mp≫ 140~141°
≪$[\alpha]_D$≫ 0°
≪Ph.≫
≪OP.≫ Ungernia trisphaera[209]
≪C. & So.≫
≪Sa. & D.≫ B–HCl: 283~285° (d) B–HBr: 272~274° (d) B–HNO_3: 197~198°
≪SD.≫ 209)
≪S. & B.≫

Insulamine $C_{16}H_{17}O_3N \equiv 271.30$

≪mp≫ 177~178° (Me_2CO)
≪$[\alpha]_D$≫ −95° ($CHCl_3$)
≪Ph.≫
≪OP.≫ Narcissous incomparabilis[15]
≪C. & So.≫ Tablets, Sol. in MeOH, $CHCl_3$, Mod. sol. in Me_2CO
≪Sa. & D.≫
≪SD.≫ 15)
≪S. & B.≫

Powellidine $C_{16}H_{17}O_3N \equiv 271.30$

≪mp≫ 207～209°

≪[α]D≫ +100° ($CHCl_3$)

≪Ph.≫

≪OP.≫ Crinum powelli (moorei X longifolium)[142][146]

≪C. & So.≫

≪Sa. & D.≫ B–$HClO_4$: 177° B–Pic: 198°

≪SD.≫ 146°

≪S. & B.≫

Annapawine $C_{16}H_{17}O_3N \equiv 271.30$

≪mp≫ 219°

≪[α]D≫ +74° ($CHCl_3$)

≪Ph.≫ Rf[33]

≪OP.≫ Hippeastrum Hybr. "Anna Pawlowna"[13]

≪C. & So.≫

≪Sa. & D.≫ B–Pic: 249°

≪SD.≫

≪S. & B.≫

Brunsvigine $C_{16}H_{17}O_4N \equiv 287.30$

≪mp≫ 243° (AcOEt)

≪[α]D≫ −77° ($CHCl_3$)

≪Ph.≫ UV[37] IR[37]

≪OP.≫ Brunsvigia cooperi[37][38], B. radalosa[38], Haemanthus amarylloides[38]

≪C. & So.≫ Needles, Sublimation at 180°/0.1

≪Sa. & D.≫ B–HCl: 220 or 245° B–Pic: 195～196° or 219° B–MeI: 252° B–O,O–DiAc: 184°

≪SD.≫ 37)

≪S. & B.≫

Daphnarcine $C_{16}H_{17}O_4N \equiv 287.30$

≪mp≫ 258～260° (d) (MeOH–Me_2CO)

≪[α]D≫ +40° (DMF)

≪Ph.≫

≪OP.≫ Narcissus poeticus[15], Nerine undulata[10]

≪C. & So.≫ Prisms, Sol. in MeOH, Spar. sol. in Me_2CO

≪Sa. & D.≫ B–Pic: 246° (d)

≪SD.≫ 15)

≪S. & B.≫

Elwesine $C_{16}H_{19}O_3N \equiv 273.32$

≪mp≫ 218～219°

≪[α]D≫ −32°

≪Ph.≫ Rf[33]

≪OP.≫ Galanthus elwesii[10]

≪C. & So.≫

≪Sa. & D.≫

≪SD.≫

≪S. & B.≫

Rudoline $C_{16}H_{19}O_4N \equiv 289.30$

≪mp≫

≪[α]D≫ +67° ($CHCl_3$)

≪Ph.≫

≪OP.≫ Narcissus odorus var. rugulosus[176]

≪C. & So.≫

≪Sa. & D.≫

≪SD.≫

≪S. & B.≫

Poetaricine $C_{16}H_{17}O_4N \equiv 287.30$

≪mp≫ 273° (d) (MeOH)

≪[α]D≫ −60° (EtOH)

≪Ph.≫

≪OP.≫ Narcissus poeticus var. ornatus[15], N. jonquilla[143]

≪C. & So.≫ Prisms

≪Sa. & D.≫ B–$HClO_4$: 228～230° (d) B–Pic: 190° (d) B–Ac: 165°

≪SD.≫

≪S. & B.≫

Macrantine $C_{16}H_{19}O_5N \equiv 305.32$
≪mp≫ 238～240° (d) (EtOH)
≪$[\alpha]_D$≫ −190° ($CHCl_3$)
≪Ph.≫
≪OP.≫ Crinum macrantherum[11)]
≪C. & So.≫
≪Sa. & D.≫ B–OAc: 222～224°
B–O,O–DiAc: 219～221°
≪SD.≫
≪S. & B.≫

Hippauline $C_{17}H_{19}O_4N \equiv 285.33$
≪mp≫ 163～164°
≪$[\alpha_D]$≫ +10° ($CHCl_3$)
≪Ph.≫
≪OP.≫ Hippeastrum aulicum var. robustum[72)]
≪C. & So.≫
≪Sa. & D.≫ B–$HClO_4$: 175° (d)
≪SD.≫
≪S. & B.≫

Hippandrine $C_{17}H_{19}O_4N \equiv 285.33$
≪mp≫ +194° (Me_2CO)
≪$[\alpha]_D$≫ −110° ($CHCl_3$)
≪Ph.≫ IR[73)]
≪OP.≫ Hippeastrum brachyandrum[73)]
≪C. & So.≫ Prisms
≪Sa. & D.≫ B–Pic: 179° (d)
≪SD.≫ 73)
≪S. & B.≫

Krelagine $C_{17}H_{19}O_4N \equiv 285.33$
≪mp≫ 202°
≪$[\alpha]_D$≫ +290° ($CHCl_3$)
≪Ph.≫ CR[33)]
≪OP.≫ Crinum powellii var. krelagei[12)57)]
≪C. & So.≫
≪Sa. & D.≫ B–Pic: 268° B–MeI: 260°
≪SD.≫ 12)
≪S. & B.≫

Brusvinine $C_{17}H_{19}O_4N \equiv 285,33$
≪mp≫ 202° (Me_2CO)
≪$[\alpha]_D$≫
≪Ph.≫
≪OP.≫ Brunsvigia cooperi[37)]
≪C. & So.≫
≪Sa. & D.≫ B–Pic: 67～69°
≪SD.≫
≪S. & B.≫

Mesembrinol $C_{17}H_{25}O_3N \equiv 291.38$
≪mp≫ 143～146°
≪$[\alpha]_D$≫ −30° ($CHCl_3$)
≪Ph.≫
≪OP.≫ Mesembryantheum tortuosum[207)]
≪C. & So.≫
≪Sa. & D.≫
≪SD.≫ 207)
≪S. & B.≫

Hippamine $C_{17}H_{19}O_4N \equiv 285.33$
≪mp≫ 162° (Me_2CO)
≪$[\alpha]_D$≫ −72° (EtOH)
≪Ph.≫
≪OP.≫ Hippeastrum vittatum[74)]
≪C. & So.≫ Prisms
≪Sa. & D.≫
≪SD.≫
≪S. & B.≫

Natalensine $C_{17}H_{19}O_4N \equiv 285.33$
≪mp≫ 203～203.5° (Me_2CO)
≪$[\alpha]_D$≫ +19.7° (MeOH)
≪Ph.≫ UV[67)]
≪OP.≫ Haemanthus natalensis[67)], H. puniceus[67)]
≪C. & So.≫
≪Sa. & D.≫ B–Pic: 224～226° (d)
B–MeI: 186～191°
≪SD.≫ 67)
≪S. & B.≫

Vallotine $C_{17}H_{19}O_5N \equiv 317.33$
≪mp≫ 217~218° (Me_2CO)
≪[α]D≫ −53° ($CHCl_3$)
≪Ph.≫ CR[44]
≪OP.≫ Vallota purpurea[44]
≪C. & So.≫ Prisms
≪Sa. & D.≫
≪SD.≫ 44)
≪S. & B.≫

Flexine $C_{17}H_{19}O_5N \equiv 317.33$
≪mp≫ 212°
≪[α]D≫ +75° ($CHCl_3$)
≪Ph.≫ Rf[33]
≪OP.≫ Nerine flexuosa[13]
≪C. & So.≫
≪Sa. & D.≫ B–$HClO_4$: 242° B–MeI: 221°
≪SD.≫
≪S. & B.≫

Suisenine $C_{17}H_{19}O_5N \equiv 317.33$
≪mp≫ 229° (EtOH)
≪[α]D≫
≪Ph.≫
≪OP.≫ Narcissus tazetta[267]
≪C. & So.≫ Needles
≪Sa. & D.≫ B–HCl 180° B–Pic: 189° B_2–H_2PtCl_6: 194° B–OBz: 196°
≪SD.≫ 267)
≪S. & B.≫

Flexamine $C_{17}H_{19}O_5N \equiv 317.33$
≪mp≫ 226~228°
≪[α]D≫ 0°
≪Ph.≫ CR[72] IR[72] Rf[33]
≪OP.≫ Nerine flexuosa[72][267]
≪C. & So.≫
≪Sa. & D.≫ B–MeI: 245° (d)
≪SD.≫ 72)
≪S. & B.≫

Luteine $C_{17}H_{19}O_5N \equiv 317.33$
≪mp≫ 185~186° (Me_2CO)
≪[α]D≫ −18.4° (EtOH)
≪Ph.≫
≪OP.≫ Sternbergia lutea[44][95], S. fischeriana[95]
≪C. & So.≫
≪Sa. & D.≫ B–$HClO_4$: 122~123° B–Pic: 201~203° B–MeI: 195~197°
≪SD.≫ 95)
≪S. & B.≫

Base III $C_{17}H_{19}O_5N \equiv 317.33$
≪mp≫ 212~213°
≪[α]D≫ +75° ($CHCl_3$)
≪Ph.≫
≪OP.≫ Nerine flexuosa[13], Sternbergia fischeriana[95]
≪C. & So.≫
≪Sa. & D.≫ B–HBr: 174~175° B–$HClO_4$: 242° (d) B–Pic: 217~219° B–MeI: 221° (d)
≪SD.≫
≪S. & B.≫

Base D $C_{17}H_{19\sim21}O_3N \equiv$
≪mp≫ 228~229° (d) (Me_2CO)
≪[α]D≫ −175° ($CHCl_3$)
≪Ph.≫
≪OP.≫ Narcissus incomparabilis[15]
≪C. & So.≫ Prisms, Sol. in MeOH, Spar. sol. in Me_2CO
≪Sa. & D.≫
≪SD.≫ 15)
≪S. & B.≫

Bodamine $C_{17}H_{21}O_3N \equiv 287.35$
≪mp≫ 208~210°
≪[α]D≫ 0°
≪Ph.≫ IR[72]
≪OP.≫ Nerine bowdeni[72]
≪C. & So.≫
≪Sa. & D.≫ B–HI: 245° B–MeI: 265°
≪SD.≫ *dl*-galanthamine(?)
≪S. & B.≫

Robecine $C_{17}H_{21}O_3N \equiv 287.35$
≪mp≫
≪[α]D≫ B–HI : −95° (DMF)
≪Ph.≫
≪OP.≫ Narcissus pseudonarcissus[15]
≪C. & So.≫
≪Sa. & D.≫ B–HI : 240~241° (d)
B–MeI : 248~249° (d)
≪SD.≫ 15)
≪S. & B.≫

Rulodine $C_{17}H_{21}O_4N \equiv 303.35$
≪mp≫ 193°
≪[α]D≫ −15° ($CHCl_3$)
≪Ph.≫
≪OP.≫ Narcissus odorus var. rugulosus[176]
≪C. & So.≫
≪Sa. & D.≫ B–HI : 135~136°
≪SD.≫
≪S. & B.≫

Nerifline $C_{17}H_{21}O_5N \equiv 319.35$
≪mp≫ 152°
≪[α]D≫
≪Ph.≫ Nerine flexuosa[72]
≪OP.≫
≪C. & So.≫
≪Sa. & D.≫ B–HI : 162~163°
≪SD.≫ 72)
≪S. & B.≫

Desacetylbowdensine $C_{17}H_{21}O_5N \equiv 319.35$
≪mp≫ 277~278°
≪[α]D≫ −44° (MeOH)
≪Ph.≫
≪OP.≫ Nerine bowdeni[56]
≪C. & So.≫
≪Sa. & D.≫
≪SD.≫
≪S. & B.≫

Petomine $C_{17}H_{21}O_6N \equiv 335.35$
≪mp≫ 253~254° (MeOH–Et_2O)
≪[α]D≫ 0° ($CHCl_3$)
≪Ph.≫
≪OP.≫ Amaryllis parkeri (A. belladonna X Brunsvigia josephinae)[3], Crinum powellii (moorei X longifolium)[146], Narcissus cyclamineus[15]
≪C. & So.≫ Plates, Mod. sol. in MeOH, Spar. sol. in Me_2CO
≪Sa. & D.≫
≪SD.≫ 15)
≪S. & B.≫

Yemensine $C_{18}H_{19}O_4N \equiv 313.34$
≪mp≫ 193° (d) (Me_2CO)
≪[α]D≫ +100° ($CHCl_3$)
≪Ph.≫
≪OP.≫ Crinum yemense (latifolium)[9]
≪C. & So.≫ Prisms, Sol. in $CHCl_3$, MeOH, Mod. sol. in Me_2CO
≪Sa. & D.≫ B–HI : 205~206° (d) B–Pic : 262~265° (d) B–MeI : 235° (d)
≪SD.≫
≪S. & B.≫

Undunerine $C_{18}H_{19}O_5N \equiv 329.34$
≪mp≫ 132~134°
≪[α]D≫ +40° ($CHCl_3$)
≪Ph.≫
≪OP.≫ Nerine undulata[44]
≪C. & So.≫
≪Sa. & D.≫
≪SD.≫
≪S. & B.≫

Distichamine $C_{18}H_{19}O_5N \equiv 329.34$

≪mp≫ 161∼162° (Me_2CO)
≪$[\alpha]_D$≫ −56° ($CHCl_3$)
≪Ph.≫ UV[8] IR[8] Rf[8]
≪OP.≫ Buphane disticha[8]
≪C. & So.≫ Prisms
≪Sa. & D.≫ B–MeI: 223∼226° (d)
≪SD.≫
≪S. & B.≫

Sternine $C_{18}H_{21}O_3N \equiv 299.36$

≪mp≫ 231∼232° (Me_2CO)
≪$[\alpha]_D$≫ +12.5° (EtOH)
≪Ph.≫
≪OP.≫ Sternbergia fischeriana[95]
≪C. & So.≫
≪Sa. & D.≫ B–HCl: 181∼182° B–HBr: 226∼226.5° B–Pic: 227∼228° B–MeI: 202∼203°
≪SD.≫
≪S. & B.≫

Manthidine $C_{18}H_{21}O_4N \equiv 315.36$

≪mp≫ 269∼272°
≪$[\alpha]_D$≫ −27° ($CHCl_3$)
≪Ph.≫ UV[4)38] IR[67]
≪OP.≫ Haemanthus amarylloides[38], H. coccineus[67]
≪C. & So.≫ Sublimation at $200°/2\times10^{-4}$
≪Sa. & D.≫
≪SD.≫ 67)
≪S. & B.≫

Vallotidine $C_{18}H_{21}O_5N \equiv 331.36$

≪mp≫ 172∼173° (Me_2CO)
≪$[\alpha]_D$≫ +57° ($CHCl_3$)
≪Ph.≫ CR[44]
≪OP.≫ Vallota purpurea[44]
≪C. & So.≫ Prisms
≪Sa. & D.≫
≪SD.≫ 44)
≪S. & B.≫

Neflexine $C_{18}H_{21}O_5N \equiv 331.36$

≪mp≫ 249∼250° (d)
≪$[\alpha]_D$≫ +65° ($CHCl_3$)
≪Ph.≫ CR[72] IR[72]
≪OP.≫ Crinum powellii[7], Nerine flexuosa[72]
≪C. & So.≫
≪Sa. & D.≫ B–Pic: 195∼196°
≪SD.≫
≪S. & B.≫

Brunsdonnine $C_{18}H_{21}O_5N \equiv 331.36$

≪mp≫ 253° (Me_2CO–MeOH)
≪$[\alpha]_D$≫ +75° ($CHCl_3$)
≪Ph.≫ CR[7] Rf[33]
≪OP.≫ Brunsdonna tubergenii (Brunsvigia josephinae X Amaryllis belladonna)[7]
≪C. & So.≫
≪Sa. & D.≫ B–HI: 248°(d) B–$HClO_4$: 232(d) B–Pic: 232°(d) B–MeI: 280∼281° (d)
≪SD.≫ 7)
≪S. & B.≫

Base NB $C_{18}H_{21}O_5N \equiv 331.36$

≪mp≫ 139∼140°
≪$[\alpha]_D$≫
≪Ph.≫
≪OP.≫ Nerine bowdenii[72]
≪C. & So.≫
≪Sa. & D.≫
≪SD.≫ 72)
≪S. & B.≫

Crispanine $C_{18}H_{21}O_5N \equiv 331.36$

≪mp≫ 139∼141°
≪$[\alpha]_D$≫ +78° ($CHCl_3$)
≪Ph.≫ CR[268] IR[268]
≪OP.≫ Nerine crispa[268]
≪C. & So.≫
≪Sa. & D.≫ B–$HClO_4$: 221°(d) B–Pic: 201° (d)
≪SD.≫ 268)
≪S. & B.≫

Neruscine $C_{18}H_{23}O_3N \equiv 301.37$
≪mp≫
≪[α]D≫ +92° ($CHCl_3$)
≪Ph.≫
≪OP.≫ Nerine corusca[81], N. flexuosa[81]
≪C. & So.≫ Oil
≪Sa. & D.≫ B-HI: 247° B-$HClO_4$: 221~223° (d)
≪SD.≫
≪S. & B.≫

Coruscine $C_{18}H_{23}O_5N \equiv 333.37$
≪mp≫ 170° (Me_2CO)
≪[α]D≫ +70° ($CHCl_3$)
≪Ph.≫ IR[81]
≪OP.≫ Brunsdonna tubergenii (Brunsvigia josephinae X Amryllis belladonna)[7], Nerine corusca[81]
≪C. & So.≫ Prisms, Sol. in MeOH, $CHCl_3$, Mod. sol. in Me_2CO
≪Sa. & D.≫ B-HI: 179~180° B-MeI: 311° (d)
≪SD.≫ 81)
≪S. & B.≫

Galanthamidine $C_{18}H_{23}O_5N \equiv 333.37$
≪mp≫ 211~213°
≪[α]D≫ −94.2° (MeOH)
≪Ph.≫
≪OP.≫ Galanthus woronovi[269]
≪C. & So.≫
≪Sa. & D.≫ B-MeI: 219° (d)
≪SD.≫ 269)
≪S. & B.≫

Nerundine $C_{18}H_{21}O_5N$ or $C_{18}H_{23}O_6N$
≪mp≫ 256° (d)
≪[α]D≫ −95° ($CHCl_3$)
≪Ph.≫
≪OP.≫ Nerine undulata[157]
≪C. & So.≫
≪Sa. & D.≫ B-$HClO_4$: 221~222° (d) B-Pic: 168° (d) B-MeI: 286°
≪SD.≫ 157)
≪S. & B.≫

Crispine $C_{18}H_{23}O_6N \equiv 349.37$
≪mp≫ 275° (d) (Me_2CO)
≪[α]D≫ −96° ($CHCl_3$)
≪Ph.≫ CR[44]
≪OP.≫ Nerine undulata[44]
≪C. & So.≫ Plates, Sol. in $CHCl_3$, MeOH, Me_2CO
≪Sa. & D.≫ B-$HClO_4$: 268~269° (d) B-$MeClO_4$: 282~283° (d) B-Pic: 258° (d)
≪SD.≫
≪S. & B.≫

Trisphaerine $C_{19}H_{17}O_6N \equiv 355.33$
≪mp≫ 209~210°
≪[α]D≫ +122.3° ($CHCl_3$)
≪Ph.≫
≪OP.≫ Ungernia trisphaera[101]
≪C. & So.≫
≪Sa. & D.≫ B-HCl: 280~281° B-HBr: 266~267° B-$HClO_4$: 256~258° B-HNO_3: 269~271° B-Pic: 242~243° B-MeI: 216~218° (d) B-O,O-DiAc: 174~175°
≪SD.≫
≪S. & B.≫

Punikathine $C_{19}H_{23}O_5N \equiv 345.38$
≪mp≫ 177~178° (Me_2CO)
≪[α]D≫ −46° ($CHCl_3$)
≪Ph.≫
≪OP.≫ Haemanthus-Hybr. "King Albert"[44]
≪C. & So.≫ Leaflets, Sol. in $CHCl_3$, MeOH, Me_2CO, AcOEt, Spar. sol. in Et_2O
≪Sa. & D.≫
≪SD.≫ 44)
≪S. & B.≫

Base F $C_{19}H_{24}O_5N^{\oplus} \equiv 346.38$

≪mp≫
≪[α]D≫
≪Ph.≫
≪OP.≫ Nerine corusca[81], N. flexuosa[81]

≪C. & So.≫
≪Sa. & D.≫ B–MeI: 258°
≪SD.≫
≪S. & B.≫

Poeticine $C_{20}H_{23}O_6N \equiv 373.39$

≪mp≫ 209～210°
≪[α]D≫ −89° ($CHCl_3$)
≪Ph.≫
≪OP.≫ Narcissus poeticus[87]

≪C. & So.≫
≪Sa. & D.≫
≪SD.≫
≪S. & B.≫

Narzettine $C_{20}H_{23}O_6N \equiv 373.39$

≪mp≫ 224～226° (MeOH)
≪[α]D≫ −90° ($CHCl_3$)
≪Ph.≫
≪OP.≫ Narcissus tazetta[87]

≪C. & So.≫ Plates
≪Sa. & D.≫ B–MeI: 256～258°
≪SD.≫
≪S. & B.≫

Ungeridine $C_{20}H_{25}O_4N \equiv 343.41$

≪mp≫ 200～201° (Me_2CO)
≪[α]D≫ +153° ($CHCl_3$)
≪Ph.≫
≪OP.≫ Ungernia (Lycoris) severtzovii[97][150][270], U. tadshikorum[96], U. trisphaera[101]

≪C. & So.≫
≪Sa. & D.≫ B–Pic: 190～191° B–MeI: 179～180° B–OAc: 184～185° Dihydro–Der.: 164～165°
≪SD.≫ 97)
≪S. & B.≫

Buphacetine $C_{20}H_{25}O_7N \equiv 391.41$

≪mp≫ 182～183° (Me_2CO–Et_2O)
≪[α]D≫ −51° (EtOH), −73° ($CHCl_3$)
≪Ph.≫ UV[8] IR[8] Rf[8]
≪OP.≫ Buphane disticha[8]

≪C. & So.≫ Needles
≪Sa. & D.≫
≪SD.≫
≪S. & B.≫

Clivatine $C_{21}H_{25}O_4N \equiv 355.42$

≪mp≫ 156～159°
≪[α]D≫ +50° ($CHCl_3$)
≪Ph.≫
≪OP.≫ Clivia miniata[210]

≪C. & So.≫
≪Sa. & D.≫ B–Picrol: 150～152°
≪SD.≫
≪S. & B.≫

Bowdensine $C_{21}H_{25}O_7N \equiv 403.42$

≪mp≫
≪[α]D B–$HClO_4$: +17°
≪Ph.≫
≪OP.≫ Nerine bowdenii[56]

≪C. & So.≫ Amorph or glass
≪Sa. & D.≫ B–$HClO_4$: 260～262° B–MeI: 284～285°
≪SD.≫
≪S. & B.≫

Coccinine

≪mp≫
≪[α]D≫
≪Ph.≫
≪OP.≫ Haemanthus albomaculates[45]

≪C. & So.≫
≪Sa. & D.≫
≪SD.≫
≪S. & B.≫

Hippacine
≪mp≫
≪[α]$_D$≫
≪Ph.≫
≪OP.≫ Hippeastrum vittatum[231]
≪C. & So.≫
≪Sa. & D.≫
≪SD.≫
≪S. & B.≫

Minatine
≪mp≫
≪[α]$_D$≫
≪Ph.≫
≪OP.≫ Clivia miniata[210]
≪C. & So.≫
≪Sa. & D.≫
≪SD.≫
≪S. & B.≫

Narcipoetine
≪mp≫ 172°
≪[α]$_D$≫ 84.4° (EtOH)
≪Ph.≫
≪OP.≫ Narcissus poeticus[17]
≪C. & So.≫
≪Sa. & D.≫ B–HCl: 271°(d) B–$HAuCl_4$: 131~132°(d) B–Pic: 261°(d)
≪SD.≫ 213)
≪S. & B.≫

Nivalidine
≪mp≫ 152~153°
≪[α]$_D$≫
≪Ph.≫
≪OP.≫ Galanthus nivalis var. gracilis[63]
≪C. & So.≫ Plates
≪Sa. & D.≫ B–HI: 343~344° B–MeI: 252~253°
≪SD.≫
≪S. & B.≫

Pancratine
≪mp≫
≪[α]$_D$≫
≪Ph.≫
≪OP.≫ Pancratium maritinum[194]
≪C. & So.≫
≪Sa. & D.≫ B–OMe=tazettine(?)
≪SD.≫ 194)
≪S. & B.≫

Spuren
≪mp≫
≪[α]$_D$≫
≪Ph.≫
≪OP.≫ Galanthus nivalis[61]
≪C. & So.≫
≪Sa. & D.≫
≪SD.≫
≪S. & B.≫

1) M. Normatov, Kh. A. Abduazimov, S. Yu. Yunusov: Uzbeksh. Khim. Zhur., **9**, 25 (1965). [C. A., **63**, 7061 (1965)].
2) H.G. Boit, H. Ehmke: Ber., **89**, 2093 (1956).
3) H.G. Boit, W. Doepke: Ber., **92**, 2578 (1959).
4) E.W. Warnhoff, W.C. Wildman: J. Am. Chem. Soc., **82**, 1472 (1960).
5) L.H. Mason, E.R. Puschett, W.C. Wildman: J. Am. Chem. Soc., **77**, 1253 (1955).
6) H. Hauth, D. Sauffacher: Helv. Chim. Acta, **45**, 1307 (1962).
7) H.G. Boit, W. Doepke: Naturwiss., **47**, 159 (1960).
8) W. Doepke: Arch. Pharm., **279**, 39 (1964).
9) H.G. Boit, W. Doepke, W. Stender: Ber., **90**, 2203 (1957).
10) H.G. Boit: "Ergebnisse der Alkaloid Chemie bis 1960," Akademie-Verlag, Berlin, 1961.
11) H. Hauth, D. Stauffacher: Helv. Chim. Acta, **47**, 185 (1964).
12) H.G. Boit, W. Doepke: Naturwiss, **47**, 498 (1960).
13) H.G. Boit, W. Doepke: Naturwiss., **47**, 470 (1960).
14) H.G. Boit, W. Doepke: Naturwiss., **47**, 323 (1960).

15) H. G. Boit, W. Doepke, A. Beitner: Ber., **90**, 2197 (1957).
16) W. C. Wildman, C. J. Kaufman: J. Am. Chem. Soc., **77**, 4807 (1955).
17) H. G. Boit, W. Doepke: Naturwiss., **45**, 85 (1958).
18) H. M. Fales, E. W. Warnhoff, W. C. Wildman: J. Am. Chem. Soc., **77**, 5885 (1955).
19) K. Takeda, K. Kotera: Chem. & Ind., **1956**, 347.
20) K. Takeda, K. Kotera: Chem. Pharm. Bull., **5**, 234 (1957).
21) E. W. Warnhoff, W. C. Wildman: J. Am. Chem. Soc., **79**, 2192 (1957).
22) K. Takeda, K. Kotera, S. Mizukami: J. Am. Chem. Soc., **80**, 2562 (1958).
23) H. M. Fales, W. C. Wildman: J. Am. Chem. Soc., **80**, 4395 (1958).
24) S. Mizukami: Tetrahedron, **11**, 89 (1960).
25) K. Kotera: Tetrahedron, **12**, 240 (1961).
26) K. Kotera, Y. Hamada, K. Tori, K. Aone, K. Kuriyama: Tetrahedron Letters, **1966**, 2009.
27) M. Normatov, Kh. A. Abduazimov, S. Yu. Yunusov: Dokl. Akad. Nauk SSR., **19**, (12), 27 (1962). [C. A., **59**, 6456 (1964)].
28) D. H. R. Barton, T. Cohen: "Festschrift Arthur Stoll," p. 117. Birkhauser, Basel (1957).
29) A. F. Sangster, K. L. Stuart: Chem. Revs., **65**, 94 (1965).
30) L. H. Briggs, L. D. Colebrook, H. M. Fales, W. C. Wildman: Anal. Chem., **29**, 904 (1957).
31) T. H. Kinstle, W. C. Wildman, C. L. Brown: Tetrahedron Letters, **1966**, 4659.
32) T. Ikuka, H. Irie, A. Kato, S. Uyeo, K. Kotera, Y. Nakagawa: Tetrahedron Letters, **1966**, 4745.
33) W. Doepke: Arch. Pharm., **295**, 605 (1962).
34) K. Gorter: Bull. Jard. Bot. Buitenzorg, [3], **2**, 331 (1920). [C., **1921**, I, 92].
35) A. J. Ewins: J. Chem. Soc., **97**, 2406 (1910).
36) A. N. Bates, J. K. Cooke, L. J. Dry, A. Goosen, N. Kriisi, F. L. Warren: J. Chem. Soc., **1957**, 2537.
37) L. J. Dry, M. Poynton, M. E. Tempson, F. L. Warren: J. Chem. Soc., **1958**, 4701.
38) Y. Inubushi, H. M. Fales, E. W. Warnhoff, W. C. Wildman: J. Org. Chem., **25**, 2153 (1960).
39) F. Tutin: J. Chem. Soc., **99**, 1240 (1911).
40) A. G. Humber, W. I. Tayler: Can. J. Chem., **33**, 1268 (1955).
41) H. Hauth, D. Stauffacher: Helv. Chim. Acta, **44**, 491 (1961).
42) J. Renz, D. Stauffacher, E. Seebeck: Helv. Chim. Acta, **38**, 1209 (1955).
43) H. G. Boit, W. Doepke: Ber., **90**, 1827 (1957).
44) H. G. Boit: Ber., **89**, 1129 (1956).
45) C. K. Briggs, P. F. Highet, R. J. Highet, W. C. Wildman: J. Am. Chem. Soc., **78**, 2899 (1956).
46) H. G. Boit: Ber., **87**, 1704 (1954).
47) G. A. Greathouse, N. E. Rigler: Am. J. Bot., **28**, 702 (1941).
48) K. Gorter: Bull. Jard. Bot. Buitenzorg, [3], **1**, 352 (1920). [C., **1920**, II, 846].
49) K. Tanaka: J. Pharm. Soc. Japan, **57**, 139 (1937).
50) S. Rangaswami, E. V. Rao: Indian J. Pharm., **12**, 67, 140 (1955). [C. A., **50**, 7403, 13375 (1956)].
51) H. M. Fales, D. H. S. Horn, W. C. Wildman: Chem. & Ind., **1959**, 1415.
52) H. M. Fales, W. C. Wildman: J. Am. Chem. Soc., **82**, 3368 (1960).
53) A. Hunger, T. Reichstein: Helv. Chim. Acta, **36**, 824 (1953).
54) S. Rangaswami, E. V. Rao: Current Sci., **24**, 55 (1955). [C. A., **49**, 9233 (1955)].
55) S. Rangaswami, M. Suryanarayana: Indian J. Pharm., **17**, 229 (1955). [C. A., **50**, 7404 (1956)].
56) R. E. Lyle, E. A. Kieler, J. R. Crowder, W. C. Wildman: J. Am. Chem. Soc., **82**, 2620 (1960).
57) W. Doepke: Arch. Pharm., **295**, 868 (1962).
58) H. G. Boit, H. Ehmke: Ber., 88, 1590(1955).
59) B. Reichart: Arch. Pharm., **276**, 328 (1938).
60) L. B. Oliveros, A. C. Santos: Univ. Philippines Natur Appl. Sci. Bull., **4**, 41 (1934). [C., **1935**, II, 858].
61) H. G. Boit: Ber., **87**, 724 (1954).
62) H. G. Boit: Ber., **87**, 1448 (1954).
63) L. Ivanova: Pharmazie [Sofia], 8, No. 4, 24 (1954). [C. A., **53**, 17422 (1959)].
64) L. Bubeva-Ivanova: Ber., **95**, 1348 (1962).
65) N. F. Proskurnina, L. Ya. Areschkina: Zhur. Obshchei Khim., **17**, 1216 (1947). [C. A., **42**, 1595 (1948)].
66) N. F. Proskurnina: Zhur. Priklado. Khim., **90**, 565 (1953). [C. A., **48**, 1025 (1954)].
67) W. C. Wildman, C. J. Kaufman: J. Am. Chem. Soc., **77**, 1248 (1955).
68) H. G. Boit, W. Doepke: Ber., **91**, 1965 (1958).
69) H. M. Fales, W. C. Wildman: J. Org. Chem., **26**, 881 (1961).
70) H. M. Fales, W. C. Wildman: J. Am.

Chem. Soc., **82**, 197 (1960).
71) H. G. Boit: Ber., **87**, 1339 (1954).
72) H. G. Boit, W. Doepke: Naturwiss., **47**, 109 (1960).
73) H. G. Boit, W. Doepke: Ber., **92**, 2582 (1959).
74) W. Doepke: Arch. Pharm., **295**, 920 (1962).
75) H. G. Boit, W. Doepke: Naturwiss., **45**, 315 (1958). [C. A., **53**, 650 (1959)].
76) W. C. Wildman, C. J. Kaufman: J. Am. Chem. Soc., **76**, 5815 (1954).
77) N. F. Proskurnina: Zhur. Obshchei Khim., **27**, 3365 (1957). [C. A., **52**, 9165 (1958)].
78) A. Gheorghiu, E. Ionescu-Matiu: Ann. Pharm. Franc., **20**, 531 (1962). [C.A., **59**, 3976 (1963)].
79) H. G. Boit: Ber., **87**, 681 (1954).
80) H. G. Boit, W. Doepke, W. Stender: Naturwiss., **45**, 390 (1958).
81) H. G. Boit, H. Ehmke: Ber., **90**, 369 (1957).
82) K. Morishima: Ar. Pth., **40**, 221 (1897). [C., **1898**, I, 254].
83) S. H. Huang, K. E. Ma: Yao Hsueh Hsueh Pao, **11**, 1 (1964). [C. A., **61**, 3154 (1964)].
84) S. Uyeo, Y. Yamato: J. Pharm. Soc. Japan, **85**, 615 (1965).
85) H. G. Boit, W. Stender, A. Boitner: Ber., **90**, 725 (1957).
86) J. W. Cook, J. D. Loudon, P. McCloskey: J. Chem. Soc., **1954**, 4176.
87) H. G. Boit, W. Doepke: Ber., **89**, 2462 (1956).
88) H. G. Boit, W. Stender: Ber., **87**, 624 (1954).
89) H. G. Boit, H. Ehmke: Ber., **89**, 163 (1956).
90) P. W. Jaffs, F. L. Warren: Chem. & Ind., **1961**, 468.
91) S. Rangaswani, R. V. K. Rao: Tetrahedron Letters, **1966**, 4481.
92) M. L. duMérac: Compt. rend., **239**, 300 (1954). [C. A., **49**, 1159 (1955)].
93) N. F. Proskurnina: Zhur. Obshchei Khim., **25**, 834 (1955). [C. **1955**, 11216].
94) F. Sandberg, S. Agurell: Svensk Farm. Tidskr., **63**, 657 (1959). [C. A., **54**, 4787 (1960)].
95) N. F. Proskurnina, N. M. Ismallow: Zhur. Obshchei Khim., **23**, 2056 (1953). [C., **1954**, 8353].
96) S. Yu. Yunusov, C. A. Abduasimov: Ber. Akad. Wiss. Uzbek. SSR, **1956**, No. 4, 7; Zhur. Obshchei Khim., **27**, 3357 (1957). [C., **1959**, 4155].
97) S. Yu. Yunusov, C. A. Abduasimov: Zhur. Obshchei Khim., **29**, 1724 (1959). [C. A., **54**, 9173 (1960)].
98) L. S. Smirnova, Kh. A. Abduazimov, S. Yu. Yunusov: Dokl. Akad. Nauk Uz. SSR., **21**, 44 (1964); Khim Prirodn. Soedin., Akad. Nauk Uz. SSR., **1965**, 332. [C. A., **64**, 5151 (1966)].
99) A. Popelak, E. Haack, G. Lettenbauer, H. Spingler: Naturwiss., **47**, 231 (1960). [C. A., **55**, 23572 (1961)].
100) N. K. Juraschewski: Zhur. Obshchei Khim., **8**, 949 (1938). [C., **1939**, I, 2988].
101) Kh. Allayarov, Kh. A. Abduazimov, S. Yu. Yunusov: Dokl. Akad. Nauk Uz. SSR., **1961**, No. 10, 25. [C. A., **58**, 2475 (1963)].
102) S. Ozeki: J. Pharm. Soc. Japan, **84**, 253 (1964).
103) S. Ozeki: J. Pharm. Soc. Japan, **84**, 1194 (1964).
104) W. Doepke: Arch. Pharm., **298**, 704 (1965).
105) K. Gorter: Bull. Jard. Bot. Buitenzorg, [3], **2**, 1 (1920) [C., **1920**, III, 842].
106) H. Kondo, K. Tomimura: J. Pharm. Soc. Japan, **48**, 36 (1928).
107) H. Kondo, K. Tomimura: J. Pharm. Soc. Japan, **48**, 323 (1928).
108) H. Kondo, S. Uyeo: Ber., **68**, 1756 (1935).
109) H. Kondo, H. Katsura, S. Uyeo: Ber., **71**, 1529 (1938).
110) H. Kondo, H. Katsura: Ber., **72**, 2083 (1939).
111) H. Kondo, H. Katsura: Ber., **73**, 112 (1940).
112) H. Kondo, H. Katsura: Ber., **73**, 1424 (1940).
113) R. B. Kelly, W. I. Taylor, K. Wiesner: J. Chem. Soc., **1953**, 2094.
114) K. Wiesner, W. I. Taylor, S. Uyeo: Chem. & Ind., **1954**, 46.
115) W. I. Taylor, B. R. Thomas, S. Uyeo: Chem. & Ind., **1954**, 929.
116) E. Wenkert: Chem. & Ind., **1954**, 1175.
117) J. W. Cook, J. D. Loudon, P. McCloskey: Chem. & Ind., **1954**, 1199.
118) B. S. Thyagarajan: Chem. & Ind., **1954**, 1299.
119) L. G. Humber, H. Kondo, K. Kotera, S. Takagi, K. Takeda, W. I. Taylor, B. R. Thomas, Y. Tsuda, K. Tsukamoto, S. Uyeo, H. Yajima, N. Yanaihara: J. Chem. Soc., **1954**, 4622.
120) H. Kondo, K. Takeda, K. Kodera: Ann. Rept. ITSUU Lab., **5**, 66 (1954).
121) T. Shingu, S. Uyeo, H. Yajima: J. Chem. Soc., **1955**, 3557.
122) S. Takagi, W. I. Taylor, S. Uyeo, H. Yajima: J. Chem. Soc., **1955**, 4003.
123) Y. Nakagawa, S. Uyeo, H. Yajima: Chem. & Ind., **1956**, 1238.
124) Y. Nakagawa, S. Uyeo: J. Chem. Soc., **1959**, 3736.
125) K. Takeda, K. Kotera, S. Mizukami, M. Kobayashi: Chem. Pharm. Bull. 8, 483 (1960).
126) R. K. Hill, J. A. Toule, L. J. Loeffer: Chem. & Ind., **1962**, 1573.

127) R.K. Hill, J.A. Toule, L.J. Loeffer: J. Am. Chem. Soc., **84**, 4951 (1962).
128) W. Doepke, H. Dalmer: Naturwiss., **52**, 61 (1965).
129) N. Ueda, T. Tokuyama, T. Sakan: Bull. Chem. Soc. Japan, **39**, 2012 (1966).
130) D.H.R. Barton, G.W. Kirby: Proc. Chem. Soc., **1960**, 392.
131) A.R. Battersby, R. Binks, W.C. Wildman: Proc. Chem. Soc., **1960**, 410.
132) A.R. Battersby, R. Binks, S.W. Breuer, H.M. Fales, W.C. Wildman: Proc. Chem. Soc., **1961**, 243.
133) R.J. Suhadolnik, A.G. Fischer, J. Zulalian: J. Am. Chem. Soc., **84**, 4348 (1962).
134) W.C. Wildman, A.R. Battersby, S.W. Breuer: J. Am. Chem. Soc., **84**, 4599 (1962).
135) W.C. Wildman, H.M. Fales, R.J.H. Highet, S.W. Breuer, A.R. Battersby: Proc. Chem. Soc., **1962**, 180.
136) D.A. Archer, S.W. Breuer, R. Binks, A.R. Battersby, W.C. Wildman: Proc. Chem. Soc., **1963**, 168.
137) A.R. Battersby, H.M. Fales, W.C. Wildman: J. Am. Chem. Soc., **83**, 4098 (1961).
138) A.R. Battersby, R. Binks, S.W. Breuer, H.M. Fales, W.C. Wildman, R.J. Highet: J. Chem. Soc., **1964**, 1595.
139) S. Uyeo, N. Yanaihara: J. Chem. Soc., **1959**, 172.
140) H. Kondo, K. Tomimura, S. Ishiwata: J. Pharm. Soc. Japan, **52**, 51 (1932).
141) H.M. Fales, L.D. Giuffrida, W.C. Wildman: J. Am. Chem. Soc., **78**, 4145 (1956).
142) H.G. Boit, H. Ehmke, S. Uyeo, H. Yajima: Ber., **90**, 363 (1957).
143) W. Doepke, H. Dalmer: Naturwiss., **52**, 60 (1965).
144) S. Ozeki: J. Pharm. Soc. Japan, **85**, 200 (1965).
145) T. Kitagawa, S. Uyeo, N. Yokoyama: J. Chem. Soc., **1959**, 3741.
146) H.G. Boit, W. Doepke: Naturwiss., **46**, 475 (1959).
147) N.F. Proskurnina, L. Ya. Areschkina: Zhur. Obshchei Khim., **17**, 1216 (1947). [C.A., **42**, 1595 (1948)].
148) H.M. Fales, W.C. Wildman: J. Am. Chem. Soc., **78**, 4151 (1956).
149) D.H.R. Barton, G.W. Kirby, J.B. Taylor, G.M. Thomas: Proc. Chem. Soc., **1961**, 254.
150) L.S. Smirnova, Kh.A. Abduazinov, S. Yu. Yunusov: Dokl. Akad. Nauk SSR., **154**, 171 (1964). [C.A., **60**, 9324 (1964)].
151) W. Doepke: Arch. Pharm., **296**, 725 (1963).
152) S. Yu. Yunusov: Zhur. Obshchei Khim., **20**, 368 (1950). [C.A., **44**, 6582 (1950)].
153) S. Yu. Yunusov: Zhur. Obshchei Khim., **20**, 1514 (1950). [C.A., **45**, 2490 (1951)].
154) K. Torssell: Acta Chem. Scand., **15**, 947 (1961).
155) F. Benington, R.D. Morin: J. Org. Chem., **27**, 142 (1962).
156) H.A. Lloyd, E.A. Kielar, R.J. Highet, S. Uyeo, H.M. Fales, W.C. Wildman: J. Org. Chem., **27**, 373 (1962).
157) H.G. Boit, W. Doepke: Naturwiss., **46**, 228 (1959).
158) K. Fragner: Ber., **24**, 1498 (1891).
159) J. Stanek: Chem. & Ind., **1955**, 1557.
160) H. Kondo, S. Ishiwata: Ber., **70**, 2427 (1937).
161) A.L. Goosen, E.V.O. John, F.L. Warren, E.C. Yates: J. Chem. Soc., **1961**, 4038.
162) A.L. Burlingame, H.M. Fales, R.J. Highet: J. Am. Chem. Soc., **86**, 4976 (1964).
163) W.C. Wildman: Chem. & Ind., **1956**, 1090.
164) W.C. Wildman: J. Am. Chem. Soc., **78**, 4180 (1956).
165) W.C. Wildman: J. Am. Chem. Soc., **80**, 2567 (1958).
166) P.W. Jeffs, F.L. Warren, W.G. Wright: J. Chem. Soc., **1960**, 1090.
167) E.W. Warnhoff, W.C. Wildman: J. Am. Chem. Soc., **82**, 1472 (1960).
168) S. Uyeo, H. Irie, A. Yoshitake, A. Ito: Chem. Pharm. Bull., **13**, 427 (1965).
169) H. Muxfeldt, R.S. Schneider, J.B. Mooberg: J. Am. Chem. Soc., **88**, 3670 (1966).
170) J.B. Hendrickson, C. Foote, N. Yoshimura: Chem. Comm., **1965**, 165.
171) H. Irie, S. Uyeo, A. Yoshitake: Chem. Comm., **1966**, 635.
172) G.L. Smith, W.H. Whitelock, Jr.: Tetrahedron Letters, **1966**, 2711.
173) H.H. Lloyd, E.H. Kielar, R.J. Highet, S. Uyeo, H.M. Fales, W.C. Wildman: Tetrahedron Letters, **1961**, 105.
174) S. Uyeo, K. Kotera, T. Okada, S. Takagi, Y. Tsuda: Chem. Pharm. Bull., **14**, 793 (1966).
175) R.E. Lyle, E.A. Kielar, J.R. Crowder, W. C. Wildman: J. Am. Chem. Soc., **82**, 2620 (1960).
176) H.G. Boit, W. Doepke, W. Stender: Naturwiss., **45**, 262 (1958).
177) A.M. Duffield, R.T. Aplin, H. Budzikiewicz, C. Djerassi, C.M. Murphy, W.C. Wildman: J. Am. Chem. Soc., **87**, 4902 (1965).
178) L. Lewin: Ar. Pth., **68**, 333 (1912). [C., **1912**, II, 525].
179) A.L. Goosen, F.L. Warren: J. Chem. Soc., **1960**, 1094.

180) W.C. Wildman: J. Am. Chem. Soc., **80**, 5815 (1958).
181) H.G. Boit, H. Ehmke: Ber., **90**, 57 (1957).
182) H.M. Fales, W.C. Wildman: Chem. & Ind., **1958**, 561.
183) F.L. Warren, W.G. Wright: J. Chem. Soc., **1958**, 4696.
184) S. Uyeo, H.M. Fales, R.J. Highet, W.C. Wildman: J. Am. Chem. Soc., **80**, 2590 (1958).
185) W.C. Wildman, H.M. Fales: J. Am. Chem. Soc., **80**, 6465 (1958).
186) W. Stender: Dissertation, Berlin, 1958.
187) W.C. Wildman, H.M. Fales, A.R. Battersby: J. Am. Chem. Soc., **84**, 681 (1961).
188) P.W. Jeffs: Proc. Chem. Soc., **1962**, 80.
189) D.H.R. Barton, G.W. Kirby, J.B. Taylor: Proc. Chem. Soc., **1962**, 340.
190) R.D. Haugwitz, P.W. Jeffs, E. Wenkert: J. Chem. Soc., **1965**, 2001.
191) S. Ozeki: J. Pharm. Soc. Japan, **85**, 206 (1965).
192) H.M. Fales, W.C. Wildman: J. Org. Chem., **26**, 181 (1961).
193) H. Irie, Y. Tsuda, S. Uyeo: J. Chem. Soc., **1959**, 1446.
194) N.F. Proskurnina: Dokl. Akad. Nauk USSR., **103**, 851 (1955). [C.A., **50**, 9434 (1956)].
195) W.C. Wildman: Chem. & Ind., **1956**, 123.
196) H.G. Boit, W. Stender: Ber., **89**, 161(1956).
197) J. Goosen, P.W. Jeffs, J. Graham, F.L. Warren, W.G. Wright: J. Chem. Soc., **1960**, 1088.
198) C.F. Murphy, W.C. Wildman: Tetrahedron Letters, **1964**, 3863.
199) R.W. King, C.F. Murphy, W.C. Wildmann: J. Am. Chem. Soc., **87**, 4912(1965).
200) W.A. Hawksworth, P.W. Jeffs, B.K. Tidd, T.P. Toube: J. Chem. Soc., **1965**, 1991.
201) A. Goosen, F.L. Warren: Chem. & Ind., **1957**, 267.
202) E.W. Warnhoff, W.C. Wildman: Chem. & Ind., **1958**, 1293.
203) H. Hauth, D. Stauffacher: Helv. Chim. Acta, **46**, 810 (1963).
204) P. Naegelli, E.W. Warnhoff, H.M. Fales, R.E. Lyle, W.C. Lyle, W.C. Wildman: J. Org. Chem., **28**, 206 (1963).
205) K. Bodendorf, W. Krieger: Arch. Pharm., **290**, 441 (1957).
206) A. Popelak, E. Haack, G. Lettenbauer, H. Spingler: Naturwiss., **47**, 156 (1960).
207) E. Smith, N. Hosansky, M. Shamma, J.B. Mors: Chem. & Ind., **1961**, 402.
208) M. Shamma, H.R. Rodriguez: Tetrahedron Letters, **1965**, 4847.
209) Kh. Allayarov, Kh.A. Abduazimov, S.Yu. Yunusov: Uzbeksk. Khim. Nauk Uz. SSR., **21**, 24(1964). [C.A., **62**, 10471(1965)].
210) B. Mehlis: Naturwiss., **52**, 33 (1965).
211) Kh.A. Abduazimov, L.S. Smirnova, S.Yu. Yunusov: Dokl. Akad. Nauk Uz. SSR., **21**, 24 (1964). [C.A., **62**, 10471(1965)].
212) H. Kondo, K. Tomimura: J. Pharm. Soc. Japan, **48**, 76 (1928).
213) F. Kolle, K. Gloppe: Pharm. Zentralhalle, **75**, 237 (1934). [C., **1934**, II, 261].
214) H.G. Boit, L. Paul, W. Stender: Ber., **88**, 133 (1955).
215) T. Kitagawa, W.I. Taylor, S. Uyeo, H. Yajima: J. Chem. Soc., **1955**, 1066.
216) S. Uyeo, T. Kitagawa, Y. Yamamoto: Chem. Pharm. Bull., **12**, 408 (1964).
217) S.Yu. Yunusov, Kh.A. Abduazimov: Ber. Akad. Wiss. Uz. SSR., **1953**, No.6, 44. [C. A., **49**, 1281 (1955)].
218) N.F. Proskurnina: Zhur. Obshchei Khim., **33**, 1686 (1963). [C.A., **59**, 14035(1963)].
219) Kh.A. Abduazimov, S.Yu. Yunusov: Dokl. Akad. Nauk SSR., **153**, 1315 (1963).[C.A., **60**, 9324 (1964)].
220) D.F.C. Garbutt, P.W. Jeffs, F.L. Warren: J. Chem. Soc., **1962**, 5010.
221) H. Kondo, T. Ikeda: Ber., **73**, 867 (1940).
222) H. Kondo, T. Ikeda: Ann. Rept. ITSUU Lab., **3**, 55 (1954).
223) E. Wenkert, J.H. Hansen: Chem. & Ind., **1954**, 1262.
224) R.J. Highet, W.C. Wildman: J. Am. Chem. Soc., **77**, 4399 (1955).
225) G.R. Clemo, R. Robinson: Chem. & Ind., **1955**, 1086.
226) S. Uyeo, H. Yajima: J. Chem. Soc., **1955**, 3392.
227) T. Ikeda, W.I. Taylor, Y. Tsuda, S. Uyeo, H. Yajima: J. Chem. Soc., **1956**, 4749.
228) P.W. Jeffs, W.A. Hawkowth: Tetrahedron Letters, **1963**, 217.
229) Kh.A. Abduazimov, L.S. Smirnova, S. Yu. Yunusov: Dokl. Akad. Nauk Uz. SSR., **20**, 19 (1963). [C.A., **60**, 10736 (1964)].
230) S. Ozeki: J. Pharm. Soc. Japan, **85**, 699 (1965).
231) W. Doepke, M. Bienert: Pharmazie, **21**, 323 (1966).
232) C.F. Murphy, W.C. Wildman: Tetrahedron Letters, **1964**, 3857.
233) G.R. Clemo, D.G.I. Felton: Chem. & Ind., **1952**, 807.
234) E. Späth, L. Kuhovec: Ber., **67**, 1501 (1934).
235) S. Norkina, A. Orechoff: Ber., **69**, 500

(1936).
236) S. Yu. Yunusov, Kh. A. Abduazimov: Ber. Akad. Wiss. Uz. SSR., **1956**, No. 5, 11; Zhur. Obshchei Khim., **29**, 1724 (1959). [C., **1960**, 1510].
237) H. Kondo, T. Ikeda, N. Okuda: Ann. Rept. ITSUU Lab., **1**, 21, 61 (1950).
238) H. Kondo, T. Ikeda: Ann. Rept. ITSUU. Lab., **2**, 55 (1951).
239) H. Kondo, T. Ikeda: Ann. Rept. ITSUU. Lab., **2**, 60 (1951).
240) H. Kondo, T. Ikeda, J. Taga: Ann. Rept. ITSUU. Lab., **3**, 65 (1952).
241) H. Kondo, T. Ikeda, J. Taga: Ann. Rept. ITSUU. Lab., **4**, 73 (1953).
242) H. Kondo, T. Ikeda, J. Taga: Ann. Rept. ITSUU. Lab., **5**, 72 (1954).
243) G. R. Clemo, M. Haggarth: Chem. & Ind., **1954**, 1046.
244) E. Wenkert: Experientia, **10**, 476 (1954).
245) T. Ikeda, W. I. Taylor, S. Uyeo: Chem. & Ind., **1955**, 1088.
246) R. J. Highet, W. C. Wildman: Chem. & Ind., **1955**, 1159.
247) W. I. Taylor, S. Uyeo, H. Yajima: J. Chem. Soc., **1955**, 2692.
248) T. Ikeda, W. I. Taylor, Y. Tsuda, S. Uyeo: Chem. & Ind., **1956**, 411.
249) K. Wiesner, Z. Valenta: Chem. & Ind., Brit. Inds., **50**, Fair. Rev. Apr. 1956, R. 36 [C. A., **50**, 15026 (1956)].
250) Y. Tsuda, S. Uyeo: J. Chem. Soc., **1961**, 1055.
251) Y. Tsuda, S. Uyeo: J. Chem. Soc., **1961**, 2485.
252) R. J. Highet, P. F. Highet: Tetrahedron Letters, **1966**, 4099.
253) S. Uyeo, H. Irie, T. Kitagawa, T. Hirose, A. Yoshitake: Chem. Pharm. Bull., **12**, 489 (1964).
254) D. H. R. Barton, G. W. Kirby: J. Chem. Soc., **1962**. 806.
255) G. W. Kirby, H. P. Tiwari: J. Chem. Soc., **1964**, 4655.
256) D. H. R. Barton, G. W. Kirby, J. B. Taylor, G. H. Thomas: Proc. Chem. Soc., **1962**, 180.
257) S. Uyeo, S. Kobayashi: Pharm. Bull., **1**, 139 (1953).
258) N. F. Proskurnina, A. P. Jakowlewa: Zhur. Obshchei Khim., **22**, 1899 (1952). [C., **1955**, 8628].
259) S. Uyeo, J. Koizumi: Pharm. Bull., **1**, 202 (1953).
260) N. F. Proskurnina, A. P. Jackowlewa: Zhur. Obshchei Khim., **25**, 1035 (1955). [C., **1956**, 1025].
261) S. Kobayashi, T. Shingu, S. Uyeo: Chem. & Ind., **1956**, 177.
262) S. Kobayashi, S. Uyeo: J. Chem. Soc., **1957**, 638.
263) D. J. Williams, D. Rogers: Proc. Chem. Soc., **1964**, 357.
264) J. Koizumi, S. Uyeo: Chem. Pharm. Bull., **12**, 696 (1964).
265) S. Minami, S. Uyeo: Chem. Pharm. Bull., **12**, 1012 (1964).
266) H. Kondo, S. Ishiwata: J. Pharm. Soc. Japan, **53**, 807 (1933).
267) Y. Kihara: Bull. Agr. Chem. Soc. Japan, **15**, 17 (1939).
268) W. Doepke: Naturwiss., **49**, 469 (1962).
269) N. F. Proskurnia, A. P. Jakowalowa: Zhur. Obshchei Khim., **26**, 172 (1956). [C., **1956**, 11436].
270) S. Yu. Yunusov, Kh. A. Abduazimov: Ber. Akad. Wiss. Uz. SSR., **1953**, No. 6, 44. [C. A., **49**, 1281 (1955)].
271) M. Shio, T. Sato, H. Koyama: Chem. & Ind., **1966**, 1229.
272) K. Kotera, Y. Hamada, R. Mitsui: Tetrahedron Letters, **1966**, 6273.

Chapter 22 Benzophenanthridine Alkaloids

Norchelidonine $C_{19}H_{17}O_5N \equiv 339.33$

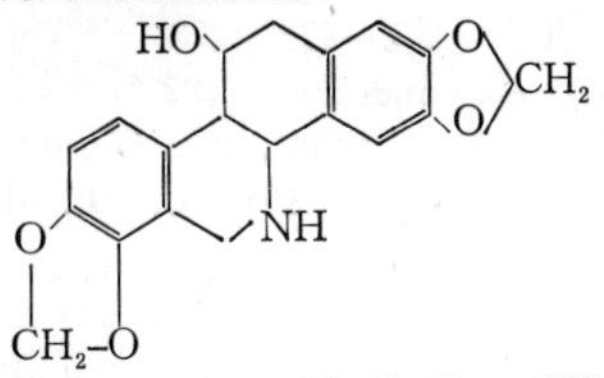

≪mp≫ 198～199° (Et_2O or $CHCl_3$–EtOH)

≪$[\alpha]_D$≫ $-112\pm3°$ (EtOH), $-100\pm3°$ ($CHCl_3$)

≪Ph.≫ UV[3] IR[3]

≪OP.≫ Glaucium flavum[1], G. flavum var. fulvum[1]

≪C. & So.≫ Prisms, Sol. in $CHCl_3$, EtOH, Spar. sol. in Et_2O

≪Sa. & D.≫ B–NAc: 195～196° B–O, N–DiAc: 140～155° B–N–NO: 221～222°

≪SD.≫ 1), 2), 3)

≪S. & B.≫

Oxysanguinarine $C_{20}H_{13}O_5N \equiv 347.31$

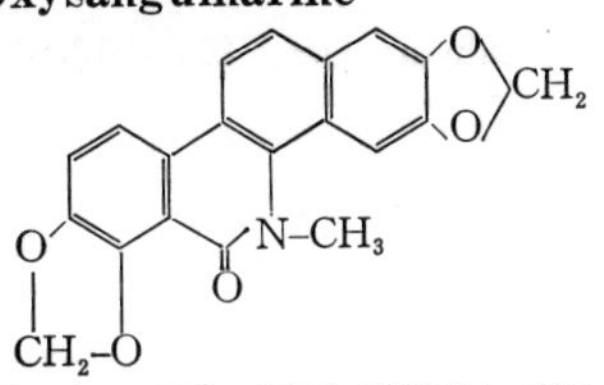

≪mp≫ 360～361° ($CHCl_3$), 356～358° (Py), cf. 223～224°

≪$[\alpha]_D$≫ 0°

≪Ph.≫

≪OP.≫ Chelidonium majus[4], Dicranostigma lactinocoides[5], Macleaya cordata[6], Sanguinaria canadensis[7][8]

≪C. & So.≫ Needles, Sol. in hot Py, Insol. in $CHCl_3$, EtOH

≪Sa. & D.≫

≪SD.≫ 7), 9)

≪S. & B.≫ 9)

Dihydrosanguinarine $C_{20}H_{15}O_4N \equiv 333.33$

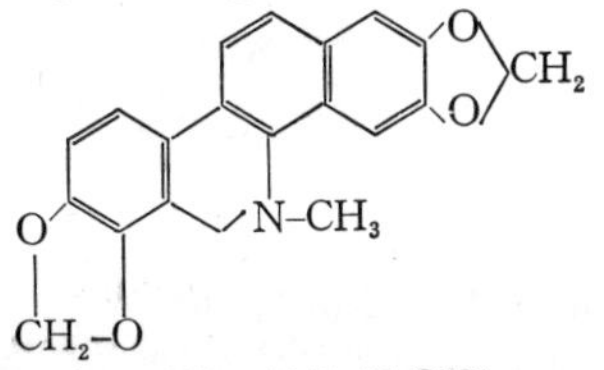

≪mp≫ 190～191° (EtOH)

≪$[\alpha]_D$≫ 0°

≪Ph.≫ Rf[11]

≪OP.≫ Argemone mericana[10][11], Chelidonium majus[12]

≪C. & So.≫

≪Sa. & D.≫ B–HCl: 192°

≪SD.≫ 7), 9), 13), 14), 15), 16)

≪S. & B.≫ 13), 14), 16)

Sanguinarine $C_{20}H_{15}O_5N \equiv 349.33$
(φ-Chelerythrine)

≪mp≫ 205~206° or 211° ($CHCl_3$–EtOH), 242~243° or 264° (Et_2O), 266~267° ($CHCl_3$)
≪$[\alpha]_D$≫ 0°

≪Ph.≫ UV[8] Rf[11,26,28,32,36,40,41,53] Polaro[54]
≪OP.≫ Argemone alba[17], A. mexicana[11], Bocconia (Macleaya) cordata[6,18,19], B. frutscens[20], B. pearcei[21], Chelidonium majus[12,16,22,23,24,25,26], Dicentra spectabilis[27,28]*
≪C. & So.≫ Sol. in most org. solv.
≪Sa. & D.≫ B⊕–Cl⊖ : 272~273°, 278°
B–NAc : 283° (d) φ–cyanide : 238° (227°)
≪SD.≫ 7), 9), 22), 23), 27), 44), 46), 47), 48)
≪S. & B.≫ 10), 13), 14), 15), 16), 49), 50), 51)
BS : 52)

Avicine $C_{20}H_{15}O_5N \equiv 349.33$

≪mp≫
≪$[\alpha]_D$≫ 0°

≪Ph.≫ UV[55,56]
≪OP.≫ Zanthoxylum avicennae[55]
≪C. & So.≫
≪Sa. & D.≫ B⊕–OAc⊖ : 160°
φ–cyanide : >340° Dihydro–Der. : 211~212.5°
≪SD.≫ 55)
≪S. & B.≫

Oxychelidonine $C_{20}H_{17}O_6N \equiv 363.34$

≪mp≫ 285° (AcOH or $CHCl_3$)
≪$[\alpha]_D$≫ +102.5° ($CHCl_3$–EtOH)

≪Ph.≫
≪OP.≫ Chelidonium majus[57]
≪C. & So.≫ Needles
≪Sa. & D.≫
≪SD.≫ 57), 58)
≪S. & B.≫

***d*-Chelidonine** $C_{20}H_{19}O_5N \equiv 353.36$

≪mp≫ 135~136° (MeOH), 160~161° (EtOH), 173~174° (AcOEt)
≪$[\alpha]_D$≫ +115° (EtOH), +117° ($CHCl_3$), +148±3° (AcOH)

≪Ph.≫ UV[73] IR[3,72,73,105] NMR[73] Rf[3,15]
≪OP.≫ Chelidonium majus[12,22,23,58,59,60,61,62,63], Dicentra spectabilis[27], Dicranostigma franchetianum[92], Glaucium flavum[1], G. flavum var. fulvum[1], Stylophorum diphyllum[22,45,61,64,92]
≪C. & So.≫ Prisms (MeOH), Plates (EtOH), Sol. in EtOH, Et_2O, $CHCl_3$, Insol. in H_2O
≪Sa. & D.≫ B–HCl : 295° B–$HAuCl_4$: 155°
B–OAc : 165~166° B–OBz : 217°
B–Pic : 220° (d) φ–cyanide : 243°
≪SD.≫ 3), 15), 23), 27), 47), 48), 58), 62), 63), 65), 66), 67), 68), 69), 70), 71), 72), 73)
≪S. & B.≫ 13), 14), 49), BS : 52), 98)

---

* Dicranostigma franchetianum[29], D. lactinocoides[5], D. lactucoides[21], Eschscholtzia californica[23,30,31], Glaucium corniculatum[15], G. elegans[32], G. fimbrilligrerum[33], G. flavum[1,34], G. flavum var. fulvum[1], G. oxylobum[35], G. vitellinum[36], Hunnemannia fumariaefolia[37], Hypecoum leptocarpum[38], H. procumbens[38], H. trilobum[39], Macleaya (Bocconia) microcarpa[40], Meconopsis aculeata[41], M. betonicifolia[41], M. horridula[41], M. latifolia[41], M. paniculata[41], M. rudis[41], Sanguinaria canadensis[8,22,30,42,43,44], Stylophorum diphyllum[45], Zanthoxylum lactucoides[21]

l-Chelidonine

≪mp≫ 135～136° (MeOH)
≪[α]D≫ −112±2° (EtOH)
≪Ph.≫ UV[2] Rf[15]
≪OP.≫ Glaucium corniculatum[15], G. flavum[1], G. flavum var. fulvum[1]
≪C. & So.≫ Prisms
≪Sa. & D.≫
≪SD.≫ 1), 2), 15)
≪S. & B.≫ 2)

Oxynitidine $C_{21}H_{17}O_5N$≡351.34

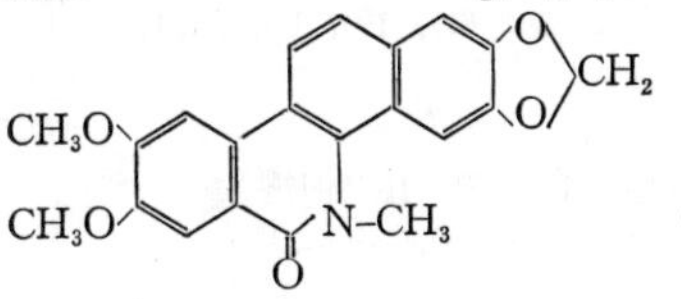

CH_2 CH_3O CH_3O N–CH_3 O

≪mp≫ 284～285° (EtOH)
≪[α]D≫ 0°
≪Ph.≫ UV[74,75]
≪OP.≫ Zanthoxylum nitidum[74,75]
≪C. & So.≫ Needles
≪Sa. & D.≫
≪SD.≫ 74), 75)
≪S. & B.≫ 34), 49)

Dihydrochelerythrine $C_{21}H_{19}O_4N$≡349.37

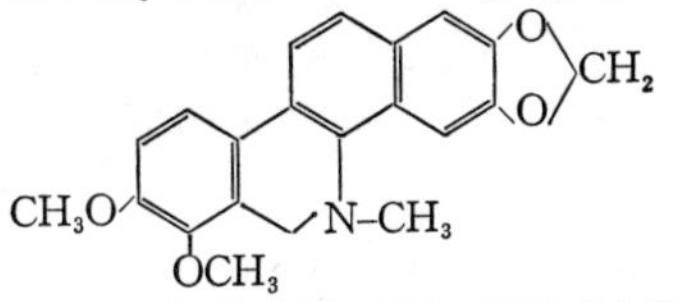

CH_2 CH_3O N–CH_3 OCH_3

≪mp≫ 143～144° (EtOH), 160～165° (MeOH)
≪[α]D≫ 0°
≪Ph.≫
≪OP.≫ Fagara semiarticulata[76], Toddalia aculeata[77]
≪C. & So.≫ Plates
≪Sa. & D.≫
≪SD.≫ 9), 13), 14), 67), 78)
≪S. & B.≫ 13), 14)

Chelerythrine (Toddaline) $C_{21}H_{19}O_5N$≡353.36

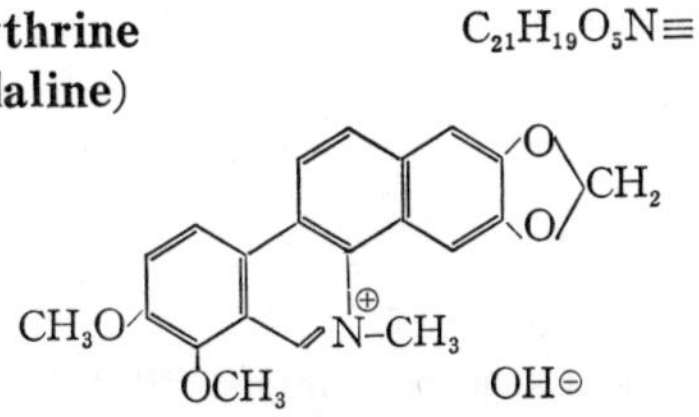

CH_2 CH_3O ⊕N–CH_3 OCH_3 OH⊖

≪mp≫ 203° (Et_2O), 207° (EtOH), 269～270° ($CHCl_3$), 282～283° (d) (Et_2O)
≪[α]D≫ 0°
≪Ph.≫ UV[23,76,88,99] IR[88] Rf[11,26,28,32,36,40,53,86,91] Polaro[54]
≪OP.≫ Argemone alba[17], A. mexicana[11], Bocconia (Macleaya) arborea[79], B. cordata[6,18,19,80], B. frutescens[81], B. pearcei[21,82], Chelidonium majus[8,9,12,22,23,26,42,58,60,83,84], Dicentra spectabilis[28]*
≪C. & So.≫ Needles, Sol. in $CHCl_3$, C_6H_6, Spar. sol. in EtOH, Me_2CO, AcOEt
≪Sa. & D.≫ B⊕–Cl⊖ : 210° (d) [cf. 203～205° (d), 213～214° (d)] B⊕–NO_3⊖ : 240° B⊕–$^1/_2SO_4$⊖ : 236～239° B–Pic : 237～238° B⊕–$AuCl_4$⊖ : 233° (d) B⊕–$HPtCl_6$⊖ : 254～256° ϕ-cyanide : 221° or 260～261° (d)
≪SD.≫ 9), 24), 25), 42), 47), 67), 78), 89), 93), 94), 95), 96)
≪S. & B.≫ 9), 13), 14), 49), 50), 51), 88)
BG : 97)

* Dicranostigma franchetianum[29], D. lactucoides[21], Eschscholtzia californica[23,30,31,81], Fagara chiloperone var. angustifolia[85], F. coco[85], F. naranjiro[85], F. naranjiro var. paraguariensis[85], F. pterota[85], F. semiarticulata[76], F. xanthoxyloides[86], Glaucium corniculatum[15], G. elegans[32], G. flavum[1,34,87], G. flavum var. fulvum[1], G. fimbrilligerum[33], G. luteum[23,83], G. oxylobum[35], G. vitellum[36], Hunnemannia fumariaefolia[37], Hypecoum leptocarpum[38], H. procumbens[38], H. trilobum[39], Macleaya (Bocconia) microcarpa[40,88], Prangos probularia[39], Sanguinaria canadensis[8,42,43], Stylophorum diphyllum[45], Toddalia aculeata[77,89], Zanthoxylum brachyacanthum (Z. veneficum)[90], Z. rhetsa[91]

Nitidine $C_{21}H_{19}O_5N \equiv 353.36$

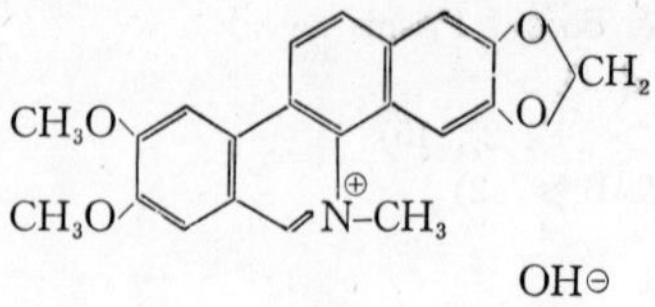

≪mp≫

≪$[\alpha]_D$≫ 0°

≪Ph.≫ UV[75)]
≪OP.≫ Fagara chiloperone var. angustifolia[85)], F. coco[85)], F. nigrescens[85)], F. pterota[85)], Zanthoxylum nitidum[74)75)]
≪C. & So.≫
≪Sa. & D.≫ B⊕–I⊖ : 285～286° B⊕–OAc⊖ : 255～260° ó–cyanide : 234° (d)
≪SD.≫ 74), 75)
≪S. & B.≫ 14), 49), 100), 101)

Corynoline $C_{21}H_{21}O_5N \equiv 355.37$

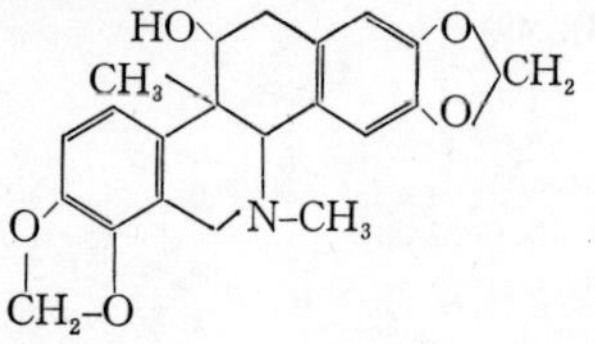

≪mp≫

≪$[\alpha]_D$≫

≪Ph.≫ UV[102)] IR[102)103)]
≪OP.≫ Corydalis incisa[102)]
≪C. & So.≫
≪Sa. & D.≫ B–OAc : 159～160°
≪SD.≫ 102), 103)
≪S. & B.≫

Methoxychelidonine $C_{21}H_{21}O_6N \equiv 385.40$

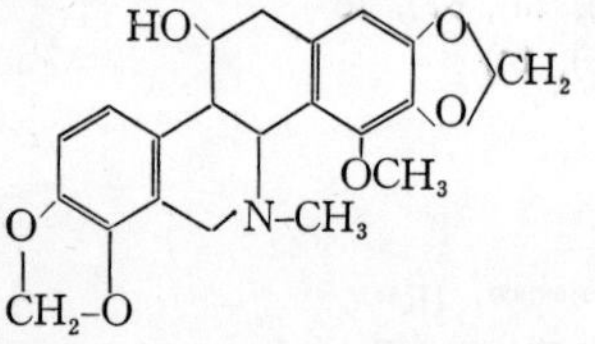

≪mp≫ 221°

≪$[\alpha]_D$≫ +115° ($CHCl_3$)

≪Ph.≫
≪OP.≫ Chelidonium majus[44)]
≪C. & So.≫
≪Sa. & D.≫ B–$HAuCl_4$: 237～238° B–OAc : 147°
≪SD.≫ 44), 47), 71)
≪S. & B.≫

α-Homochelidonine $C_{21}H_{23}O_5N \equiv 369.40$

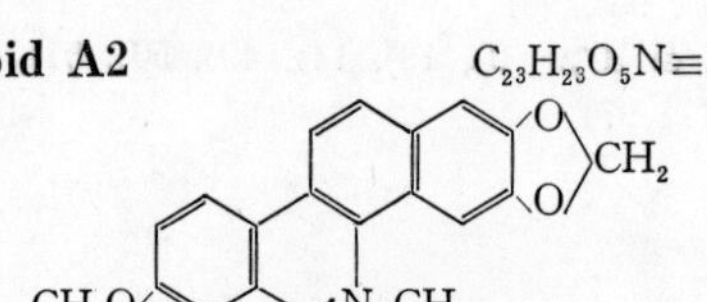

≪mp≫ 181～182° (AcOEt or Et_2O)

≪$[\alpha]_D$≫ +118° (EtOH), +116° ($CHCl_3$)

≪Ph.≫ IR[3)]
≪OP.≫ Chelidonium majus[12)23)42)44)84)], Eschscholtzia californica[30)]
≪C. & So.≫ Prisms, Sol. in EtOH, $CHCl_3$, AcOEt, AcOH, Spar. sol. in Et_2O
≪Sa. & D.≫
≪SD.≫ 3), 47), 62), 64), 65), 66), 67), 71), 96)
≪S. & B.≫

Alkaloid A2 $C_{23}H_{23}O_5N \equiv 393.42$

≪mp≫ 200～202° and 239～242° (Double mp)

≪$[\alpha]_D$≫

≪Ph.≫ UV[86)] IR[86)] NMR[86)] Mass[86)]
≪OP.≫ Fagara xanthoxyloides[86)104)]
≪C. & So.≫
≪Sa. & D.≫
≪SD.≫ 86), 104)
≪S. & B.≫

Toddalinine $C_{19}H_{15}O_4N$≡321.32
≪mp≫ 185° (MeOH), cf. 180~200° (d)
≪$[\nu]_D$≫
≪Ph.≫
≪OP.≫ Toddalia aculeata[77,89]
≪C. & So.≫ Amorph
≪Sa. & D.≫ B–HCl: 283~285° or 245° (d)
B–$HAuCl_4$: 231~232° (d) B–Pic: 230~235°
≪SD.≫ 89)
≪S. & B.≫

Chelilubine $C_{20}H_{19}O_6N$≡369.36
≪mp≫ 211~213° or 217~218° (Et_2O),
230~231° ($CHCl_3$–EtOH), 257~258° (Et_2O)
≪$[\alpha]_D$≫ 0°
≪Ph.≫ Rf[8,28,32,40]
≪OP.≫ Chelidonium majus[12,26], Dicentra spectabilis[28], Dicranostigma franchetianum[29], D. leaticoides[5]*
≪C. & So.≫ Sol. in $CHCl_3$
≪Sa. & D.≫ B⊕–Cl⊖: 282~283°
ψ–cyanide: 269~270°
≪SD.≫
≪S. & B.≫

Sanguirubine $C_{23}H_{21}O_6N$≡407.41
≪mp≫ 224~225° ($CHCl_3$–EtOH),
252~253° (Et_2O)
≪$[\alpha]_D$≫ 0°
≪Ph.≫ Rf[8]
≪OP.≫ Sanguinaria canadensis[8]
≪C. & So.≫ Prisms
≪Sa. & D.≫ B⊕–Cl⊖: 275~276°
ψ–cyanide: 237~238°
≪SD.≫
≪S. & B.≫

Sanguirutine $C_{23}H_{25}O_6N$≡411.44
≪mp≫ 211~212° ($CHCl_3$–EtOH),
246~247° (Et_2O)
≪$[\alpha]_D$≫ 0°
≪Ph.≫ Rf[8]
≪OP.≫ Sanguinaria canadensis[8]
≪C. & So.≫ Sol. in $CHCl_3$
≪Sa. & D.≫ B⊕–Cl⊖: 163~164° (d)
ψ–cyanide: 232~233° (d)
≪SD.≫
≪S. & B.≫

Chelilutine $C_{23}H_{25}O_6N$≡411.44
≪mp≫ 202~203° ($CHCl_3$–EtOH),
229~230° (Et_2O)
≪$[\alpha]_D$≫ 0°
≪Ph.≫ Rf[8,40]
≪OP.≫ Chelidonium majus[8,12,26], Dicentra spectabilis[28], Eschscholtzia californica[31], Glaucium oxylobum[35]**
≪C. & So.≫ Needles, Sol. in $CHCl_3$, Et_2O
≪Sa. & D.≫ B⊕–Cl⊖: 197~198° (d)
ψ–cyanide: 270.5~271°
≪SD.≫
≪S. & B.≫

Bocconia Alkaloid P-61 $C_{21}H_{19}O_5N$≡365.37
≪mp≫ 210°
≪$[\alpha]_D$≫
≪Ph.≫
≪OP.≫ Bocconia arborea[79]
≪C. & So.≫
≪Sa. & D.≫
≪SD.≫ 79)
≪S. & B.≫

---

* Eschscholtzia californica[31], Glaucium canadensis[8], G. corniculatum[15], G. elegans[32], G. flavum[1,34], G. flavum var. fulvum[1], G. oxylobum[35], G. vitellinum[36], Hunnemannia fumariaefolia[37], Macleaya (Bocconia) microcarpa[40]

** Hunnemannia fumariaefolia[37], Macleaya microcarpa[40], Sanguinaria canadensis[8]

Bocconia Alkaloid A $C_{20}H_{17}O_4N \equiv 335.34$
≪mp≫ 302°
≪$[\alpha]_D$≫
≪Ph.≫
≪OP.≫ Bocconia arborea[79]
≪C. & So.≫
≪Sa. & D.≫
≪SD.≫
≪S. & B.≫

Bocconia Alkaloid B $C_{20}H_{15}O_4N \equiv 333.33$
≪mp≫ 191°
≪$[\alpha]_D$≫
≪Ph.≫
≪OP.≫ Bocconia arborea[79]
≪C. & So.≫
≪Sa. & D.≫
≪SD.≫
≪S. & B.≫

Bocconia Alkaloid C $C_{31}H_{33}O_5N \equiv 499.58$
≪mp≫ 327°
≪$[\alpha]_D$≫
≪Ph.≫
≪OP.≫ Bocconia arborea[79]
≪C. & So.≫
≪Sa. & D.≫
≪SD.≫
≪S. & B.≫

β-Homochelidonine $C_{21}H_{21}O_5N$ or $C_{21}H_{23}O_5N$
≪mp≫ 159° ($EtOH-Et_2O$), 169~170° (EtOH)
≪$[\alpha]_D$≫
≪Ph.≫
≪OP.≫ Bocconia cordata[19], Chelidonium majus[22][23][58], Eschscholtzia californica[23], Sanguinaria canadensis[22], Zanthoxylum brachyacanthun (Z. veneficum)[90]
≪C. & So.≫ Needles
≪Sa. & D.≫
≪SD.≫
≪S. & B.≫

γ-Homochelidonine $C_{21}H_{21}O_5N$ or $C_{21}H_{23}O_5N$
≪mp≫ 160~161° (Et_2O)
≪$[\alpha]_D$≫
≪Ph.≫
≪OP.≫ Chelidonium majus[58], Eschscholtzia californica[23], Sanguinaria canadensis[42]
≪C. & So.≫
≪Sa. & D.≫
≪SD.≫
≪S. & B.≫

1) J. Slavik, L. Slavikova: Coll. Czech. Chem. Comm., **24**, 3141 (1959).
2) J. Slavik: Coll. Czech. Chem. Comm., **24**, 3601 (1959).
3) F. Santávy, M. Horák, M. Maturová, J. Brabenec: Coll. Czech. Chem. Comm., **25**, 1344 (1960).
4) J. Slavik: Ceskoslow. Farm., **4**, 15 (1955).
5) J. Slavik, L. Slavikova: Coll. Czech. Chem. Comm., **26**, 1839 (1961).
6) C. Tani, N. Takao: J. Pharm. Soc. Japan, **82**, 755 (1962).
7) E. Späth, F. Schlemmer, G. Schenck, A. Gempp: Ber., **70**, 1677 (1937).
8) J. Slavik, L. Slavikova: Coll. Czech. Chem. Comm., **25**, 1667 (1960).
9) E. Späth, F. Kuffner: Ber., **64**, 2034 (1931).
10) S.N. Sarkar: Nature, **162**, 265 (1948).
11) L. Slavikova, J. Slavik: Coll. Czech. Chem. Comm., **21**, 211 (1956).
12) E. Seoana: Annales Reat. Soc. Espan. Fris. Quim. Ser., **B**, **61**, 747 (1965). [C.A., **63**, 18187 (1965)].
13) A.S. Bailey, R. Robinson: J. Chem. Soc., **1950**, 1375.
14) A.S. Bailey, R. Robinson, R.S. Staunton: J. Chem. Soc., **1950**, 2277.
15) J. Slavik, L. Slavikova: Coll. Czech. Chem. Comm., **22**, 279 (1957).
16) D. Beke, M.B. Bárczaí, L. Toke: Magyar Kém. Folyoírat, **64**, 125 (1958). [C.A., **54**, 17437 (1960)].
17) L. Slavikova, T. Shun, J. Slavik: Coll. Czech. Chem. Comm., **25**, 756 (1960).
18) J. F. Eijkman: Rec. trav. chim., **3**, 182 (1884); Pharm. J., [3], **13**, 87 (1882).
19) P. Murrill, J.O. Schlotterbeck: Ber., **33**, 2802 (1900).
20) Greshoff: Mededeel. Land. Plantent., **25**, 28 (1898).

21) R.H.F. Manske: Can. J. Chem., **32**, 83 (1954).
22) E. Schmidt: Arch. Pharm., **231**, 136, 177 (1893).
23) E. Schmidt: Arch. Pharm., **239**, 395 (1901).
24) J. Gadamer, A. Stichel: Arch. Pharm., **262**, 488 (1924).
25) J. Gadamer, K. Winterfeld: Arch. Pharm., **262**, 483 (1924).
26) J. Slavik, L. Slavikova: Coll. Czech. Chem. Comm., **20**, 21 (1954).
27) P.W. Danckwortt: Arch. Pharm., **260**, 94 (1922).
28) J. Slavik: Coll. Czech. Chem. Comm., **24**, 2506 (1959).
29) L. Slavikova, J. Slavik: Chem. Listy, **51**, 1923 (1957).
30) R. Fischer: Arch. Pharm., **239**, 421 (1901).
31) J. Slavik, L. Slavikova: Coll. Czech. Chem. Comm., **20**, 27 (1954).
32) J. Slavik: Coll. Czech. Chem. Comm., **25**, 1698 (1960).
33) R.A. Konovalova, S. Yu. Yunusov, A.P. Orechov: Zhur. Obshchei Khim., USSR, **9**, 1939 (1939).
34) J. Slavik: Coll. Czech. Chem. Comm., **20**, 32 (1955).
35) J. Slavik, L. Slavikova: Coll. Czech. Chem. Comm., **28**, 2530 (1963).
36) J. Slavik: Coll. Czech. Chem. Comm., **24**, 3999 (1959).
37) L. Slavikova, J. Slavik: Coll. Czech. Chem. Comm., **31**, 1355 (1966).
38) J. Slavik, L. Slavikova: Coll. Czech. Chem. Comm., **26**, 1472 (1961).
39) S. Yu. Yunusov, S. T. Akramov, G. P. Sidyakin: Dokl. Akad. Nauk Uz. SSR., **1957**, No. 7, 23. [C.A., **53**, 3606 (1959)].
40) J. Slavik, L. Slavikova: Coll. Czech. Chem. Comm., **20**, 356 (1955).
41) J. Slavik: Coll. Czech. Chem. Comm., **25**, 1663 (1960).
42) E. Schmidt: Arch. Pharm., **231**, 136, 145 (1893).
43) R. Fischer: Arch. Pharm., **239**, 409 (1901).
44) J. Gadamer, K. Winterfeld: Arch. Pharm., **262**, 589 (1924).
45) J.O. Schlotterbeck, H.C. Watkins: Ber., **35**, 7 (1902).
46) J. Gadamer, A. Stichel: Arch. Pharm., **262**, 248 (1924).
47) F.v. Bruchhausen, H.W. Bersch: Ber., **63**, 2520 (1930).
48) E. Späth, F. Kuffner: Ber., **64**, 370 (1931).
49) T. Richardson, R. Robinson, E. Seijo: J. Chem. Soc., **1937**, 835.
50) A.S. Bailey, R. Robinson: Nature, **164**, 402 (1949).
51) A.S. Bailey, R. Robinson, R. S. Staunton: Nature, **165**, 235 (1950).
52) E. Leete: J. Am. Chem. Soc., **85**, 473 (1963).
53) R.H.F. Manske: "The Alkaloids", Vol. VII, p. 431, edited by R.H.F. Manske, Academic Press, New York (1960).
54) J. Bartek, F. Santavy: Chem. Listy, **47**, 461 (1953).
55) H. R. Arthur, W. H. Hui, Y. L. Ng: J. Chem. Soc., **1959**, 4007.
56) G.E. Foster, J. Macdonald: J. Pharm. Pharmcol., **3**, 127 (1951).
57) J. Gadamer, M. Theissen: Arch. Pharm., **262**, 578 (1924).
58) M. Wington: Arch. Pharm., **239**, 438 (1901).
59) Godefrog: J. Pharm., **10**, 635 (1824).
60) J.M. Probst: Ann., **29**, 113 (1839).
61) E. Schmidt: Arch. Pharm., **228**, 96 (1890).
62) J. Gadamer: Arch. Pharm., **262**, 249 (1924).
63) J. Gadamer, K. Winterfeld: Arch. Pharm., **262**, 452 (1924).
64) F. Selle: Arch. Pharm., **228**, 96 (1890).
65) J. Gadamer: Arch. Pharm., **257**, 298 (1919).
66) J. Gadamer: Arch. Pharm., **258**, 148 (1920).
67) J. Gadamer: Arch. Pharm., **258**, 160 (1920).
68) J. Gadamer, A. Stichel: Arch. Pharm., **262**, 499 (1924).
69) Kling: Dissertation, Marburg, (1927).
70) Schwarz: Dissertation, Marburg, (1928).
71) F.v. Bruchhausen, H.W. Bersch: Ber., **63**, 2520 (1930).
72) L. Marion, D.P. Ramsay, R.N. Jones: J. Am. Chem. Soc., **73**, 305 (1951).
73) E. Sevane: Annales Real Soc. Espan. Fris. Quim., Ser., **B**, **61**, 755 (1965). [C.A. **63**, 16395 (1965)].
74) H. R. Arthur, W.H. Hygi, Y.L. Ng: Chem. & Ind., **1958**, 1514.
75) H.R. Arthur, W.H. Hygi, Y.L. Ng: J. Chem. Soc., **1959**, 1840.
76) P.J. Scheuer, M.Y. Chang, C.E. Swanholn: J. Org. Chem., **27**, 1472 (1962).
77) T.R. Govindachari, R.S. Thyagarajan: J. Chem. Soc., **1956**, 769.
78) P. Karrer: Ber., **50**, 212 (1917).
79) R.H.F. Manske: Can. J. Res., **21 B**, 140 (1943).
80) E. Schmidt: Arch. Pharm., **239**, 401 (1901).
81) J.A. Battandier: Compt. rend., **120**, 1276 (1895).
82) I. Maccio: Arch. farm. y. bioquim. Tucumám., **3**, 27 (1946). [C.A., **41**, 3507 (1947)].
83) J.M. Probst: Ann., **29**, 120 (1839).

84) R. H. F. Manske: Can. J. Res., **21 B**, 140 (1943).

85) A. M. Kuck, S. M. Albonico, V. Deulofeu: Chem. & Ind., **1966**, 945.

86) F. G. Torto, P. Sefcovic, B. A. Dadson: Tetrahedron Letters, **1966**, 181.

87) J. M. Probst: Ann., **31**, 241 (1839).

88) A. S. Bailey, C. R. Worthing: J. Chem. Soc., **1956**, 4535.

89) B. B. Dey, P. P. Pillay: Arch. Pharm., **271**, 477 (1933).

90) J. R. Canon, G. K. Hughes, E. Ritchie, W. C. Taylor: Austral. J. Chem., **6**, 86 (1953).

91) A. Chatterjee: Tetrahedron, **7**, 257 (1959).

92) R. H. F. Manske: Can. J. Res., **20 B**, 53 (1942).

93) K. H. Bauer, K. J. Hedinger: Arch. Pharm., **258**, 167 (1920).

94) P. Karrer: Helv. Chim. Acta, **6**, 232 (1923).

95) J. Gadamer, A. Stichel: Arch. Pharm., **262**, 498 (1924).

96) E. Späth, F. Kuffner: Ber., **64**, 1123 (1931).

97) R. Robinson: "The Structural Relations of Natural Products", Clarendon Press, Oxford (1955)

98) A. R. Battersby, R. J. Francis, E. A. Ruveda, J. Staunton: Chem. Comm., **1965**, 89.

99) S. F. Mason: "Physical Methods in Heterocyclic Chemistry", Vol. II, p. 7. edited by A. R. Katritsky, Academic Press, 1963.

100) H. R. Arthur, Y. L. Ng: J. Chem. Soc., **1959**, 4010.

101) K. W. Gopinath, T. R. Govindachari, P. C. Pathasarathy, N. Viswanathan: J. Chem. Soc., **1959**, 4012.

102) N. Takao: Chem. Pharm. Bull., **11**, 1306 (1963).

103) N. Takao: Chem. Pharm. Bull., **11**, 1312 (1963).

104) R. Paris, H. Moyse-Mignox: Ann. Pharm. France, **5**, 410 (1947).

105) H. W. Bersch: Arch. Pharm., **291**, 491 (1958).

Chapter 23 Dibenzopyrrocoline Alkaloids

Cryptaustoline $C_{20}H_{24}O_4N^{\oplus} \equiv 342.39$

≪mp≫

≪$[\alpha]_D$≫ B⊕–I⊖ : −151° (EtOH)

≪Ph.≫

≪OP.≫ Cryptocarya bowiei[1]

≪C. & So.≫

≪Sa. & D.≫ B⊕–I⊖ : 214° (d)

B⊕–OMe–I⊖ : 153∼155° B⊕–OEt–I⊖ : 90∼95°

≪SD.≫ 1), 2)

≪S. & B.≫ 3), cf. 4), 5), 6), 8), BG: 7)

Cryptowoline $C_{19}H_{20}O_4N^{\oplus} \equiv 326.35$

≪mp≫

≪$[\alpha]_D$≫ B⊕–I⊖ : −186° (EtOH)

≪Ph.≫

≪OP.≫ Cryptocarya bowiei[2]

≪C. & So.≫

≪Sa. & D.≫ B⊕–I⊖ : 245∼246°

B⊕–OMe–I⊖ : 227° B⊕–OEt–I⊖ : 215°

≪SD.≫ 2)

≪S. & B.≫ 8)

---

1) J. Ewing, G. K. Hughes, E. Ritchie, W. C. Taylor: Nature, **169**, 618 (1952). [C. A., **47**, 6954 (1953)].
2) J. Ewing, G. K. Hughes, E. Ritchie, W. C. Taylor: Austral. J. Chem., **6**, 78 (1953). [C. A., **47**, 12399 (1953)].
3) G. K. Hughes, E. Ritchie, W. C. Taylor: Austral. J. Chem., **6**, 315 (1953). [C. A., **48**, 6446 (1954)].
4) R. Robinson, S. Sugasawa: J. Chem. Soc., **1932**, 789.
5) C. Schöpf, K. Thierfelder: Ann., **497**, 22 (1932).
6) J. Harley-Mason: J. Chem. Soc., **1953**, 1465.
7) R. Robinson: "The Structural Relations of Natural Products", Clarendon Press, Oxford (1955).
8) T. Kametani, K. Ogasawara: J. Chem. Soc. (C), **1967**, 2208

Chapter 24 Phenethylisoquinoline and Colchicine Alkaloids

Androcymbine $C_{21}H_{25}O_5N \equiv 371.42$

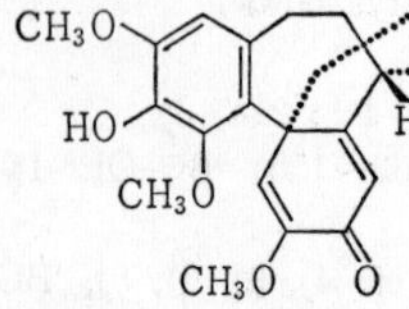

≪mp≫ 199~201° (EtOAc or Me_2CO)
≪$[\alpha]_D$≫ 260±2° ($CHCl_3$)

≪Ph.≫ CR[1] UV[1)2)] IR[1)2)] NMR[2] Mass[2] ORD[2] Rf[1]
≪OP.≫ Androcymbium melanthioides var. stricta[1]
≪C. & So.≫ Colourless prisms
≪Sa. & D.≫
≪SD.≫ 2)
≪S. & B.≫ BG: 2)

Melanthioidine $C_{38}H_{42}O_6N_2 \equiv 622.73$

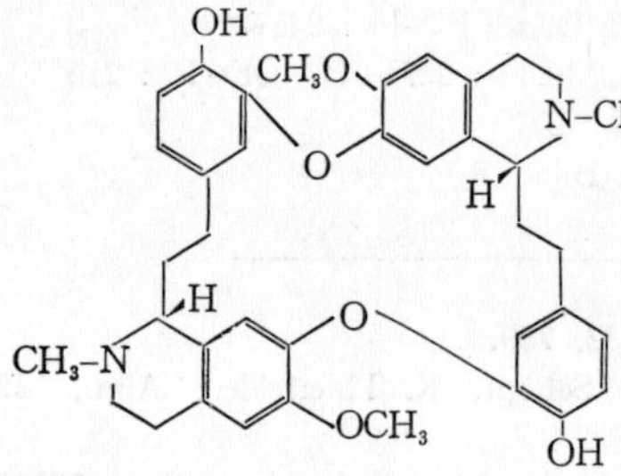

≪mp≫ 142~144° (AcOEt or Me_2CO)
≪$[\alpha]_D$≫ −63±3° ($CHCl_3$)
≪Ph.≫ UV[1)3)] IR[1)3)] NMR[3] Mass[3] ORD[3] Rf[1]
≪OP.≫ Androcymbium melanthioides var. stricta[1]
≪C. & So.≫ Colourless prisms
≪Sa. & D.≫
≪SD.≫ 3)
≪S. & B.≫

Colchicine $C_{22}H_{25}O_6N \equiv 399.43$

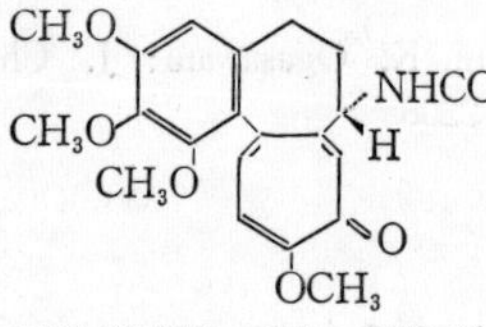

≪mp≫ 140° (C_6H_6), 155~157° (AcOEt)
≪$[\alpha]_D$≫ −121° ($CHCl_3$), −219° (MeOH), −429° (H_2O)
≪Ph.≫ CR[8)16)18)52)]
UV[81)98)105)107)111)116)122)123)124)148)]
IR[90)94)98)99)104)105)107)111)116)124)] NMR[125)126)]
Mass[127] Rf[6] pKa[122)123)] Polaro[81)84)] X-ray[97)128)]

≪OP.≫ Androcymbium gramineum[4)5)6)], A. melanthioides[1)7)], A. sphodelus[8], Bulbocodium tofieldia[8], Colchicum agrippium[9], C. alpinum[10], C. arenarium[11], C. autumnale[1)4)6)11)~31)], C. bornmnelleri[11], C. byzanthium[32)33)]*
≪C. & So.≫ Colourless needles, Sol. in EtOH, $CHCl_3$, H_2O, Very spar. sol. in Et_2O, C_6H_6. cf purification[49]
≪Sa. & D.≫ B-$HAuCl_4$: 209° B-Bz: 207~209°
≪SD.≫ 16), 23), 50)~115)
≪S. & B.≫ 116), 117), 118), 119), 120), 121) BG: 120), 129)~135), BS: 7), 32), 33), 136), 137), 138), 139), 140), 141), 142), 143)

---

* C. cilicium[11], C. comigerum[34], C. crociflorum[35], C. cupani[18], C. hierosolymitanum[36], C. lusitanum[6], C. luteum[6)37)38)], C. monatum[10], C. multiflorum[10], C. neapolitanum[10], C. speciosum[24)29)30)39)], C. variegatum[28], C. vernum[9], Fritillaria[8], Gloriosa rothschildiana[40)41)], G. simplex[1)41)], G. superba[1)37)42)43)], G. virescens[1)44)], Hemerocallis[8], Littonia modesta[1)45)], Lloydia[8], Merendera attica[9], M. bulbocodium (Bulbocodium ruthenicum)[8)46)], M. caucasica[8)38)], M. kesselringii[47], M. persica[37], M. robusta[47)48)], M. sobolifera[9)18)], M. trigyna[35], Muscari[8], Ornithogalum tulipa[8], Sandersonia aurantiaca[1], Veratrum anthericum[8]

Colchiceine $C_{21}H_{23}O_6N \equiv 385.40$

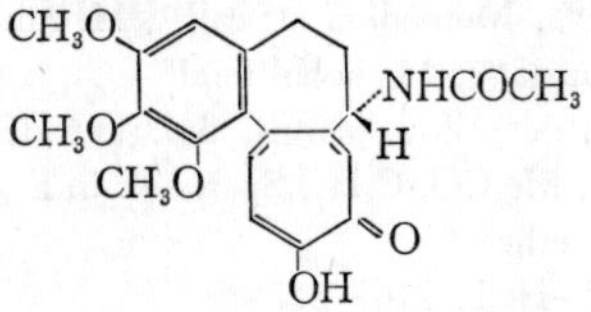

≪mp≫ 139～140° or 178～179° (Dioxane-Et_2O)
≪[α]D≫ −253° ($CHCl_3$)

≪Ph.≫ CR[16,52] UV[68,81,111,123] IR[90,94,99,104,111] Mass[127] Polaro[81,83,84]
≪OP.≫ Colchicum autumnale[15,28,30], C. cornigerum[34], C. robusta[48], C. speciosum[30,144]
≪C. & So.≫ Prisms, Sol. in EtOH, $CHCl_3$, Spar. sol. in H_2O, Insol. in Et_2O, C_6H_6
≪Sa. & D.≫ B-OAc: 124° B-OBz: 205～207°
≪SD.≫ 53), 59), 68), 69), 71), 73), 81), 83), 84), 85), 86), 87), 93), 94), 96), 99), 100), 104), 113), 145)
→colchicine
≪S. & B.≫ 146), 147)

2-Desmethyl desacetylcolchicine (Compound U) $C_{19}H_{21}O_5N \equiv 343.37$

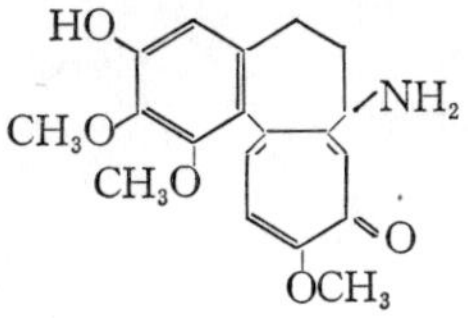

≪mp≫ 224～226°

≪[α]D≫ −93°
≪Ph.≫
≪OP.≫ Colchicum autumnale[30,149], C. speciosum[30]
≪C. & So.≫
≪Sa. & D.≫ B-DiAc: 233°
≪SD.≫ 149)
≪S. & B.≫

Demecolceine (Colchameine, Compound Ta) $C_{20}H_{23}O_5N \equiv 357.39$

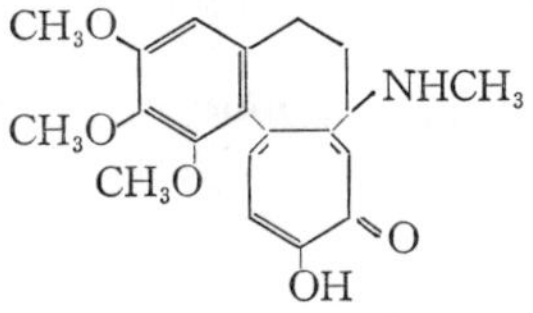

≪mp≫ 133～135° (MeOH)

≪[α]D≫ −220° ($CHCl_3$), −99° (EtOH)
≪Ph.≫ UV[111] IR[111]
≪OP.≫ Colchicum autumnale[30,150,151], Merendera robusta[48]
≪C. & So.≫
≪Sa. & D.≫ B-TriAc: 231～232°
≪SD.≫ 151), 152)
→demecolcine
≪S. & B.≫

3-Desmethyldemecolcine (Compound S) $C_{20}H_{23}O_5N \equiv 357.39$

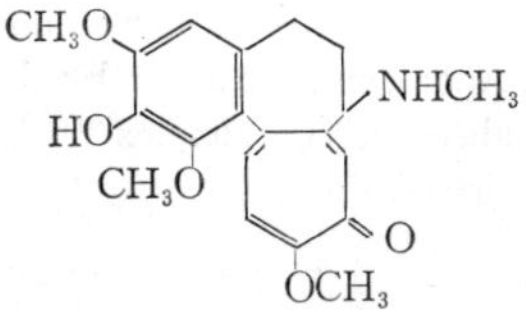

≪mp≫ 136～138° (MeOH-Et_2O)
≪[α]D≫ −119° ($CHCl_3$)

≪Ph.≫
≪OP.≫ Colchicum autumnale[28,29,30,149,150,151], C. hierosolymitanum[35], C. speciosum[29,30]
≪C. & So.≫ Sol. in $CHCl_3$, H_2O, MeOH, Me_2CO, Spar. sol. in Et_2O, AcOEt, Insol. in Petr. ether
≪Sa. & D.≫ B-OAc: 200～202° B-DiAc: 224°
≪SD.≫ 150), 151), 153)
≪S. & B.≫

2-Demethyldemecolcine $C_{20}H_{23}O_5N \equiv 357.39$

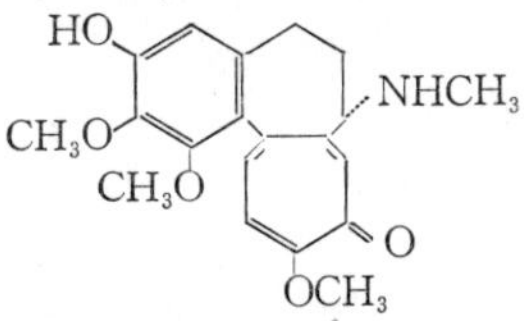

≪mp≫ 220～222°

≪[α]D≫ −128±3° ($CHCl_3$)
≪Ph.≫ Rf[154]
≪OP.≫ Colchicum cornigerum[154]
≪C. & So.≫
≪Sa. & D.≫
≪SD.≫ cf. 154)
≪S. & B.≫

Demecolcine (Colchamine[144], Compound F[155]) $C_{21}H_{25}O_5N \equiv 371.42$

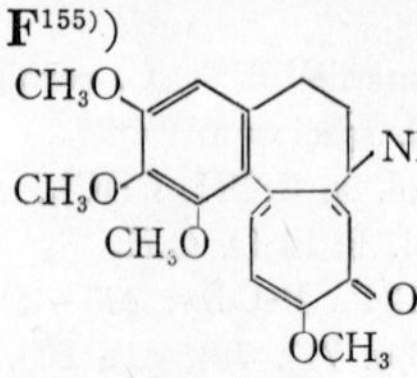

≪mp≫ 184～186°
≪$[\alpha]_D$≫ $-130 \pm 3°$ ($CHCl_3$)
≪Ph.≫ CR[162] UV[148)152)161] IR[148] Rf[162]
≪OP.≫ Colchicum arenarium[11], C. autumnale[6)26)27)28)30)149)150)156], C. bommuelleri[11], C. byzanthium[32], C. cilicium[11], C. hierosolymitanum[38], C. speciosum[11)29)30)144)157], Merendera attical[35], M. caucasica[38], M. robusta[48], M. sobolifera[38]
≪C. & So.≫ Pale yellow prisms, Sol. in $CHCl_3$, EtOH, EtOAc, Me_2CO, C_6H_6, Spar. sol. in Et_2O, Insol. in Petr. ether
≪Sa. & D.≫ B-HCl: 216～217°
B-HI: 230～232° B-$HClO_4$: 264°
B-NMe: 201～202° B-NAc: 227～229°
B-NBz: 209～211° cf. Deriv[158]
≪SD.≫ 25), 27), 144), 148), 151), 152), 153), 155), 157), 158), 159), 160)
≪S. & B.≫ cf. 161), BG: 33)

***N*-Methyldemecolcine** $C_{22}H_{27}O_5N \equiv 385.44$

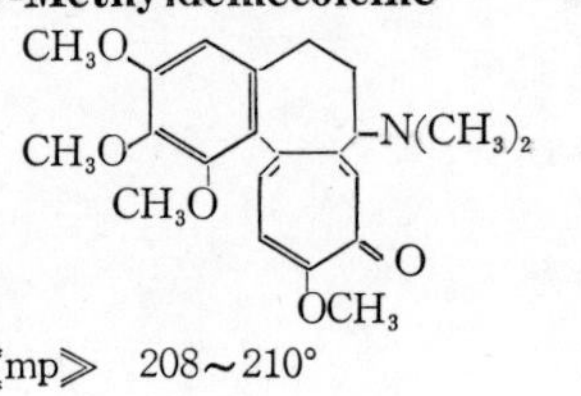

≪mp≫ 208～210°
≪$[\alpha]_D$≫ $-104° \pm 4°$ ($CHCl_3$)
≪Ph.≫ Rf[154]
≪OP.≫ Colchicum cornigerum[154]
≪C. & So.≫
≪Sa. & D.≫
≪SD.≫ cf. 154)
≪S. & B.≫

Cornigerine $C_{21}H_{21}O_6N \equiv 383.39$

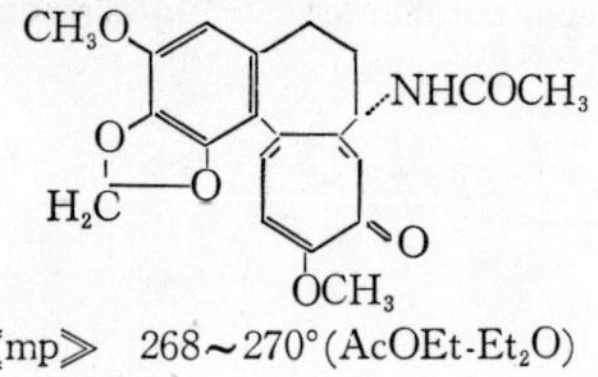

≪mp≫ 268～270° ($AcOEt$-Et_2O)
≪$[\alpha]_D$≫ $-150 \pm 2°$ ($CHCl_3$), $-233 \pm 3°$ (MeOH)
≪Ph.≫ CR[162] UV[162] IR[162] NMR[154] Mass[164] Rf[162]
≪OP.≫ Colchicum cornigerum[154)162]
≪C. & So.≫
≪Sa. & D.≫
≪SD.≫ 164)
≪S. & B.≫

***N*-Formyldesacetylcolchicine (Compound B)** $C_{21}H_{23}O_6N \equiv 385.40$

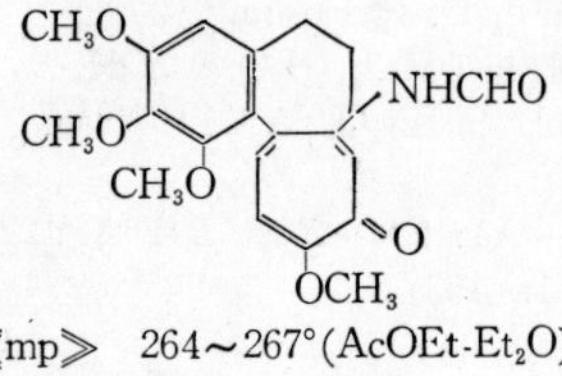

≪mp≫ 264～267° ($AcOEt$-Et_2O)
≪$[\alpha]_D$≫ $-171°$ ($CHCl_3$)
≪Ph.≫ UV[165] Rf[6]
≪OP.≫ Colchicum arenarium[25)28], C. autumnale[1)6)25)26)29)30)31], C. hierosolymitanum[38], C. speciosum[30)38], Androcymbium melanthioides[1], Gloriosa rothschildiana[41], G. simplex[1)41], G. superba[41)43], G. virescens[1]*
≪C. & So.≫
≪Sa. & D.≫
≪SD.≫ 100), 156), 165)
≪S. & B.≫

* Littonia modesta[45], Merendera caucasica[41], M. robusta[48], M. sobolifera[41], Sandersonia aurantiaca[1]

2-Desmethylcolchicine (Compound C)

$C_{21}H_{23}O_6N \equiv 385.40$

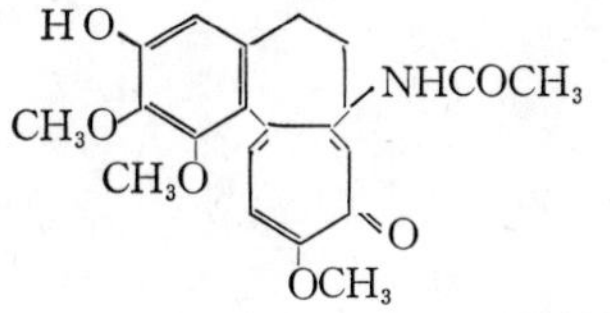

≪mp≫ 189∼190° or 275∼280° (AcOEt-Et_2O)

≪$[\alpha]_D$≫ −130° ($CHCl_3$), −231° (MeOH)

≪Ph.≫ CR[162] UV[81] IR[81] Rf[6)162] Poloro[81]

≪OP.≫ Androcymbium gramineum[6], A. melanthioides[1], Colchicum arenarium[6], C. autumnale[1)6)25)30)149], C. crociflorum[36], C. hierosolymitanum[38], C. luteum[38], C. speciosum[30], Gloriosa rothschildiana[41], G. simplex[1)41] *

≪C. & So.≫

≪Sa. & D.≫ B–OAc: 231∼233°

≪SD.≫ 81), 108), 156), 166)

≪S. & B.≫

3-Desmethylcolchicine (Compound E_1)

$C_{21}H_{23}O_6N \equiv 385.40$

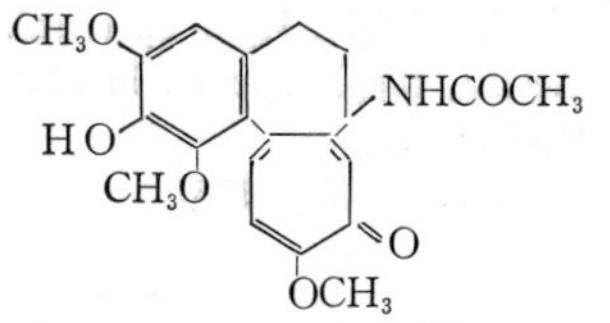

≪mp≫ 140° or 178∼180°

≪$[\alpha]_D$≫ −133° ($CHCl_3$)

≪Ph.≫

≪OP.≫ Androcymbium gromineum[38], A. melanthioides[1], Colchicum arenarium[6], C. autumnale[1)27)28)29)30)156], C. speciosum[29)30)38], Gloriosa superba[31], Merendera robusta[48]

≪C. & So.≫

≪Sa. & D.≫ B–OAc: 194°

≪SD.≫ 27), 156), 166), 167)

≪S. & B.≫

Lumi-Compound E (Compound D)

$C_{21}H_{23}O_6N \equiv 397.41$

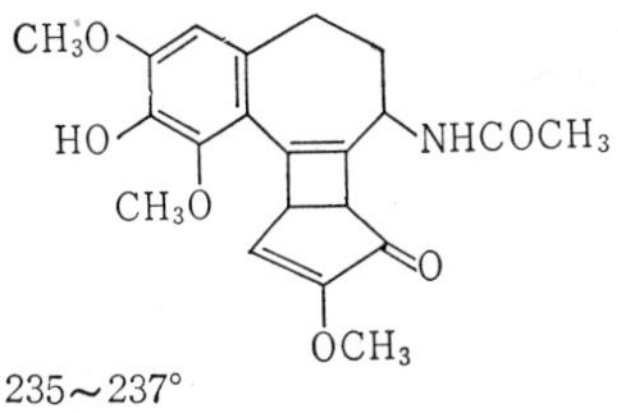

≪mp≫ 235∼237°

≪$[\alpha]_D$≫ +294° ($CHCl_3$)

≪Ph.≫

≪OP.≫ Colchicum autumnale[25)26)27)28)30], C. speciosum[30], Merendera robusta[48]

≪C. & So.≫

≪Sa. & D.≫ B–OAc: 230∼232°

B–Oxime: 299∼301°

≪SD.≫ 25), 27), 167)

≪S. & B.≫ cf. 167)

β-Lumicolchicine (Lumicolchicine-I, Compound I)

$C_{22}H_{25}O_6N \equiv 399.43$

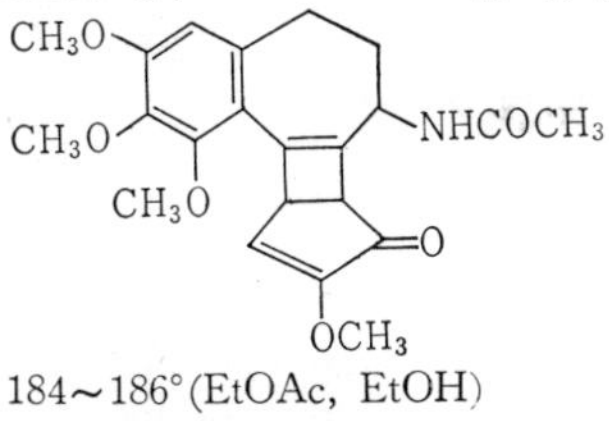

≪mp≫ 184∼186° (EtOAc, EtOH)

≪$[\alpha]_D$≫ +309° ($CHCl_3$)

≪Ph.≫ NMR[162] Rf[6]

≪OP.≫ Colchicum autumnale[25)26)27)28)30)149)167)168], C. luteum[6], C. speriosum[30], C. vernum[9], Gloriosa rothschildiana[41], G. simplex[41], G. superba[41], Merendera attica[9], M. robusta[48]

≪C. & So.≫ Sol. in $CHCl_3$, MeOH, Spar. sol. in Et_2O, Insol. in Petr. ether

≪Sa. & D.≫ B–Ac: 227∼229°

B–Oxime: 263° or 274∼276°

≪SD.≫ 25), 167), 168), 169), 170)

≪S. & B.≫ cf. 169)

γ-Lumicolchicine (Lumicolchicine-II, Compound J)

$C_{22}H_{25}O_6N \equiv 399.43$

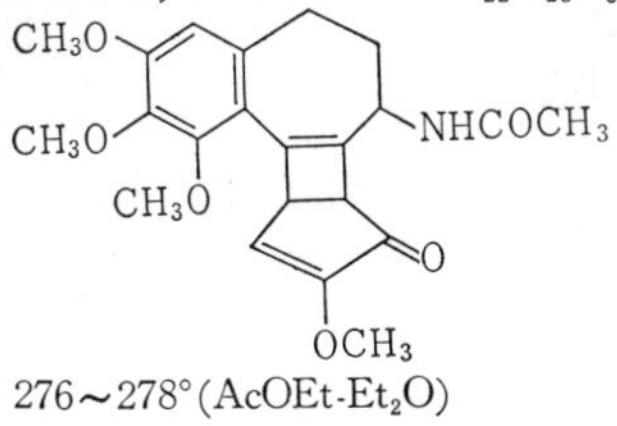

≪mp≫ 276∼278° (AcOEt-Et_2O)

≪$[\alpha]_D$≫ −450° ($CHCl_3$)

≪Ph.≫ NMR[162] Mass[127]

≪OP.≫ Colchicum autumnale[25)26)28)30)167)168], C. speciosum[30], C. vernum[9], Merendera attica[9], M. robusta[48]

≪C. & So.≫

≪Sa. & D.≫ B–Oxime: 284∼286° or 309∼311°

≪SD.≫ 26), 167), 168), 169), 170)

≪S. & B.≫ cf. 169)

---

* G. superba[41], G. virescens[1], Littonia modesta[1], Merendera attica[35], M. caucasia[38], M. robusta[48], M. sobolifera[38], Sandersonia aurantiaca[1]

Alkaloid CC-12 $C_{22}H_{25}O_7N \equiv 415.43$

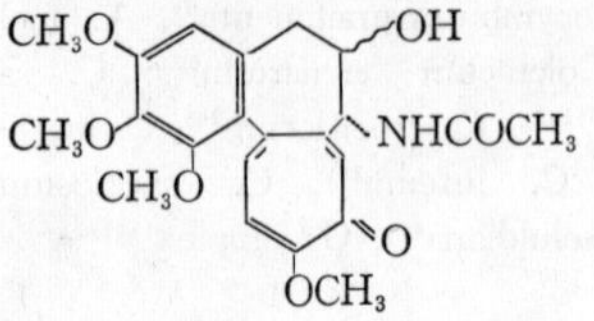

≪mp≫ 197～199°
≪[α]$_D$≫ $-45\pm4°$ or $-83\pm4°$ ($CHCl_3$)
≪Ph.≫ UV[34] IR[34] NMR[34] Mass[34]
≪OP.≫ Colchicum cornigerum[34]
≪C. & So.≫
≪Sa. & D.≫
≪SD.≫ 34)
≪S. & B.≫

Colchicoside $C_{27}H_{33}O_{11}N \equiv 547.54$

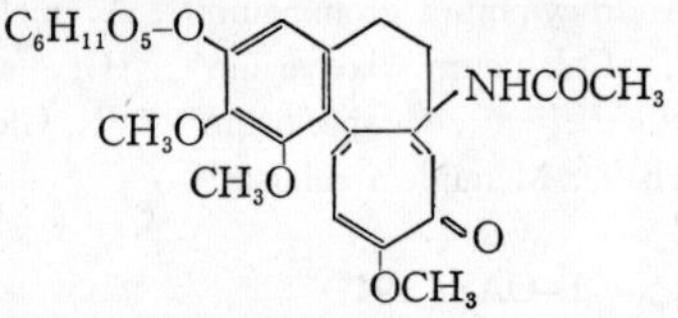

≪mp≫ 216～218° (EtOH)
≪[α]$_D$≫ $-360°$ (H_2O)
≪Ph.≫ UV[111)124)] IR[111)124)] ORP[173)]
≪OP.≫ Colchicum autumnale[171)]
≪C. & So.≫ Sol. in MeOH, H_2O, Spar. sol. in EtOH, Et_2O, AcOEt, Me_2CO, Insol. in $CHCl_3$
≪Sa. & D.≫ Deriv.: 172)
≪SD.≫ 101), 171), 173), 174), 175)
≪S. & B.≫

Base G 4 $C_{16}H_{23}O_4N \equiv 293.35$

≪mp≫ 260～262°
≪[α]$_D$≫ $-78°$ (MeOH)
≪Ph.≫
≪OP.≫ Gloriosa superba[43)]
≪C. & So.≫
≪Sa. & D.≫
≪SD.≫
≪S. & B.≫

Base G 2 $C_{19}H_{25-7}O_4N$

≪mp≫ 228～230°
≪[α]$_D$≫ $-110°$ (MeOH)
≪Ph.≫
≪OP.≫ Gloriosa superba[31)43)]
≪C. & So.≫
≪Sa. & D.≫
≪SD.≫
≪S. & B.≫

Compound O $C_{22}H_{25}O_6N \equiv 399.43$

≪mp≫ 254～256°
≪[α]$_D$≫ $-114°$ ($CHCl_3$)
≪Ph.≫
≪OP.≫ Colchicum autumnale[28)]
≪C. & So.≫
≪Sa. & D.≫
≪SD.≫
≪S. & B.≫

No Name $C_{22}H_{26}O_4NS_2 \equiv 431.43$

≪mp≫ 265～267° (MeOH)
≪[α]$_D$≫ $-366°$ ($CHCl_3$)
≪Ph.≫
≪OP.≫ Colchicum autumnale[150)176)]
≪C. & So.≫ Yellow crystals, Sol. in $CHCl_3$, dioxane, DMF, C_6H_6, Me_2CO, Insol. in H_2O, Et_2O, EtOH
≪Sa. & D.≫
≪SD.≫
≪S. & B.≫

Compound M $C_{27}H_{33}O_{11}N \equiv 547.54$

≪mp≫ 310~314°
≪[α]D≫
≪Ph.≫
≪OP.≫ Colchicum autumnale[28]

≪C. & So.≫
≪Sa. & D.≫ B-TetraAc: 302~304°
≪SD.≫
≪S. & B.≫

Speciosine $C_{28}H_{31}O_6N \equiv 477.54$

≪mp≫ 209~211°(Me_2CO)
≪[α]D≫ −22°($CHCl_3$)
≪Ph.≫
≪OP.≫ Colchicum speciosum[177]

≪C. & So.≫
≪Sa. & D.≫
≪SD.≫
≪S. & B.≫

Compound Ka

≪mp≫ 212~214°
≪[α]D≫ −140°($CHCl_3$)
≪Ph.≫
≪OP.≫ Colchicum autumnale[149]

≪C. & So.≫
≪Sa. & D.≫
≪SD.≫
≪S. & B.≫

Compound N

≪mp≫ 227~229°(AcOEt)
≪[α]D≫ B-Ac: −110°($CHCl_3$)
≪Ph.≫
≪OP.≫ Colchicum autumnale[28]

≪C. & So.≫
≪Sa. & D.≫ B-Ac: 229°
≪SD.≫
≪S. & B.≫

Compound P

≪mp≫ 228~229°
≪[α]D≫ −226°($CHCl_3$)
≪Ph.≫
≪OP.≫ Colchicum autumnale[30]

≪C. & So.≫
≪Sa. & D.≫
≪SD.≫
≪S. & B.≫

Compound R

≪mp≫ 196~198°
≪[α]D≫
≪Ph.≫
≪OP.≫ Colchicum autumnale[30]

≪C. & So.≫
≪Sa. & D.≫
≪SD.≫
≪S. & B.≫

Compound To

≪mp≫ 236~238°
≪[α]D≫ −65°(Py)
≪Ph.≫
≪OP.≫ Colchicum autumnale[30]

≪C. & So.≫
≪Sa. & D.≫
≪SD.≫
≪S. & B.≫

Compound G

≪mp≫ 187~189°
≪[α]D≫ −139.2°($CHCl_3$)
≪Ph.≫
≪OP.≫ Colchicum autumnale[11][25][26]

≪C. & So.≫
≪Sa. & D.≫
≪SD.≫
≪S. & B.≫

Compound E_2

≪mp≫ 140～180°
≪$[\alpha]_D$≫ −110°
≪Ph.≫
≪OP.≫ Colchicum autumnale[27]

≪C. & So.≫
≪Sa. & D.≫
≪SD.≫
≪S. & B.≫

Compound H_3

≪mp≫ 183～185°
≪$[\alpha]_D$≫
≪Ph.≫
≪OP.≫ Colchicum autumnale[30]

≪C. & So.≫
≪Sa. & D.≫
≪SD.≫
≪S. & B.≫

Compound G_3

≪mp≫
≪$[\alpha]_D$≫
≪Ph.≫
≪OP.≫ Gloriosa superba[43]

≪C. & So.≫
≪Sa. & D.≫
≪SD.≫
≪S. & B.≫

Alkaloid CC-1

≪mp≫ 115°
≪$[\alpha]_D$≫ −74°($CHCl_3$)
≪Ph.≫ CR[162] UV[162] IR[162] Rf[162]
≪OP.≫ Colchicum cornigerum[162]

≪C. & So.≫
≪Sa. & D.≫
≪SD.≫
≪S. & B.≫

Alkaloid CC-2

≪mp≫ 170°
≪$[\alpha]_D$≫ +38°($CHCl_3$)
≪Ph.≫ CR[162] UV[162] IR[162] Rf[162]
≪OP.≫ Colchicum cornigerum[162]

≪C. & So.≫
≪Sa. & D.≫
≪SD.≫
≪S. & B.≫

Alkaloid CC-3 $C_{20}H_{25}O_5N$≡359.41

≪mp≫ 197～199°(AcOEt-Et_2O)
≪$[\alpha]_D$≫ 155±3°($CHCl_3$)
≪Ph.≫ CR[162] UV[162] IR[162] Rf[162]
≪OP.≫ Colchicum cornigerum[162]

≪C. & So.≫
≪Sa. & D.≫
≪SD.≫
≪S. & B.≫

Alkaloid CC-4

≪mp≫ 126°
≪$[\alpha]_D$≫
≪Ph.≫ CR[162] Rf[162]
≪OP.≫ Colchicum cornigerum[162]

≪C. & So.≫
≪Sa. & D.≫
≪SD.≫
≪S. & B.≫

Alkaloid CC-5

≪mp≫ 208～210°
≪$[\alpha]_D$≫
≪Ph.≫ CR[162] Rf[162]
≪OP.≫ Colchicum cornigerum[162]

≪C. & So.≫
≪Sa. & D.≫
≪SD.≫
≪S. & B.≫

Alkaloid CC-6

≪mp≫
≪[α]D≫
≪Ph.≫ CR[162] Rf[162]
≪OP.≫ Colchicum cornigerum[162]

≪C. & So.≫
≪Sa. & D.≫ B–Ac: 180°
≪SD.≫
≪S. & B.≫

No Name $C_{22}H_{25}O_6N$≡399.43

≪mp≫ 168～170°(Et_2O-AcOEt)
≪[α]D≫ +38±4°($CHCl_3$)
≪Ph.≫
≪OP.≫ Colchicum cornigerum[154]

≪C. & So.≫
≪Sa. & D.≫
≪SD.≫
≪S. & B.≫

No Name $C_{23}H_{27}O_8N$≡445.45

≪mp≫ 178～181°
≪[α]D≫ −115±4° MeOH
≪Ph.≫
≪OP.≫ Colchicum cornigerum[154]

≪C. & So.≫
≪Sa. & D.≫
≪SD.≫
≪S. & B.≫

No Name $C_{25}H_{35}O_6N$≡445.54

≪mp≫ 123～125°(Et_2O-Petr. ether)
≪[α]D≫ −70±4°($CHCl_3$)
≪Ph.≫
≪OP.≫ Colchicum cornigerum[154]

≪C. & So.≫
≪Sa. & D.≫
≪SD.≫
≪S. & B.≫

---

1) J. Hrbek, F. Santavy: Coll. Czech. Chem. Comm., **27**, 225 (1962).
2) A.R. Battersby, R.B. Herbert, L. Pijewska, F. Santavy: Chem. Comm., **1965**, 228.
3) A.R. Battersby, R.B. Herbert F. Santavy: Chem. Comm., **1965**, 415; T. Kametani, et al: Chem. Pharm. Bull., **16**, 663 (1968).
4) E. Perrot: Compt. rend., **202**, 1086 (1936).
5) E. Perrot: Bull. sci. pharmacol., **43**, 257 (1936).
6) V. Delong, J. Havrliková, F. Santavy: Ann. pharm. franc., **13**, 449 (1955). [C. A., **50**, 1266 (1956)].
7) A.R. Battersby, R.B. Herbert, E. McDonald, R. Ramage, J.H. Clements: Chem. Comm., **1966**, 603.
8) G. Klein, G. Pollauf: Oesterr. Botan. Z., **78**, 251 (1929). [C., **1930**, II, 1104].
9) F. Santavy, E. Confalik: Coll. Czech. Chem. Comm., **16**, 198 (1951).
10) Rochette: Union farm., **17**, 200 (1876).
11) F. Santavy, M. Cernoch, J. Malinsky, B. Lang, A. Zajikova: Ann. pharm. franc., **9**, 50 (1951).
12) P.J. Pelletier, J. Caventou: Ann. chim. phys., **14**, 69 (1820).
13) Ph. L. Geiger: Ann. chim. pharm., **7**, 274 (1833).
14) A. Aschoff: Arch. Pharm., **89**, 14 (1857).
15) M.L. Oberlin: Ann. chim. phys., **50**, 108 (1857).
16) S. Zeisel: Monatsh., **4**, 162 (1883).
17) A. Houdes: Compt. rend., **98**, 1442 (1884).
18) A. Albo: Arch. sci phys. nat., **12**, 227 (1901).
19) J. Grier: Yearbook Pharm., **1923**, 611.
20) P. Lipták: Ber. ungar. pharm. Ges., **3**, 346 (1927).
21) F. Chemritius: J. prakt. Chem., **118**, 29 (1928).
22) E. Niemann: Pharm. Acta Helv., 8, 92 (1933).
23) A. Cohen, J.W. Cook, E.M.F. Roe: J. Chem. Soc., **1940**, 194.
24) E.N. Taran: Pharmazie (russ), **1940**, No. 9/10, 38. [C. A., **35**, 5255 (1941)].
25) F. Santavy, J. Reichstein: Helv. Chim. Acta, **33**, 1606 (1950).
26) F. Santavy: Pharm. Acta Helv., **25**, 248 (1950).
27) F. Santavy: Coll. Czech. Chem. Comm., **15**, 552 (1950).
28) F. Santavy, V. Macak: Coll. Czech. Chem. Comm., **19**, 805 (1954).
29) V. Masinova, F. Santavy: Coll. Czech. Chem. Comm., **19**, 1283 (1954).
30) F. Santavy, Z. Hoscalkova, P. Podivinsky, H. Potesilova: Coll. Czech. Chem. Comm., **19**, 1289 (1954).

31) F. Santavy, F. A. Kincl, A. R. Shinde: Arch. Pharm., **290**, 376 (1957).
32) E. Leete, P. E. Nemeth: J. Am. Chem. Soc., **82**, 6055 (1960).
33) E. Leete, P. E. Nemeth: J. Am. Chem. Soc., **83**, 2192 (1961).
34) A. D. Cross, A. Bl-Namidi, L. Pijewska, F. Santavy: Coll. Czech. Chem. Comm., **31**, 377 (1966).
35) F. Santavy, D. V. Zajicek, A. Nemeckova: Coll. Czech. Chem. Comm., **22**, 1482 (1957).
36) C. Weizmann: Bull. Res. Council, Israel, **2**, 21 (1952). [C. A., **47**, 11661 (1953)].
37) P.M. Mehra, J.N. Khoshoo: J. Pharm. Pharmacol., **3**, 486 (1951).
38) H. Potesilova, I. Bortosova, F. Santavy: Ann. pharm. franc., **12**, 616 (1954). [C. A., **49**, 5594 (1955)].
39) A. A. Beer, Sh. A. Karapetjan, A. I. Kolesnikow, D. P. Snegirew: Ber. Akad. Wiss., USSR., **67**, 883 (1949).
40) J. T. Bayan, W. M. Lauter: J. Am. Pharm. Assoc., **40**, 253 (1951).
41) F. Santavy, J. Bartex: Pharmazie, **7**, 595 (1952).
42) H. W. B. Clewer, S. J. Green, F. Tutin: J. Chem. Soc., **107**, 835 (1915).
43) M. Maturova, B. Lang, T. Reichstein, F. Santavy: Plant med., **7**, 298 (1959). [C. A., **54**, 5011 (1960)].
44) E. H. W. J. Burden, D. N. Grindley, G. A. Prowse: J. Pharm. Pharmacol., **7**, 1063 (1955). [C. A., **50**, 4453 (1956)].
45) F. Santavy: Coll. Czech. Chem. Comm., **22**, 652 (1957).
46) P. Fourment, H. Roques: Bull. soc. pharm. Bordeaux, **65**, 26 (1927). [C., **1927**, II 1062].
47) G. V. Lasurjewski, V. A Masslennikowa: Ber. Akad. Wiss. USSR., **63**, 449 (1948).
48) A. S. Sadykov, M. K. Yusupov, Nauchn. Jr.: Tashkentsk. Gos. Univ., **1962**, No. 203, 15. [C. A., **59**, 6451 (1963)].
49) J. N. Ashley, J. O. Harris: J. Chem. Soc., **1944**, 677.
50) H. G. Boit: "Ergebnisse der Alkaloid-Chemie bis 1960", Akademie-Verlag, Berlin, 1961
51) S. Zeisel: Manatsh., **6**, 989 (1885).
52) S. Zeisel: Manatsh., **7**, 557 (1886).
53) S. Zeisel: Monatsh., **9**, 1 (1888).
54) G. Johanny, S. Zeisel: Monatsh., **9**, 865 (1888).
55) A. Windaus: Sitzber, heidelbelg Akad. Wiss., Math. Naturwiss. Klasse, A, **1910**, 2 Abh.
56) A. Windaus: Sitzber, heidelberg Akad. Wiss Math. Naturwiss. Klasse, A. **1911**, 2. Abh.
57) S. Zeisel, A. Friedlich: Monatsh., **34**, 1181 (1913).
58) S. Zeisel, K. Ritter, V. Stockert: Monatsh., **34**, 1327 (1913).
59) S. Zeisel, K. Ritter, V. Stockert: Monatsh., **34**, 1339 (1913).
60) A. Windaus: Sitzber, heidelberg Akad. Wiss., Math. Naturwiss. Klasse, A. **1914**, 18. Abh.
61) A. Windaus: Sitzber, heidelberg Akad. Wiss., Math. Naturwiss. Klasse, A., **1919**, 16. Abh.
62) A. Windaus, H. Schiele: Nachr. Ges. Wiss. Göttingen, Jahresber. Geschäftjahr, Math. physik. Klasse, Fachgruppen, **1923**, 17.
63) A. Windaus: Ann., **439**, 59 (1924).
64) A. Windaus, H. Schiele: Ann., **439**, 71 (1924).
65) A. Windaus, W. Eickel: Ber., **57**, 1871(1924).
66) A. Windaus, H. Jensen, A. Schramme: Ber., **57**, 1875 (1924).
67) T. M. Sharp: J. Chem. Soc., **1936**, 1234.
68) K. Bursion: Ber., **71**, 245 (1938).
69) R. Grewe: Ber., **71**, 907 (1938).
70) E. C. Horning: Chem. Revs., **33**, 107 (1943).
71) J. W. Cook, W. Graham, A. Cohen, R. W. Lapsley, C. A. Lawrence: J. Chem. Soc., **1944**, 322.
72) G. L. Buchanan, J. W. Cook, J. D. Loudon: J. Chem. Soc., **1944**, 325.
73) K. Meyer, T. Reichstein: Pharm. Acta Helv., **19**, 127 (1944).
74) N. Barton, J. W. Cook, J. D. Loudon: J. Chem. Soc., **1945**, 176.
75) M. J. S. Dewar: Nature, **155**, 141 (1945).
76) D. S. Tarbell, H. R. Franck, P. E. Fanta: J. Am. Chem. Soc., **68**, 502 (1946).
77) M. Sorkin: Helv. Chim. Acta, **29**, 146 (1946).
78) F. Santavy: Compt. rend. soc. biol., **140**, 932 (1946).
79) J. W. Cook, G. T. Dickson, J. D. Loudon: J. Chem. Soc., **1947**, 746.
80) G. L. Buchanan, J. W. Cook, J. D. Loudon, J. MacMillan: Nature, **162**, 629 (1948).
81) F. Santavy: Helv. Chim. Acta, **31**, 821 (1948).
82) H. R. V. Arnstein, D. S. Tarbell, H. T. Huang, G. P. Scott: J. Am. Chem. Soc., **70**, 1669 (1948).
83) R. Brdicka: Arkiv. kemi Mineral. Geol., **26**B, No. 19 (1948).
84) F. Santavy: Coll. Czech. Chem. Comm., **14**, 145 (1949).
85) J. Cech, F. Santavy: Coll. Czech. Chem. Comm., **14**, 532 (1949).
86) J. W. Cook, G. T. Dickson, D. Ellis, J. D. Loudon: J. Chem. Soc., **1949**, 1704.

87) H. R. V. Arnstein, D. S. Tarbell, G. P. Scott, H. T. Huang: J. Am. Chem. Soc., **71**, 2448 (1949).
88) H. Fernholz: Ann., **568**, 63 (1950), cf. Angew. Chem., **65**, 319 (1953).
89) J. W. Cook, T. Y. Johnston, J. D. Loudon: J. Chem. Soc., **1950**, 537.
90) G. P. Scott, D. S. Tarbell: J. Am. Chem. Soc.,**72**, 240 (1950).
91) A. D. Kemp, D. S. Tarbell: J. Am. Chem. Soc., **72**, 243 (1950).
92) H. Rapoport, A. W. Williams, M. Cisney: J. Am. Chem. Soc., **72**, 3324 (1950); **73**, 1414, 3654 (1951).
93) E. C. Horning, M. G. Horning, J. Koo, M. S. Fish, J. P. Parker, G. N. Walker, R. M. Hornwitz, G. E. Ullyot: J. Am. Chem. Soc., **72**, 4840 (1950).
94) H. P. Koch: J. Chem. Soc., **1951**, 512.
95) J. W. Cook, J. Jack, J. D. Loudon: J.Chem. Soc., **1951**, 1397.
96) H. Rapoport, A. W. Williams: J. Am. Chem. Soc., **73**, 1896 (1951).
97) M. V. King, J. L. DeVris, R. Pepeinsky: Acta Cryst., **5**, 437 (1952). [C. A., **48**, 3950 (1954)].
98) R. M. Horowitz, G. E. Ullyot: J. Am. Chem. Soc., **74**, 587 (1952).
99) A. Uffer: Helv. Chim. Acta, **35**, 2135 (1952).
100) F. Santavy: Chem. Listy, **46**, 280 (1952).
101) P. Bellet, G. Amiard, M. Pesez, A. Petit: Ann. pharm. franc., **10**, 241 (1952). [C. A., **47**, 3323 1953)].
102) J. W. Cook, J. Jack, J. D. Loudon: J. Chem. Soc., **1952**, 607.
103) J. Koo: J. Am. Chem. Soc., **75**, 720 (1953).
104) G. A. Nicholls, D. S. Tarbell: J. Am. Chem. Soc., **75**, 1104 (1953).
105) R. F. Raffauf, A. L. Farren, G. E. Ullyot: J. Am. Chem. Soc., **75**, 5292 (1953).
106) T. Nozoe, T. Ikemi, S. Ito: Science Repts. Tohoku Univ., 1st. Ser., [1], **38**, 117 (1954).
107) H. Rapoport, A. W. Williams, J. E. Campion, D. E. Pack: J. Am. Chem. Soc., **76**, 3693 (1954).
108) M. M. P. Bellet, P. Régnier: Bull. Soc. Chim. France, **1945**, 408.
109) H. Rapoport, J. B. Lavigne: J. Am. Chem. Soc., **77**, 667 (1955).
110) H. Rapoport, J. E. Campion, J. E. Gordon: J. Am. Chem. Soc., **77**, 2389 (1955).
111) J. Fabian, V. Delaroff, P. Poirier, M. Legrand: Bull. Soc. Chim. France, **1955**, 1455.
112) H. Corrodi, H. Hardegger: Helv. Chim. Acta, **78**, 2030 (1955).
113) H. Corrodi, H. Hardegger: Helv. Chim. Acta, **40**, 193 (1957).
114) K. Ahmad, G. L. Buchanan, J. W. Cook: J. Chem. Soc., **1957**, 3278.
115) H. J. E. Loewenthal, P. Rona: Proc. Chem. Soc., **1958**, 114.
116) J. Schreiber, W. Leimgruber, M. Pesaro, P. Schudel, A. Eschenmoser: Angew. Chem., **71**, 637 (1959).
117) E. E. van Tamelen, T. A. Spencer, D. S. Allen, R. L. Orvis: J. Am. Chem. Soc., **81**, 6341 (1959).
118) J. Schreiber, W. Leimgruber, M. Pesaro, P. Schudel, T. Threlfall, A. Eschenmoser: Helv. Chim. Acta, **44**, 540 (1961).
119) E. E. van Tamelen, T. A. Spencer, Jr., D. S. Allen, Jr., R. L. Orvis: Tetrahedron, **14**, 8 (1961).
120) A. I. Scott, F. McCapra, J. Nabney, D. W. Young, A. J. Baker, T. A. Davidson, A. C. Day: J. Am. Chem. Soc., **85**, 3040 (1963).
121) A. I. Scott, F. McCapra, R. L. Buchanan, A. C. Day, D. W. Young: Tetrahedron, **21**, 3605 (1965).
122) J. M. Kolthoff: Biochem. Z., **162**, 348 (1925).
123) H. Schuhler: Compt. rend., **210**, 490 (1940).
124) L. Velluz, G. Muller: Bull. Soc. Chim. France, **1954**, 755.
125) V. Delaroff, P. Rathle: Bull. Soc. Chim. France, **1965**, 1621.
126) G. L. Buchanan, A. L. Porte, J. K. Sutherland: Chem. & Ind., **1962**, 859.
127) J. M. Wilson, M. Ohashi, H. Budzikiewicz, F. Santavy, C. Djerassi: Tetrahedron, **19**, 2225 (1963).
128) E. J. Forbes: Chem. & Ind., **1956**, 192.
129) F. A. L. Anet, R. Robinson: Nature, **166**, 924 (1950).
130) B. Belleau: Experientia, **9**, 178 (1953).
131) H. G. Griesebach: Z. Naturforsch., **12b**, 227, 597 (1957); **13b**, 335 (1958).
132) T. A. Geissman, T. Swan: Chem. & Ind., **1957**, 984.
133) H. Erdtman, C. A. Wachtmeister: "Festschrift Arthur Stoll," p. 144. Birkhauser, Basel. 1957.
134) E. Wenkert: Experientia, **15**, 165 (1957).
135) A. I. Scott: Nature, **186**, 556 (1960).
136) E. J. Walaszek, F. E. Kelsey, E. M. K. Gelling: Science, **116**, 225 (1952).
137) A. R. Battersby, J. J. Reynolds: Proc. Chem. Soc., **1960**, 346.
138) J. H. Richards, L. D. Ferretti: Proc. Natl. Acad. Sci., **46**., 1438 (1960).
139) E. Leete: J. Am. Chem. Soc., **85**, 3666

(1963).
140) A.R. Battersby, R. Binks, J.J. Reynolds, D.A. Yeowell: J. Chem. Soc., **1964**, 4257.
141) A.R. Battersby, R. Binks, D.A. Yeowell: Proc. Chem. Soc., **1964**, 86.
142) A.R. Battersby, R.B. Herbert: Proc. Chem. Soc., **1964**, 260.
143) R. Leete: Tetrahedron Letters, **1965**, 333.
144) V.V. Kisselew, G.P. Menschikow: Ber. Akad. Wiss. USSR., **87**, 227 (1952).
145) H. Lettre, H. Fernholz, E. Hartaig: Ann., **576**, 147 (1952).
146) T. Nakamura, Y. Murase, R. Hayashi, Y. Endo: Chem. Pharm. Bull., **10**, 251 (1962).
147) G. Sunagawa, T. Nakamura, J. Nakazawa: Chem. Pharm. Bull., **10**, 291 (1962).
148) Y. Ueno: J. Pharm. Soc. Japan, **73**, 1230, 1232, 1235 (1953).
149) F. Santavy, M. Talas: Coll. Czech. Chem. Comm., **19**, 141 (1954).
150) G. Muller, P. Bellet: Ann. pharm. franc., **13**, 81 (1955).
151) F. Santavy: Chem. Listy, **52**, 957 (1958).
152) A. Uffer, O. Schindler, F. Santavy, T. Reichstein: Helv. Chim. Acta, **37**, 18 (1954).
153) F. Santavy: Coll. Czech. Chem. Comm., **24**, 2237 (1959).
154) M. Saleh, S. El-Gangihi, A. El-Hamidi, F. Santavy: Coll. Czech. Chem. Comm., **28**, 3413 (1963).
155) V.V. Kisselew: Ber. Akad. Wiss. USSR., **96**, 527 (1954).
156) F. Santavy, M. Talas: Chem. Listy, **46**, 373 (1952).
157) V.V. Kisselew, G.P. Menschikow: Ber. Akad. Wiss. USSR., 88, 825 (1953).
158) A. Uffer: Experientia, **10**, 76 (1954).
159) S. Moeschlin, H. Meyer, A. Lichtman: Schweiz. Med. Wschr., **83**, 990 (1953). [C. A., **48**, 4107 (1954)].
160) F. Santavy: Coll. Czech. Chem. Comm., **16**, 676 (1951).
161) A. Uffer, O. Schindler, F. Santavy, T. Reichstein: Helv. Chim. Acta, **37**, 18 (1954).
162) O.L. Chapman, H.G. Smith, R.W. King: J. Am. Chem. Soc., **85**, 803 (1963).
163) F. Santavy, R. Winkler, T. Reichstein: Helv. Chim. Acta, **36**, 1319 (1953).
164) A.D. Cross, A. El-Hamidi, J. Hrbek, F. Santavy: Coll. Czech. Chem. Comm., **29**, 1187 (1964).
165) R.F. Raffauf, A.L. Farren, G.E. Ullyot: J. Am. Chem. Soc., **75**, 3854 (1953).
166) F. Santavy, M. Tales, O. Telupilova: Coll. Czech. Chem. Comm., **18**, 710 (1953).
167) F. Santavy: Coll. Czech. Chem. Comm., **16**, 665 (1951).
168) E.J. Forbes: J. Chem. Soc., **1955**, 3864.
169) R. Grewe, W. Wulf: Ber., **84**, 621 (1951).
170) P.D. Gordner, R.L. Brandon, G.R. Haynes: J. Am. Chem. Soc., **79**, 6334 (1957).
171) P. Bellet: Ann. pharm. franc., **10**, 81, 241 (1952).
172) P. Bellet, P. Regnier: Bull. Soc. Chim. France, **1953**, 756.
173) P. Bellet, P. Regnier: Ann. pharm. franc., **10**, 340 (1952).
174) L. Velluz, G. Muller: Bull. Soc. Chim. France, **1955**, 194.
175) J. Stanek: Chem. & Ind., **1956**, 488.
176) P. Bellet, G. Muller: Ann. pharm. france, **13**, 84 (1955).
177) V. Kisselew: Zhur. Obshchei Khim., **26**, 3218 (1956).

Appendix*

Chapter 2 Simple Isoquinoline Alkaloids

Hydrohydrastinine
cf. p. 25.
≪S. & B.≫ 1)

Corypalline
cf. p. 25.
≪S. & B.≫ 1)

***d*-Salsoline**
cf. p. 25.
≪S. & B.≫ 2)

Carnegine
cf. p. 26.
≪OP.≫ Carnegia gigantea[3)]

Anhalamine
cf. p. 26.
≪S. & B.≫ 4)

Lophophorine
cf. p. 27.
≪S. & B.≫ BS: 5)

Pellotine
cf. p. 28.
≪S. & B.≫ BS: 5)

Anhalonidine
cf. p. 27.
≪S. & B.≫ BS: 5)

Gigantine $C_{13}H_{19}O_3N \equiv 237.29$

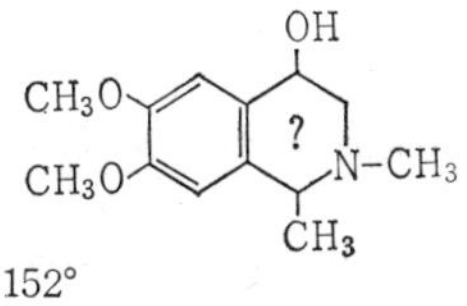

OH
CH_3O
CH_3O
?
$N-CH_3$
CH_3

≪mp≫ 151～152°
≪$[\alpha]_D$≫ 27.1° (MeOH)
≪Ph.≫ IR[3)] NMR[3)] Mass[3)]
≪OP.≫ Carnegia gigantea[3)]
≪C. & So.≫
≪Sa. & D.≫
≪SD.≫ 3), 6)
≪S. & B.≫ 6)

1) J. M. Bobbitt, D. N. Roy, A. Marchand, C. W. Allen: J. Org. Chem., **32**, 2225 (1967).
2) J. Kametani, S. Takano, F. Sasaki: J. Pharm. Soc. Japan, **87**, 191 (1967).
3) J. E. Hodgkins, S. D. Brown, J. L. Massingill: Tetrahedron Letters, **1967**, 1321.
4) T. Kametani, N. Wagatsuma, F. Sasaki: J. Pharm. Soc. Japan, **86**, 913 (1966).
5) A. R. Battersby, R. Binks, R. Huxtable: Tetrahedron Letters, **1967**, 563.
6) G. Grethe, M. Uskokovic, T. Williams, A. Brossi: Helv. Chim. Acta, **50**, 2397 (1967).

Chapter 3 Benzylisoquinoline Alkaloids

***d*-Coclaurine**
cf. p. 32.
≪Ph.≫ NMR[1)]
≪OP.≫ Alseodaphne archboldiana[1)], Cassytha racemosa[2)], Xylopia papuana[1)]

Papaverine
cf. p. 31.
≪S. & B.≫ BS: 3)

***N*-Norarmepavine**
cf. p. 33.
≪Ph.≫ NMR[1)]
≪OP.≫ Alseodaphne archboldiana[1)]

Armepavine
cf. p. 33.
≪Ph.≫ UV[4)]
≪OP.≫ Papaver persicum[4)5)], P.caucasicum[5)]
≪S. & B.≫ 6)

Romeine
cf. p. 34.
≪OP.≫ Romneya coulteri[7)]
≪SD.≫ 7)

***N*-Methylpapaveraldinium iodide**
cf. papaveraldine p. 32.
≪OP.≫ Stephania sasaki[12)]

* The data in this appendix were collected from the papers and Chemical Abstracts reported in 1967.

d-Reticuline

cf. p. 34.

≪Ph.≫ UV[7] IR[7] NMR[1)7]

≪OP.≫ Alseodaphne archboldiana[1], Romneya coulteri[7]

Laudanosine

cf. p. 35.

≪Ph.≫ NMR[8]

≪S. & B.≫ 9)

d-Magnocurarine

cf. p. 36.

≪Ph.≫ CR[10] UV[10)11] NMR[11] Rf[11]

≪OP.≫ Colletia spinosissima[11], Litsea cubeba[10]

l-Coclaurine $C_{17}H_{19}O_3N \equiv 285.33$

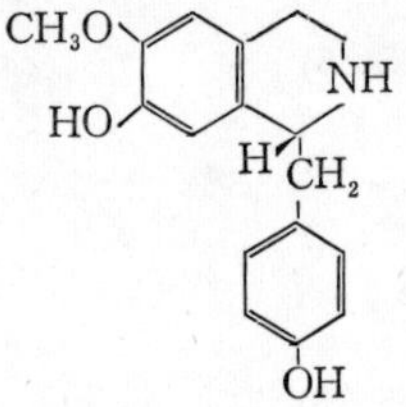

≪Ph.≫ NMR[1]

≪OP.≫ Alseodaphne arch boldiana[1]

≪C. & So.≫

≪Sa. & D.≫

≪SD.≫ 1)

≪S. & B.≫

≪mp≫

≪$[\alpha]_D$≫

Colletine $C_{20}H_{26}O_3N^{\oplus} \equiv 328.42$

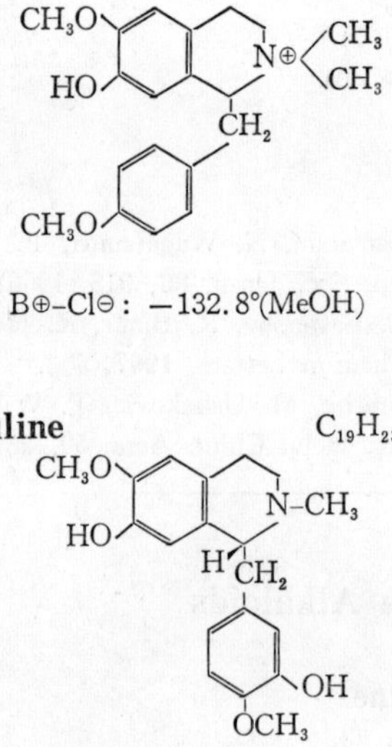

≪Ph.≫ UV[11] NMR[11] Rf[11]

≪OP.≫ Colletia spinosissima[11]

≪C. & So.≫

≪Sa. & D.≫ $B^{\oplus}$-$Cl^{\ominus}$: 130～132° $B^{\oplus}$-$I^{\ominus}$: 169～173°

≪SD.≫ 11)

≪S. & B.≫ 11)

≪mp≫

≪$[\alpha]_D$≫ $B^{\oplus}$-$Cl^{\ominus}$: −132.8°(MeOH)

l-Reticuline $C_{19}H_{23}O_4N \equiv 329.38$

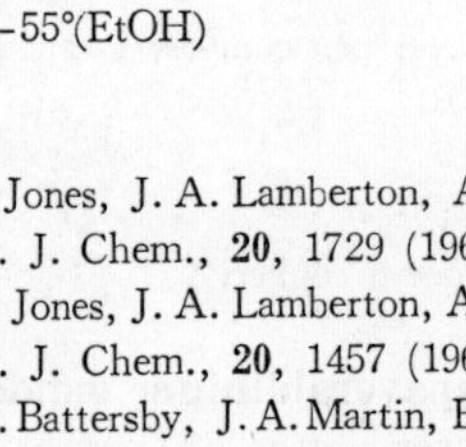

≪Ph.≫ UV[7] IR[7] NMR[7)8]

≪OP.≫ Hernandia ovigera[13], Romneya couteri[7]

≪C. & So.≫ Amorph

≪Sa. & D.≫

≪SD.≫ 7)

≪S. & B.≫

≪mp≫ 78～90°

≪$[\alpha]_D$≫ −55°(EtOH)

1) S. R. Jones, J. A. Lamberton, A. A. Sioumis: Aust. J. Chem., **20**, 1729 (1967).
2) S. R. Jones, J. A. Lamberton, A. A. Sioumis: Aust. J. Chem., **20**, 1457 (1967).
3) A. R. Battersby, J. A. Martin, E. Brochmann-Hanssen: J. Chem. Soc. (C), **1967**, 1785.
4) S. Pfeifer, L. Kuehn: Pharmazie, **22**, 221 (1967).
5) V. Pleininger, J. Appelt, L. Slavikova, J. Slavik: Coll. Czech. Chem. Comm., **32**, 2682 (1967).
6) J. Sam, A. J. Bej: J. Pharm. Sci., **56**, 906 (1967).
7) F. R. Stermitz, L. C. Teng: Tetrahedron Letters, **1967**, 1601.
8) A. H. Jackson, J. A. Martin: J. Chem. Soc. (C), **1966**, 2061.
9) B. K. Cassels: Chem. & Ind., **1966**, 1635.

10) S.-T. Lu, F.-M. Lin: J. Pharm. Soc. Japan, **87**, 878 (1967).
11) E. Sánchez, J. Comin: Tetrahedron, **23**, 1139 (1967).
12) J. Kunitomo, T. Nagai, E. Yuge: J. Pharm. Soc. Japan, **87**, 1010 (1967).
13) H. Furukawa, S.-T. Lu: J. Pharm. Soc. Japan, **86**, 1143 (1966).

Chapter 4 Pavine and Isopavine Alkaloids

Eschscholtzine
cf. p. 41.
≪SD.≫ 1)
≪S. & B.≫ 1)

***l*-Caryathine**
cf. p. 41.
≪SD.≫ 1)

Bisnorargemonine
cf. p. 41.
≪SD.≫ 1)

Eschscholtzidine
cf. p. 42.
≪Ph.≫ ORD[1]
≪SD.≫ 1)

Norargemonine
cf. p. 42.
≪SD.≫ 1)

Argemonine
cf. p. 42.
≪Pb.≫ UV[2] ORD[1][2][3][4][5] CD[2][3][5]
≪SD.≫ 1), 2), 3), 4), 5), 6), 7)
≪S. & B.≫ BS: 4)

Remrefine (Roemrefine) $C_{21}H_{24}O_4N^{\oplus}$≡354.41
cf. p. 43.

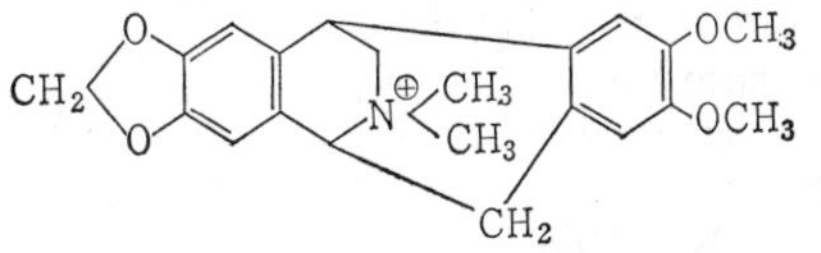

≪mp≫
≪$[\alpha]_D$≫
≪Ph.≫ NMR[8]
≪OP.≫ Roemeria refracta[8]
≪C. & So.≫
≪Sa. & D.≫
≪SD.≫ 8)
≪S. & B.≫

1) A. C. Barker, A. R. Battersby: J. Chem. Soc. (C), **1967**, 1317.
2) R. P. K. Chan, J. C. Craig, R. H. F. Manske, T. O. Soine: Tetrahedron, **23**, 4209 (1967).
3) O. Červinka, A. Fábryová, V. Novák: Tetrahedron Letters, **1966**, 5375.
4) A. C. Barker, A. R. Battersby: Tetrahedron Letters, **1967**, 135.
5) S. F. Mason, G. W. Vane, J. S. Whitehurst: Tetrahedron, **23**, 4087 (1967).
6) M. J. Martell, Jr., T. O. Soine, L. B. Kier: J. Pharm. Sci., **56**, 973 (1967).
7) M. M. Abdel-Monem, T. O. Soine: J. Pharm. Sci., **56**, 976 (1967).
8) M. S. Yunusov, S. T. Skramov, S. Yu. Yunusov: Khim. Prir. Soedin., **3**, 68 (1967). [C. A., **67**, 1114 (1967)].

Chapter 5 *N*-Benzylisoquinoline Alkaloids

Sendaverine
cf. p. 44.
≪Ph.≫ UV[1] IR[1] NMR[1] Mass[1] Rf[1]
≪SD.≫ 1)
≪S. & B.≫ 1)

1) T. Kametani, K. Ohkubo: Chem. Pharm. Bull., **15**, 608 (1967).

Chapter 6 Bisbenzylisoquinoline Alkaloids

Magnoline
cf. p. 45.
≪Ph.≫ IR[1] NMR[1]
≪S. & B.≫ 1); BS: 2)

Cuspidaline
cf. p. 46.
≪mp≫
≪$[\alpha]_D$≫ −48°($CHCl_3$)
≪Ph.≫ UV[3] IR[3] NMR[3]
≪OP.≫ Limacia cuspidata[3]
≪Sa. & D.≫ B-O,O-DiMe·2MeI: 181~182°
≪SD.≫ 3)

Magnolamine
cf. p. 46.
≪S. & B.≫ 4)

Liensinine
cf. p. 47.
≪Ph.≫ IR[5] NMR[6]
≪S. & B.≫ 6), 7)

Isoliensinine
cf. p. 47.
≪Ph.≫ NMR[6][8]
≪S. & B.≫ 6), 8)

Neferine
cf. p. 47.
≪Ph.≫ NMR[6]

Thalmelatine
cf. p. 47.
≪SD.≫ 9)

Dehydrothalicarpine
cf. p. 48.
≪Ph.≫ Rf[10]
≪OP.≫ Thalictrum minus[10]
≪SD.≫ 10)

Thalicarpine
cf. p. 48.
≪Ph.≫ UV[11] IR[11] NMR[11] Rf[10]
≪OP.≫ Thalictrum minus[10]
≪Sa. & D.≫ 9)
≪SD.≫ 11)
≪S. & B.≫ 11)

Thalicrine
cf. p. 49.
≪S. & B.≫ BG: 2)

Oxyacanthine
cf. p. 49.
≪Ph.≫ UV[2] IR[12] Rf[12][13]
≪S. & B.≫ 12); BG: 2)

Repandine
cf. p. 50.
≪Ph.≫ Rf[13]
≪SD.≫ 13)

Epistephanine
cf. p. 51.
≪Ph.≫ UV[2] IR[2] NMR[2] Mass[2]
≪OP.≫ Stephania japonica[2]
≪S. & B.≫ 2)

Stebisimine
cf. p. 51.
≪mp≫ 233~235°
≪$[\alpha]_D$≫ 0°
≪Ph.≫ UV[2] IR[2] NMR[2] Mass[2]
≪OP.≫ Stephania japonica[2]
≪Sa. & D.≫ B–2HCl: >290°(d) B–2Pic: 254~256° B–2MeI: >290°(d)
≪SD.≫ 2)
≪S. & B.≫ 14); BG: 2)

Berbamine
cf. p. 52.
≪Ph.≫ IR[12] Rf[12][13]
≪S. & B.≫ 12); BG: 2)

Pycnamine
cf. p. 52.
≪Ph.≫ Rf[13]

Fangchinoline
cf. p. 52.
≪Ph.≫ NMR[15]
≪OP.≫ Stephania tetrandra[15]

Limacine
cf. p. 53.
≪mp≫ 154~156°(C_6H_6–Me_2CO)
≪$[\alpha]_D$≫ −212°($CHCl_3$)
≪Ph.≫ UV[3] NMR[3]
≪OP.≫ Limacia cuspidata[3]
≪C. & So.≫ Colourless needles
≪Sa. & D.≫ B–2Pic: 184~186°
≪SD.≫ 3)
≪S. & B.≫ BG: 3)

Limacusine
cf. p. 53.
≪mp≫ 235~237° (MeOH–Me_2CO)
≪$[\alpha]_D$≫ +110° ($CHCl_3$)
≪Ph.≫ UV[3] NMR[3] Mass[3]
≪OP.≫ Limacia cuspidata[3]
≪C. & So.≫ Colourless powder
≪Sa. & D.≫ B–OMe: 210~212°
≪SD.≫ 3)
≪S. & B.≫ BG: 3)

Tetrandrine

cf. p. 53.

≪OP.≫ Stephania tetrandra[15]

Isotetrandrine

cf. p. 55.

≪OP.≫ Cyclea barbata[16]

Cepharanthine

cf. p. 55.

≪OP.≫ IR[17] NMR[17] Mass[17]

≪S. & B.≫ 17)

Repanduline $C_{37}H_{36}O_7N_2$=620.67

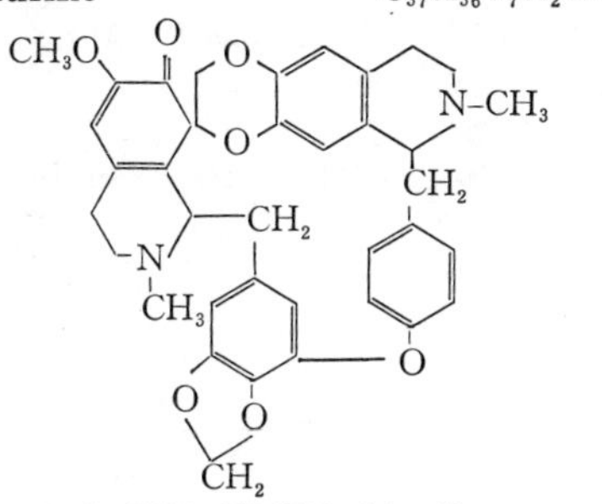

≪Ph.≫ UV[18] IR[18] NMR[19] Mass[19]

≪SD.≫ 18), 19), 20)

Thalicberine

cf. p. 56.

≪S. & B.≫ BG: 2)

Isochondodendrine

cf. p. 58.

≪$[\alpha]_D$≫ 114° (0.1 N-HCl)

≪Ph.≫ UV[21][22] IR[22] NMR[21] Mass[23] Rf[21][25]

≪OP.≫ Cissampelos mucronata[22], Epinetrum cordifolium[21], E. mangenotii[21]

≪SD.≫ 22)

≪S. & B.≫ 24)

Norcycleanine

cf. p. 58.

≪Ph.≫ UV[21] NMR[21] Rf[21]

≪OP.≫ Epinetrum cordifolium[21], E. mangenotii[21]

Cycleanine

cf. p. 59.

≪Ph.≫ UV[21] NMR[21] Rf[21][24]

≪OP.≫ Epinetrum cordifolium[21], E. mangenotii[21]

Curine (Chondodendrine)

cf. p. 59.

≪Ph.≫ Mass[23]

≪S. & B.≫ 25); BG: 2)

***d*-Tubocurarine**

cf. p. 61.

≪Ph.≫ UV[26] Mass[23]

≪S. & B.≫ 25), 26), 27)

Hayatine

cf. p. 62.

≪Ph.≫ Mass[23]

≪OP.≫ Cissampelos pareira[28]

≪Sa. & D.≫ B-O, O-DiEt: 160°

≪SD.≫ 23), 28)

Hayatinine

cf. p. 66.

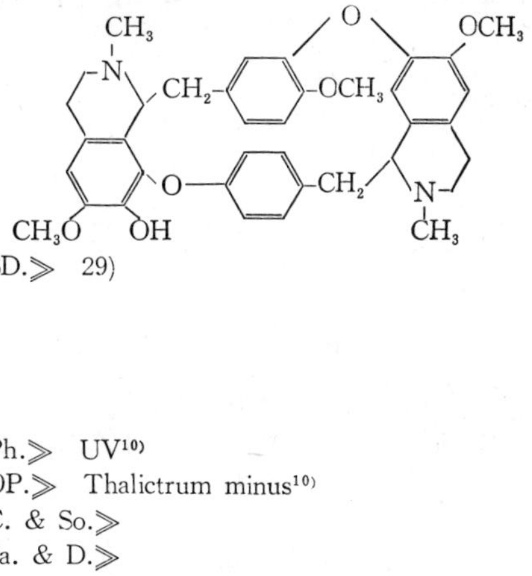

≪SD.≫ 29)

Dehydrothalmelatine $C_{40}H_{44}O_8N_2$=680.77

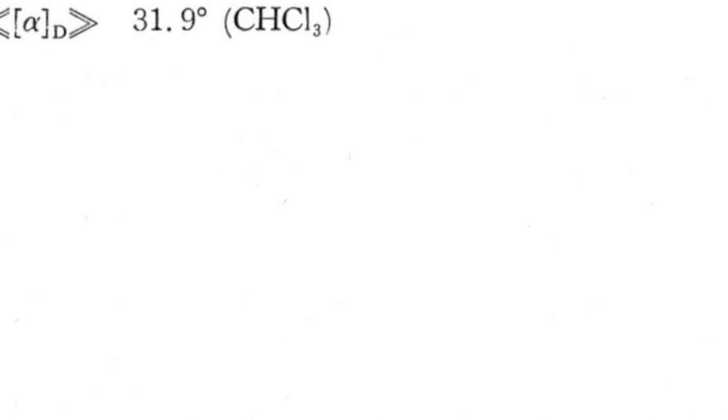

≪mp≫ 126~128° (MeOH)

≪$[\alpha]_D$≫ 31.9° ($CHCl_3$)

≪Ph.≫ UV[10]

≪OP.≫ Thalictrum minus[10]

≪C. & So.≫

≪Sa. & D.≫

≪SD.≫ 10)

≪S. & B.≫

Homoaromoline $C_{37}H_{40}O_6N_2 \equiv 608.71$

OCH_3 CH_3O
CH_3–N N–CH_3
OH
H CH_2 O H
CH_2
O
CH_3O

≪mp≫ 238~240°
≪$[\alpha]_D$≫ 416° ($CHCl_3$)

≪Ph.≫ UV[16] NMR[16]
≪OP.≫ Cyclea barbata[16]
≪C. & So.≫ Needles
≪Sa. & D.≫ B–OMe=O–Me–oxyacanthine
≪SD.≫ 16)
≪S. & B.≫

Thalidezine $C_{38}H_{42}O_7N_2 \equiv 638.73$

HO
OCH_3 CH_3O
CH_3–N N–CH_3
OCH_3
CH_2 O CH_2
O
OCH_3

≪mp≫ 158~159° (Me_2CO)
≪$[\alpha]_D$≫ 235° ($CHCl_3$)

≪Ph.≫ UV[30] NMR[30] Mass[30]
≪OP.≫ Thalictrum fendleri[30]
≪C. & So.≫
≪Sa. & D.≫ B–OMe=hernandezine B–OEt : 117~120°
≪SD.≫ 30)
≪S. & B.≫

Thalidasine $C_{39}H_{44}O_7N_2 \equiv 652.76$

OCH_3 CH_3O
O
CH_3–N OCH_3 CH_3O N–CH_3
H H
CH_2 O CH_2
OCH_3

≪mp≫ 105~107°
≪$[\alpha]_D$≫ −70° (MeOH)

≪Ph.≫ UV[31] NMR[31] Mass[31]
≪OP.≫ Thalictrum dasycarpum[31]
≪C. & So.≫ Amorph
≪Sa. & D.≫ B–2MeI : 182~183° B–2Oxal : 160~162° B–2Pic : 175~177°
≪SD.≫ 31)
≪S. & B.≫

Hayatidine $C_{37}H_{40}O_6N_2 \equiv 608.71$

CH_3 O OCH_3
N CH_2 OCH_3
O CH_2 N
CH_3O OH CH_3

≪mp≫ 179~180°
≪$[\alpha]_D$≫ −109° (Py)

≪Ph.≫ UV[32]
≪OP.≫ Cissampelos pareira[32]
≪C. & So.≫
≪Sa. & D.≫
≪SD.≫ 32)
≪S. & B.≫

1) T. Kametani, R. Yanase, S. Kano, K. Sakurai: Chem. Pharm. Bull., **15**, 56 (1967).
2) D. H. R. Barton, G. W. Kirby, A. Wiechers: J. Chem. Soc. (C), **1966**, 2313.
3) M. Tomita, H. Furukawa, K. Fukagawa: J. Pharm. Soc. Japan, **87**, 793 (1967).
4) T. Kametani, H. Yagi, S. Kaneda: Chem. Pharm. Bull., **14**, 974 (1966).
5) T. Kametani, S. Takano, F. Sasaki: J. Pharm. Soc. Japan, **87**, 191 (1967).
6) H. Furukawa: J. Pharm. Soc. Japan, **86**, 883 (1966).
7) L. V. Volkova, V. E. Kosmacheva, V. G. Voronin, O. N. Tolkachev, N. A. Preobrazhenskii:

Khim. Geterotsikl. Soedin., **1967**, 522. [C. A., **68**, 294 (1968)].
8) T. Kametani, S. Takano, K. Satoh : J. Heterocyclic Chem., **3**, 546 (1966).
9) M. A. Haimova, N. M. Mollov, H. B. Dutschewska, I. R. Petrova, D. L. Kasseva, N. G. Koitscheva : C. R. Acad. Bulg. Sci., **19**, 921 (1966). [C. A., **66**, 6192 (1967)].
10) H. B. Dutschewska, N. M. Mollov : Ber., **100**, 3135 (1967).
11) M. Tomita, H. Furukawa, S.-T. Lu, S. M. Kupchan : Chem. Pharm. Bull., **15**, 959 (1967).
12) T. Kametani, H. Iida, S. Kano, S. Tanaka, K. Fukumoto, S. Shibuya, H. Yagi : J. Heterocyclic Chem., **4**, 85 (1967).
13) J. Knabe, P. Horn : Arch. Pharm., **300**, 726 (1967).
14) T. Kametani, O. Kusama, K. Fukumoto : Chem. Comm., **1967**, 1212.
15) M. Tomita, M. Kozuka, S.-T. Lu : J. Pharm. Soc. Japan, **87**, 316 (1967).
16) M. Tomita, M. Kozuka, M. Satomi : J. Pharm. Soc. Japan, **87**, 1012 (1967).
17) M. Tomita, K. Fujitani, Y. Aoyagi : Tetrahedron Letters, **1967**, 1201.
18) J. Harley-Mason, A. S. Howard, W. I. Taylor, M. J. Vernengo, I. R. C. Bick, P. S. Clezy : J. Chem. Soc. (C), **1967**, 1948.
19) I. R. C. Bick, J. H. Bowie, J. Harley-Mason, D. H-Williams : J. Chem. Soc. (C), **1967**, 1951.
20) K. Aoki, J. Harley-Mason : J. Chem. Soc. (C), **1967**, 1957.
21) M. Debray, M. Plat, J. Le. Men : Ann. Pharm. Fr., **24**, 551 (1966).
22) M. A. Ferreira, L. N. Prista, A. C. Alves, A. S. Roque : Garcia Orta, **13**, 395 (1965). [C. A., **67**, 9444 (1967)].
23) G. W. A. Milne, J. R. Plimmer : J. Chem. Soc. (C), **1966**, 1966.
24) J. P. Sheth, O. N. Tolkachev : Tetrahedron Letters, **1967**, 1161.
25) O. N. Tolkachev : J. Sci. Ind. Res., **26**, 209 (1967).
26) O. N. Tolkachev, L. P. Kvashnina, N. A. Preobrazhenskii : Zh. Obshch. Khim., **37**, 1764 (1966). [C. A., **66**, 8052 (1967)].
27) O. N. Tolkachev, V. P. Chernova, F. L. Pao, E. V. Kuznetsova, N. A. Preobrazhenski : Zh. Obshch. Khim., **36**, 1767 (1966). [C. A., **66**, 8052 (1967)].
28) A. K. Bhatnagar, S. Bhattacharji, A. C. Roy, S. P. Popli, M. L. Dhar : J. Org. Chem., **32**, 819 (1967).
29) A. K. Bhatnagar, S. P. Popli : Indian J. Chem., **5**, 102 (1967).
30) M. Shamma, R. J. Shine, B. S. Dudock : Tetrahedron, **23**, 2887 (1967).
31) S. M. Kupchan, T.-H. Yang, G. S. Vasilikiotis, M. H. Barnes, M. L. King : J. Am. Chem. Soc., **89**, 3075 (1967).
32) A. K. Bhatnagar, S. P. Popli : Experientia, **23**, 242 (1967).

Chapter 7 Cularine and Related Alkaloids

Cularicine

cf. p. 74.

≪S. & B.≫ 1)

Cularimine

cf. p. 74.

≪Ph.≫ NMR[2,3] Mass[3]

≪S. & B.≫ 2), 3)

Cularine

cf. p. 74.

≪Ph.≫ NMR[3] Mass[3]

≪S. & B.≫ 4) ; BG : 4)

1) T. Kametani, S. Shibuya, H. Uyeno : J. Pharm. Soc. Japan, **87**, 238 (1967).
2) T. Kametani, S. Shibuya : J. Pharm. Soc. Japan, **87**, 196 (1967).
3) T. Kametani, S. Shibuya, S. Sasaki : J. Pharm. Soc. Japan, **87**, 198 (1967).
4) T. Kametani, T. Kikuchi, K. Fukumoto : Chem. Comm., **1967**, 546.

Chapter 8 Proaporphine Alkaloids

Crotonosine

cf. p. 76.

≪Ph.≫ Mass[1]

≪S. & B.≫ BS: 2)

Fugapavine (Mecambrine)

≪Ph.≫ UV[3] IR[3] NMR[4]

≪OP.≫ Meconopsis cambrica[4], Papaver persicum[3]

Stepharine

cf. p. 76.

≪Ph.≫ IR[5] Mass[1]

≪OP.≫ Laurelia novaezelandia[6], Pericampylus formosanus[5]

≪C. & So.≫ Colourless prisms

Glaziovine

cf. p. 77.

≪Ph.≫ UV[7,8] IR[7,8] NMR[7,8] Mass[7,8]

≪SD.≫ 3)

≪S. & B.≫ 7), 8)

Pronuciferine

cf. p. 77.

≪Ph.≫ IR[3] NMR[8] Mass[1]

≪OP.≫ Papaver persicum[3]

≪S. & B.≫ 7), 8)

Orientalinone

cf. p. 77.

≪OP.≫ Papaver bracteatum[9], P. caucasicum[10], P. persicum[10]

≪S. & B.≫ 11)

Linearsine

cf. p. 78.

≪Ph.≫ Mass[1]

Litsericine

cf. p. 79.

≪Ph.≫ UV[12] NMR[12] ORD[12] Rf[12]

≪SD.≫ 12)

Oreoline (**Oridine**[13]) $C_{17}H_{23}O_3N \equiv 289.36$

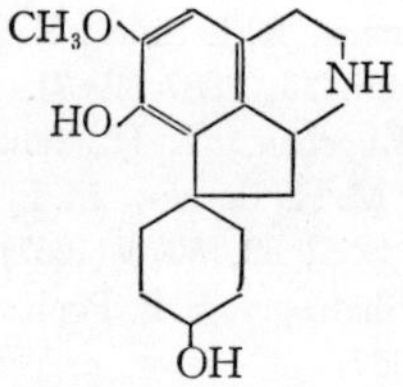

CH_3O NH HO OH

≪mp≫ 234~236°

≪$[\alpha]_D$≫ B-MeI: −87±3° (EtOH)

≪Ph.≫ UV[13,14] IR[13,14] Rf[13,14]

≪OP.≫ Papaver oreophilum[14]

≪C. & So.≫

≪Sa. & D.≫ B-MeI: 259~261°

≪SD.≫ 14)

≪S. & B.≫

***N*-Methyloreoline** $C_{18}H_{25}O_3N \equiv 303.39$

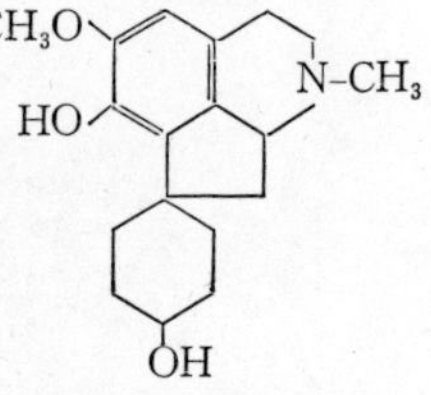

CH_3O $N\text{-}CH_3$ HO OH

≪mp≫ 192~193°

≪$[\alpha]_D$≫

≪Ph.≫ UV[14] Rf[14]

≪OP.≫ Papaver oreophilum[14]

≪C. & So.≫

≪Sa. & D.≫

≪SD.≫ 14)

≪S. & B.≫

1) M. Baldwin, A. G. Loudon, A. Maccoll, L. J. Haynes, K. L. Stuart: J. Chem. Soc. (C), **1967**, 154.
2) D. H. R. Barton, D. S. Bhakuni, G. M. Chapman, G. W. Kirby, L. J. Haynes, K. L. Stuart: J. Chem. Soc. (C), **1967**, 1295.
3) S. Pfeifer, L. Kuehn: Pharmazie, **22**, 221 (1967).
4) V. A. Mnatsakanyan, A. R. Mkrtchyan: Armyansk. Khim. Zh., **19**, 466 (1966). [C. A., **66**, 3646 (1967)].
5) M. Tomita, M. Kozuka, S.-T. Lu: J. Pharm. Soc. Japan, **87**, 315 (1967).
6) K. Bernauer: Helv. Chim. Acta, **50**, 1583 (1967).
7) T. Kametani, H. Yagi: Chem. Comm., **1967**,

366.
8) T. Kametani, H. Yagi: J. Chem. Soc. (C), **1967**, 2182.
9) K. Heydenreich, S. Pfeifer: Pharmazie, **22**, 124 (1967).
10) V. Pleininger, J. Appelt, L. Slavikova, J. Slavik: Coll. Czech. Chem. Comm., **32**, 2682 (1967).
11) A.H. Jackson, J.A. Martin: J. Chem. Soc. (C), **1966**, 2222.
12) T. Nakasato, S. Asada: J. Pharm. Soc. Japan, **86**, 1205 (1966).
13) M. Muratova, D. Davlaskova, F. Santavy: Planta Med., **14**, 22 (1966). [C.A., **65**, 1040 (1966)].
14) I. Mann, S. Pfeifer: Pharmazie, **22**, 124 (1967).

Chapter 9 Aporphine Alkaloids

_O_-Demethylnuciferine
cf. Caaverine. p. 81.
≪mp≫ 214～215°
≪OP.≫ Papaver persicum[1]
≪Sa. & D.≫ B-HI: 256～257° B-Oxal: 225～227°(d)
≪SD.≫ 1)

Roemerine
cf. p. 82.
≪Ph.≫ UV[2]
≪OP.≫ Laurelia novae-zelandiae[3], Papaver feddei[4]

_l_-Nuciferine
cf. p. 82.
≪Ph.≫ UV[5,6,7] IR[6,7] Rf[6]
≪OP.≫ Lysichiton camtschatcense var. japonicum[5], Papaver orientale[7]
≪C. & So.≫ Colourless needles

L-(+)-Nuciferine
cf. p. 82.
≪mp≫ 165～167° (MeOH)
≪$[\alpha]_D$≫ 165±5° (EtOH)
≪Ph.≫ UV[2]
≪OP.≫ Papaver persicum[2]
≪SD.≫ 2)

Mecambroline
cf. p. 83.
≪Ph.≫ UV[3]
≪OP.≫ Laurelia novae-zelandiae[3]

Tuduranine
cf. p. 84.
≪Ph.≫ CR[8] UV[8] NMR[8]
≪OP.≫ Stephania rotunda[8]
≪C. & So.≫ Amorph
≪S. & B.≫ 8)

Pukateine
cf. p. 85.
≪OP.≫ Laurelia novae-zelandiae[3]

Isothebaine
cf. p. 85.
≪Ph.≫ UV[7] IR[7]
≪OP.≫ Papaver orientale[7]

Actinodaphnine
cf. p. 86.
≪OP.≫ Cassytha melantha[9]

Laurolitsine
cf. p. 86.
≪Ph.≫ Mass[3]
≪OP.≫ Laurelia novae-zelandiae[3]

Laurelliptine
cf. p. 86.
≪OP.≫ Cassytha racemosa[10]

Laurotetanine
cf. p. 87.
≪OP.≫ Cassytha racemosa[10], Lindera pipericarpa[11]

Isoboldine
cf. p. 88.
≪Ph.≫ UV[3,12] NMR[12,13] Mass[14]
≪OP.≫ Cassytha racemosa[10], Laurelia novae-zelandiae[3]
≪S. & B.≫ 12)

Boldine
cf. p. 88.
≪Ph.≫ UV[12] NMR[12,13] Mass[14]
≪OP.≫ Laurelia novae-zelandiae[3]

Dicentrine
cf. p. 88.
≪Ph.≫ Mass[14]
≪OP.≫ Stephania dinklagei[15]

Nantenine
cf. p. 88.
≪Ph.≫ CR[16]
≪OP.≫ Cassytha racemosa[10]

Glaucentrine
cf. p. 89.
≪SD.≫ 17) =corydine (?)

Thalicmidine
cf. p. 89.
≪SD.≫ 18)

Glaucine

cf. p. 90.

≪Ph.≫ UV[12)19)] IR[19)] NMR[12)13)19)] Mass[14)]

≪OP.≫ Papaver feddei[4)]

≪S. & B.≫ 12), 19), 20)

Hernandaline

cf. p. 90.

≪Ph.≫ IR[21)] NMR[21)]

≪S. & B.≫ 21)

Hernangerine

cf. p. 93.

≪Ph.≫ UV[22)] NMR[22)]

≪OP.≫ Hernandia ovigera[22)]

≪SD.≫ 22)

Bulbocapnine

cf. p. 93.

≪Ph.≫ NMR[23)] Mass[14)]

Corytuberine

cf. p. 94.

≪Ph.≫ NMR[13)] Mass[14)]

Isocorydine

cf. p. 94.

≪Ph.≫ Mass[14)]

≪OP.≫ Papaver oreophilum[4)], Stephania dinklagei[15)]

Corydine

cf. p. 95.

≪Ph.≫ NMR[12)13)24)] Mass[14)]

≪OP.≫ Stephania venosa[24)]

≪S. & B.≫ 13)

Hernandine

cf. p. 96.

≪Ph.≫ CR[23)] NMR[23)]

≪OP.≫ Hernandia bivalvis[23)]

≪SD.≫ 23)

Ocoteine

cf. p. 97.

≪Ph.≫ UV[25)] NMR[23)]

Liriodenine (Spermatheridine)

cf. p. 97.

≪Ph.≫ Mass[5)26)]

≪OP.≫ Lysichiton camtschatcense var. japonicum[5)]

≪SD.≫ 5)

Moschatoline

cf. p. 98.

≪Ph.≫ Mass[26)]

≪SD.≫ 26)

Cassythicine $C_{19}H_{19}O_4N \equiv 325.35$

≪mp≫

≪[α]$_D$≫

≪Ph.≫ NMR[9)]

≪OP.≫ Cassytha glabella[9)], C. melantha[9)]

≪C. & So≫

≪SD.≫ 9)

≪S. & B.≫

***d*-Nuciferoline** $C_{19}H_{21}O_3N \equiv 311.37$

≪mp≫

≪[α]$_D$≫

≪Ph.≫

≪OP.≫ Papaver caucasicum[2)]

≪C. & So.≫

≪Sa. & D.≫

≪SD.≫ 2)

≪S. & B.≫ 2)

Pukateine Methyl Ether $C_{19}H_{19}O_3N \equiv 309.35$

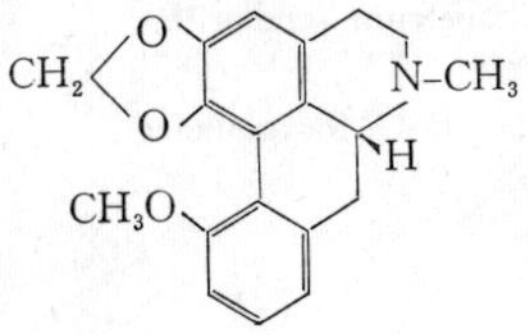

CH_2 O O N-CH_3 H CH_3O

≪Ph.≫ UV[3]
≪OP.≫ Laurelia novae-zelandiae[3]
≪C. & So.≫
≪Sa. & D.≫
≪SD.≫ 3)
≪S. & B.≫

≪mp≫ 136～138° (Et_2O)
≪[α]D≫ −293±3° ($CHCl_3$), −271±3° (EtOH)

Lindecarpine $C_{18}H_{19}O_4N \equiv 313.34$

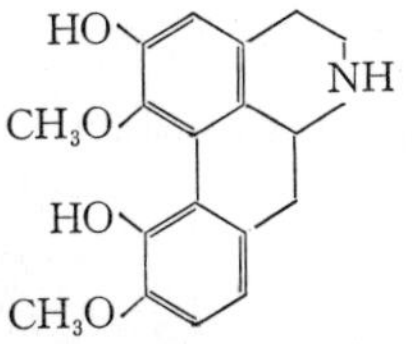

HO CH_3O NH HO CH_3O

≪Ph.≫ CR[11] UV[11] IR[11] NMR[11]
≪OP.≫ Lindera pipericarpa[11]
≪C. & So.≫ Colourless needles, unstable in air, light.
≪Sa. & D.≫ B-HCl: 200°(d) B-NMe: 196°(d) B-NMe·MeI: 187° B-N, O-DiMe=isocorydine
≪SD.≫ 11)
≪S. & B.≫

≪mp≫ 195°(d) (AcOEt)
≪[α]D≫ 166° (EtOH)

No Name $C_{19}H_{19}O_4N \equiv 325.35$

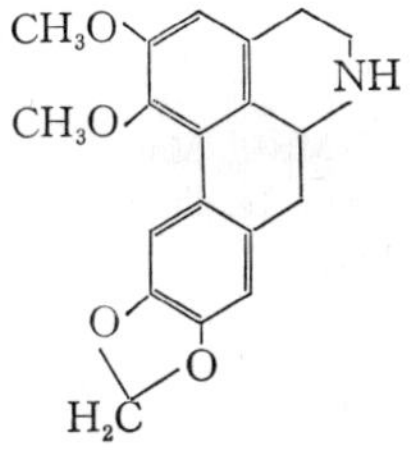

CH_3O CH_3O NH O O H_2C

≪Ph.≫
≪OP.≫ Cassytha racemosa[10]
≪C. & So.≫
≪Sa. & D.≫
≪SD.≫ 11)
≪S. & B.≫

≪mp≫
≪[α]D≫

Bracteoline $C_{19}H_{21}O_4N \equiv 327.37$

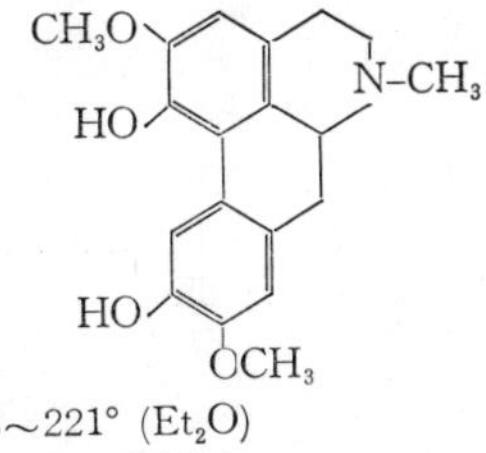

CH_3O HO N-CH_3 HO OCH_3

≪Ph.≫ UV[27] IR[27] Mass[27] Rf[27]
≪OP.≫ Papaver bracteatum[27]
≪C. & So.≫ Sol. in MeOH, Spar. sol. in Et_2O
≪Sa. & D.≫
≪SD.≫ 27)
≪S. & B.≫

≪mp≫ 218～221° (Et_2O)
≪[α]D≫ 35±8° ($CHCl_3$)

***N*-Methyllaurotetanine** $C_{20}H_{23}O_4N \equiv 341.39$

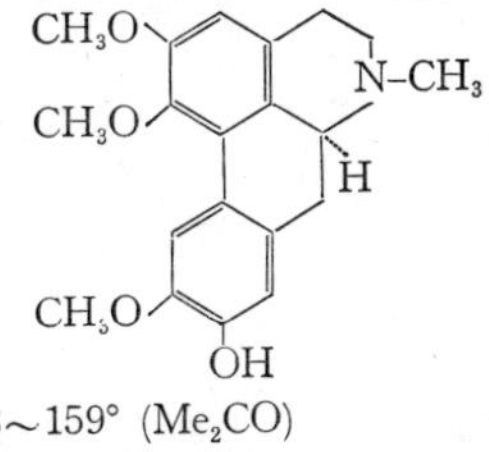

CH_3O CH_3O N-CH_3 H CH_3O OH

≪Ph.≫ UV[28] NMR[28]
≪OP.≫ Cassytha racemosa[10], Litsea cubeba[28]
≪C. & So.≫ Colourless prisms
≪Sa. & D.≫
≪SD.≫ 28)
≪S. & B.≫

≪mp≫ 158～159° (Me_2CO)
≪[α]D≫ 126° (EtOH)

Thaliporphine $C_{20}H_{23}O_4N \equiv 341.39$

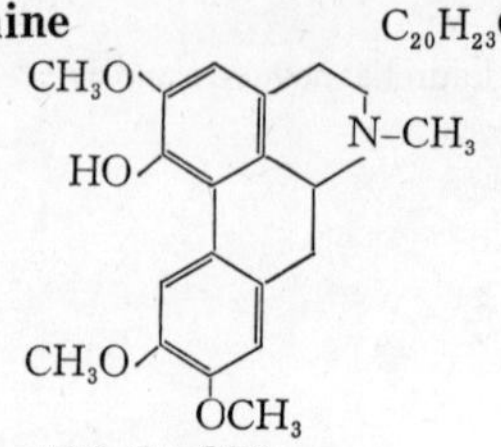

≪Ph.≫ UV[25] NMR[25] Mass[25]
≪OP.≫ Thalictrum fendleri[25]
≪C. & So.≫
≪Sa. & D.≫ B–OMe=glaucine
≪SD.≫ 25)
≪S. & B.≫

≪mp≫ 170~172° (MeOH)
≪[α]D≫

No Name $C_{21}H_{26}O_4N^{\oplus} \equiv 356.43$

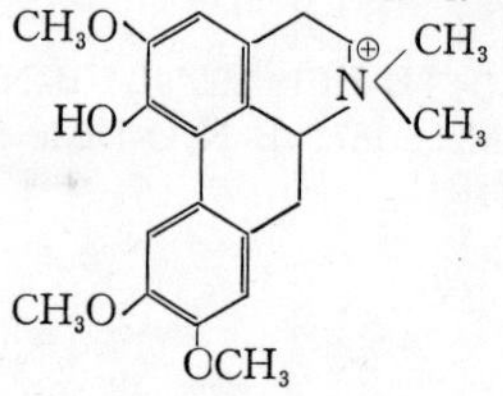

≪Ph.≫
≪OP.≫ Fagara tinguassoiba[17]
≪C. & So.≫
≪Sa. & D.≫
≪SD.≫ 17)
≪S. & B.≫ 17)

≪mp≫
≪[α]D≫

Preocoteine $C_{21}H_{25}O_5N \equiv 371.42$

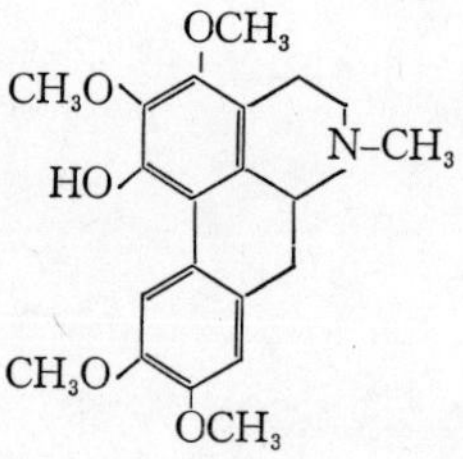

≪Ph.≫ UV[25] NMR[25] Mass[25]
≪OP.≫ Thalictrum fendleri[25]
≪C. & So.≫ Oil
≪Sa. & D.≫
≪SD.≫ 25)
≪S. & B.≫

≪mp≫
≪[α]D≫

Lysicamine $C_{18}H_{13}O_3N \equiv 291.29$

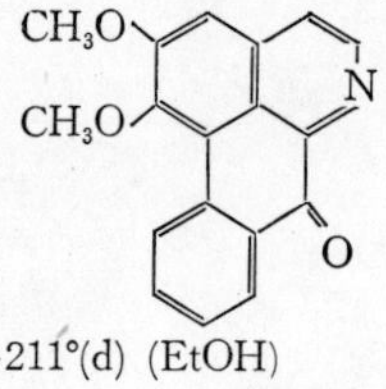

≪Ph.≫ UV[5] IR[5] NMR[5]
≪OP.≫ Lysichiton camtschatcense var. japonicum[5]
≪C. & So.≫ Yellow needles
≪Sa. & D.≫
≪SD.≫ 5)
≪S. & B.≫ 5)

≪mp≫ 210~211°(d) (EtOH)
≪[α]D≫

No Name $C_{19}H_{13}O_5N \equiv 335.30$

CH_3O CH_3O N O O O H_2C

≪mp≫
≪[α]D≫

≪Ph.≫
≪OP.≫ Cassytha racemosa[10]
≪C. & So.≫
≪Sa. & D.≫
≪SD.≫ 10)
≪S. & B.≫

No Name

HO CH_3O $N(CH_3)_2$

≪mp≫
≪[α]D≫

≪Ph.≫ UV[29] NMR[29]
≪OP.≫ Aristolochia argentina[29]
≪C. & So.≫
≪Sa. & D.≫
≪SD.≫ 29)
≪S. & B.≫

1) V. Pleininger, J. Appelt, L. Slavikova, J. Slavik: Coll. Czech. Chem. Comm., **32**, 2682 (1967).
2) S. Pfeifer, L. Kuehn: Pharmazie, **22**, 221 (1967).
3) K. Bernauer: Helv. Chim. Acta, **50**, 1583 (1967).
4) S. Pfeifer, I. Mann: Abh. Deut. Akad. Wiss. Berlin, Kl. Chem., Geol. Biol., **1966**, 315. [C. A., **67**, 3123 (1967)].
5) N. Katsui, K. Sato, S. Tobinaga, N. Takeuchi: Tetrahedron Letters, **1966**, 6257.
6) M. Muratova, D. Pavlaskova, F. Santavy: Planta Med., **14**, 22 (1966). [C. A., **65**, 1040 (1967)].
7) V. L. Pleininger, F. Santavy: Acta Univ. Palacki. Olomuc., Fac. Med., **1966**, 5. [C. A., **67**, 5115 (1967)].
8) M. Tomita, M. Kozuka: J. Pharm. Soc. Japan, **86**, 871 (1966).
9) S. R. Jones, J. A. Lamberton, A. A. Sioumis: Aust. J. Chem., **19**, 2239 (1966).
10) S. R. Jones, J. A. Lamberton, A. A. Sioumis: Aust. J. Chem., **20**, 1457 (1967).
11) A. K. Kiang, K. Y. Sim: J. Chem. Soc. (C), **1967**, 282.
12) A. H. Jackson, J. A. Martin: J. Chem. Soc. (C), **1966**, 2061.
13) A. H. Jackson, J. A. Martin: J. Chem. Soc. (C), **1966**, 2222.
14) A. H. Jackson, J. A. Martin: J. Chem. Soc. (C), **1966**, 2181.
15) A. Quevauviller, G. Sarrazin: Ann. Pharm. Fr., **25**, 371 (1967). [C. A., **67**, 10090 (1967)].
16) K. Kotera, Y. Hamada, R. Mitsui: Tetrahedron Letters, **1966**, 6273.
17) M. Shamma, W. A. Slusarchyk: Tetrahedron, **23**, 2563 (1967).
18) Kh. G. Pulatova, Z. F. Ismailov, S. Yu. Yunusov: Khim. Prir. Soedin., **3**, 67 (1967). [C. A., **67**, 1114 (1967)].
19) T. Kametani, I. Noguchi: J. Chem. Soc. (C), **1967**, 1440.
20) B. Franck, L. F. Tietz: Ang. Chem. Int. Ed. Engl., **6**, 799 (1967).
21) H. B. Dutschewska, N. M. Mollov: Ber., **100**, 3135 (1967).
22) H. Furukawa, S.-T. Lu: J. Pharm. Soc. Japan, **86**, 1143 (1966).
23) K. S. Soh, F. N. Lahey, R. Greenhalgh: Tetrahedron Letters, **1966**, 5279.
24) M. Tomita, H. Furukawa, T. Ikeda: J. Pharm. Soc. Japan, **87**, 793 (1967).
25) M. Shamma, R. J. Shine, B. S. Dudock: Tetrahedron, **23**, 2887 (1967).
26) I. R. C. Bick, J. H. Bowie, G. K. Douglas: Aust. J. Chem., **20**, 1403 (1967).
27) K. Heydenreich, S. Pfeifer: Pharmazie, **22**, 124 (1967).
28) S.-T. Lu, F.-M. Lin: J. Pharm. Soc. Japan, **87**, 878 (1967).
29) H. A. Priestap, E. A. Ruveda, S. M. Albónico, V. Deulofeu: Chem. Comm., **1967**, 754.

Chapter 10 Protoberberine Alkaloids

Coptisine
cf. p. 109.
≪Ph.≫ UV[1] IR[1] Rf[1]
≪OP.≫ Papaver caucasicum[2], P. persicum[2]
≪S. & B.≫ 3)

Berberine
cf. p. 110.
≪OP.≫ Papaver lapponicum[4], P. radicatum[4]
≪S. & B.≫ 3)

Dehydrocorydalmine
cf. p. 110.
≪Ph.≫ UV[5] IR[5] NMR[5]
≪OP.≫ Stephania glabra[5]
≪Sa. & D.≫ B⊕-Cl⊖ : 220～221°(d)
≪SD.≫ 5)

Palmatine
cf. p. 111.
≪Ph.≫ IR[6] NMR[5]
≪OP.≫ Papaver caucasicum[2], P. persicum[2]
≪S. & B.≫ 6)

Stylopine
cf. p. 112.
≪S. & B.≫ BS: 7)

Scoulerine
cf. p. 113.
≪Ph.≫ IR[6]
≪S. & B.≫ 6)

Canadine
cf. p. 114.
≪Ph.≫ UV[8]
≪OP.≫ Thalictrum minus[8]
≪Sa. & D.≫ B-MeCl: 191～193°(d)

Sinactine
cf. p. 114.
≪OP.≫ Fumaria officinalis[9]

***l*-Isocorypalmine**
cf. p. 115.
≪Ph.≫ CR[10] UV[10] NMR[10] Mass[10]
≪OP.≫ Tinomiscium petolare[10]

Cyclanoline
cf. p. 117.
≪Ph.≫ UV[11] IR[11]
≪OP.≫ Stephania tetrandra[11]

Steponine
cf. p. 117.
≪$[\alpha]_D$≫ B⊕-I⊖ : −120° (EtOH)
≪Ph.≫ CR[12] NMR[12] Rf[12]
≪OP.≫ Pericampylus formosanus[13], Stephania sasakii[12]
≪Sa. & D.≫ B⊕-O, O-DiMe·I⊖ : 208° B⊕-O, O-DiEt·I⊖ : 225°

Coreximine
cf. p. 117.
≪Ph.≫ UV[14] IR[15]
≪S. & B.≫ 14), 15), 16), 17)

Xylopinine
cf. p. 118.
≪Ph.≫ UV[16] IR[15][16] NMR[15] Rf[16]
≪S. & B.≫ 15), 16), 17)

Stepharotine
cf. p. 118.
≪OP.≫ Stephania rotunda[18]

Alborine (Alkaloid R-K[1])
cf. p. 122.
≪mp≫ 238～240° (Et_2O) or >360° (MeOH)
≪OP.≫ Papaver alboroseum[19]

Stephanarine $C_{19}H_{18}O_4N \equiv 324.34$

CH3O, HO, N⊕, OCH3, OH

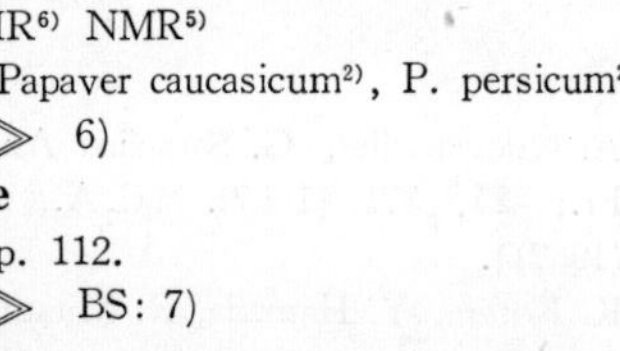

≪mp≫
≪$[\alpha]_D$≫
≪Ph.≫ NMR[5] Rf[5]
≪OP.≫ Stephania glabra[5]
≪C. & So.≫
≪Sa. & D.≫ B⊕-Cl⊖ : 274～275°(d)
≪SD.≫ 5)
≪S. & B.≫

Coramine

$C_{19}H_{21}O_4N \equiv 327.37$

≪mp≫ 247～252°

≪[α]$_D$≫

≪Ph.≫ Corydalis pseudoadunca[21)]

≪OP.≫

≪C. & So.≫

≪Sa. & D.≫ B–O,O–DiMe=xylopinine B–O, O–DiMe·MeI: 243～246° B–O,O–DiEt·MeI: 226～227° B–O, O–DiAc: 191～193°

≪SD.≫ 20)

≪S. & B.≫

Mecambridine

$C_{22}H_{25}O_6N \equiv 399.43$

≪mp≫ 178°

≪[α]$_D$≫ −236° ($CHCl_3$)

≪Ph.≫ UV[1) 22) 23)] IR[1) 22) 23)] NMR[22)] Rf[1) 22)]

≪OP.≫ Papaver alboroseum[19)], P. bracteatum[19)], P. nudicaule var. xanthopetalum[19)], P. nudicaule var. leiocarpum[19)], P. oreophilum[22)], P. orientale[23)]

≪C. & So.≫

≪Sa. & D.≫ B–OAc: 119～120°

≪SD.≫ 22)

≪S. & B.≫ BS: 22)

1) M. Muratova, D. Pavlaskova, F. Santavy: Planta Med., **14**, 22 (1966). [C.A., **65**, 1040 (1966)].
2) V. Pleininger, J. Appelt, L. Slavikova, J. Slavik: Coll. Czech. Chem. Comm., **32**, 2682 (1967).
3) X.A. Dominguez, J.A. Delgado, W.P. Reeves, P.D. Gardner: Tetrahedron Letters, **1967**, 2493.
4) H. Boehm: Abh. Deut. Akad. Wiss. Berlin, Kl. Chem., Geol. Biol., **1966**, 325. [C.A., **66**, 10769 (1967)].
5) R.W. Doskotch, M.Y. Malik, J.L. Beal: J. Org. Chem., **32**, 3253 (1967).
6) T. Kametani, M. Ihara: J. Chem. Soc. (C), **1967**, 530.
7) A.R. Battersby, R.J. Francis, M. Hirst, R. Southgate, J. Staunton: Chem. Comm., **1967**, 602.
8) K.I. Kuchkova, I.V. Terentéva, G.V. Lazuŕevskii: Khim. Prir. Soedin., **3**, 141 (1967). [C.A., **67**, 5121 (1967)].
9) N.M. Mollov, G.I. Yakimov, P.P. Panov: C. R. Acad. Bulg. Sci., **20**, 557 (1967). [C.A., **67**, 11029 (1967)].
10) M. Tomita, H. Furukawa: J. Pharm. Soc. Japan, **87**, 881 (1967).
11) M. Tomita, M. Kozuka, S.-T. Lu: J. Pharm. Soc. Japan, **87**, 316 (1967).
12) J. Kunitomo, T. Nagai, E. Yuge: J. Pharm. Soc. Japan, **87**, 1010 (1967).
13) M. Tomita, M. Kozuka, S.-T. Lu: J. Pharm. Soc. Japan, **87**, 315 (1967).
14) T. Kametani, M. Ihara: J. Chem. Soc. (C), **1966**, 2010.
15) T. Kametani, M. Ihara: J. Pharm. Soc. Japan, **87**, 174 (1967).
16) T. Kametani, I. Noguchi, S. Nakamura, Y. Konno: J. Pharm. Soc. Japan, **87**, 168 (1967).
17) T. Kametani, M. Satoh: J. Pharm. Soc. Japan, **87**, 179 (1967).
18) M. Tomita, M. Kozuka: J. Pharm. Soc. Japan, **86**, 871 (1966).
19) S. Pfeifer, D. Thomas: Pharmazie, **21**, 701 (1967).
20) M.S. Yunusov, S.T. Akramov, S. Yu. Yunusov: Khim. Prir. Soedin., **2**, 340 (1966). [C.A., **66**, 8050 (1967)].
21) M.S. Yunusov, S.T. Akramov, S. Yu. Yunusov: Dokl. Akad. Nauk SSSR, **162**, 607 (1965). [C.A., **63**, 5695 (1966)].
22) S. Pfeifer, I. Mann, L. Dolejs, Š.V. Hanuš, A.D. Cross: Tetrahedron Letters, **1967**, 83.
23) V.L. Pleininger, F. Santavy: Acta Univ. Palacki. Olomuc., Fac. Med., **1966**, 5. [C.A., **67**, 5115 (1967)].

Chapter 11 Protopine Alkaloids

Protopine
cf. p. 129.
≪Ph.≫ UV[1] IR[1] Mass[2] Rf[1]
≪OP.≫ Bocconia cordata[3], B. frutescens[4], Fumaria officinalis[5], Papaver caucasicum[6], P. oreophilum[7], P. persicum[6], P. radicatum[8]

Hunnemannine
cf. p. 129.
≪Ph.≫ UV[9] NMR[9]
≪S. & B.≫ 9)

α-Allocryptopine
cf. p. 130.
≪OP.≫ Bocconia cordata[3], B. frutescens[4]

β-Allocryptopine (Thalictrimine)
cf. p. 131.
≪Ph.≫ UV[1)10)] IR[1)4)10)] Mass[2] Rf[1]
≪OP.≫ Thalictrum minus[10)11)]
≪Sa. & D.≫ B-HBr: 178～179°(d) B_2-H_2SO_4: 218～219° B-MeI: 187～188°(d)
≪SD.≫ 10)

Muramine
cf. p. 131.
≪Ph.≫ Mass[2]

Oreophiline $C_{23}H_{25}O_6N$≡411.44

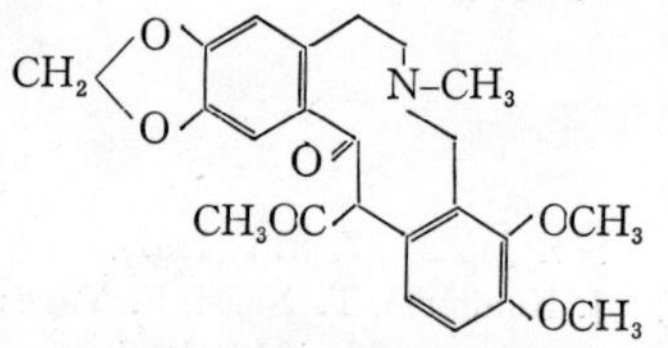

≪mp≫
≪[α]$_D$≫
≪Ph.≫
≪OP.≫ Papaver oreophilum[7]
≪C. & So.≫
≪Sa. & D.≫
≪SD.≫ 7)
≪S. & B.≫

1) M. Muratova, D. Pavlaskova, F. Santavy: Planta Med., **14**, 22 (1966). [C. A., **65**, 1040 (1966)].
2) V. Hanus, L. Dolejs, B. Mueller, W. Doepke: Coll. Czech. Chem. Comm., **32**, 1759 (1967).
3) N. G. Kiryakov, M. S. Kitova, A. V. Georgieva: C. R. Acad. Bulg. Sci., **20**, 189 (1967). [C. A., **67**, 3123 (1967)].
4) C. Tani, S. Takao: J. Pharm. Soc. Japan, **87**, 699 (1967).
5) N. M. Mollov, G. I. Yakinov, P. P. Panov: C. R. Acad. Bulg. Sci., **20**, 557 (1967). [C. A., **67**, 11029 (1967)].
6) V. Pleininger, J. Appelt, L. Slavikova, J. Slavik: Coll. Czech. Chem. Comm., **32**, 2682 (1967).
7) S. Pfeifer, I. Mann: Abh. Deut. Akad. Wiss. Berlin, Kl. Chem., Geol. Biol., **1966**, 315. [C. A., **67**, 3123 (1967)].
8) H. Boehm: Abh. Deut. Akad. Wiss. Berlin, Kl. Chem., Geol. Biol., **1966**, 325. [C. A., **66**, 10769 (1966)].
9) D. Giacopello, V. Deulofeu: Tetrahedron, **23**, 3265 (1967).
10) K. I. Kuchkova, G. V. Lazurévskii: Izv. Akad. Nauk Mold. SSR, Ser. Khim. Biol., No.11. 43 (1965). [C. A., **66**, 8932 (1967)].
11) K. I. Kuchkova, I. V. Terenteva, G. V. Lazurévskii: Khim. Prior. Soedin., **3**, 141 (1967). [C. A., **67**, 5121 (1967)].

Chapter 12 Phthalide Alkaloids

Narcotine
cf. p. 138.
≪S. & B.≫ BS: 1)

1) A. R. Battersby, R. J. Francis, M. Hirst, R. Southgate, J. Staunton: Chem. Comm., **1967**, 602.

Chapter 13 Rhoeadine and Related Alkaloids

Papaverrubine A (Alkaloid R-5[1])
cf. p. 144.
≪Ph.≫ Mass[2] Rf[1]
≪SD.≫ 3), 4)

Papaverrubine D
cf. p. 144.
≪Ph.≫ Mass[2]
≪OP.≫ Papaver alboroscum[5]

Rhoeadine
cf. p. 145.
≪Ph.≫ UV[6][7] IR[7] Mass[2] Rf[1][7]
≪OP.≫ Papaver oreophilum[8]

Rhoeagenine
cf. p. 145.
≪Ph.≫ UV[7] IR[7] Mass[2] Rf[7]
≪OP.≫ Papaver oreophilum[8]

Isorhoeadine
cf. p. 145.
≪Ph.≫ UV[7] IR[7] Mass[2] Rf[1][7]
≪OP.≫ Papaver oreophilum[8]
≪SD.≫ 3)

Papaverrubine B
cf. p. 146.
≪Ph.≫ Mass[2] Rf[9]
≪SD.≫ 4)

Glaudine
cf. p. 146.
≪Ph.≫ NMR[9] Mass[2]

Glaucamine
cf. p. 146.
≪Ph.≫ Mass[2]

Oreodine
cf. p. 146.
≪Ph.≫ NMR[9] Mass[2]
≪OP.≫ Papaver oreophilum[8]
≪SD.≫ 8)

Oreogenine
cf. p. 147.
≪Ph.≫ Mass[2]
≪OP.≫ Papaver oreophilum[8]
≪SD.≫ 8)

Papaverrubine E $C_{20}H_{19}O_6N \equiv 369.36$

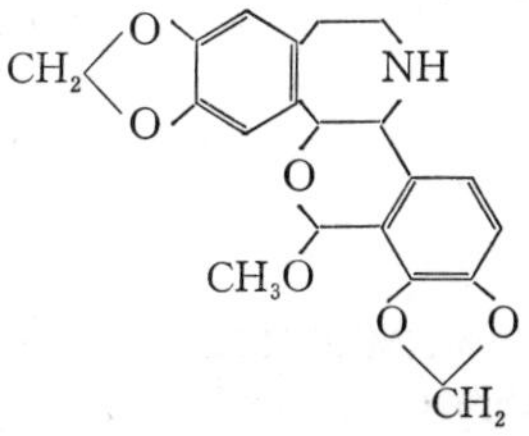

≪mp≫
≪$[\alpha]_D$≫

≪Ph.≫ Mass[2]
≪OP.≫
≪C. & So.≫
≪Sa. & D.≫
≪SD.≫ 4)
≪S. & B.≫

Papaverrubine C (Epiporphyroxine)
$C_{20}H_{21}O_6N \equiv 371.38$

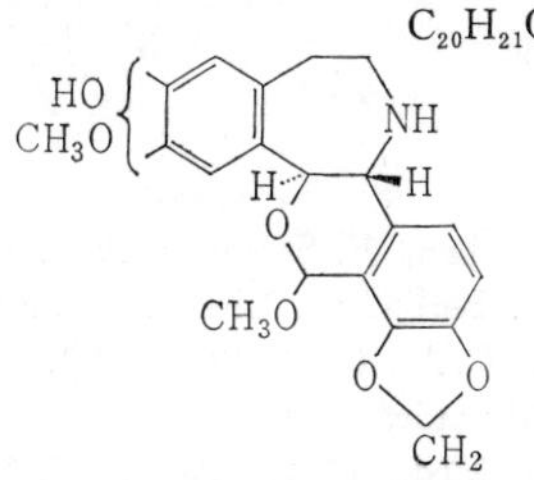

≪mp≫ 187～188° (C_6H_6–Petr. ether, AcOEt–Ligroin, EtOH)
≪$[\alpha]_D$≫ 283° ($CHCl_3$)

≪Ph.≫ UV[9] IR[9] NMR[9] Mass[9] Rf[9]
≪OP.≫ Opium[9], Papaver alboroseum[5], P. caucasicum[9]
≪C. & So.≫
≪Sa. & D.≫ B–OMe: 67～70°
≪SD.≫ 4), 9)
≪S. & B.≫

Papaverrubine F $C_{21}H_{23}O_6N \equiv 385.40$

<Ph.>
<OP.> Papaver oreophilum[4)]
<C. & So.>
<Sa. & D.> B–MeI: 186~189°
<SD.> 4)
<S. & B.>

<mp> 223~225° (Me_2CO)
<$[\alpha]_D$>

Alkaloid E $C_{22}H_{27}O_6N \equiv 401.44$

<Ph.> UV[10)] IR[10)] NMR[10)] Mass[2)10)]
<OP.> Papaver bracteatum[10)]
<C. & So.>
<Sa. & D.> B–HCl: 161~163° (d) **B–OMe: 105°**
B–N→O: 167~169°
<SD.> 10)
<S. & B.>

<mp> 186.5~187°
<$[\alpha]_D$> 306° (MeOH)

Alpinigenine $C_{22}H_{27}O_6N \equiv 401.44$

<Ph.> IR[11)] NMR[11)] Mass[11)]
<OP.> Papaver alpinum[11)]
<C. & So.>
<Sa. & D.>
<SD.> 3), 11)
<S. & B.>

<mp> 193~195° (AcOEt)
<$[\alpha]_D$> 286±5° (MeOH)

Alpinine $C_{23}H_{29}O_6N \equiv 415.47$

<Ph.> IR[11)] NMR[11)] Mass[2)11)]
<OP.> Papaver alpinum[11)]
<C. & So.>
<Sa. & D.>
<SD.> 11)
<S. & B.>

<mp>
<$[\alpha]_D$> 288±10° ($CHCl_3$)

1) F. Santavy, A. Nemeckova: Coll. Czech. Chem. Comm., **32**, 461 (1967).
2) L. Dolejš, V. Hanuš: Tetrahedron, **23**, 3265 (1967).
3) I. Mann, H. Doehnert, S. Pfeifer: Pharmazie, **21**, 494 (1966).
4) I. Mann, S. Pfeifer: Pharmazie, **21**, 700 (1966).
5) S. Pfeifer, D. Thomas: Pharmazie, **21**, 701 (1966).
6) L. Hruban, F. Santavy: Abh. Deut. Akad. Wiss. Berlin, Kl. Chem., Geol. Biol., **1966**, 627. [C. A., **67**, 2112 (1967)].
7) M. Muratova, D. Pavlaskova, F. Santavy: Planta Med., **14**, 22 (1966). [C. A., **65**, 1040 (1966)].
8) S. Pfeifer, I. Mann: Abh. Deut. Akad. Wiss. Berlin, Kl. Chem., Geol. Biol., **1966**, 315. [C. A., **67**, 3123 (1967)].
9) D. W. Hughes, L. Kuehn, S. Pfeifer: J.

Chem. Soc. (C), **1967**, 444.
10) A. Guggisberg, M. Hesse, H. Schmid, H. Boehm, H. Roensch, K. Mothes: Helv. Chim. Acta, **50**, 621 (1967).
11) M. Maturova, H. Potesilova, F. Santavy, A. D. Cross, V. Hanus. , L. Dolejs: Coll. Czech. Chem. Comm., **32**, 419 (1967).

Chapter 14 Morphine and Related Alkaloids

Norsinoacutine

cf. p. 148.

≪mp≫ 113～115° [B+AcOEt]
≪[α]$_D$≫ −107° (EtOH)
≪Ph.≫ IR[1] UV[1] NMR[1]
≪OP.≫ Croton balsamifera[1], C. flavens[1]
≪Sa. & D.≫ B–N, O–DiMe·MeI: 163～166°

Amurine

cf. p. 148.

≪Ph.≫ NMR[1]
≪OP.≫ Papaver feddei[2], P. radicatum[3]

Nudaurine (Amurinol-I)

cf. p. 148.

≪Ph.≫ Rf[4]
≪SD.≫ 4)

Salutaridine

cf. p. 149.

≪Ph.≫ UV[5] IR[5] NMR[1] Mass[6]
≪OP.≫ Papaver bracteatum[7], P. orientale[5]
≪Sa. & D.≫ B-OMe: 147～148°
≪SD.≫ 8)
≪S. & B.≫ 8)

Sinomenine

cf. p. 149.

≪Ph.≫ Mass[6]

Isosinomenine

cf. p. 149.

≪Ph.≫ Mass[6]

Morphine

cf. p. 150.

≪Ph.≫ Mass[6] ORD[9][10] CD[11]
≪SD.≫ 8)
≪S. & B.≫ 8); BS: 13)

Codeine

cf. p. 150.

≪Ph.≫ Mass[6] ORD[10]
≪S. & B.≫ BS: 13)

Neopine

cf. p. 150.

≪Ph.≫ Mass[6]

Thebaine

cf. p. 151.

≪Ph.≫ UV[12] IR[12] Mass[6] Rf[12]
≪OP.≫ Papaver feddei[2], P. radicatum[3]
≪S. & B.≫ BS: 13)

Flavinine $C_{18}H_{19}O_4N \equiv 313.34$

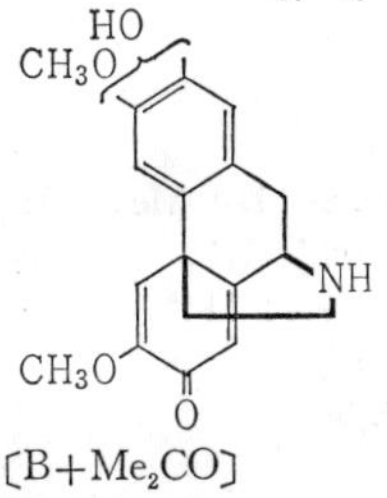

≪mp≫ 130～132° [B+Me$_2$CO]
≪[α]$_D$≫ −6° (EtOH)
≪Ph.≫ UV[1] IR[1] NMR[1] CD[1]
≪OP.≫ Croton flavens[1]
≪C. & So.≫ Sol. in CHCl$_3$
≪Sa. & D.≫ B–N, O–DiMe·MeI: 250～252°(d)
≪SD.≫ 1)
≪S. & B.≫

8,14-Dihydronorsalutaridine $C_{18}H_{21}O_4N \equiv 315.36$

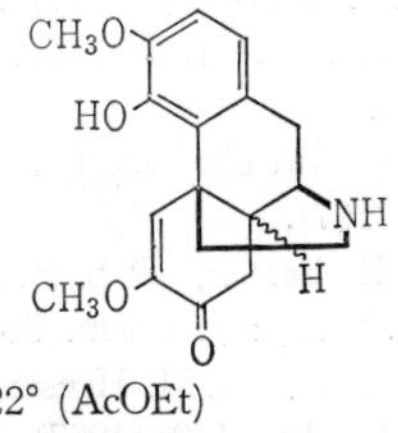

≪mp≫ 208～222° (AcOEt)
≪[α]$_D$≫ −69.1° (MeOH)
≪Ph.≫ IR[14] NMR[14] Mass[14]
≪OP.≫ Croton linearis[14]
≪C. & So.≫
≪Sa. & D.≫
≪SD.≫ 14)
≪S. & B.≫

8,14-Dihydrosalutaridine $C_{19}H_{23}O_4N \equiv 329.38$

≪mp≫ 198～203°
≪[α]$_D$≫ −76.1° (MeOH)
≪Ph.≫ UV[14] IR[14] NMR[14] Mass[14]
≪OP.≫ Croton linearis[14]
≪C. & So.≫
≪Sa. & D.≫ B–OAc: 210°
≪SD.≫ 14)
≪S. & B.≫

Acutumidine $C_{18}H_{22}O_6NCl \equiv 383.83$

≪mp≫ 239～241°(d)
≪[α]$_D$≫ −212°(Py)
≪Ph.≫ NMR[15] CD[15] X-ray[16] pKa[16]
≪OP.≫ Menispermum dauricum[16], Sinomenium acutum[16]
≪C. & So.≫
≪Sa. & D.≫
≪SD.≫ 15), 16)
≪S. & B.≫

Acutumine $C_{19}H_{24}O_6NCl \equiv 396.84$

≪mp≫ 238～240°(d)
≪[α]$_D$≫ −206°(Py)
≪Ph.≫ UV[15)16)] IR[15)16)] Mass[16)17)] CD[15] pKa[16)17)] NMR[16]
≪OP.≫ Menispermum dauricum[16], Sinomenium acutum[16]
≪C. & So.≫
≪Sa. & D.≫ B–$HAuCl_4$: 199～200° B–Semicarb: >290°
≪SD.≫ 15), 16), 17)
≪S. & B.≫

Hernandoline $C_{20}H_{25}O_5N \equiv 359.41$

≪mp≫
≪[α]$_D$≫
≪Ph.≫
≪OP.≫ Stephania hernandifolia[18]
≪C. & So.≫
≪Sa. & D.≫ B–OMe: 104° B–OMe·MeI: 190～191° B–OMe·HI: 186° B–OMe·$HClO_4$: 225～226° B–OAc: 114°
≪SD.≫ 18)
≪S. & B.≫

1) K. L. Stuart, C. Chambers: Tetrahedron Letters, **1967**, 2879.
2) S. Pfeifer, I. Mann: Abh. Deut. Akad. Wiss. Berlin., Kl. Chem., Geol. Biol., **1966**, 315. [C. A., **67**, 3123 (1967)].
3) H. Boehm: Abh. Deut. Akad. Wiss. Berlin., Kl. Chem., Geol. Biol., **1966**, 325. [C. A., **66**, 10769 (1967)].
4) D. H. R. Barton, R. James, G. W. Kirby, W. Doepke, H. Flentje: Ber., **100**, 2457 (1967).
5) V. L. Pleininger, F. Santavy: Acta Univ. Palacki. Olomuc., Fac. Med., **1966**, 5. [C. A., **67**, 5115 (1967)].
6) D. M. S. Wheeler, T. H. Kinstle, K. L. Rinehart, Jr.: J. Am. Chem. Soc., **89**, 4494 (1967).
7) K. Heydenreich, S. Pfeifer: Pharmazie, **22**, 124 (1967).
8) D. H. R. Barton, D. S. Bhakuni, R. James, G. W. Kirby: J. Chem. Soc. (C), **1967**, 128.
9) A. F. Casy, M. M. Hansan: J. Pharm. Pharmacol., **19**, 132 (1967). [C. A., **66**, 7159

(1967)].
10) G. Bernath, J. A. Szabo, K. Koczka, P. Vinkler: Acta Chim. Acad. Sci. Hung., **51**, 339 (1967). [C. A., **66**, 10770 (1967)].
11) H. Corrodi, E. Hardegger: J. Pharm. Pharmacol., **19**, 623 (1967). [C. A., **67**, 11027 (1967)].
12) M. Muratova, D. Pavlaskova, F. Santavy: Planta Med., **14**, 22 (1966). [C. A., **65**, 1040 (1966)].
13) A. R. Battersby, J. A. Martin, E. Brochmann-Hanssen: J. Chem. Soc. (C), **1967**, 1785.
14) L. J. Haynes, G. E. M. Husbands, K. L. Stuart: Chem. Comm., **1967**, 15.
15) M. Tomita, Y. Okamoto, T. Kikuchi, K. Osaki, M. Nishikawa, K. Kamiya, Y. Sasaki, K. Matoba, K. Goto: Tetrahedron Letters, **1967**, 2425.
16) M. Tomita, Y. Okamoto, T. Kikuchi, K. Osaki, M. Nishikawa, K. Kamiya, Y. Sasaki, K. Matoba, K. Goto: Tetrahedron Letters, **1967**, 2421.
17) K. Goto, M. Tomita, Y. Okamoto, Y. Sasaki, K. Matoba: Proc. Japan Acad., **42**, 1181 (1966). [C. A., **67**, 5121 (1967)].
18) I. I. Fadelva, A. D. Kuzovkov, T. N. Ilinskaya: Khim. Prir. Soedin., **3**, 106 (1967). [C. A., **67**, 4142 (1967)].

Chapter 15 Hasubanonine and Related Alkaloids (Hasubanan Alkaloids)

Metaphanine
cf. p. 158.
≪Ph.≫ UV[1] IR[1] NMR[1] Mass[1]
≪OP.≫ Stephania abyssinica[1], S. japonica[2]
≪SD.≫ 1)
≪S. & B.≫ BG: 1)

Prometaphanine
cf. p. 158.
≪Ph.≫ IR[2] NMR[2]
≪OP.≫ Stephania japonica[2]
≪SD.≫ 2)

Hasubanonine
cf. p. 158.
≪SD.≫ 3)

Cepharamine $C_{19}H_{23}O_4N \equiv 329.38$

CH_3O, HO, $N-CH_3$, O, OCH_3

≪mp≫ 186～187°
≪$[\alpha]_D$≫ −248° ($CHCl_3$)
≪Ph.≫ CR[4] UV[4] IR[4] NMR[4] Mass[4] ORD[4]
≪OP.≫ Stephania cepharantha[4]
≪C. & So.≫
≪Sa. & D.≫
≪SD.≫ 4)
≪S. & B.≫

1) H. L. deWaal, B. J. Prinsloo, R. R. Arndt: Tetrahedron Letters, **1966**, 6169.
2) M. Tomita, Y. Inubushi, T. Ibuka: J. Pharm. Soc. Japan, **87**, 381 (1967).
3) T. Ibuka, M. Kitano, Y. Watanabe, M. Matsui: J. Pharm. Soc. Japan, **87**, 1014 (1967).
4) M. Tomita, M. Kozuka: Tetrahedron Letters, **1966**, 6229.

Chapter 16 Emetine and Related Alkaloids

Dihydroprotoemetine
cf. p. 160.
≪Ph.≫ Mass[1]
≪OP.≫ Alangium lamarckii[1]
≪SD.≫ 1)
≪S. & B.≫ BG: 1)

Ankorine

cf. p. 160.

≪Ph.≫ UV[1] IR[1] MNR[1] Mass[1]

≪OP.≫ Alangium lamarckii[1]

≪SD.≫ 1)

≪S. & B.≫ BG: 1)

Ipecoside

cf. p. 160

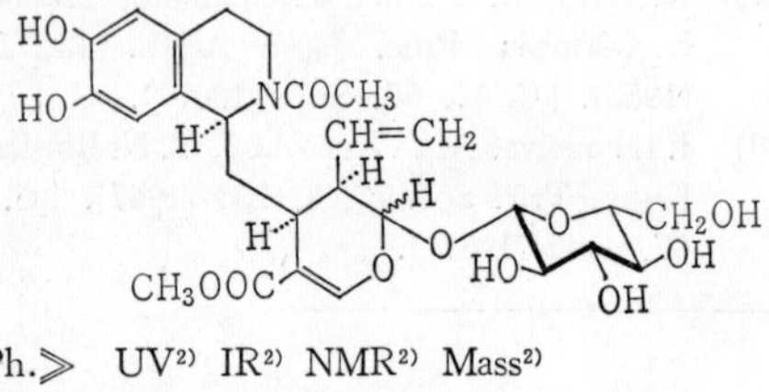

≪Ph.≫ UV[2] IR[2] NMR[2] Mass[2]

≪Sa. & D.≫ B–Dihydro-Der.: 160～161°

≪SD.≫ 2)

≪S. & B.≫ BG: 2)

Psychotrine

cf. p. 161.

≪Ph.≫ Rf[3]

≪OP.≫ Alangium lamarckii[3]

Emetine

cf. p. 162.

≪S. & B.≫ 4), 5)

Deoxytubulosine

cf. p. 163.

≪OP.≫ Alangium lamarckii[1]

Tubulosine

cf. p. 163.

≪Ph.≫ UV[6] IR[6] NMR[6] Mass[6] Rf[6]

≪S. & B.≫ 7)

Isotubulosine

cf. p. 163.

≪Ph.≫ CR[6] UV[6] IR[6] NMR[6] Mass[6] Rf[6]

≪OP.≫ Alangium lamarckii[6]

≪Sa. & D.≫ B–OMe: 163°

≪SD.≫ 6)

Alangimarckine

cf. p. 163.

≪Ph.≫ UV[1] IR[1] NMR[1] Mass[1]

≪OP.≫ Alangium lamarckii[1]

≪SD.≫ 1)

≪S. & B.≫ BG: 1)

Demethylpsychotrine $C_{27}H_{34}O_4N_2 \equiv 450.56$

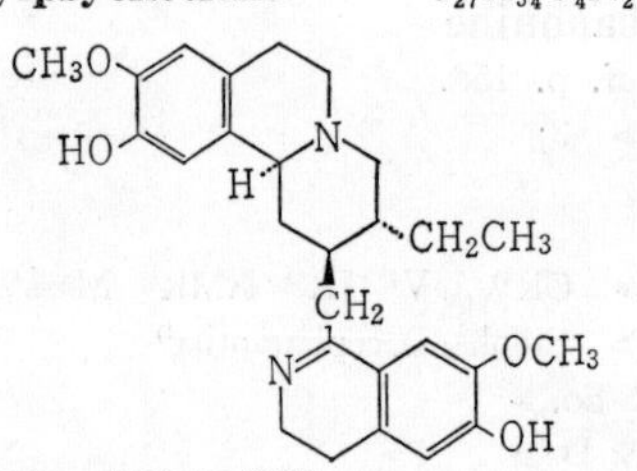

≪mp≫ 166～168° (EtOH)

≪$[\alpha]_D$≫ 67.9° (MeOH)

≪Ph.≫ UV[3] Mass[3]

≪OP.≫ Alangium lamarckii[3]

≪C. & So.≫ Dark yellow granules

≪Sa. & D.≫

≪SD.≫ 3)

≪S. & B.≫

Alangicine $C_{28}H_{36}O_5N_2 \equiv 480.58$

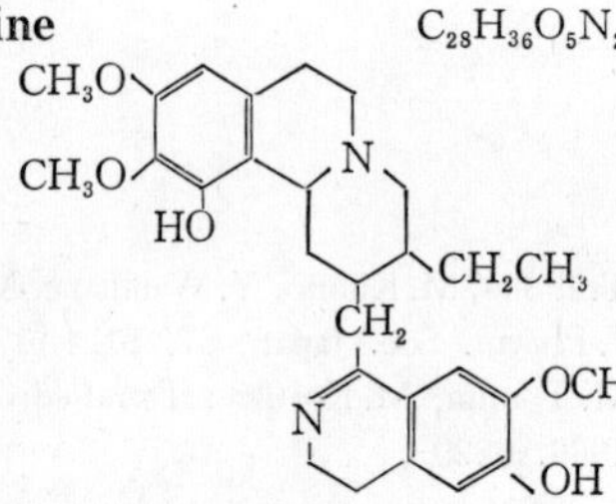

≪mp≫ 147～148° (EtOH)

≪$[\alpha]_D$≫ 64.1° (MeOH)

≪Ph.≫ UV[3] Mass[3]

≪OP.≫ Alangium lamarckii[3]

≪C. & So.≫ Yellow granules

≪Sa. & D.≫

≪SD.≫ 3)

≪S. & B.≫

1) A. R. Battersby, R. S. Kapil, D. S. Bhakuni, S. P. Popli, J. R. Merchant, S. S. Salgar: Tetrahedron Letters, **1966**, 4965.
2) A. R. Battersby, B. Gregory, H. Spencer, J. C. Turner, M. M. Janot, P. Potier, P. Francois, J. Levisalles: Chem. Comm., **1967**, 219.
3) S. C. Pakrashi, E. Ai: Tetrahedron Letters, **1967**, 2143.
4) Chinoin Gyogyszer es Vegyeszeti Termerker Gyara Rt. Australian, 252,467. [C. A., **67**, 2114 (1967)].
5) Cs. Szantay, L. Toke, P. Kolonits: Magy. Kem.

Foly., **73**, 293 (1967). 〔C. A., **68**, 295 (1968)〕.
6) A. Popelak, E. Haack, H. Spingler: Tetrahedron Letters, **1966**, 5077.
7) Cs. Szantay, Gy. Kalaus: Acta Chim. Acad. Sci. Hung., **49**, 427 (1966). 〔C. A., **67**, 2114 (1967)〕.

Chapter 17 Erythrina Alkaloids

Cocculolidine
cf. p. 167.
≪Ph.≫ UV[1] IR[1] NMR[1] Mass[1]
≪OP.≫ Cocculus trilobus[1]
≪C. & So.≫ B–Dihydro-Der.: 121～121.5° B–Tetrahydro-Der.: 187～189° B–NAc: 173～175° B–MeI 261～264°(d) B–HCl: 247～251°(d)
≪SD.≫ 1)

β-Erythroidine
cf. p. 167.
≪Ph.≫ ORD[2]

Erythraline
cf. p. 168.
≪Ph.≫ Mass[3]
≪S. & B.≫ BS: 3)

Erythramine
cf. p. 168.
≪Ph.≫ NMR[3] Mass[3]
≪SD.≫ 3)

Dihydroerysodine
cf. p. 169.
≪Ph.≫ Mass[3]

Erythratine
cf. p. 169.
≪Ph.≫ Mass[3]
≪S. & B.≫ BS: 3)

1) K. Wada, S. Marumo, K. Munakata: Agr. Biol. Chem., **31**, 452 (1967).
2) U. Weiss, H. Zeffer: Experientia, **19**, 660 (1963).
3) D. H. R. Barton, R. James, G. W. Kirby, D. A. Widdowson: Chem. Comm., **1967**, 266.

Chapter 18 Protostephanine Alkaloids

Protostephanine
cf. p. 173.
≪Ph.≫ Mass[1]
≪OP.≫ Stephania japonica[2]
≪SD.≫ 3)
≪S. & B.≫ 1)

1) B. Pecherer, A. Brossi: J. Org. Chem., **32**, 1053 (1967).
2) M. Tomita, Y. Inubushi, T. Ibuka: J. Pharm. Soc. Japan, **87**, 381 (1967).
3) B. Pecherer, A. Brossi: Helv. Chim. Acta, **49**, 2261 (1966).

Chapter 19 Amaryllidaceae Alkaloids

Caranine
cf. p. 176.
≪Ph.≫ UV[1] Mass[2,3] ORD[1] CD[1]
≪SD.≫ 4)

Lycorine
cf. p. 176.
≪Ph.≫ UV[1] Mass[2,3] NMR[4] ORD[1,4] CD[1,4] X-ray[4]
≪SD.≫ 4)
≪S. & B.≫ BS: 5)

Pluviine
cf. p. 178.
≪SD.≫ 4)

Methylpseudolycorine
cf. p. 178.
≪Ph.≫ Mass[3]

Narcissidine
cf. p. 181.
≪Ph.≫ Mass[3]

Bellamarine
cf. p. 181.
≪Ph.≫ Mass[3]

Aulamine
cf. p. 182.
≪Ph.≫ Mass[3]

***O*¹-Acetyllycorine**
cf. p. 182.
≪Ph.≫ Mass[3]

Nartazine
cf. p. 182.
≪Ph.≫ Mass[3]

Crinine
cf. p. 183.
≪S. & B.≫ 6)

Haemanthamine
cf. p. 185.
≪Ph.≫ UV[1] ORD[1] CD[1]

Crinamine
cf. p. 186.
≪Ph.≫ UV[1] ORD[1] CD[1]

Masonine
cf. p. 190.
≪Ph.≫ NMR[7]
≪Sa. & D.≫ B–Tetrahydro: 189～192°
≪SD.≫ 7)

Hippeastrine
cf. p. 190.
≪Ph.≫ Mass[2,8] ORD[1] CD[1]

Clivonine
cf. p. 191.
≪Ph.≫ NMR[9] Mass[9] ORD[9] CD[9]
≪SD.≫ 9)

Homolycorine
cf. p. 191.
≪Ph.≫ Mass[2]

Ungerine
cf. p. 191.
≪Ph.≫ Mass[8]

Clivimine
cf. p. 193.
≪Ph.≫ NMR[9] Mass[9]
≪SD.≫ 9)

Oduline
cf. p. 193.
≪Ph.≫ NMR[7]
≪SD.≫ 7)

Lycorenine
cf. p. 194.
≪Ph.≫ Mass[2]

Unsevine
cf. p. 194.
≪Ph.≫ Mass[8]

Tazettine
cf. p. 196.
≪Ph.≫ UV[1] ORD[1] CD[1]
≪SD.≫ 10)

Galanthamine
cf. p. 198.
≪S. & B.≫ 11)

1) K. Kuriyama, T. Iwata, M. Moriyama, K. Kotera, Y. Hamada, R. Mitsui, K. Takeda: J. Chem. Soc. (B), **1967**, 46.
2) T. Ibuka, H. Irie, A. Kato, S. Uyeo, K. Kotera, Y. Nakagawa: Tetrahedron Letters, **1966**, 4745.
3) T.H. Kinstle, W.C. Wildman, C.L. Brown: Tetrahedron Letters, **1966**, 4659.
4) K. Kotera, Y. Hamada, R. Mitsui: Tetrahedron Letters, **1966**, 6273.
5) W.C. Wildman, N.E. Heimer: J. Am. Chem. Soc., **89**, 5265 (1967).
6) H.W. Whitlock. Jr., G.L. Smith: J. Am. Chem. Soc., **89**, 3600 (1967).
7) W. Doepke, M. Bienert: Arch. Pharm., **299,** 994 (1966).
8) R. Razakov, Kh. A. Abduazimov, N.S. Vulfson, S. Yu. Yunusov: Khim. Prir. Soedin., **3**, 23 (1967). 〔C.A., **67**, 4142 (1967)〕.
9) W. Doepke, M. Bienert, A.L. Burlingame, H. K. Schnoes, P.W. Jeffs, D.S. Farrier: Tetrahedron Letters, **1967**, 451.
10) Y. Tsuda, Y. Sasaki, S. Uyeo: Chem. Pharm. Bull., **15**, 980 (1967).
11) S. Uyeo, H. Shirai, A. Koshiro, T. Yashiro, K. Kagei: Chem. Pharm. Bull., **14**, 1033 (1967).

Chapter 20 Benzophenanthridine Alkaloids

Oxysanguinarine

cf. p. 213.

≪Ph.≫ IR[1,2] UV[1,2] Rf[1]

≪OP.≫ Papaver orientale[2]

Sanguinarine

cf. p. 214.

≪Ph.≫ UV[1] IR[1,3] Rf[1]

≪OP.≫ Bocconia cordata[4], B. frutescens[3], Papaver radicatum[5]

Chelidonine

cf. p. 214.

≪S. & B.≫ BS: 6)

Chelerythrine

cf. p. 215.

≪Ph.≫ IR[3]

≪OP.≫ Bocconia frutescens[3], Toddalia aculeata[7]

No Name $C_{20}H_{15}O_4N \equiv 333.33$

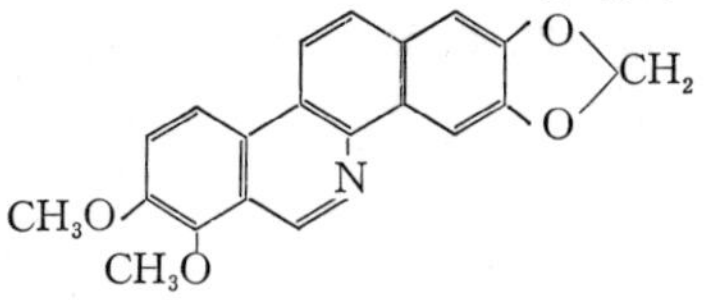

≪mp≫ 220～221°

≪$[\alpha]_D$≫

≪Ph.≫ UV[7] NMR[7] Mass[7]

≪OP.≫ Toddalia aculeata[7]

≪C. & So.≫

≪Sa. & D.≫

≪SD.≫ 7)

≪S. & B.≫

1) M. Muratova, D. Pavlaskova, F. Santavy: Planta Med., **14**, 22 (1966). [C. A., **65**, 1040 (1966)].
2) V. L. Pleininger, F. Santavy: Acta Univ. Palacki. Olomuc., Fac. Med., **1966**, 5. [C. A., **67**, 5115 (1967)].
3) C. Tani, S. Takao: J. Pharm. Soc. Japan, **87**, 699 (1967).
4) N. G. Kiryakov, M. S. Kitova, A. V. Georgieva: C. R. Akad. Bulg. Sci., **20**, 189 (1967). [C. A., **67**, 3123 (1967)].
5) H. Boehm: Abh. Deut. Akad. Wiss. Berlin., Kl. Chem., Geol. Biol., **1966**, 325. [C. A., **66**, 10769 (1967)].
6) A. R. Battersby, R. J. Francis, M. Hirst, R. Southgate, J. Staunton: Chem. Comm., **1967**, 602.
7) T. R. Govindachari, N. Viswanathan: Indian J. Chem., **5**, 280 (1967).

Chapter 21 Dibenzopyrrocoline Alkaloids

Cryptaustoline

cf. p. 221.

≪Ph.≫ Mass[1], cf. NMR[1]

≪S. & B.≫ 1), 2)

Cryptowoline

cf. p. 221.

≪Ph.≫ Mass[1], cf. NMR[1]

≪S. & B.≫ 1), 3)

1) T. Kametani, K. Ogasawara: J. Chem. Soc. (C), **1967**, 2208.
2) B. Franck, L. F. Tietz: Ang. Chem. Int. Ed. Engl., **6**, 799 (1967).
3) F. Benington, R. D. Morin: J. Org. Chem., **23**, 1050 (1967).

Chapter 22 Phenethylisoquinoline and Colchicine Alkaloids

Melanthioidine

cf. p. 222.

≪Ph.≫ UV[1] IR[1] NMR[1] Mass[1]

≪SD.≫ 1)

≪S. & B.≫

Colchicine

cf. p. 222.

≪OP.≫ Kreysigia multiflora[2]

Deacetylcolchicine

≪OP.≫ Kreysigia multiflora[2]

Kreysiginone $C_{20}H_{23}O_4N \equiv 341.39$

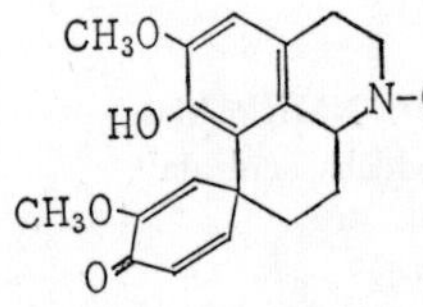

≪mp≫

≪[α]D≫

≪Ph.≫ UV[3,4] IR[3,4] NMR[3,4] Mass[3,4]

≪OP.≫ Kreysigia multiflora[3]

≪C. & So.≫

≪Sa. & D.≫

≪SD.≫ 3)

≪S. & B.≫ 3), 4), 5)

Dihydrokreysiginone $C_{20}H_{25}O_4N \equiv 343.41$

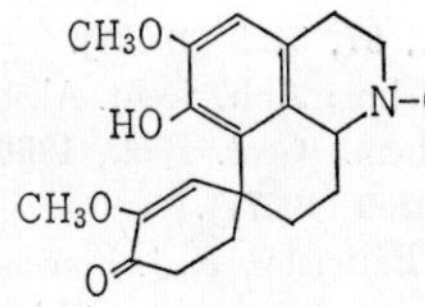

≪mp≫ 217～222°(d)

≪[α]D≫

≪Ph.≫ UV[3] IR[3] NMR[3] Mass[3]

≪OP.≫ Kreysigia multiflora[3]

≪C. & So.≫

≪Sa. & D.≫

≪SD.≫ 3)

≪S. & B.≫

Floramultine $C_{21}H_{25}O_5N \equiv 371.42$

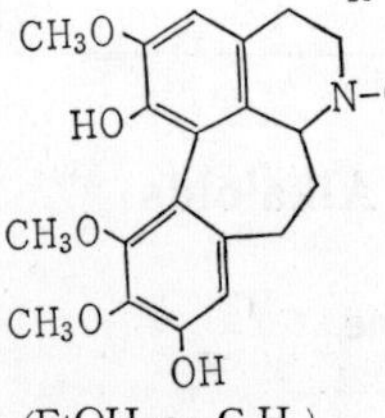

≪mp≫ 230°(d) (EtOH or C_6H_6)

≪[α]D≫ −77° ($CHCl_3$)

≪Ph.≫ UV[6] IR[6] NMR[2] Mass[2] Rf[6]

≪OP.≫ Kreysigia multiflora[2]

≪C. & So.≫

≪Sa. & D.≫ B–2OMe·HBr: 243°

≪SD.≫ 2), 6)

≪S. & B.≫

Multifloramine $C_{21}H_{25}O_5N \equiv 371.42$

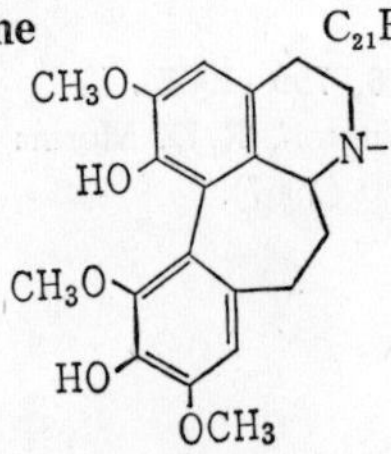

≪mp≫

≪[α]D≫

≪Ph.≫ NMR[2]

≪OP.≫ Kreysigia multiflora[2]

≪C. & So.≫

≪Sa. & D.≫

≪SD.≫ 2)

≪S. & B.≫ 2)

Kreysigine

$C_{22}H_{27}O_5N \equiv 385.44$

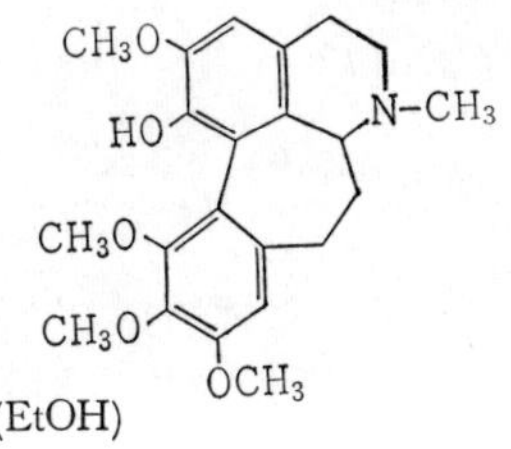

CH_3O
HO
N–CH_3
CH_3O
CH_3O
OCH_3

≪mp≫ 188° (EtOH)
≪$[\alpha]_D$≫ 0°

≪Ph.≫ CR[6] IR[2)6)] UV[2)6)] NMR[2] Mass[2] Rf[6]
≪OP.≫ Kreysigia multiflora[6]
≪C. & So.≫ Insol. in NaOH
≪Sa. & D.≫ B-MeI: 265~266°
≪SD.≫ 2), 6)
≪S. & B.≫ 2)

1) A.R. Battersby, R.B. Herbert, L. Mo, F. Santavy: J. Chem. Soc. (C) **1967**, 1739.
2) A.R. Battersby, R.B. Bradbury, R.B. Herbert, M.H.G. Munro, R. Ramage: Chem. Comm., **1967**, 450.
3) A.R. Battersby, E. McDonald, M.H.G. Munro, R. Ramage: Chem. Comm., **1967**, 934.
4) T. Kametani, K. Fukumoto, H. Yagi, F. Satoh: Chem. Comm., **1967**, 878.
5) T. Kametani, F. Satoh, H. Yagi, K. Fukumoto: Chem. Comm., **1967**, 1103.
6) G.M. Badger, R.B. Bradbury: J. Chem. Soc., **1960**, 445.

Alkaloid Name Index

A

B

C

O

P

R

S

T

U

V

W

X

Y

Z

Subject Index

A, B, C

D, E, F

G, H, I

M, N, O

P

R, S, T, W

7954962
3 1378 00795 4962